Advanced Materials for Wastewater Treatment

Scrivener Publishing
100 Cummings Center, Suite 541J
Beverly, MA 01915-6106

Advanced Materials Series

The Advanced Materials Series provides recent advancements of the fascinating field of advanced materials science and technology, particularly in the area of structure, synthesis and processing, characterization, advanced-state properties, and applications. The volumes will cover theoretical and experimental approaches of molecular device materials, biomimetic materials, hybrid-type composite materials, functionalized polymers, supramolecular systems, information- and energy-transfer materials, biobased and biodegradable or environmental friendly materials. Each volume will be devoted to one broad subject and the multidisciplinary aspects will be drawn out in full.

Series Editor: Ashutosh Tiwari
Institute of Advanced Materials
Linköping University
SE-581 83 Linköping
Sweden
E-mail: ashutosh.tiwari@liu.se

Publishers at Scrivener
Martin Scrivener (martin@scrivenerpublishing.com)
Phillip Carmical (pcarmical@scrivenerpublishing.com)

Advanced Materials for Wastewater Treatment

Edited by
Shahid-ul-Islam

Scrivener Publishing

WILEY

This edition first published 2017 by John Wiley & Sons, Inc., 111 River Street, Hoboken, NJ 07030, USA and Scrivener Publishing LLC, 100 Cummings Center, Suite 541J, Beverly, MA 01915, USA

For more information about Scrivener publications please visit www.scrivenerpublishing.com.

Wiley Global Headquarters
111 River Street, Hoboken, NJ 07030, USA

For details of our global editorial offices, customer services, and more information about Wiley products visit us at www.wiley.com.

Library of Congress Cataloging-in-Publication Data
ISBN 978-1-119-40776-8

Cover image: The Editor
Cover design by: Russell Richardson

Set in size of 11pt and Minion Pro by Exeter Premedia Services Private Ltd., Chennai, India

10 9 8 7 6 5 4 3 2 1

Contents

Preface **xv**

1 Arsenic: Toxic Effects and Remediation **1**
Sharf Ilahi Siddiqui and Saif Ali Chaudhry
1.1 Introduction 1
1.2 Arsenic Concentration in Water 2
1.3 Exposure of Arsenic in Human Body 3
1.4 Metabolism and Excretion of Arsenious Compounds 4
1.5 Arsenic Toxicity and Mechanism 6
1.5.1 Oxidative Stress 6
1.5.2 Binding to Sulfhydryl Group 7
1.5.3 Replacement of Phosphate Group 8
1.5.4 Alternation in the Gene Expression 9
1.5.5 Arsenic Impairs Glucose Catabolism 9
1.6 Detoxification of Arsenic 10
1.6.1 Antioxidants Agents 10
1.6.2 Chelating Agents 11
1.7 Arsenic Remediation Technologies 12
1.8 Adsorption and Recent Advancement 15
1.9 Conclusion 16
Acknowledgment 17
Abbreviations 17
References 18

2 Recent Trends in Textile Effluent Treatments: A Review **29**
Shumaila Kiran, Shahid Adeel, Sofia Nosheen, Atya Hassan, Muhammad Usman and Muhammad Asim Rafique
2.1 Introduction 30
2.2 Industrial Dyes, Dying Practices, and Associated Problems 31
2.3 Wastewater Remediation 31

2.4 Physical Methods 33
2.4.1 Adsorption 35
2.4.2 Coagulation and Flocculation 35
2.4.3 Membrane Processes 35
2.4.4 Ultra Filtration 36
2.4.5 Micellar-Enhanced Ultrafiltration (MEUF) 36
2.4.6 Reverse Osmosis 36
2.4.7 Nanofiltration 37
2.5 Chemical Methods 37
2.5.1 Photo Catalytic Degradation of Dyes 37
2.5.2 Oxidation and Photocatalysis with Hydrogen Peroxide 38
2.5.3 Ozonation 39
2.5.4 Degradation of Dyes Using Sodium Hypochlorite (NaOCl) 39
2.5.5 Electrochemical Method 39
2.6 Bioremediation 40
2.7 Products Recognition and Mechanisms of Dye Degradation 40
2.8 Conclusion 42
2.9 Future Outlook 43
References 43

3 Polyaniline as an Inceptive Dye Adsorbent from Effluent 51
Raminder Kaur and Monika Duhan
3.1 Introduction 52
3.1.1 Effluent from the Industries 53
3.2 Pollution Due to Dyes 56
3.2.1 Lethal Effects of Dyes 57
3.3 Methods Used for the Dye Removal 58
3.3.1 Removal of Dyes by Adsorption 59
3.3.1.1 Factors Affecting Adsorption 62
3.4 Adsorption Kinetics 71
3.4.1 Adsorption Isotherms 72
3.5 Polyaniline: An Emerging Adsorbent 74
3.5.1 Polyaniline in Dye Removal 74
3.5.2 Polyaniline in Metal Ions Removal 81
3.5.3 Polyaniline in Phenols Removal 83
3.5.4 Polyaniline in Acid Removal 83
3.6 Conclusion 84
References 84

4 Immobilized Microbial Biosorbents for Wastewater Remediation **101**
Mohammad Asaduddin Laskar, Rajeev Kumar and Mohamed A. Barakat
4.1 Introduction 102
4.2 Immobilized Microbial Biosorbent 103
4.2.1 Algae Biosorbent 103
4.2.2 Fungi Biosorbent 106
4.2.3 Bacteria Biosorbent 111
4.3 Biosorption Mechanism 114
4.3.1 Algae-Based Biocomposite 114
4.3.2 Bacteria-Based Bio-Composite 116
4.3.3 Fungi-Based Biocomposite 119
4.4 Conclusion 120
References 122

5 Remediation of Cr (VI) Using Clay Minerals, Biomasses and Industrial Wastes as Adsorbents **129**
Rashmi Acharya, Satyabadi Martha and K. M. Parida
5.1 Introduction 130
5.2 Isotherm Models 133
5.2.1 Langmuir Isotherm Model 133
5.2.2 Freundlich Isotherm Model 134
5.2.3 Dubnin–Radushkevich Isotherm Model 135
5.3 Thermodynamics of Adsorption 135
5.4 Kinetics of Adsorption 136
5.4.1 Pseudo-First-Order Kinetics 136
5.4.2 Pseudo-Second–Order Kinetics 137
5.5 Solution pH 137
5.6 Clay Minerals 139
5.6.1 Natural Clay Minerals 139
5.6.2 Natural Clay Minerals Along with Reducing Agents 140
5.6.3 Modified Clay Minerals 140
5.7 Biomasses 146
5.8 Industrial Wastes 159
5.9 Conclusion 161
References 163

6. Microbial Diversity as a Tool for Wastewater Treatment **171**
Sadia Ilyas and Haq Nawaz Bhatti
6.1 Overview of Wastewater; Sources, Pollutants, and Characteristics 171

6.1.1 Biodiversity of Wastewater Plants 175
6.2 Role of Dominant Wastewater Treatment Communities in Biodegradation 179
6.2.1 Hydrolytic Microbial Community 179
6.2.2 Acetogenic, Coliforms, and Cyanobacterial Community 181
6.2.3 Denitrifying, Fecal Coliforms, and Fermentative Microbial Community 182
6.2.4 Floc-Forming and Gram-Negative Microbial Community 183
6.2.5 Nocardioforms and Methane-Forming Microbial Community 183
6.2.6 Nitrifying Microbial Community 184
6.2.7 Denitrifying Microbial Community 187
6.2.8 Phosphorous Solubilizing Microbial Community 190
6.2.9 Sulfur Oxidizing and Reducing Microbial Community 197
6.3 Methods for the Treatment of Wastewater 200
6.3.1 Preliminary Treatments 200
6.3.2 Primary Treatments 204
6.3.3 Secondary/Biological Treatments 205
6.3.3.1 Aerated Lagoons and Bioaugmentation 207
6.3.3.2 Trickling Filter Process 211
6.3.3.3 Activated Sludge Process 213
6.3.3.4 Oxidation Ditch and Oxidation Pond Process 214
6.3.3.5 Anaerobic Digestion Process 216
6.3.3.6 Biogenic Enzymatic Wastewater Treatment 216
6.4 Conclusion 218
References 218

7 Role of Plant Species in Bioremediation of Heavy Metals from Polluted Areas and Wastewaters 223
Mayerly Alexandra Oyuela Leguizamo
7.1 Introduction 224
7.2 Heavy Metals (HM) Worldwide 225
7.3 Allochthonous and Autochthonous Plants 227
7.4 Phytoremediation of Heavy Metals (HM) 231

7.4.1 Phytoremediation 231
7.4.2 Phytoremediation Approaches and Technologies 231
7.5 Methodology 238
7.6 Analysis of Research on Heavy Metals (HM) and Native and Endemic Plant Species 238
7.7 Results 249
7.8 Conclusion 249
References 252

8 Bioremediation: A Green, Sustainable and Eco-Friendly Technique for the Remediation of Pollutants 263
Munawar Iqbal, Arif Nazir, Mazhar Abbas, Qudsia Kanwal and Dure Najaf Iqbal
8.1 Introduction 264
8.2 Immobilization 264
8.3 Enzyme Immobilization Strategies 265
8.4 Adsorption 265
8.5 Entrapment 267
8.6 Encapsulation 268
8.7 Covalent Binding 269
8.8 Self-Immobilization 270
8.9 Properties of Immobilized Enzymes 271
8.9.1 Immobilized LiP 271
8.9.2 Immobilized MnP 273
8.9.3 Immobilized Lac 274
8.10 Enzymes Sources 276
8.11 Conditions for Lipid Degradation 276
8.12 Environmental Applications of Ligninolytic Enzymes 279
8.12.1 Degradation and Decolorization of Industrial (Textile) Dyes 279
8.12.2 Dye Decolorization with Free Ligninolytic Enzymes 280
8.12.3 Dye Removal by Immobilized Ligninolytic Enzymes 286
8.12.4 Degradation of Lipids 291
8.12.5 Degradation of Miscellaneous Compounds 292
8.12.6 Xenobiotics and Industrial Effluents 295
8.12.7 Degradation of Aromatic Compounds 296
8.13 Conclusions 299
References 300

9 Role of Plant-Based Biochar in Pollutant Removal: An Overview **313**
D.S. Malik, C.K. Jain, Anuj K. Yadav and Sushmita Banerjee
9.1 Introduction 313
9.2 Preparation Methods of Biochar 315
9.2.1 Pyrolysis 315
9.2.2 Slow Pyrolysis 315
9.2.3 Fast Pyrolysis 315
9.2.4 Gasification 315
9.2.5 Hydrothermal Carbonization 315
9.3 Physico-chemical Characterization of Plant-Based Biochar 316
9.3.1 pH 317
9.3.2 Ash Content 317
9.3.3 Moisture Content 317
9.3.4 Bulk Density 317
9.3.5 Elemental Analysis 320
9.3.6 BET (Brunauer, Emmett, and Teller) 320
9.3.7 SEM and EDX 320
9.3.8 FTIR 320
9.4 Biochar for Heavy Metal Removal 320
9.5 Biochar for Dye Removal 321
9.6 Biochar for Fluoride Removal 322
9.7 Biochar for Persistent Organic Pollutant Removal 323
9.8 Biochar for Other Pollutant Removal 323
9.9 Biochar for Soil Treatment/Improvement 324
9.10 Conclusion 324
Acknowledgments 325
References 325

10 A Review on Ferrate(VI) and Photocatalysis as Oxidation Processes for the Removal of Organic Pollutants in Water and Wastewater **331**
Kyriakos Manoli, Malini Ghosh, George Nakhla and Ajay K. Ray
10.1 Introduction 332
10.2 Ferrate(VI) 335
10.2.1 Introduction 335
10.2.2 Synthesis 336
10.2.2.1 Electrochemical Synthesis 336
10.2.2.2 Wet Chemical Method 338
10.2.2.3 Dry Thermal Method 338
10.2.3 Characterization 338

10.2.4 Oxidation 340
10.2.4.1 Kinetics of the Oxidation of Organics by Ferrate(VI) 340
10.2.4.2 Stoichiometry 341
10.2.4.3 Application and Performance of Ferrate(VI) in Wastewater Treatment 357
10.2.5 Future Directions 359
10.3 Photocatalysis 360
10.3.1 Introduction 360
10.3.1.1 General Concept of Photocatalysis 360
10.3.1.2 Basic Principle of Photocatalysis 361
10.3.2 Design Parameters of Photocatalysis 363
10.3.2.1 Different Aspects of Design Parameters 364
10.3.2.2 Reactor Design Limitations Along with Proposed Solution 365
10.3.3 Photocatalysts 367
10.3.3.1 Doping of TiO_2 369
10.3.3.2 Coupled Semiconductors 371
10.3.3.3 Dye-Sensitized Catalyst 373
10.3.4 Challenges and Future Prospects of Photocatalysis 376
10.4 Combination of Photocatalysis (UV/TiO_2) and Ferrate(VI) 376
10.5 Conclusion 378
References 379

11 Agro-Industrial Wastes Composites as Novel Adsorbents 391
Haq Nawaz Bhatti, Amina Kamal and Munawar Iqbal
11.1 Introduction 392
11.2 Material and Methods 400
11.2.1 Chemical, Reagent and Instruments 400
11.2.2 Biomass Collection and Preparation 401
11.2.3 Composites Preparation 401
11.2.4 Dye Solution Preparation 402
11.2.5 Adsorption Experiments 402
11.3 Results and Discussion 402
11.3.1 Screening of Adsorbents 402
11.3.2 Effect of pH 403
11.3.3 Effect of Composites Dose 405
11.3.4 Effect of Contact Time 406

11.3.5 Effect of Initial Concentration 406
11.3.6 Effect of Temperature 408
11.3.7 Kinetic Study 409
11.3.8 Intraparticle Diffusion Model 412
11.3.9 Isotherm Modelling 412
11.3.10 Thermodynamic Study 417
11.4 Conclusion 421
References 421

12 A Review on the Removal of Nitrate from Water by Adsorption on Organic-Inorganic Hybrid Biocomposites 433
Wondalem Misganaw Golie, Kaisar Ahmad and Sreedevi Upadhyayula
12.1 Introduction 433
12.1.1 Risks Associated to High Level of Nitrate in Water 434
12.1.2 Technologies for the Removal of Nitrate from Water 435
12.2 Adsorbents for the Removal of Nitrate from Water 437
12.3 Models for Adsorption Process 445
12.3.1 Batch Adsorption Models 445
12.3.1.1 Adsorption Isotherms and Models 446
12.3.1.2 Langmuir Isotherm 446
12.3.1.3 Freundlich Isotherm 447
12.3.1.4 Temkin Isotherm 448
12.3.1.5 Dubinin-Radushkevich (D-R) Isotherm 448
12.3.1.6 Sips Isotherm 449
12.3.1.7 Redlich-Peterson Isotherm 450
12.3.1.8 Thermodynamic Parameters 450
12.3.1.9 Adsorption Kinetics 452
12.4 Column Study 454
12.4.1 Breakthrough Curve Analysis 455
12.4.2 Models of Column Studies 457
12.4.2.1 Adams-Bohart Model 457
12.4.2.2 Thomas Model 458
12.4.2.3 Yoon and Nelson Model 459
12.4.2.4 Clark Model 460
12.4.2.5 The Wolborska Model 461
12.4.2.6 Bed Depth Service Time (BDST) Model 462

12.5 Conclusion 463
Nomenclatures 464
References 467

13 Nitrate Removal and Nitrogen Sequestration from Polluted Waters Using Zero-Valent Iron Nanoparticles Synthesized under Ultrasonic Irradiation 479
Mohammadreza Kamali, Maria Elisabete Costa and Isabel Capela
13.1 Introduction 480
13.2 Materials and Methods 483
13.2.1 Experimental 483
13.2.1.1 Reagents 483
13.2.1.2 Synthesis Protocol 483
13.2.2 Characterization 484
13.2.3 Taguchi Design and Reactivity Analysis 485
13.3 Results and Discussion 486
13.3.1 Characterization 486
13.3.2 Reactivity of nZVI 489
13.3.2.1 Statistical Analysis 489
13.3.2.2 Nitrate Removal Reaction: Mechanisms and Pathways 492
13.4 Conclusion 497
Acknowledgments 498
References 498

Index 507

Preface

Water is an essential component for living organisms on planet earth and its pollution is one of the critical global environmental issues today. The influx of significant quantities of organic and inorganic waste, sediments, surfactants, synthetic dyes, sewage, and heavy metals into all types of water bodies has been increasing substantially over the past century due to rapid industrialization, population growth, agricultural activities, and other geological and environmental changes. These pollutants are very dangerous and are posing serious threat to us all.

Currently, a number of methods including ion exchange, membrane filtration, advanced oxidation, biological degradation, photocatalytic degradation, electro-coagulation, and adsorption are in operation for removing or minimizing these wastes. This book on *Advanced Materials for Wastewater Treatment* brings together innovative methodologies and research strategies that remove toxic effluents from wastewaters through fourteen important chapters written by leading scientists working in this field. I have no doubt that readers of this book will benefit from its comprehensive coverage of the current literature, up-to-date overviews of all aspects of toxic chemical remediation, including the role of nanocomposites. Together they showcase in a very lucid manner an array of technologies that complement the traditional as well as advanced treatment practices of textile effluents. I would also like to thank all the authors who contributed chapters to this book and provided their valuable ideas and knowledge. I am also very thankful to the publishers and, in particular, Martin Scrivener, for their generous cooperation at every stage of the book's compilation and production.

Shahid-ul-Islam
Indian Institute of Technology Delhi (IITD),
Hauz Khas, New Delhi, India
August 2017

1

Arsenic: Toxic Effects and Remediation

Sharf Ilahi Siddiqui and Saif Ali Chaudhry*

Environmental Chemistry Research Laboratory, Department of Chemistry, Jamia Millia Islamia, New Delhi, India

Abstract

Arsenic is associated with cancerous and non-cancerous human diseases. Arsenic from drinking water is the most common source of human exposure and it has becomes major calamity for the world. Pentavalent arsenic, As(V), can be reduced to trivalent arsenic, As(III), in the blood, which is transferred to the liver and metabolized. Arsenic produces various toxic intermediates during the metabolism, and is generally excreted from lever via urine. But on the high exposure, it retains in body and binds to soft or hard tissue. Arsenic replaces the phosphate group, which is involved in various biological pathways, inhibits glucose transporters, alters expression of genes, and can stimulate oxidative stress. This chapter enlighten the toxicity of arsenic toward living cell. Previous literatures evaluated the toxicity profiles of inorganic arsenate, arsenite and methylated metabolites, pentavalent monomethylarsonic acid (MMAV), and dimethylarsinic acid (DMAV). This chapter discusses the recently identified toxic trivalent forms of methylated metabolites. Several detoxifying nutritional supplements are also highlighted. The remediation of arsenic from drinking water is also depicted.

Keywords: Arsenic exposure, metabolism and excretion, toxicity, de-toxification, adsorption

1.1 Introduction

Industrial wastewater released in freshwater without proper treatment causes the contamination in freshwater which responds to various aquatic problems. Pesticides, fertilizers, suspended solids, color stuffs, and toxic

**Corresponding author*: saifchaudhry09@gmail.com

Shahid-ul-Islam (ed.) Advanced Materials for Wastewater Treatment, (1–28)

metals, etc., are well-known water pollutants that change the quality of freshwater [1]. Natural activities are also involved in the water contamination. Heavy metals, particularly arsenic in water, is creating more serious environmental problems for various continents, particularly Asia. Millions of people from Asian countries are living in the zone of arsenic poisoning [2].

The regular exposure to arsenic containing water is associated with toxicity and harzardous effects toward human health. It is a major calamity for various countries viz Bangladesh, India, Nepal, China, Taiwan, Thailand, Mexico, Japan, and Argentina [3]. In Bangladesh alone, millions of people have died due to arsenic poisoning and are living under the same threat. Over all, more than 140 million people from 70 countries are living in these conditions. [4].

Poisoning and toxicity of arsenic is the resulting effect of regular and high exposure of arsenic contaminated water. When arsenic becomes concentrated in human body, it accumulates in the tissues, binds to sulfhydryl sites, changes their functioning, and causes various damages. These damages have been discussed in literature [5].

To overcome these major environmental problems, concerning agencies from various countries are spending large amounts of money to control the arsenic discharge in freshwater and are collecting data of arsenic concentration in their aquatic environment. This step provides the proper way of controlling the high concentration of arsenic in water and to save the water from arsenic contamination. Various environmental agencies gave the maximum limit of arsenic 0.01 mg/L in water for safe drinking [3].

1.2 Arsenic Concentration in Water

Human activities such as mining, smelting, fossil fuel combustion, pesticides, and fertilizers are the major cause of high arsenic concentration in water [6]. Tailing is a waste that comes from mining which responds to 200 mg/L concentration of arsenic in water, whereas the pesticides and fertilizers formulated with arsenic responds to 612×10^8 g/year and 2380×10^8 g/year of arsenic discharge in water through soil erosion and leaching, respectively [7, 8]. Moreover, more than 200 arsenic containing minerals, particularly sulfide minerals contribute to direct or indirect releasing of arsenic in water [9]. Therefore, nearby areas of mining and mineralizing are under arsenic threat, where the concentration of arsenic rises up to 100–5000 mg/L in water [10]. Generally, natural water acquires arsenic in the range 1 and 2 mg/L, although it may be high up to 12 mg/L in areas containing natural sources [10]. Therefore, the concentration of arsenic in water becomes much higher than the maximum limit of arsenic in water fixed by WHO. More than 100 million of people are drinking arsenious water with a concentration of more than 0.01 mg/L.

High concentration of arsenic beyond WHO guidelines maximum permissible value (0.01 mg/L) has been reported from number of countries such as Indo-Bangladesh region (0.8 mg/L), Argentina (0.2 mg/L), Mexico (0.4 mg/L), and Taiwan (0.05–2.0 mg/L) [11]. −3, 0, +3, and +5 are the major oxidation states of arsenic. Last two oxidation states of arsenic named arsenite (As(III)) and arsenate (As(V)) are generally isolated from the water, which are stable in reducing and oxidizing environments, respectively [12].

Previous reports show that As(III) strongly binds to sulfhydryl sites of proteins and is considered to be the most toxic. As(V) also causes the large poisoning in body. As(III) and As(V) are the arsenic species found in water in the form of either oxy ions or organic and inorganic molecules [13]. Inorganic forms of arsenic are 100 times more toxic than organic forms. Inorganic and organic forms of arsenic in water are the result of pH and redox potential of water [14].

Reports from various regions suggest that the high level exposure of arsenic is associated with various adverse health effects such as cancer, diabetes, hypertension, neurological arteriosclerosis, and cardiovascular diseases [15, 16] (Figure 1.1). Arsenic induces the alteration in the cell calcium signalling, oxidative stress, impairment of cell mitochondrial function, and cell cycle progression where these effects ultimately lead to cancer [17].

1.3 Exposure of Arsenic in Human Body

Arsenic can enter into the human body via ingestion, inhalation, and skin absorption [18]. The ingestion of arsenic through drinking water is considered as a major source of arsenic concentration into body and their toxicity [19]. Aresnic has normal behavior toward body and is easily absorbed by the blood stream from gastrointestinal tract or lungs on ingestion into the body [20]. As(V) molecules are less reactive with membranes of the gastrointestinal tract than As(III), hence, As(V) completely absorbed by blood stream from gastrointestinal tract. In blood stream (erythrocytes), arsenic bounds to the globin, and circulates in various parts of human body viz bones, muscles, lungs, kidneys, across the placenta, and keratin-rich tissues such as skin nails and hair [21].

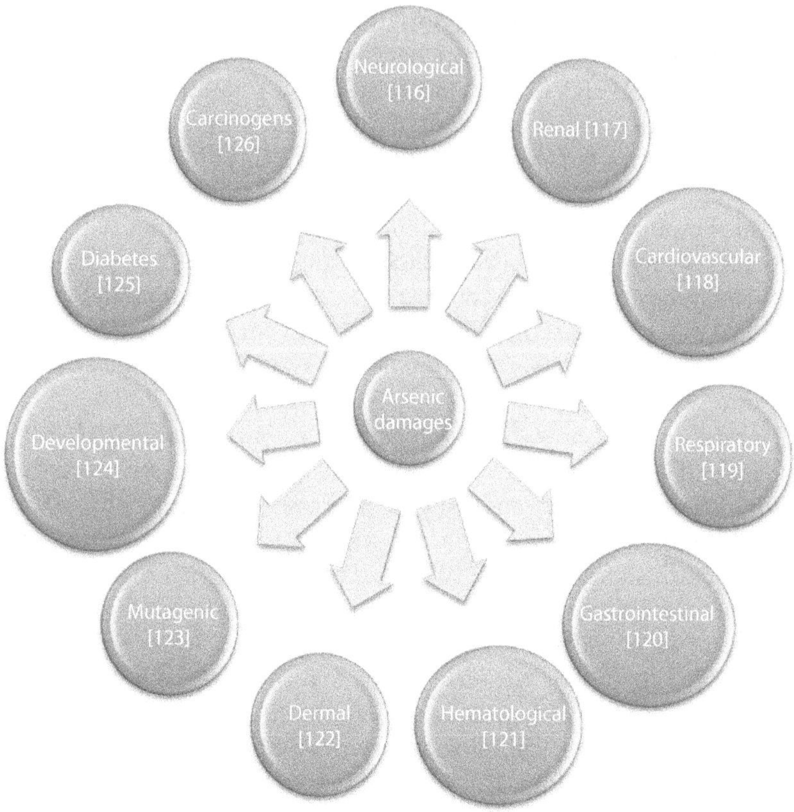

Figure 1.1 Effects of arsenic.

1.4 Metabolism and Excretion of Arsenious Compounds

The liver is the major part of the body where arsenic metabolism occurs. Primarily, metabolism of arsenic is to be considered as normal way of arsenic detoxification but recent studies suggest that intermediates of metabolism induce the toxicity [22]. Briefly, in the arsenic metabolism, ingested As(III) or As(V) molecules convert into the methylated metabolite and inorganic arsenicals [23]. Arsenic metabolism is an enzyme-induced biochemical reaction, where As(V) first reduced to As(III) by glutathione enzyme then methylation of arsenic takes place, and S-adenosylmethionine (SAM) works as methyl donor and glutathione sulfhydryl works as a vital co-factor [24].

The monomethylarsenic acid (MMA) and dimethylarsenic acid (DMA) are the resulting products of methylation of As(III) formed through the

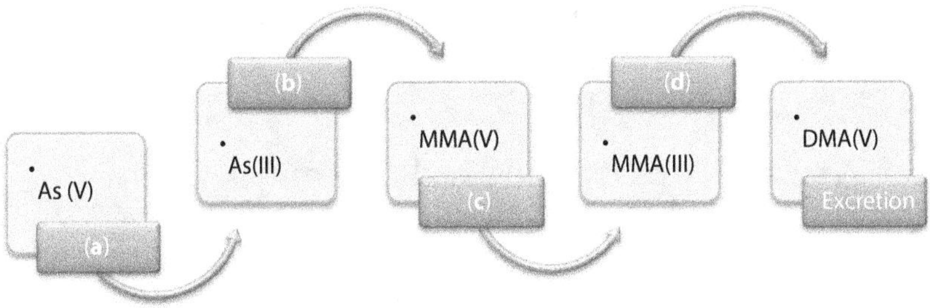

Figure 1.2 Arsenic methylation pathway in the human body (a): Arsenate reductase or purine nucleoside phosphorylase (PNP), (b): Arsenite methyl transferase (As3MT), (c): Glutathione S-transferase omega 1 or 2 (GSTO1, GSTO2), and (d): Arsenite methyl transferase (As3MT), MMA(V): Monomethylarsenic acid, MMA(III): Monomethylarsonous acid, DMA(V): Dimethylarsenic acid.

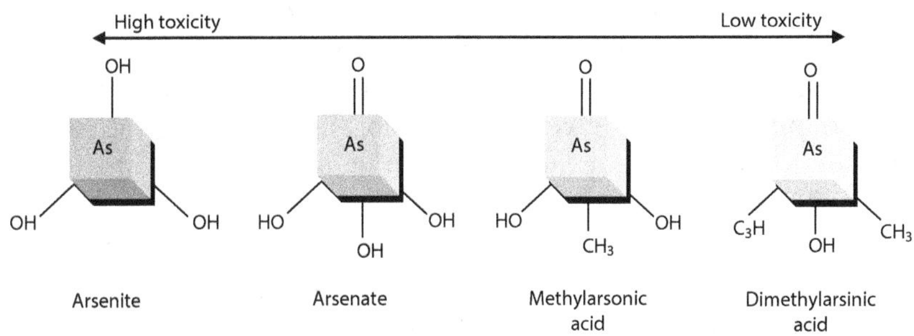

Figure 1.3 Toxicity trends of arsenic.

enzymatic transfer of the methyl group from SAM to methyl arsenate and dimethyl arsenate [20] (Figure 1.2). MMA is more toxic intermediate than DMA [25]. MMA and DMA are more toxic than other inorganic or organic arsenic molecules [26] (Figure 1.3). Dimethylmonothioarsenic acid ($DMMTA^V$), Dimethyldithioarsenic acid ($DMDTA^V$), arsenosugars, and arsenobetaines are other reported intermediate metabolites of arsenic [26]. The distribution and metabolism of $DMMTA^V$ and $DMDTA^V$ are similar to DMA^{III} and DMA^V, respectively [27]. $DMMTA^V$ is reported as more toxic than $DMDTA^V$ [28]. Arsenosugars and arsenobetaines are the products of inorganic arsenic consumed by marine organism [29].

Generally, the ingested arsenic molecules excrete from liver through urine either as as-ingested form or as methylated intermediate [30]. Skin arsenic excretes in lower rate than other organs. Blood arsenic excretes most rapidly from the human body, where 50–90% of arsenic excretes within 2–4 days, while remainder excretes slowly [20]. The excess level of arsenic in the blood stream is associated with the retention of arsenic in tissues and its toxicity [13].

1.5 Arsenic Toxicity and Mechanism

Arsenic induces various types of target-based toxicity such as arsenic-induced cardiovascular dysfunction, diabetes mellitus, neurotoxicity, nephrotoxicity, hepatotoxicity, and carcinogenicity [31]. The mechanism of arsenic toxicity is discussed below.

1.5.1 Oxidative Stress

Arsenic causes various adverse health effects by inducing high oxidative stress which affect the antioxidant enzymes found in the body. Arsenic stimulates the production of reactive oxygen species (ROS) and induces the toxicity [32]. The abnormal electron transfer through respiratory organ to mitochondrion of cell is responsible for generating the ROS in mitochondrion followed by the production of hydrogen peroxide (H_2O_2), superoxide anion (O_2^-), and hydroxyl radicals (OH^-) [33].

The electrons passed through the respiratory organ to mitochondrion of cell trigger the molecular oxygen (O_2) to form superoxide anion (O_2^-) and then dismutate to H_2O_2. H_2O_2 is the result of production of methylated metabolites such as dimethylarsinic radicals [$(CH_3)_2As\bullet$] and dimethylarsinic peroxyl [$(CH_3)_2AsOO\bullet$], during the oxidation of As(III) to As(V) [34]. Therefore, the free radicals generation during inorganic arsenic metabolism is responsible for oxidative stress.

The oxidative stress directly depend on the ingestion level of arsenic in the body, the excess level of arsenic in the cell, consumed oxygen by the cell, resulting the increased ROS production and oxidative stress [32]. Excess level of ROS is responsible for the oxidative damages in cellular and metabolism system which causes the physiological abnormalities and deleterious chronic disorders. Hemeoxygenase-1 (HO-1) is also responsible for the ROS generation which produces the free iron. The resulting free iron takes part in the Fenton reaction and forms the hydroxyl free radical ($^{\bullet}OH$) [35]. This free radical may attack DNA and impart the adverse effect to health [36].

Recently, Zhao *et al.* [37] investigated the effect of arsenic exposure on the nervous system of Gallus Gallus in response to oxidative stress and heat shock proteins (Hsps). Histological changes in the antioxidant enzyme activity, and the expressions of Hsps on arsenic exposure were observed. The malondialdehyde (MDA) content was increased on increasing arsenic dose while the activities of Glutathione peroxidase (GSH-Px) and catalase (CAT) were decreased. Moreover, the change in the expression of Hsps and Hsp60 and Hsp70 were also observed. Therefore, they suggested that sub-chronic exposure to arsenic-induced neurotoxicity in chickens was due to the disturbance in oxidative stress.

The human endothelial cell apoptosis, inflammation, oxidative stress, and nitric oxide (NO) production were also affected by the excessive amount of arsenic (5 μM of As_2O_3) [38]. Result showed that arsenic induced the significant enhancement in endothelial cell apoptosis and inflammation as indicated by the increase of mRNA and protein expression of vascular cell adhesion molecule-1, intercellular adhesion molecule-1, and pentraxin 3. Moreover, the exposure of arsenic also increased the intracellular ROS. The change in the activity of NADPH oxidase (NOX) and up-regulated mRNA expression of NOX subunits p22phox were also investigated. Oxidative stress and impaired NO production are involved in their pro-inflammatory and pro-apoptotic effects. Similarly, arsenic-induced oxidative stress was also observed in human embryonic kidney (HEK) cells and HaCaT cells [39]. ROS can alter the expression of atherosclerosis-related genes and stimulate the various cell signals by the oxidation of sulfhydryl groups and by changing the intracellular redox status [40–43].

1.5.2 Binding to Sulfhydryl Group

Arsenic shows the high binding affinity for vicinal thiol sites of enzyme, which inhibits their catalytic activities and induce the toxicity (Figure 1.4). The complexes between arsenic and vicinal thiol are generally stable [44]. Generally,

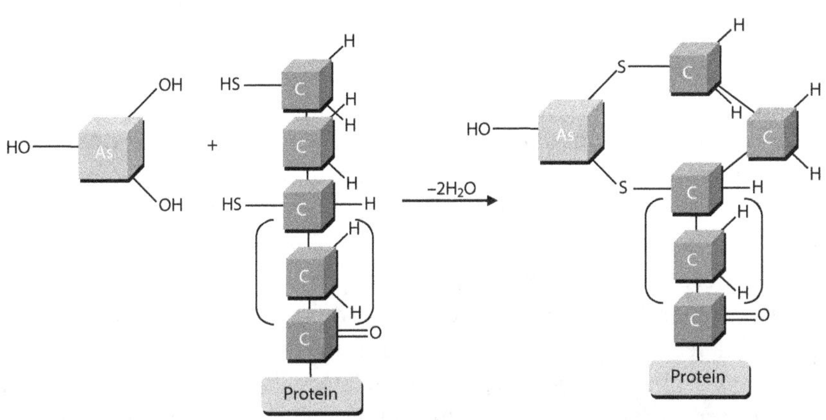

Figure 1.4 Representation of sulfhydryl-arsenic bonding.

arsenic decreases the cellular-reduced glutathione (GSH) level through the reduction of As(V) to As(III), where GSH functions as an electron donor or through arsenic-induced free radical oxidize GSH. Arsenic also decreases the cellular GSH level through binding to their sulfhydryl sites. In comparison to As(V), As(III) easily and strongly bind to sulfhydryl sites of reduced glutathione (GSH) [45]. Monomethylarsonous acid, MMA(III) also affect the functioning of GSH and thioredoxin reductase on thiol binding [46].

The arsenic exposure also reduces the generation of ATP in Kreb's cycle due to their high binding affinity for vicinal thiols sites of enzymes such as GSH resulting the cell damage and death [46]. Moreover, arsenic may bind to the thiol sites of pyruvate dehydrogenase and ketoglutarate dehydrogenase enzyme. The presence of arsenic also disturbs the cellular redox condition which leads to cytotoxicity, synthetic peptides based on the Zn finger region, and the estrogen binding region of the human estrogen receptor-α which contributes to carcinogenicity, tubulin, poly(ADP-ribose) polymerase (PARP-1), thioredoxin reductase, estrogen receptor-alpha, arsenic(+3)methyltransferase, and Keap-1 which leads to the several genetic effects and human breast cancer cell line MCF-7 [47–50].

1.5.3 Replacement of Phosphate Group

Structure and the properties of As(V) resembles to the phosphate anion thus the presence of As(V) disturb the various biochemical reactions viz glycolysis, glycogenesis, gluconeogenesis and glycogenolysis, and pentose phosphate pathway (PPP), which involves the glucose-6-phosphate and 6-phosphogluconate as essential mediator [51]. In vitro, As(V) replaces the phosphate group of glucose-6-phosphate and 6-phosphogluconate during the biochemical reactions and form glucose-6-arsenate and 6-arsenogluconate. Simply, arsenic competes with the phosphate binding sites [52]. The phosphate group involved in the sodium pump and anion exchange transport system of human erythrocytes is also replaced by arsenate [53]. Studies reported that the production of adenosine-tri-phosphate (ATP) during the glycolysis is stopped due to the production of adenosine-tri-arsenate on the replacement of phosphate by As(V) [20]. Generally, 1,3-diphospho-D-glycerate is formed during the glycolysis process by enzymatical addition of phosphate anion to D-gylceraldehyde-3-phosphate. However, in the presence of As(V), one of the phosphate group of 1,3-diphospho-D-glycerate is replaced by the As(V) and form the unstable anhydride named 1-arsenato-3-phospho-D-glycerate, which later easily hydrolyze into As(V) and 3-phosphoglycerate due to the longer As-O bond length [51]. The replacement of phosphate group by As(V) is known to be as arsenolyis.

Decrease in the ATP generation on As(V) exposure was also observed from human and rabbit erythrocytes [54, 55]. Similarly, in mitochondrion of cell, during the oxidative phosphorylation, adenosine-5′-diphosphate (ADP-phosphate) is replaced by ADP-arsenate in the presence of succinate [52, 54]. Moreover, the production of nicotinamide adenine dinucleotide phosphate (NADPH) and glucose-6-phosphate dehydrogenase (G6PDH), an enzyme of pentose phosphate pathway (PPP), are also reduced on exposure to arsenic [20].

1.5.4 Alternation in the Gene Expression

It has been investigated that arsenic can induce the alteration in gene expression [55]. The ingestion of 5 mol/L of As(III) in rat pancreatic cells decreased the mRNA expression. 6 mol/L of As(III) exposed in mouse adipocytes cell altered the gene expression of peroxisome proliferative-activated receptor (PPAR), and high level exposure of As(III) in human adipocyte cells, decreased the expression of AKT genes [56–58]. AKT gene expression is also altered on the ingestion of As(III) into 3T3-L1 adipocytes cell [59]. It also has been reported that As(III) can inhibit the activation of AKT gene. Similarly, exposure of 0.1 and 5mol/L As(III) in human GM847 fibroblast cells caused the alteration of expression of c-fos and c-jun genes, respectively [60].

Moreover, increased expression was also observed in phosphoenol pyruvate carboxykinase (PEPCK) gene on the As(III) ingestion in chick embryos [61, 62]. The immunomodulatory effect of arsenic on cytokine and HSP gene expression in Labeo rohita fingerlings were also investigated [63]. Furthermore, down regulates in the gene expression at the postsynaptic density in mouse cerebellum on arsenic exposure were also observed [64]. Arsenic ingestion is also associated with decreased gene expression and increased DNA methylation in peripheral blood cells in women [65]. Moreover, the low dose of inorganic arsenic, 50 μg/L arsenic trioxide for 90 days, changed the antioxidant genes expression and also triggered the oxidative stress in Zebrafish brain [66].

1.5.5 Arsenic Impairs Glucose Catabolism

It has been reported that arsenic may disturb the glucose metabolism pathway and insulin signalling [67]. Numerous enzyme complexes such as succinyl Co-A synthase, ketoglutarate dehydrogenase, and pyruvate dehydrogenase (PDH) are involved in the glucose metabolism [68]. As(III) may inhibit their function on binding [22].

Generally, PDH enzyme complex viz dihydrolipoyl transacetylase, dihydrolipoyl dehydrogenase, pyruvate decarboxylase, thiamine pyrophosphate, lipoic acid, CoASH, FAD, and NAD^+ are most sensitive to As(III) [69]. PDH inhibition in the presence of As(III) was reported on the result of binding between As(III) and lipoic acid moiety [70]. However, MMA^{III} were reported as stronger inhibitor of PDH than As(III) [69]. Moreover, phenylarsine oxide (PAO), an organic arsenic species, inhibits the basal or insulin stimulated glucose uptake by canine kidney cells, adipocytes and intact skeletal muscle [71–73].

1.6 Detoxification of Arsenic

To control the arsenic effect, detoxification of arsenic through the nutrients and chelation therapy has become meaningful.

1.6.1 Antioxidants Agents

The generation of ROS, on arsenic exposure, reduces the cellular antioxidant which increases the oxidative stress in human body. To prevent the ROS generation or decrease the oxidative stress, numerous endogenous antioxidants such as superoxide dismutase (SOD), glutathione reductase (GR), catalase, glutathione peroxidase (GPx), and reduced glutathione (GSH) are naturally generated in the human body which trigger the antioxidant system [74]. However, the high exposure of arsenic decreases the generation of these antioxidants thus external antioxidants such as vitamin C and E, quercetin, *N*-acetylcysteine (NAC), lipoic acid, and thiol-based antioxidant are injected in body to scavenge the ROS [75].

Vitamins A, C, and E work as antioxidant and decrease the oxidative stress on resulting arsenic exposure through scavenging of ROS [76]. It has been reported that vitamin C may trap the arsenic and alleviate the arsenic-induced oxidative stress. It may scavenge the ROS by electron transfer to prevent the lipid pre-oxidation. Furthermore, it binds to free radicals to protect the membrane from oxidative damages [77]. Similarly, vitamin E also has ability to trap the free radicals, to protect the membrane from arsenic toxicity and oxidative damages [78]. It has been reported that administration of vitamin C and E could effectively reduce the fragmentation of DNA in the presence of arsenic [79]. Moreover, vitamin A, B, B_{12}, and folic acid may also reduce the arsenic toxicity to reduce the oxidative damages [80].

Quercetin is a bioflavonoid, has also been reported as antioxidant, which protect the cell from oxidative damage through trapping the ROS. Quercetin inhibit the cytotoxicity due to low-density lipoprotein [81].

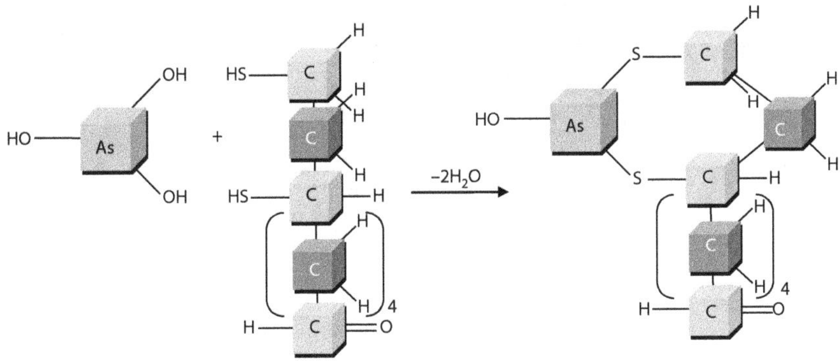

Figure 1.5 Detoxification of arsenite oxy anions by lipoic acid.

The *N*-acetylcysteine (NAC) also shows protective effect against arsenic toxicity. It may also trap arsenic on chelation and recovered the hepatic malondialdehyde level [82]. It may reduce the arsenic-induced hepatic, however, showed renal toxicity on co-administration with zinc [83]. Lipoic acid has also the free radical scavenging properties, which leads to reduction in arsenic toxicity and oxidative damage [84] (Figure 1.5).

These are the some reported antioxidants that are externally injected in the body and showed efficient result against the arsenic toxicity due to their free radical scavenging and chelation properties. Most of the antioxidants may be obtained from naturals sources. Antioxidants agents obtained from naturals sources can be better antioxidants than synthetic agents due to easily available, low cost, eco-friendly nature, and no further toxicity. Plant and plant extracts have strong antioxidant activity [85]. Hippophae rhamnoides [86], Moringa oleifera [87], Spirulina [88], Centella asiatica [89], Curcumin [90], Mentha piperita [91], and Aloe vera barbadensis [92] have strong antioxidant properties to protect the cell from arsenic-induced oxidative stress.

1.6.2 Chelating Agents

The complexation between the metal ions and multi-dentate ligand is known as chelation, and the multi-dentate ligand referred to as chelating agent. Chelating agents are organic compounds that are able to donate their electron to metal ions and form the chelate complex. Similarly, to trap and detoxify the arsenic, chelation therapy is being used to generate the chemically inert arsenic-ligand complex [93]. This chemically inert arsenic-ligand complex is further excreted from body without any interaction within body (Figure 1.6).

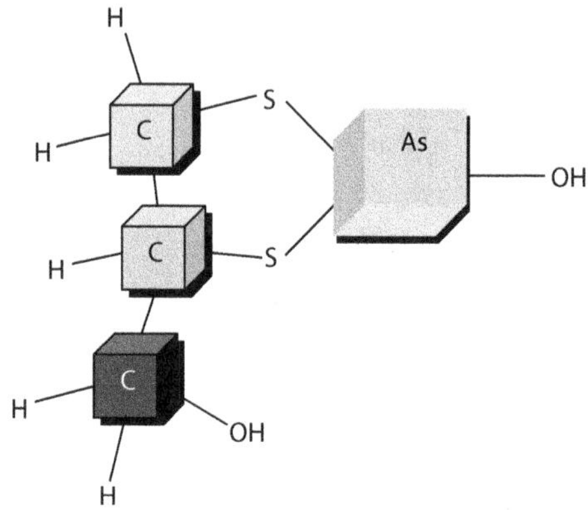

Figure 1.6 Excreted arsenite chelate complex.

Meso-2,3-dimercaptosuccinic acid (DMSA) and 2,3-dimercapto-1-propanesulphonic acid (DMPS) are the most commonly used chelating agents which could detoxify the arsenic through complex formation [94, 95]. Moreover, numerous derivatives of DMSA viz mono isoamyl DMSA (MiADMSA), mono *n*-amyl DMSA (MnDMSA), mono *n*-butyl DMSA (MnBDMSA), mono *i*-butyl DMSA (MiBDMSA), dimethyl DMSA (DMDMSA), diethyl DMSA (DEDMSA), diisoamyl DMSA, and diisopropyl DMSA (DiPDMSA) have also been reported that could also be effective chelating agents to reduce the arsenic concentration from different parts of body [96]. However, there are large drawbacks of chelating agents such as non-specificity, low therapeutic index, and failure to permeate the plasma membrane. Despite of this, metal redeployment and binding of chelating agents to other sites can also induce the side effects and toxicity [97]. Moreover, the use of antioxidants and chelating agents is not mass effective and limited to the particular body systems, therefore, making arsenic free water for human consumption is the only solution.

1.7 Arsenic Remediation Technologies

The concentration of arsenic in water can be maintained at WHO recommended maximum limits through various treatment processes such as oxidation-coagulation [98], electro-coagulation and co-precipitation [99, 100], oxidation-precipitation [101], reverse osmosis [102], electro dialysis [103], and ion exchange technology [104] (Table 1.1). However, these technologies are hardly handling and are very costly. Besides, adsorption

Table 1.1 Techniques utilized for arsenic removal.

Removal techniques	Advantages	Disadvantages	Cost	pH Dependency	Removal efficiency		Ref.
					As(III)	As(V)	
Precipitation	Simple, low-cost	Slow process and Produce large sludge	Lower	Independent	Less effective	Less effective	[144, 145]
Coagulation	Low costs, simple chemicals used, No monitoring of breakthrough is required	Required Pre-oxidation step, produced toxic sludge and hard to operate	Comparatively low	Dependent	Less effective	>90%	[146, 147]
Ion exchange	Eco-friendly, used for industrial and municipal water, provides high flow rate, removes dissolved inorganics effectively	High-cost medium, high-tech operation and maintenance, need high operation skill, sludge disposal problem	High	Dependent	Very less effective	>90%.	[148, 149]
Membrane filtration	No chemicals required and does not influence water taste and color	Temperature limitation, effective for minor stream sizes, and fouling of the membranes	High	Independent	Less effective	>80%	[150, 151]

(*Continued*)

Table 1.1 Cont.

Removal techniques	Advantages	Disadvantages	Cost	pH Dependency	Removal efficiency		Ref.
					As(III)	As(V)	
Reverse osmosis	Very effective at removing inorganic constituents, little maintenance, and no addition of chemicals	Pre-oxidation of As(III) to As(V) required and high tech operation and maintenance	Very high	Independent	>60%	>90%	[152, 153]
Electro dialysis	Easy to handling	Interference by oxidizing agent's	High	Independent	25–60%	>70%	[154]

technology is inexpensive; does not involve sophisticated instrumentation and do not require long procedure. The process is simple, safe to handle, and effectively work at low and high arsenic concentration in water [105, 106]. Therefore, adsorption of arsenic can be the better option for cleaning the arsenic contaminated water at different scales ranging from household module to community plants.

1.8 Adsorption and Recent Advancement

Being surface phenomena, adsorption process is based on interaction between the solute (adsorbate) and solid surface (adsorbent). Low particle diameter, high surface area, high active sites, and magnetic character of adsorbent are responsible for the higher removal capacity for arsenic [107]. Numerous adsorbents with above characteristics have been utilized. Sometimes, pre-oxidation step is preferred to remove As(III), which makes the process costly [108]. Moreover, these steps enhance the chance of formation of un-healthy by-products [109]. Numerous adsorbents having oxidative properties have been utilized for simultaneous oxidation of As(III) to As(V) and adsorption of As(V) [110].

Recently, metal-based adsorbents, metal oxides and nanocomposites are utilized for arsenic cleanup from pollutant sites, followed by easily oxidation of As(III) to As(V) [111].

Nanosized metal oxides and nanocomposites cleanup water under the various water quality constraints such as pH and competing ions [112]. In addition, generation and regeneration of adsorbent make the process significant in respect to cost and removal capacity. Activated carbon (AC) is one of the most utilized highly amorphous and porous adsorbent, however, the generation and regeneration of AC activated carbon is very difficult [113].

These drawbacks of AC generate the large sludge in cleaning sites and made the process costly. This drawback of AC could be avoided in the era of magnetic adsorbent. Metals like iron, titanium, and cobalt-based adsorbent respond to magnet. Moreover, nanosized magnetic adsorbent respond to low gradient magnet [114]. The impregnation or doping of magnetic NPs into AC or organic framework imparts their magnetic characteristics to the AC or organic framework, which makes adsorbent suitable for magnetic separation from water [115]. Recently, various magnetic organic–inorganic hybrid adsorbents have been successfully utilized in the field of water cleaning. This recent development in adsorbents provide a variety of eco-friendly and cost effective adsorbents, having remarkable

Table 1.2 Adsorbents utilized for arsenic removal.

Adsorbent	Removal capacity (mg/g)		Ref.
	As(III)	As(V)	
IHB	51.9	59.6	[127]
Fe (III)-BSX	54.35	–	[128]
IOCSp	4.2	4.6	[129]
CCB	–	96.46	[130]
GO–$ZrO(OH)_2$ nanocomposite	–	84.89	[131]
α-Fe_2O_3	95.0	47.0	[132]
γ-Fe_2O_3 NPs	74.83	105.25	[133]
AAC- Fe_3O_4	46.06	16.56	[134]
Fe-Cu BO	122.3	82.7	[135]
Fe-Ce MO	86.29	55.51	[136]
MIO-GO	54.18	27.76	[137]
GO-$MnFe_2O_4$ MNH	97.0	136.0	[138]
$MnFe_2O_4$	146.0	207.0	
Fe_3O_4-RGO-MnO_2 Ns	14.0	12.0	[139]
Fe-MnO_x/RGO	47.0	49.0	[140]
β-FeOOH-GONs	77.50	45.70	[141]
Diatom-FeO_x composite	10.0	12.5	[142]
HCO NPs	170.0	107.0	[143]

potential for arsenic remediation from water and wastewater [13]. Various adsorbents utilized for arsenic remediation has been depicted in Table 1.2.

1.9 Conclusion

This chapter shows that arsenic can cause a major calamity as well as be a threat for water dependent bodies. Arsenic shows toxicity and carcinogenicity towards human body on bonding with binding sites available on various working enzyme. This chapter is associated with the brief biochemical metabolic pathway mechanism of arsenic ingestion and risk of toxicity. Further, this study attracted the attention of scientists to search ways of detoxifying the arsenic. Although, various therapeutic and nutritional strategies have been incorporated to discard the arsenic toxicity. This study reveals that the adsorptive remediation of arsenic from water is better option instead of detoxification of arsenic. In general, improved adsorption capacity of adsorbents probably is due to higher number of active binding sites on their surface, therefore more number of adsorbents with new functional groups are required to search out. This chapter will help the young scientist to quick understand the toxicity, detoxification, and remediation of arsenic from water.

Acknowledgment

The authors appreciate the Jamia Millia Islamia, New Delhi, India, for equipping the Environmental Chemistry Research Laboratory where this research work was carried out.

Abbreviations

AAC-Fe_3O_4 (ascorbic acid-coated Fe_3O_4)
CCB (chitosan-coated biosorbent)
Fe (III)-BSX (Fe (III)-treated biomass of *Staphylococcus xylosus*)
Fe-MnO_x/G (dispersed graphene matrix)
α-Fe_2O_3 (ultrafine iron oxide)
β-FeOOH/GONs (akaganeite [β-FeOOH] decorated graphene oxide nano composite)
γ-Fe_2O_3 (saturated magnetic γ-Fe_2O_3 NPs)
GO (graphene oxide)
GO-$MnFe_2O_4$ MNH (graphene oxide-$MnFe_2O_4$ magnetic nanohybrid)
ICBO (Fe-Cu binary oxide)
ICF (iron chitosan flakes)
IOCSp (iron oxide-coated sponge)
IHB (inonotus hispidus biomass),
MIO-GO (magnetic iron oxide-loaded graphene oxide)
NPs (nanoparticles)
Ns (nanocomposite)

References

1. Wang, Q., Yang, Z., Industrial water pollution, water environment treatment, and health risks in China. *Environ. Pollut.,* 218, 358–365, 2016.
2. Singh, N., Kumar, D., Sahu, A., Arsenic in the environment: effects on human health and possible prevention. *J. Environ. Biol.,* 28, 359–365, 2007.
3. Jadhav, S.V., Bringas, E., Yadav, G.D., *et al.*, Arsenic and fluoride contaminated groundwaters: a review of current technologies for contaminants removal. *J. Environ. Manage.,* 162, 306–325, 2015.
4. Ng, J.C., Wang, J., Shraim, A., Global health problems caused by arsenic from natural sources. *Chemosphere,* 52, 1353–1359, 2003.
5. Mazumder, G., Chronic arsenic toxicity & human health. *Indian J. Med. Res.,* 128, 436–447, 2008.
6. Chaudhry, S.A., Ahmed, M., Siddiqui, S.I., Ahmed, S., Fe(III)–Sn(IV) mixed binary oxide-coated sand preparation and its use for the removal of As(III) and As(V) from water: application of isotherm, kinetic and thermodynamics. *J. Mol. Liq.,* 224, 431–441, 2016.
7. Mohan, D., Pittman, J., Arsenic removal from water/wastewater using adsorbents-a critical review. *J. Hazard. Mater.,* 142, 1–53, 2007.
8. Mackenzie, E.T., Lamtzy, RJ., Petorson, V., Global trace metals cycles and predictions. *J. Int. Assoc. Math. Geol.,* 6, 99–142, 1979.
9. Mandal, B.K., Suzuki, K.T., Arsenic round the world: a review. *Talanta,* 58, 201–235, 2002.
10. McIntyre, D.O., Linton, T.K., *6 - Arsenic. Fish Physiol.,* 31, 297–349, 2011.
11. Flora, S.J.S., Arsenic-induced oxidative stress and its reversibility. *Free Radical Biol. Med.,* 51, 257–281, 2011.
12. Hoang, T.H., Ju-Yong, K., Sunbaek, B., Kyoung, W.K., Source and fate of as in the environment. *Geo. system Eng.,* 13, 35–42, 2010.
13. Siddiqui, S.I., Chaudhry, S.A., Arsenic removal from water using nanocomposites: a review. *Cur. Environ. Eng.,* 2017, In Press. DOI: 10.2174/2212717804666161214l437-15.
14. Bowell, R.J., Alpers, C.N., Jamieson, H.E., Nordstrom, D.K., Majzlan, J., The environmental geochemistry of arsenic-an overview. *Rev. Mineral. Geochem.,* 79, 1–16, 2014.
15. Shameem, K., Abdul, M., Jayasinghe, S.S., *et al.*, Arsenic and human health effects: a review. *Environ. Toxicol. Pharmacol.,* 40, 828–846, 2015.
16. Sanchez, T.R., Perzanowski, M., Graziano, J.H., Inorganic arsenic and respiratory health, from early life exposure to sex-specific effects: a systematic review. *Environ. Res.,* 147, 537–555, 2016.
17. Yunus, M., Sohel, N., Hore, S.K., Rahman, M., Arsenic exposure and adverse health effects: a review of recent findings from arsenic and health studies in Matlab, Bangladesh. *The Kaohsiung J. Med. Sci.,* 27, 371–376, 2011.
18. Davis, M.A., Signes-Pastor, A.J., Argos, M., *et al.*, Assessment of human dietary exposure to arsenic through rice. *Sci. Total Environ.,* 586, 1237–1244, 2017.

19. Pompili, M., Vichi, M., Dinelli, E., *et al.*, Arsenic: association of regional concentrations in drinking water with suicide and natural causes of death in Italy. *Psychiatry Res.,* 249, 311–317, 2017.
20. Kulshrestha, A., Jarouliya, U., Prasad, G.B.K.S., Flora, S.J.S., Bisen, P.S., Arsenic-induced abnormalities in glucose metabolism: biochemical basis and potential therapeutic and nutritional interventions. *World J. Transl. Med.,* 3, 96–111, 2014.
21. Ghosh, A., Evaluation of chronic arsenic poisoning due to consumption of contaminated ground water in West Bengal, India. *Int. J. Prev. Med.,* 4, 976–979, 2013.
22. Wei, B., Yu, J., Wang, J., *et al.,* The relationships between arsenic methylation and both skin lesions and hypertension caused by chronic exposure toarsenic in drinking water. *Environ. Toxicol. Pharmacol.,* 53, 89–94, 2017.
23. Naranmandura, H., Iwata, K., Suzuki, K.T., Ogra, Y., Distribution and metabolism of four different dimethylated arsenicals in hamsters. *Toxicol. Appl. Pharmacol.,* 245, 67–75, 2010.
24. Hsueh, Y.M., Chung, C.J., Shiue, H.S, *et al.,* Urinary arsenic species and CKD in a Taiwanese population: a case-control study. *Am. J. Kidney Dis.,* 54, 859–870, 2009.
25. Islam, K., Haque, A., Karim, R., Dose-response relationship between arsenic exposure and the serum enzymes for liver function tests in the individuals exposed to arsenic: a cross sectional study in Bangladesh. *Environ. Health,* 10, 64, 2011.
26. Kumagai, Y., Sumi, D., Arsenic: signal transduction, transcription factor, and biotransformation involved in cellular response and toxicity. *Annu. Rev. Pharmacol. Toxicol.,* 47, 243–262, 2007.
27. Tseng, C.H., Arsenic methylation, urinary arsenic metabolites and human diseases: current perspective. *J. Environ. Sci. Health C. Environ. Carcinog. Ecotoxicol. Rev.,* 25, 1–22, 2007.
28. Suzuki, K.T., Iwata, K., Naranmandura, H., Suzuki, N., Metabolic differences between two dimethylthioarsenicals in rats. *Toxicol. Appl. Pharmacol.* 218, 166–173, 2007.
29. Kahn, M., Raml, R., Schmeisser, E., *et al.,* Two movel thio-arsenosugars in scallops identified with HPLC-ICPMS and HPLC-ESMS. *Environ. Chem.,* 2, 171–176, 2005.
30. Raml, R., Goessler, W., Traar, P., Ochi, T., Francesconi, K.A., Novel thioarsenic metabolites in human urine after ingestion of an arsenosugar, 2′,3′-dihydroxypropyl 5-deoxy-5-dimethylarsinoyl-beta-D-riboside. *Chem. Res. Toxicol.,* 18, 1444–1450, 2005.
31. Flora, S.J.S., Agrawal, S., Chapter 31- Arsenic, Cadmium, and Lead. *Reprod. Dev. Toxicol.,* 537–566, 2017.
32. Barchowsky, A., Klei, L.R., Dudek, E.J., Swartz, H.M., James, P.E., Stimulation of reactive oxygen, but not reactive nitrogen species, in vascular endothelial cells exposed to low levels of arsenite. *Free Radical Biol. Med.,* 27, 1405–1412, 1999.

33. Naranmandura, H., Xu, S., Sawata, T., *et al.*, Mitochondria are the main target organelle for trivalent monomethylarsonous acid (MMA(III))-induced cytotoxicity. *Chem. Res. Toxicol.*, 24, 1094–1103, 2011.
34. Yamanaka, K., Okada, S., Induction of lung-specific DNA damage by metabolically methylated arsenics via the production of free radicals. *Environ. Health Perspect.*, 102, 37–40, 1994.
35. Liu, S.X., Athar, M., Lippai, I., Waldren, C., Hei, T.K., Induction of oxyradicals by arsenic: implication for mechanism of genotoxicity. *Proc. Natl. Acad. Sci.*, 98, 1643–1648, 2001.
36. Kitchin, K.T., Ahmad, S., Oxidative stress as a possible mode of action for arsenic carcinogenesis. *Toxicol. Lett.*, 137, 3–13, 2003.
37. Zhao, P., Guo, Y., Zhang, W., Neurotoxicity induced by arsenic in Gallus Gallus: Regulation of oxidative stress and heat shock protein response. *Chemosphere*, 166, 238–245, 2017.
38. Ma, Y., Ma, Z., Yin, S., Yan, X., Wang, J., Arsenic and fluoride induce apoptosis, inflammation and oxidative stress in cultured human umbilical vein endothelial cells. *Chemosphere*, 167, 454–461, 2017.
39. Ma, L., Li, J., Zhan, Z., *et al.*, Specific histone modification responds to arsenic-induced oxidative stress. *Toxicol. Appl. Pharmacol.*, 302, 52–61, 2016.
40. Rezaei, M, Khodayar, M.J., Seydi, E., Soheila, A., Parsi, I.K., Acute, but not chronic, exposure to arsenic provokes glucose intolerance in rats: possible roles for oxidative stress and the adrenergic pathway. *Can. J. Diabetes*, 41, 273–280, 2017.
41. Escudero-Lourdes, C., Toxicity mechanisms of arsenic that are shared with neurodegenerative diseases and cognitive impairment: role of oxidativestress and inflammatory responses. *Neuro.Toxicol.*, 53, 223–235, 2016.
42. Bandyopadhyay, A., Role of oxidative stress in arsenic(III) induced genotoxicity in cells of meristematic tissue of Allium cepa: an in vivo study. *Mater. Today*, 3, Part A. 3194–3199, 2016.
43. Ogun, M., Ozcan, A., Karaman, M., *et al.*, Oleuropein ameliorates arsenic induced oxidative stress in mice. *J. Trace Elem. Med. Biol.*, 36, 1–6, 2016.
44. Aposhian, H.V., Biochemical toxicology of arsenic. In: Hodgson E, Bend JR, and Philpot RM, editors. *Biochem. Toxicol. Elsevier Sci.*, 265–299, 1989.
45. Styblo, M., Serves, S.V., Cullen, W.R., Thomas, D.J., Comparative inhibition of yeast glutathione reductase by arsenicals and arsenothiols. *Chem. Res. Toxicol.*, 10, 27–33, 1997.
46. Lin, S, Cullen, W.R, Thomas, D.J., Methylarsenicals and arsinothiols are potent inhibitors of mouse liver thioredoxin reductase. *Chem. Res. Toxicol.*, 12, 924–930, 1999.
47. Habib, G.M., Shi, Z.Z., Lieberman, M.W., Glutathione protects cells against arsenite-induced toxicity. *Free Radic. Biol. Med.*, 42, 191–201, 2007.
48. Kitchin, K.T., Wallace, K., Arsenite binding to synthetic peptides based on the Zn finger region and the estrogen binding region of the human estrogen receptor-α. *Toxicol. Appl. Pharmacol.*, 206, 66–72, 2005.

49. Kitchin, K.T., Wallace, K., The role of protein binding of trivalent arsenicals in arsenic carcinogenesis and toxicity. *J. Inorg. Biochem.*, 102, 532–539, 2008.
50. Zhang, X., Yang, F., Shim, J.Y., *et al.*, Identification of arsenic-binding proteins in human breast cancer cells. *Cancer Lett.*, 255, 95–106, 2007.
51. Dixon, H.B.F., The biochemical action of arsenic acids especially as phosphate analogues. *Adv. Inorg. Chem.*, 44, 191–227, 1997.
52. Gresser, M.J., ADP-arsenate. Formation by submitochondrial particles under phosphorylating conditions. *J. Biol. Chem.*, 256, 5981–5983, 1981.
53. Kenny, L.J., Kaplan, J.H., Arsenate substitutes for phosphate in the human red cell sodium pump and anion exchanger. *J. Biol. Chem.*, 263, 7954–7960, 1988.
54. Crane, R.L., Lipmann, F., The effect of arsenate on aerobic phosphorylation. *J. Biol. Chem.*, 201, 235–243, 1953.
55. Sun, X., Li, B., Li, X., *et al.*, Effects of sodium arsenite on catalase activity, gene and protein expression in HaCaT cells. *Toxicol. In. Vitro.*, 20, 1139–1144, 2006.
56. Díaz-Villaseñor, A., Burns, A.L., Hiriart, M., Cebrián, M.E., Ostrosky-Wegman, P.. Arsenic-induced alteration in the expression of genes related to type 2 diabetes mellitus. *Toxicol. Appl. Pharmacol.*, 225, 123–133, 2007.
57. Wauson, E.M., Langan, A.S., Vorce, R.L., Sodium arsenite inhibits and reverses expression of adipogenic and fat cell-specific genes during in vitro adipogenesis. *Toxicol. Sci.*, 65, 211–219, 2002.
58. Paul, D.S., Hernández-Zavala, A., Walton, F.S., *et al.*, Examination of the effects of arsenic on glucose homeostasis in cell culture and animal studies: development of a mouse model for arsenic-induced diabetes. *Toxicol. Appl. Pharmacol.*, 222, 305–314, 2007.
59. Wang, Z.X., Jiang, C.S., Liu, L., *et al.*, The role of Akt on arsenic trioxide suppression of 3T3-L1 preadipocyte differentiation. *Cell Res.*, 15, 379–386, 2005.
60. Hu, Y., Jin, X., Snow, E.T., Effect of arsenic on transcription factor AP-1 and NF-kappaB DNA binding activity and related gene expression. *Toxicol. Lett.*, 133, 33–45, 2002.
61. Hamilton, J.W., Kaltreider, R.C., Bajenova, O.V., *et al.*, Molecular basis for effects of carcinogenic heavy metals on inducible gene expression. *Environ. Health Perspect.*, 106 Suppl. 4, 1005–1015, 1998.
62. Kaltreider, R.C., Davis, A.M., Lariviere, J.P., Hamilton, J.W., Arsenic alters the function of the glucocorticoid receptor as a transcription factor. *Environ. Health Perspect.*, 109, 245–251, 2001.
63. Banerjee, S., Mitra, T., Purohit, G.K., Mohanty, S., Mohanty, B.P., Immunomodulatory effect of arsenic on cytokine and HSP gene expression in Labeo rohita fingerlings. *Fish Shellfish Immunol.*, 44, 43–49, 2015.
64. Zhang, C., Li, S., Sun, Y., et al., Arsenic downregulates gene expression at the postsynaptic density in mouse cerebellum, including genes responsible for long-term potentiation and depression. *Toxicol. Lett.*, 228, 260–269, 2014.

65. Ameer, S.S., Engström, K., Hossain, M.B., et al., Arsenic exposure from drinking water is associated with decreased gene expression and increased DNA methylation in peripheral blood. *Toxicol. Appl. Pharmacol.*, 321, 57–66, 2017.
66. Sarkar, S., Mukherjee, S., Chattopadhyay, A., Bhattacharya, S., Low dose of arsenic trioxide triggers oxidative stress in zebrafish brain: Expression of antioxidant genes. *Ecotoxicol. Environ. Safety,* 107, 1–8, 2014.
67. Tseng, C.H., Tseng, C.P., Chiou, H.Y., et al., Epidemiologic evidence of diabetogenic effect of arsenic. *Toxicol. Lett.,* 133, 69–76, 2002.
68. Paul, D.S., Walton, F.S., Saunders, R.J., Stýblo, M., Characterization of the impaired glucose homeostasis produced in C57BL/6 mice by chronic exposure to arsenic and high-fat diet. Environ. *Health Perspect.,* 119, 1104–1109, 2011.
69. Boquist, L., Boquist, S., Ericsson, I., Structural beta-cell changes and transient hyperglycemia in mice treated with compounds inducing inhibited citric acid cycle enzyme activity. *Diabetes,* 37, 89–98, 1988.
70. Petrick, J.S., Jagadish, B., Mash, E.A., Aposhian, H.V., Monomethylarsonous acid (MMA(III)) and arsenite: LD(50) in hamsters and in vitro inhibition of pyruvate dehydrogenase. *Chem. Res. Toxicol.,* 14, 651–656, 2001.
71. Chouhan, S., Flora, S.J., Arsenic and fluoride: two major ground water pollutants. *Indian J. Exp. Biol.,* 48, 666–678, 2010.
72. Douen, A.G., Kacem, R., Jones, M.N., Direct interactions of phenylarsine oxide with hexose transporters in isolated rat adipocytes. *Biochimica. Et. Biophysica. Acta,* 944, 444–450, 1988.
73. Frost, S.C., Kohanski, R.A., Lane, M.D., Effect of phenylarsine oxide on insulin-dependent protein phosphorylation and glucose transport in 3T3-L1 adipocytes. *J. Biol. Chem.,* 262, 9872–9876, 1987.
74. Liebl, B., Muckter H, Doklea, E., Fichtl, B., Forth, W., Influence of organic and inorganic arsenicals on glucose uptake in Madin-Darby canine kidney (MDCK) cells. *Analyst,* 117, 681–684, 1992.
75. Reddy, P.S., Rani, G.P., Sainath, S.B., Meena, R., Supriya, C., Protective effects of N-acetylcysteine against arsenic-induced oxidative stress and reprotoxicity in male mice. *J. Trace Elem. Med. Biol.,* 25, 247–253, 2001.
76. Shay, K.P., Moreau, R.F., Smith, E.J., Smith, A.R., Hagen, T.M., α-lipoic acid as a dietary supplement: molecular mechanisms and therapeutic potential. *Biochim. Biophys. Acta,* 1790, 1149–1160, 2009.
77. Khandker, S., Dey, R.K., Maidul Islam, A.Z.M., Ahmad, S.A., Al-Mahmud, I., Arsenic-safe drinking water and antioxidants for the management of arsenicosis patients. *Bangladesh J. Pharmacol.,* 1, 42–50, 2006.
78. Banerjee, P., Bhattacharyya, S.S., Bhattacharjee, N., Ascorbic acid combats arsenic-induced oxidative stress in mice liver. *Ecotoxicol. Environ. Safety,* 72, 639–649, 2009.
79. Birben, E., Sahiner, U.M., Sackesen, C., Erzurum, S., Kalayci, O., Oxidative stress and antioxidant defense. *World Allergy Organ. J.,* 5, 9–19, 2012.

80. Bhattacharjee, S., Sarkar, C., Pal, S., Additive beneficial effect of folic acid vitamin B12 coadministration on arsenic induced oxidative damage in cardiac tissue in vivo. *Asian J. Pharm. Clin. Res.*, 6, 64–69, 2013.
81. Mishra, D., Flora S.J.S., Quercetin administration during chelation therapy protects arsenic-induced oxidative stress in mice. *Biol. Trace Elem. Res.*, 122, 137–147, 2008.
82. Hemalatha, P., Reddy, A.G., Reddy, Y.R., Shivakumar, P., Evaluation of protective effect of N-acetylcysteine on arsenic induced hepatotoxicity. *J. Nat. Sci. Biol. Med.*, 4, 393–395, 2013.
83. Modi, M., Kaul, R.K., Kannan, G.M., Flora, S.J.S., Co-administration of zinc and n-acetylcysteine prevents arsenic-induced tissue oxidative stress in male rats. *J. Trace Elem. Med. Biol.*, 20, 197–204, 2006.
84. Gonzalez-Pereza, O., Gonzaleza-Castaneda, R.E., Therapeutic perspectives on the combination of alpha-lipoic acid and vitamin E. *Nutr. Res.*, 26, 1–5, 2006.
85. Suryakumar, G., Gupta, A., Medicinal and therapeutic potential of Sea buckthorn (Hippophae rhamnoides L.). *J. Ethnopharmacol.*, 138, 268–278, 2011.
86. Pintea, A., Varga, A., Stepnowski, P., et al., Chromatographic analysis of carotenol fatty acid esters in Physalis alkekengi and Hippophae rhamnoides. *Phytochem. Anal.*, 16, 188–195, 2005.
87. Gupta, R., Kannan, G.M., Sharma, M., Flora, S.J.S., Therapeutic effects of *Moringa oleifera* on arsenic-induced toxicity in rats. *Environ. Toxicol. Pharmacol.*, 20, 456–464, 2005.
88. Dartsch, P.C., Antioxidant potential of selected Spirulina platensis preparations. *Phytother. Res.*, 22, 627–633, 2008.
89. Gupta, R., Flora, S.J.S., Effect of Centella asiatica on arsenic induced oxidative stress and metal distribution in rats. *J. Appl. Toxicol.*, 26, 213–222, 2006.
90. El-Demerdash, F.M., Yousef, M.I., Radwan, F.M., Ameliorating effect of curcumin on sodium arsenite-induced oxidative damage and lipid peroxidation in different rat organs. *Food Chem. Toxicol.*, 47, 249–254 2009.
91. Sharma, A., Sharma, M.K., Kumar, M., Protective effect of Mentha piperita against arsenic-induced toxicity in liver of awiss albino mice. *Basic Clin. Pharmacol. Toxicol.*, 100, 249–257, 2006.
92. Gupta, R., Flora, S.J.S., Protective value of Aloe vera against some toxic effects of arsenic in rats. *Phytother. Res.*, 19, 23–28, 2005.
93. Guha Mazumder, D.N., Chronic arsenic toxicity & amp; human health. *Indian J. Med. Res.*, 128, 436–447, 2008.
94. Flora, S.J.S., Flora, G., Saxena, G., Mishra, M., Arsenic and lead induced free radical generation and their reversibility following chelation. *Cell Mol. Biol.*, 53, 26–47, 2007.
95. Andersen, O., Principles and recent developments in chelation treatment of metal intoxication. *Chem. Rev.*, 99, 2683–2710, 1999.
96. Flora, S.J.S., Pant, B.P., Tripathi, N., Kannan, G.M., Jaiswal, D.K., Therapeutic efficacy of a few diesters of meso 2,3- dimercaptosuccinic acid during subchronic arsenic intoxication in rats. *J. Occup. Hlth.*, 39, 119–123, 1997.

97. Aaseth, J., Skaug, M.A., Cao, Y., Andersen, O., Chelation in metal intoxication-Principles and paradigms. *J. Trace Elem. Med. Biol.*, 31, 260–266, 2015.
98. Bora, A.J., Gogoi, S., Baruah, G., Dutta, R.K., Utilization of co-existing iron in arsenic removal from groundwater by oxidation-coagulation at optimized pH. *J. Environ. Chem. Eng.*, 4, 2683–2691, 2016.
99. Banerji, T., Chaudhari, S., Arsenic removal from drinking water by electrocoagulation using iron electrodes- an understanding of the process parameters. *J. Environ. Chem. Eng.*, 4, 3990–4000, 2016.
100. Park, J.H., Han, Y.S., Ahn, J.S., Comparison of arsenic co-precipitation and adsorption by iron minerals and the mechanism of arsenic natural attenuation in a mine stream. *Water Res.*, 106, 295–303, 2016.
101. Tresintsi, S., Simeonidis, K., Vourlias, G., Stavropoulos, G., Mitrakas, M., Kilogram-scale synthesis of iron oxy-hydroxides with improved arsenic removal capacity: study of Fe(II) oxidation–precipitation parameters. *Water Res.*, 46, 5255–5267, 2012.
102. Garea, A.A., Irabien, A., Arsenic removal from drinking water by reverse osmosis: minimization of costs and energy consumption. *Sep. Purif. Technol.*, 144, 46–53, 2015.
103. Van der Bruggen, B., 6 - Advances in electrodialysis for water treatment. *Adv. Membr. Technol. Water Treat.*, 185–203, 2015.
104. Lee, C.G., Alvarez, P.J.J., Nam, A., *et al.*, Arsenic(V) removal using an amine-doped acrylic ion exchange fiber: kinetic, equilibrium, and regeneration studies. *J. Hazard. Mater.*, 325, 223–229, 2017.
105. Khatamian, M., Khodakarampoor, N, Saket-Oskou, M., Efficient removal of arsenic using graphene-zeolite based composites. *J. Colloid Interface Sci.*, 498, 433–441, 2017.
106. Hu, Q., Liu, Y., Gu, X., Zhao, Y., Adsorption behavior and mechanism of different arsenic species on mesoporous MnFe2O4 magnetic nanoparticles. *Chemosphere,* 181, 328–336, 2017.
107. Martínez-Cabanas, M., López-García, M., Barriada, J.L., Herrero, R., Sastre de Vicente, M.E., Green synthesis of iron oxide nanoparticles. Development of magnetic hybrid materials for efficient As(V) removal. *Chem. Eng. J.*, 301, 83–91, 2016.
108. Bernaurdshaw, N., Evrim, C., Heechul, C., Photochemical oxidation of arsenic(III) to arsenic(V) using peroxydisulfate ions as an oxidizing agent. *Environ. Sci. Technol.*, 42, 6179–6184, 2008.
109. Brian, B., Reaction of chlorine with organic poly electrolytes in water treatment: a review. *J. Water Supply Res. Technol. AQUA.*, 54, 531–544, 2005.
110. Pinakidou, F., Katsikini, M., Simeonidis, K., Paloura, E.C., Mitrakas, M., An X-ray absorption study of synthesis- and As adsorption-induced micro structural modifications in Fe oxy-hydroxides. *J. Hazard. Mater.*, 298, 203–209, 2015.
111. Su, H., Ye, Z., Hmidi, N., High-performance iron oxide–graphene oxide nanocomposite adsorbents for arsenic removal. *Colloids Surf. A. Physicochem. Eng. Asp.*, 522, 161–172, 2017.

112. Kumar, S.K., Jiang, S.J., Synthesis of magnetically separable and recyclable magnetic nanoparticles decorated with β-cyclodextrin functionalized graphene oxide an excellent adsorption of As(V)/(III). *J. Mol. Liq.*, 237, 387–401, 2017.
113. Natale, F.D., Erto, A., Lancia, A., Musmarra, D., A descriptive model for metallic ions adsorption from aqueous solutions onto activated carbons. *J. Hazard. Mater.*, 169, 360–369, 2009.
114. Mayo, J.T., Yavuz, C., Yean, S. *et al.*, The effect of nanocrystalline magnetite size on arsenic removal. *Sci. Technol. Adv. Mater.*, 8, 71–75, 2007.
115. Chen, B., Zhu, Z., Ma, J., *et al.*, One-pot, solid-phase synthesis of magnetic multiwalled carbon nanotube/iron oxide composites and their application inarsenic removal. *J. Colloid Interface Sci.*, 434, 9–17, 2014.
116. Dannan, M., Dally, S., Conso, F., Arsenic induced encephalopathy. *Neurology*, 34, 1524–1529, 1984.
117. Winship, K.A., Toxicity of inorganic arsenic salts. *Adverse Drug React. Acute Poisoning Rev.*, 3, 129–160, 1984.
118. Axelson, O., Dahlgren, E., Jansson, C.D., Rehnuland, S.O., Arsenic exposure and mortality, A case Ref. study from a Swedish copper smelter. *Br. J. Ind. Med.*, 35, 8–15, 1978.
119. Saha, K.C., Chronic arsenical dermatoses from tube-well water in West Bengal during 1983-87. *Indian J. Dermatol.*, 40, 1–12, 1995.
120. Goebel, H.H., Schmidt, P.E., Bohl, J., Tettenborn, B., Kramer, G., Guttman, L., Poly neuropathy due to arsenic intoxication: biopsy studies. *J. Neuropathol. Exp. Neurol.*, 49, 137–149, 1990.
121. Franzblau, A., Lilis, R., Acute arsenic intoxication from environmental arsenic exposure. *Arch. Environ. Health*, 44, 385–390, 1989.
122. Bhowmick, S., Chakraborty, S., Mondal, P., Renterghem, W.V., Berghe, S.V.D., Ross, G.R., Bickley, L.K., Papa, C.M., Montmorillonite-supported nanoscale zero-valent iron for removal of arsenic from aqueous solution: kinetics and mechanism. *N.J. Med.*, 86, 377–380, 1989.
123. Barrett, J.C., Lamb, P.W., Wang, T.C., Lee, T.C., Mechanisms of arsenic induced cell transformation. *Biol. Trace. Elem. Res.*, 21, 421–429, 1989.
124. Nordstrom, S., Beckman, L., Nordenson, I., Occupational and environmental risks in and around a smelter in Northern Sweden, spontaneous abortion among female employers and decreased birth weight in their offspring. *Hereditas*, 90, 291–296, 1979.
125. Rahman, M., Tondel, M., Ahmad S.A., Axelson, O., Diabetes mellitus associated with arsenic exposure in Bangladesh. *Am. J. Epidemiol.*, 148, 198–203, 1989.
126. Michael, H.F., Beck, B.D., Chen, Y., Lewis, A.S., Thomas, D.J., Arsenic exposure and toxicology: a historical perspective. *Toxicol. Sci.*, 123, 305–332, 2011.
127. Sari, A., Tuzen. M., Biosorption of As (III) and As (V) from aqueous solution by macrofungus (Inonotus hispidus) biomass: Equilibrium and kinetic studies. *J. Hazard. Mater.*, 164, 1372–1378, 2009.

128. Aryal, M., Ziagova, M., Kyriakides, M.L., Study on arsenic biosorption using Fe (III)-treated biomass of Staphylococcus xylosus. *Chem. Eng. J.*, 162, 178–185, 2010.
129. Nguyen, T.V., Vigneswaran, S., Ngo, H.H., Kandasamy, J., Arsenic removal by iron oxide coated sponge: experimental performance and mathematical models. *J. Hazard. Mater.*, 182, 723–729, 2010.
130. Boddu, V., Abburi, K., Talbott, J., Smith, E., Haasch, R., Removal of arsenic (III) and arsenic (V) from aqueous medium using chitosan-coated biosorbent. *Water Res.*, 42, 633–642, 2008.
131. Luo, X., Wang, C., Wang, L., Deng, F., Luo, S., Tu, X., Au, C., Nano-composites of graphene oxide-hydrated zirconium oxide for simultaneous removal of As (III) and As (V) from water. *Chem. Eng. J.*, 220, 98–106, 2013.
132. Tang, W., Li, Q., Gao, S., Shang, J.K., Arsenic (III, V) removal from aqueous solution by ultrafine α-Fe_2O_3 nanoparticles synthesized from solvent thermal method. *J. Hazard. Mater.*, 192, 131–138, 2011.
133. Lin, S., Lu, D., Liu, Z., Removal of arsenic contaminants with magnetic γ-Fe_2O_3 nanoparticles. *Chem. Eng. J.*, 211–212, 46–52, 2012.
134. Feng, L., Cao, M., Ma, X., Zhu, Y., Hu, C., Super paramagnetic high-surface-area Fe_3O_4 nanoparticles as adsorbents for arsenic removal. *J. Hazard. Mater.*, 217, 439–446, 2012.
135. Zhang, G., Ren, Z., Zhang, X., Chen, J., Nanostructured iron(III)-copper(II) binary oxide: A novel adsorbent for enhanced arsenic removal from aqueous solutions. *Water Res.*, 47, 4022–4031, 2013.
136. Basu, T., Nandi, D., Sen, P., Ghosh, U.C., Equilibrium modeling of As(III,V) sorption in the absence/presence of some groundwater occurring ions by iron(III)-cerium(IV) oxide nanoparticle agglomerates: a mechanistic approach of surface interaction. *Chem. Eng. J.*, 228, 665–680, 2013.
137. Fei, Y., Sainan, S., Jie, M., Sheng, H., Enhanced removal performance of arsenate and arsenite by magnetic graphene oxide with high iron oxide loading. *Phys. Chem. Chem. Phys.*, 17, 4388–4397, 2015.
138. Kumar, S., Nair, R.R., Pillai, P.B., *et al.*, Graphene oxide-$MnFe_2O_4$ magnetic nano-hybrids for efficient removal of lead and arsenic from water. *ACS Appl. Mater. Interfaces*, 6, 17426–17436, 2014.
139. Xubiao, L., Chengcheng, W., Shenglian, L., Ruizhi, D., Xinman, T., Guisheng, Z., Adsorption of As(III) and As(V) from water using magnetite Fe_3O_4-reduced graphite oxide-MnO_2 nano composites. *Chem. Eng. J.*, 187, 45–52, 2012.
140. Jin, Z., Zimo, L., Yu, L., Ruiqi, F., Shams, AB., Xinhua, X., Adsorption behavior and removal mechanism of arsenic on graphene modified by iron-manganese binary oxide ($FeMnO_x$/RGO) from aqueous solutions. *RSC Adv.*, 5, 67951–67961, 2015.
141. Ming, L.C., Yan, S., Chun, B.H., Chen, L., Jian, H.W., Akaganeite decorated graphene oxide composite for arsenic adsorption/removal and its pro concentration at ultra-trace level. *Chemosphere*, 130, 52–58, 2015.

142. Thakkar, M., Randhawa, V., Mitra, S., Wei., L., Synthesis of diatom-FeOx composite for removing trace arsenic to meet drinking water standards. *J. Colloid. Interface Sci.,* 457, 169–173, 2015.
143. Li, R., Li, Q., Gao, S., Shang, J.K., Exceptional arsenic adsorption performance of hydrous cerium oxide nanoparticles: Part A. Adsorption capacity and mechanism. *Chem. Eng. J.,* 185–186, 127–135, 2012.
144. Dambies, L., Existing and prospective sorption technologies for the removal of arsenic in water. *Sep. Sci. Technol.,* 39, 603–627, 2005.
145. Choong, T.S.Y., Chuah, T.G., Robiah, Y., Koay, F.L.G., Azni, I., Arsenic toxicity, health hazards and removal techniques from water: an overview. *Desalination,* 217, 139–166, 2007.
146. Vasudevan, S, Kannan, B.S., Lakshmi, J., Mohanraj, S., Sozhan, G., Effects of alternating and direct current in electrocoagulation process on the removal of fluoride from water. *J. Chem. Technol. Biotechnol.,* 86, 428–436, 2011.
147. Zhao, X., Zhang, B., Liu, H., Qu, J., Simultaneous removal of arsenite and fluoride via an integrated electro-oxidation and electrocoagulation process. *Chemosphere,* 83, 726–729, 2011.
148. An, B., Liang, Q., Zhao, D., Removal of arsenic (V) from spent ion exchange brine using a new class of starch-bridged magnetite nano particles. *Water Res.,* 45, 1961–1972, 2011.
149. Baciocchi, R., Chiavola, A., Gavasci, R., Ion exchange equilibria of arsenic in the presence of high sulfate and nitrate concentrations. *Water Sci. Technol. Water Supply,* 5, 67–74, 2005.
150. Harisha, R.S., Hosamani, K.M., Keri, R.S., Natarajm, S.K., Aminabhavi, T.M., Arsenic removal from drinking water using thin film composite nanofiltration membrane. *Desalination,* 252, 75–80, 2010.
151. Floch, J., Hideg, M., Application of ZW-1000 membranes for arsenic removal from water sources. *Desalination,* 162, 75–83, 2004.
152. Slater, C., Ahlert, R., Uchrin, C., Applications of reverse osmosis to complex industrial wastewater treatment. *Desalination,* 48, 171, 1983.
153. Xu, P., Capito, M., Cath, T.Y., Selective removal of arsenic and monovalent ions from brackish water reverse osmosis concentrate. *J. Hazard. Mater.,* 260, 885–891, 2013.
154. Schoeman, J.J., Evaluation of electro dialysis for the treatment of a hazardous leachate. *Desalination,* 224, 178–182, 2008.

2

Recent Trends in Textile Effluent Treatments: A Review

Shumaila Kiran[1]*, Shahid Adeel[2], Sofia Nosheen[3], Atya Hassan[4], Muhammad Usman[2] and Muhammad Asim Rafique[5]

[1]*Department of Applied Chemistry, Govt. College University, Faisalabad, Pakistan*
[2]*Department of Chemistry, Govt. College University, Faisalabad, Pakistan*
[3]*Department of Environmental Science, Lahore College for Women University, Lahore, Pakistan*
[4]*Department of Chemistry, Federal Urdu University Arts Science & Technology, Gulshan Campus, Karachi, Pakistan*
[5]*Department of Public Administration, Govt. College University Faisalabad 38000 Pakistan*

Abstract

The proper use of dyes, auxiliaries, energy, and water is requisite in order to attain the preferred color, handle, wearing comfort, and design for textiles. For many years, achieving the ultimate result was the chief concern, but recently, owing to the change in the values in our society, security of the environment has turn out to be the central community concern. The significance and the requirement to boost the competence of cleaning procedure of the residual water from textile industry are the main thrust of this chapter. In order to discover the best condition and parameters for textile effluent treatment, different methods of treatment of the residual wastewaters are discussed. At the present time, domestic, farming waste and industrial processes pollute water resources. Due to toxic, carcinogenic, and mutagenic by-products, community apprehension more than the ecological shock of wastewater contamination has amplified. A number of conventional techniques are in use to remove these pollutants but the advanced techniques have proven more eco-friendly, rapid, cost, and time effective in nature. This chapter focuses on the use of such novel technologies to remove contaminants from effluent load.

Keywords: Industrial wastewater, treatment methods

**Corresponding author*: shumaila.asimch@gmail.com

Shahid-ul-Islam (ed.) Advanced Materials for Wastewater Treatment, (29–50)

2.1 Introduction

Many man-made dyes are used every day depending on the demand of customer. About 30 million tons is the annual world production of textiles and they need 700,000 tons of dyes per year [1]. It is used for different purposes such as fabric sprinting, leather coloring, farming investigate, photo electrochemical cells, and hair complexion. The use of such huge amount of dyes produces considerable environmental pollution through the release of their effluent load usage. Dyes possess diverse chemical chromophoric groups, and amongst the majority normally used structural groups are the subsequent: azo, anthraquinone, triarylmethane, and phthalocyanine [2].

Most of the textile dyes possess azo group in its own structural formula which plays a significant role in dyestuffs chemistry. During textile processing, the textile dyeing production utilizes huge amounts of hose down and releases heavy load of waste matter [3, 4]. Most of the industries dispose of the waste material such as colors, chemicals derivative and products into the river. The category of dye being used, nature of cloth being processed and the strength of the dyeing material greatly influence the uniqueness of wastewater releasing from dyeing factory (e.g., textile industry) that vary from day-to-day [5–7]. Primarily, the dyes present in industrial wastewaters are matter of great concern because of having toxic and potentially hazardous moeties as well as being the source of aesthetic problems [8]. The problems offered by such effluent are the hot issue for the most of developing countries including Pakistan. These effluents cause various problems and generating environmental issues such as it destroy the food chain, food web, health risk, etc. Furthermore, presence of insignificant concentration of the dyes is lethal to some microorganisms as well as to aquatic life and possibly will grounds straight destruction or inhibition of their catalytic actions. It is common observation that the most of dyes are not easily degradable [9–11]. Generally, it was observed that many dyes are not only carcinogenic and mutagenic but also cause the skin irritation. The above mentioned problem is the serious issue in Pakistan as well as others foreign countries.

A large numbers of researches have been conducted for wastewater treatment because a large section of industrial dyes are left over in wastewater due to low removal efficiency of the conventional wastewater treatment plants [12]. Therefore, treating effluent that is contaminated with synthetic dyes is deemed necessary before it is discharged into wastewater bodies. Hence there is a need to treat wastewater to make it viable for agriculture, domestic usage, and other purposes[13].

2.2 Industrial Dyes, Dying Practices, and Associated Problems

There is a variety of synthetic dyes which are in common use. It was seen that reactive dyes have extraordinary properties (viz., they chemically bind textile fibers, which significantly improves product color stability and washability). The dyeing procedure used is reliant on the chemical configuration and properties of the specific dye; water is used as a medium to smooth the progress of the dyeing procedure for all dyes [14].

There are three common dying processes:

1. In exhaust dyeing, the fabric stuff comes in contact with dye fluid and it picks the dyes. Consequently, the amount of dye in the container steadily decreases [15].
2. In constant dyeing, in which the dye is set against the yarn by means of warmth or chemicals, subsequent to which surplus dye is waste out. Continuous dyeing is mainly appropriate for use with natural fibers [16].
3. Bunch dyeing is the mainly accepted and frequent way used for dyeing fabric supplies that slowly transferred the dye from a relatively huge quantity dyeing container to the substrate.

By comparison to the exhaust method, the constant and bunch dyeing approaches usually use much lower quantities of chemicals and aqueous media [17]. Waste effluent load from the continuous dyeing process needs such methodology which are cost, time and energy effective. To formulate waste matter troubles extra comprehensible being faced by the cloth manufacturing, it seems to believe essential to have apparent figure of processes carried out in textile industry (Figure 2.1) which result in effluent production. A description of individual operations that are performed on cotton textiles and the main pollutants that result from each operation has been shown in Figure 2.2.

Although there are a lot of dyes being used to color the textiles, here are some dyes which are regularly used in textile sector [18].

2.3 Wastewater Remediation

During synthesis and dyeing processes about 10–15% of the whole manufactured colorants is vanished [19]. Moreover, the color imparted to wastewater from the dyes must be removed before being discharged to water

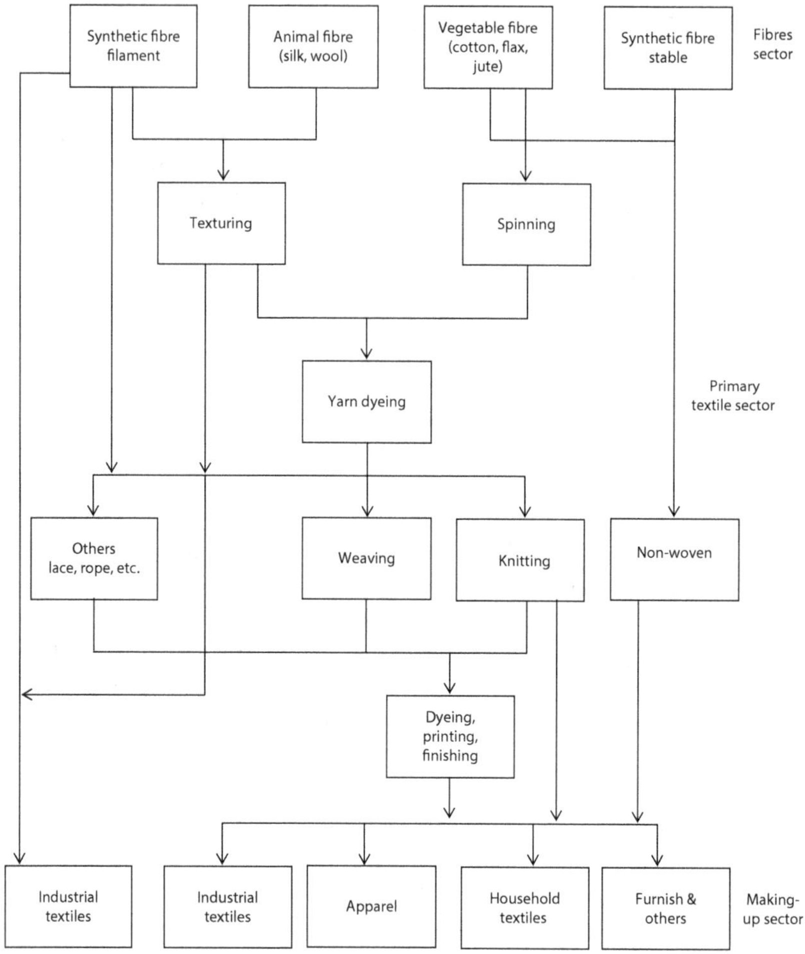

Figure 2.1 Flow chart showing the process carried in textile industry.

bodies, because it affects water transparency and is regarded to be a pollutant. Removing color from wastewater during dye processing represents a challenge and major concern for industry researchers [20]. It is deemed necessary to minimize the quantity of dyes or their alteration into a lesser amount of poisonous or yet less risk-free products in effluent because discharge standards are becoming more rigorous [21]. Comparative expenditure, levels of handling requisite, or location limitations are the important factors that influence the wastewater treatment methods.

Generally, following methods are in use for the removal of color from wastewaters.

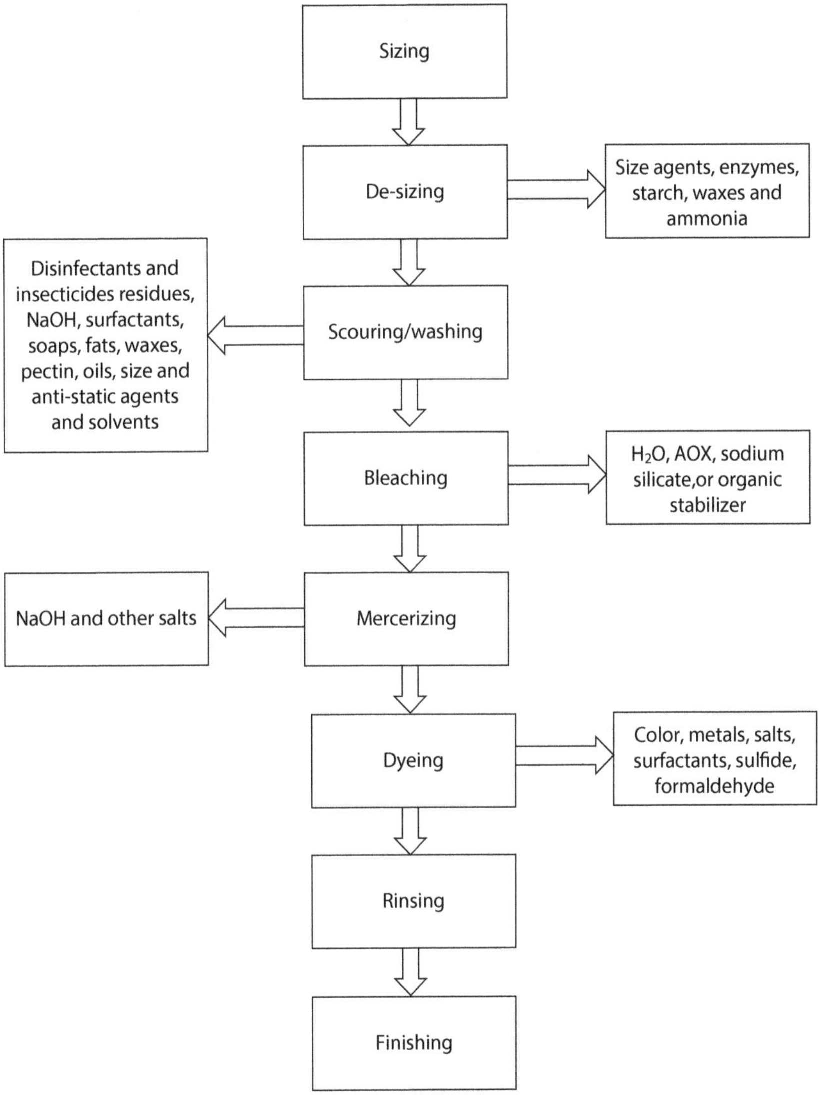

Figure 2.2 A description of the individual operations that are performed on cotton textile and the main pollutants that result from each operation.

2.4 Physical Methods

Physical methods for effluent management consist of exclusion of substances by means of logically happening forces, such as gravity, electrical attraction, van der Waal forces, and substantial barriers. Usually, substantial

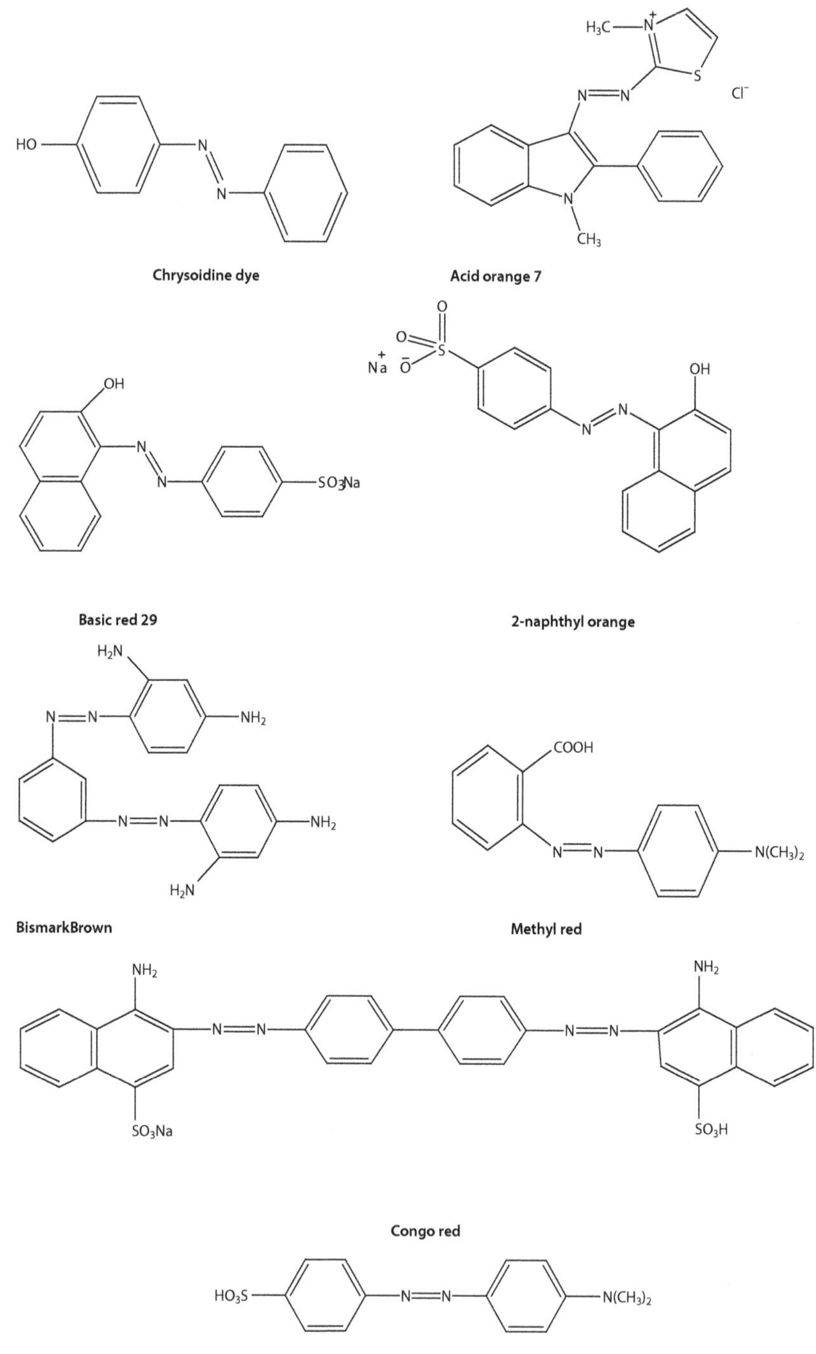

Figure 2.3 The chemical composition of man-made dyes mostly used in the textile industry.

treatments do not change the chemical constitution of the required materials. Although in the case of vaporization, physical state changes and frequently detached substances are caused to agglomerate [22]. So for initial treatment physical methods are used which include:

2.4.1 Adsorption

Amongst physical methods, adsorption process is extremely efficient for treating industrial wastewater. Diverse types of adsorbents are commonly used. These include synthetic and natural adsorbents, for example, ferrierite, laumontite, silica, alumina. Activated carbon is proved to be an efficient adsorbent for exclusion of organic compounds from the industrial effluent [24] that must be regenerated or disposed-off after use. Some precautionary measures for disposing of organics as they can trickle away more the time and may originate contamination at a later stage. Some additional adsorbents include silica gel, zeolites, and a variety of clays. Polymeric adsorbents are eco-friendly, occur in nature, and act as ion-exchangers or contain structures that permit adsorption of substances inside them [25]. Cellulose bio-adsorbents also been developed synthetically and show promising results in removal of reactive dyes.

2.4.2 Coagulation and Flocculation

Coagulation and flocculation are different ways for water purification as compared to adsorption process. To treat dye wastewater for color removal, Inorganic (alum, lime and iron salts) and organic (polymers) coagulants are used, either unaided or in mixture with one another [26]. Alum is the valuable inorganic salt for dye removal from textile effluents [27]. Coagulation process combined with filtration is most commonly used to remove dissolved organic carbon, iron (Fe), manganese (Mn), additives causing turbidity and color, and it can also remove heavy metals [28].

2.4.3 Membrane Processes

Membrane technologies are used to remove the dyestuff or have potential to reuse the materials used for dyeing. Membrane pressure is the driving force for treatment of dyeing wastewater. This technique is helpful in elucidation, concentration and most important separation of dyes from wastewater [29]. However, this methodology is still not widely encouraged due to some limitation as it requires special equipment, high investment, and membrane fouling, etc. [30].

2.4.4 Ultra Filtration

Ultra filtration is the unique process that permits the recovery of dyes from water. It excludes macromolecules and particles from wastewater but the elimination of color is not comprehensive (between 31 and 76%) [31]. Membrane transport properties depend upon membrane thickness. Rate of solvent evaporation gradually increases which favors to increase the dyes separation process [32]. This work is used as pretreatment for reverse osmosis or in conjunction with a biological one [33].

2.4.5 Micellar-Enhanced Ultrafiltration (MEUF)

Although ultrafiltration has been recognized as the best technique available for removal of impurities but problem arises due to low permeability and high pressure requirement. Many industries have been used ultrafiltration but it is not extensively accepted because it needs further filtration and does not have direct reuse [34]. Micellar-enhanced ultrafiltration (MEUF) is modified version of ultrafiltration which is used to remove dyes as well as multivalent metal ions from water stream effectively [35]. It needs a lesser amount of power owing to little operating pressures. It is the uniqueness of this process [36]. Contaminants are incorporated into micelles [37] which are larger in size and difficult to pass through pores of membrane and, thus, are rejected by the membrane during ultrafiltration [38].

Positive charge at the surface of cationic micelles, for example, Cetylpyridinium chloride (CPC), Octadecyltrimethyl ammonium bromide (OTAB), Cetyltrimethyl ammonium bromide (CTAB), binds anionic impurities [39, 40] while anionic micelles can attract metal ions as well as cationic dyes. The use of sodium dodecyl sulfate (SDS) and linear alkylbenzenesulfonate (LAS) is among the least expensive and mostly easily available anionic surfactant for elimination of heavy metal ion from industrial waste such Pb^{2+}, Cd^{2+}, Cu^{2+},Ni^{2+}, and Zn^{2+} [36]. Nonionic surfactants, although more expensive than anionics, but more effective as they are more tolerant to water hardness [40]. They, however, suffer from big drawback because of being not suitable for removal of ionic impurities [41, 42]. Mixed micellar solution of anionic, cationic, and nonionic surfactants may prove very useful for simultaneous removal of all types of anionic, cationic as well as nonionic impurities [40, 42].

2.4.6 Reverse Osmosis

Reverse osmosis membranes produce an elevated value of permeate and have a rejection coefficient of 90% or even greater for the majority of ionic substances [43]. This method is also considered as hyper filtration; in which water run powerfully through a partially permeable membrane [44]. Amount of dissolved salt depends upon the osmotic stress and energy required for the separation process [45]. Nanofiltration can be employed either as a pretreatment or as a posttreatment, to lessen the lethal compounds in the ultimate product water and thus permit direct employ of the treated water for beneficial purposes without any further extra treatment [46, 47].

2.4.7 Nanofiltration

The efficiency of this separation process lies between reverse osmosis and ultrafiltration. Reactive dyes can be successfully detached from waste matter by using nanofiltration [48]. A dye maintenance speed of 85–90% was found to be acceptable for the recycling of the water [43]. Nanofiltration characterizes one of the unusual applications likely for the handling of solutions in treatment of dyeing wastewater [49] with complex and highly concentrated solutions [50, 51].

2.5 Chemical Methods

Chemical oxidation of the pollutants present in wastewater might lead to the elimination or reduction in their toxicity. Generally oxidizing agents are used to facilitate chemical oxidation [52].

2.5.1 Photo Catalytic Degradation of Dyes

Photocatalytic degradation of dyes is a tool relatively common in recent era. It was observed that dyes undergo degradation more frequently in the presence of some oxidant and nanoparticles [53] and this degradation rate depends on the photolytic properties of dyes. Photo-mediated conditions are required to be optimized in order to get optimum results [54]. Oxidation processes are extensively used for effective degradation and removal of dyes thus permitting reuse of wastewater [55, 56]. The use of UV/H_2O_2 along with other radiation sources is efficient for removal of dyes. These reactions are accelerated by OH° free radical. Numerous additional catalysts such as TiO_2, ZnO nanoparticles, beneath specific irradiation situations are recognized to decolorize a variety of man-made dyes [57].

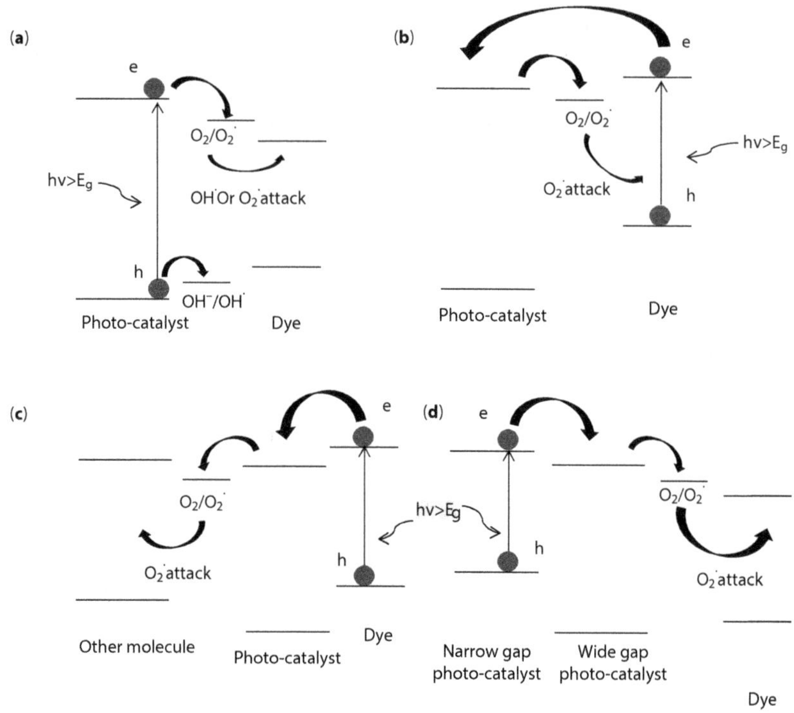

Figure 2.4 The mechanisms of light induced degradation of dyes (a) photocatalysis, (b) dye sensitization followed by dye degradation, (c) dye sensitization followed by reduction of a second molecule, and (d) degradation by coupled semiconductors under visible light [10].

2.5.2 Oxidation and Photocatalysis with Hydrogen Peroxide

Using hydrogen peroxide (H_2O_2) oxidation procedures are usually investigated as alternatives for wastewater handling in two systems:

1. Homogenous systems (accompanied using visible or UV light, ozone, and peroxidase etc. [58]).
2. Heterogeneous systems (accompanied using semiconductors, zeolites, clays with or without UV light, TiO2, stable customized zeolites along with iron and aluminum [59]).

Hydrogen peroxide has ability to decolorize industrial waste matter in the presence of Fe(II) sulphate and titanium oxide [60]. A relative study was conducted on photo-assisted removal of methylene blue dye-loaded waste matter by advanced oxidation methodising TiO_2/ZnO photo catalyst

[9]. These methods can be effectively useful to decolorize a number of areas, but they are, however, unsuccessful whilst employed with vat and disperse dyes [61].

2.5.3 Ozonation

Ozonation has found significant application for decolorizing synthetic dyes. Various dye effluents counter in a different way with ozone, depending on their chemical composition [62]. Researchers determined that ozone effectively decolorizes azo dyes in wastewaters. Additionally, it is originated that the speed of dye photo-degradation by ozonation process relies on the chemical structure of the dye being treated [60]. This treatment has advantage that it increases neither the quantity of wastewater nor the sludge accumulation [62]. The major drawback of this method is that it releases carcinogenic nitrogenous aromatics and some poisonous molecules therefore it is not recommended for use [63].

2.5.4 Degradation of Dyes Using Sodium Hypochlorite (NaOCl)

Sodium hypochlorite (NaOCl) is widely used as a cleaning and disinfection agent that can be easily transported or stored [64]. Dyes that contain amino groups in their structure may be successfully degraded by this treatment. The Cl^{-1} generated from NaOCl action destructs amino groups, which eventually result in breakage of azo bond [65]. The disadvantage of using NaOCl is the strong corrosive material that releases toxic chlorine gas when mixed with acids. Thus, safety precautions must be considered [66].

2.5.5 Electrochemical Method

Electrochemical treatment involves the redox reactions for exclusion of metals under the impudence of exterior shortest current in the presence of electrolytic medium. It is an assured waste matter management tool [67] that has become popular since last two decades. It includes electro-coagulation, electro-flotation, electro-oxidation, electro-disinfection, andelectro-deposition processes, etc. [68]. This process offers an extensive array of viable advantages than other methods, such as elevate delimination effectiveness, sparkling energy alteration simple procedure, and low ecological impact [69]. In count, it is a flexible and benevolent skill that is capable of be useful in diverse types of industrial effluents detoxification because the key reagent is electron as distinctive reagent and they do not produce solid residues [70].

2.6 Bioremediation

The issues associated with the discussed physicochemical treatment methods make biological methods more preferable due to their cost-effective and environmental friendly nature [71]. These are considered the most effective tool for removal of wastes from wastewater loaded with organic constituents. In the mineralization of multifaceted organic molecules and recalcitrant nature of molecules, microorganisms play an important role [72]. Biological treatment has advantage over physicochemical treatments methods thus about 70% of the organic materials are degraded by the biological treatment [73]. To biodegrade synthetic dyes, bioremediation is a smart and easy operational way but found to be complex [74]. Many microbial types have been evaluated for their aptitude to either decolorize or mineralize a range of dyes [75]. Various fungal strains have been well reported to effect the dye concentration on rate of color removal [76]. Elevated amounts of dyes may hinder cell development leading to a decrease in the speed of decolorization [77, 78]. Reports showed that degradation speed of dyes decreased owing to manufacture and gathering of poisonous products in the growth media [79].

The employment of microbes to treat industrial waste matter polluted with dyes comprises two apparent advantages:

These are

1. Proficient in dyes mineralization [76]
2. The final products are either less noxious or non-hazardous [62]

2.7 Products Recognition and Mechanisms of Dye Degradation

Competence of the management scheme is reliant on pH of wastewater [88]. The most favorable pH depends on the type of microbes and constitution of dyes [89]. Cell permeability is another very significant aspect not linked with the construction of dye although having a chief task in dye decolorization method and level [90]. Chemical composition of azo dyes has an important consequence on the biodegradation speed [91]. Depending on the placement of azo group and their numbers, various dyes degrade in a different way [92]; more quantity of azo groups on dye molecule would result in the reduction of degradation rate as it would need

Table 2.1 Advantages and disadvantages of dye treatment technologies.

Sr. no.	Methods	Advantages	Disadvantages	References
1	Activated carbon	Effective in removing a variety of dyes	Ineffective when used with disperse and vat dyes	[80]
2	Silica gel	Effective for removing (alkaline) dyes	Least effective when used at commercial scale	[81]
3	Membrane filtration	Effective for all dyes	High pressure, expensive	[82]
4	Ion exchange	Adsorbents can be regenerated	Ineffective against disperse dyes	[83]
5	Electro kinetic	Economically viable	Sludge production disposal constraints	[84]
6	Fenton's reagent	Removing soluble and insoluble dyes	Formation of by-products	[21]
7	Photo-Fenton's	High-quality prospective for dye detoxification	May construct lethal by-products	[85]
8	Ozonation	Functional in gaseous processes	Little half-life, constant furnish is obligatory	[86]
9	NaOCl	Effective for cleaving azo linkages	Formation of by products	[86]
10	Electrochemical	Products are not harmful	High price of electrical energy	[84]
11	Biodegradation	Price efficient, valuable for all dyes	Non-speedy method; Successful at optimal conditions	[87]

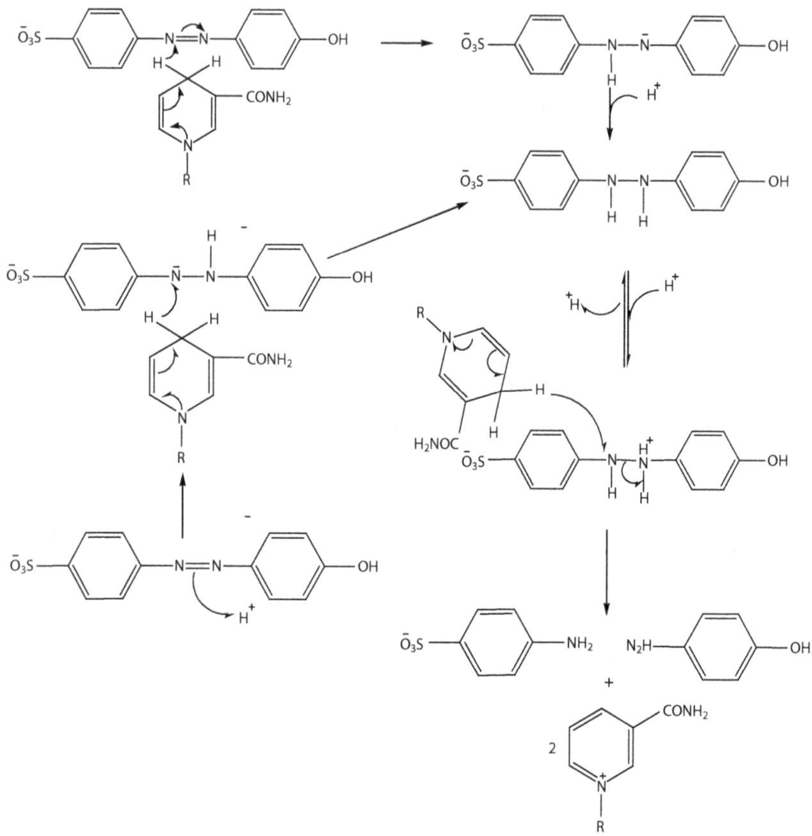

Figure 2.5 Mechanism for reduction of azo dye.

additional time to lessen additional azo groups [93]. The production of lethal dye metabolites might suppress the degradation of dye molecule. Textile wastewater has eminent quantities of salts, dispersing and solubilizing agents which create unfavorable effects on biodegradation [94].

2.8 Conclusion

This review is mainly aimed to assemble current technical achievements made in remediating the synthetic dyes in wastewater. In addition, we have categorized and explained the technological processes that are used to remove dyes from water and compare their advantages and disadvantages. However, due to high cost involved, disposal problems, and fewer compliant to extensive variety of dye wastewaters, the most of these methods have not been broadly applied. Bioremediation of dye-based wastewaters using effective dye degrading microorganisms is still seen as an striking alternative elucidation owing to its inexpensiveness, environmentally friendly, and widely suitable treatment tool.

2.9 Future Outlook

Many present literatures for the removal of synthetic dyes from wastewater have been concerning with different aspects of applications of physical, chemical, and microbiological methods and techniques, by the look for new microbes given that privileged decaying speed and with the explanation of the chief biochemical and biophysical processes underlying the decolorization of dyes. This inclination definitely proves the vital function of microbiological processes in the potential technologies used for the elimination of dyes from waters. The extensive application of joint techniques via microbiological disintegration and chemical or physical treatments to improve the efficiency of the microbiological disintegration can be anticipated in future.

References

1. Gugnami, A., Mishra, A., *Textile & apparel compendium*, Technopak, 2012.
2. Singh, H.B., Bharati, K.A., *Handbook of natural dyes and pigments*, Woodhead Publishing, New Delhi, 2014.
3. Ding, S., Li, Z., Wang, R., Overview of dyeing wastewater treatment technology. *Water Resour. Prot.*, 26, 73, 2010.
4. Khandare, R., Kabra, A., Kadam, A., Govindwar, S., Treatment of dye containing wastewaters by a developed lab scale phytoreactor and enhancement of its efficacy by bacterial augmentation. *Int. Biodeterior. Biodegrad.*, 78, 89, 2013.
5. Ong, S.A., Uchiyama, K., Inadama, D., Ishida, Y., Yamagiwa, K., Treatment of azo dye Acid Orange 7 containing wastewater using up-flow constructed wetland with and without supplementary aeration. *Bioresour. Technol.*, 101, 9049, 2010.
6. He, Y., Yi, Y., Comparison of different coke quenching methods on water environment. *J. Kunming Univ. Sci. Technol.*, 34(6), 82, 2011.
7. Kadam, A., Kamatkar, J., Khandare, R., Jadhav, J., Govindwar, S., Solid state fermentation: Tool for bioremediation of adsorbed textile dyestuff on distillery industry waste-yeast biomass using isolated *Bacillus cereus* strain EBT1. *Environ. Sci. Pollut. Res.*, 20, 1009, 2013.

8. Javed, M., Usmani, N., Assessment of heavy metal (Cu, Ni, Fe Co, Mn, Cr, Zn) pollution in effluent dominated rivulet water and their effect on glycogen metabolism and histology of *Mastacembelusarmatus*. *Springer Plus*, 2, 390, 2013.
9. Mohabansi, N.P., Patil, V.B., Yenkie, N., A comparative study on photo degradation of methylene blue dye effluent by advanced oxidation process by using TiO_2/ZNO photo catalyst. *Rasayan J. Chem.*, 4(4), 814, 2011.
10. Mehta, R., Surana, M., Comparative study of photo-degradation of dye Acid Orange -8 by Fenton reagent and Titanium Oxide- A review. *Der. Pharma. Chemica.*, 4(1), 311, 2012.
11. Montazerozohori, M., Nasr-esfahani, M., Joohari, S., Photocatalytic degradation of an organic dye in some aqueous buffer solutions using nano titanium dioxide a kinetic study. *Environ. Prot. Eng.*, 38(3), 45, 2012.
12. Giles, D.E., Mohapatra, M., Issa, T.B., Anand, S., Singh, P., Iron and aluminium based adsorption strategies for removing arsenic from water. *J. Environ. Manag.*, 92, 3011, 2011.
13. Qiao, J., Jiang, Z., Sun, B., Sun, Y., Wang, Q., Guan, X., Arsenate and arsenite removal by FeCl3: effects of pH, As/Fe ratio, initial As concentration and co-existing solutes. *Sep. Purif. Technol.*, 92, 106, 2012.
14. Fang, L., Zhang, X., Sun, D., Chemical modification of cotton fabrics for improving utilization of reactive dyes. *Carbohydr. Polym.*, 91(1), 363, 2013.
15. Teng, X., Ma, W., Zhang, S., Application of tertiary amine cationic poly acryl amide with high cationic degree in salt-free dyeing of reactive dyes. *Chin. J. Chem. Eng.*, 18(6), 1023, 2010.
16. Swamy, V.N., Gowda, K.N., Sudhakar, R., Extraction and dyeing conditions of natural dye from flowers of *Plumeriarubra* L. on textiles and fastness properties. *Indian J. Tradit.Know.*, 15, 278, 2016.
17. Tarbuk, A., Grancaric, A.M., Leskovac, M., Novel cotton cellulose by cationization during mercerization—part 2: the interface phenomena. *Cellulose*, 21, 2167, 2014.
18. Bin, N., Zhipeng, W., Dandan, S., Two methods of fast-growing poplar veneer dyeing processing. *J. Northeast Forest. Univ.*, 33(6), 6, 41, 2011.
19. Lee, C.S., Robinson, J., Chong, M.F., A review on application of flocculants in wastewater treatment. *Process Saf. Environ. Prot.*, 92, 489, 2014.
20. Lai, T.L., Yong, K.F., Yu, J.W., Chen, J.H., Wang, C.B., High efficiency degradation of 4-nitrophenol by microwave-enhanced catalytic method. *J. Hazard. Mater.*,185, 366, 2011.
21. Hao, O.J., Kim, H., Chiang, P.C., Decolorization of wastewater. *Crit. Rev. Environ. Sci. Technol.*, 30, 449, 2000.
22. Sacco, O., Stoller, M., Vaiano, V., Ciambelli, P., Chianese, A., Sannino, D., Photocatalytic degradation of organic dyes under visible light on n-doped photocatalysts. *Int. J. Photoenergy*, Article ID 626759, 2012, DOI:10.1155/2012/626759.

23. Ali, A., Zafar, M.Z.H., Haq, I., Phull, A.R., Ali, J.S., Hussain, A., Synthesis, characterization, applications, and challenges of iron oxide nanoparticles. *Nanotechnol. Sci. Appl.*, 9, 49, 2016.
24. Zhou, L., Jin, J., Liu, Z., Liang, X., Shang, C., Adsorption of acid dyes from aqueoussolutions by the ethylenediamine-modified magnetic chitosan nanoparticles. *J. Hazard. Mater.*, 185(2), 1045, 2011.
25. Lin, Z.Y., Zhang, Y.X., Chen, Y.L., Qian, H., Extraction and recycling utilization of metal ions (Cu^{2+}, Co^{2+} and Ni^{2+}) with magnetic polymer beads. *Chem. Eng. J.*, 200, 104, 2012.
26. Chen, D., Li, Y., Zhang, J., Li, W., Zhou, J., Shao, L., Efficient removal of dyes by a novel magnetic Fe_3O_4/ZnCr-layered double hydroxide adsorbent from heavy metal wastewater. *J. Hazard. Mater.*, 243, 152, 2012.
27. Lukasiewicz, E., Post-coagulation sludge management for water and wastewater treatment with focus on limiting its impact on the environment. *Econ. Environ. Studies*, 16(4), 831, 2016.
28. Teh, C.Y., Budiman, P.M., Shak, K.P.Y., Wu, T.Y., Recent advancement of coagulation–flocculation and its application in wastewater treatment. *Res. Chem. Eng. Ind.*, 5(16), 4363, 2016.
29. Fras, Z.L., Ristic, T., Tkavc,T., Adsorption and antibacterial activity of soluble and precipitated chitosan on cellulose viscose fibres. *J. Eng. Fiber Fabr.*, 7(1), 50, 2011.
30. Monsalvo, V.M., Mohedano, A.F., Rodriguez, J.J., Activated carbons from sewage sludge: application to aqueous-phase adsorption of 4-chlorophenol. *Desalination*, 277, 377, 2011.
31. Benhadji, A., Ahmed, M.T., Maachiv, R., Electrocoagulation and effect of cathode materials on the removal of pollutants from tannery wastewater of Rouiba. *Desalination*, 277, 128, 2011.
32. Ciardelli, G., Ranieri, N., The treatment and reuse of wastewater in the textile industry by means of ozonation and electroflocculation. *Water Res.*, 35, 567, 2001.
33. Zodi, S., Louvet, J.N., Michon, C., Potier, O., Pons, M.N., Lapicque, F., Leclerc, J.P., Electrocoagulation as a tertiary treatment for paper mill wastewater: removal of non-biodegradable organic pollution and arsenic. *Sep. Purif. Technol.*, 81, 62, 2011.
34. Baek, K., Yang, T.W., Competitive bind of anionic metals with cetylpyridinium chloride micelle in micellar enhanced ultrafiltration. *Desalination*, 167, 101, 2004.
35. Zeng, G.M., Micellar enhanced ultrafiltration of phenol in synthetic wastewater using polysulfone spiral membrane. *J. Membr. Sci.*, 310, 149, 2008.
36. Huang, J.H., Removal of cadmium ions using micellar-enhanced ultrafiltration with mixed anionic-nonionic surfactants. *J. Membr. Sci.*, 326, 303, 2009.
37. Zaghbani, N., Removal of Eriochrome Blue Black R from wastewater using micellar-enhanced ultrafiltration. *J. Hazard. Mater.*, 168, 1417, 2009.

38. Misra S.K., Studies on the simultaneous removal of dissolved DBP and TBP as well as uranyl ions from aqueous solutions by using Micellar-Enhanced Ultrafiltration Technique. *Hydrometallurgy*, 96, 47, 2009.
39. Ahmad, A.L., Micellar-enhanced ultrafiltration for removal of reactive dyes from an aqueous solution. *Desalination*, 191, 153, 2006.
40. Ahmad, A.L., Puasa, S.W., Reactive dyes decolourization from an aqueous solution by combined coagulation/micellar-enhanced ultrafiltration process. *J. Chem. Eng.*, 132, 257, 2007.
41. Chung, Y.S., Effects of membrane hydrophilicity on the removal of a trihalomethane via micellar-enhanced ultrafiltration process. *J. Membr. Sci.*, 326, 714, 2009.
42. Iqbal, J., Kim, H.J., Yang, J.S., Baek, K., Removal of arsenic from groundwater by micellar-enhanced ultrafiltration (MEUF). *Chemosphere*, 66, 970, 2007.
43. Muthumareeswaran, M.R., Alhoshan, M., Agarwal, G.P., Ultrafiltration membrane for effective removal of chromium ions from potable water. *Scientific Reports*, 7, 41423, 2017. DOI: 10.1038/srep41423.
44. Rahman, F., The treatment of industrial effluents for the discharge of textile dyes using by techniques and adsorbents. *Text. Sci. Eng.*, 6, 1, 2016.
45. Mehdipour, G.M., Azhdari, F., Photocatalytic degradation of textile dye direct orange 26 by using $CoFe_2O_4/Ag_2O$. *Adv. Environ. Technol.*, 2(2), 77, 2016.
46. Malekian, F., Farhadian, M., Sohrabi, M., Razmjou, A., Application of nanofiltration as a tertiary treatment in a polyester production industry for wastewater reuse. *Desalin. Water Treat.*, 57(16), 7175, 2016.
47. Phuntsho, S., Hong, S., Elimelech, M., Shon, H.K., Forward osmosis desalination of brackish groundwater: meeting water quality requirements for fertigation by integrating nanofiltration. *J. Membr. Sci.*, 436, 1, 2013.
48. Stoyko, P.P., Pencho, A.S., Color and COD retention by reverse osmosis. *Desalination*, 154, 247, 2003.
49. Chollom, M.N., Rathilal, S., Pillay, V.L., Alfa, D., The applicability of nanofiltration for the treatment and reuse of textile reactive dye effluent. *Water S. A.*, 41(3), 398, 2015.
50. Le, N.L., Nunes, S.P., Materials and membrane technologies for water and energy sustainability. *Sust. Mat. Technol.*, 7, 1, 2016.
51. Malekian, F., Farhadian, M., Sohrabi, M., Razmjou, A., Application of nanofiltration as a tertiary treatment in a polyester production industry for wastewater reuse. *J. Membr. Sci.*, 57(16), 7175, 2016.
52. Ranganathan, K., Karunagaran, K., Sharma, D.C., Recycling of wastewaters of textile dyeing industries using advanced treatment technology and cost analysis-Case studies. *Res. Conserv. Recycl.*, 50, 306, 2007.
53. Fartode, A.P., Parwate, D.V., UV photolytic decolorization study of synthetic waste water containing indigo carmine dye in presence of H_2O_2. *Inter. J. Chem. Phys. Sci.*, 3(3), 23, 2014.

54. Jafari, T., Moharreri, E., Amin, A.S., Miao, R., Song, W., Suib, S.L., Photocatalytic water splitting—the untamed dream: a review of recent advances. *Molecules*, 21(7), 900, 2016.
55. Chatha, S.A.S., Kiran, S., Gulzar, T., Kamal, S., Ghaffar, A., Chatha, M.N., Comparative study on decolorisation and mineralisation of synthetic and real textile effluents using advanced oxidation processes. *Oxid. Commun.*, 39(2), 1604, 2016.
56. Rashid, A., Nosheen, S., Kiran, S., Bhatti, H.N., Kamal, S., Shamim, F., Akram, W., Rafique, M.A., Combination of oxidation and coagulation processes for wastewater decontamination on batch scale. *Oxid. Commun.*, 39(2), 1716, 2016.
57. Aminia, M., Ashrafi, M., Photocatalytic degradation of some organic dyes under solar light irradiation using TiO_2 and ZnO nanoparticles. *Nanochem. Res.*, 1(1), 79, 2016.
58. Jadhav, V.V., Dhabbe, R.S., Sabale, S.R., Nikam, G.H., Tamhankar, B.V., Degradation of dyes using high temperature stable anatase nanosphere TiO_2 photocatalyst. *Universal J. Environ. Res. Technol.*, 3(6), 667, 2013.
59. Ananthashankar, R., Ghaly, A., Photocatalytic decolourization of textile effluent containing reactive red 120 dye with Uv/TiO_2. *Am. J. Engg. Appl. Sci.*, 6(3), 252, 2013.
60. Zaharia, C., Suteu, D., Muresan, A., Muresan, R., Popescu, A., Textile wastewater treatment by homogenous oxidation with hydrogen peroxide. *Environ. Engng. Manag. J.*, 8(6), 1359, 2009.
61. Mohabansi, N.P., Satone, A.K., Solar photocatalytic degradation of textile effluents by using titanium dioxide. *Int. J. Pharm. Chem. Biol. Sci.*, 5(2), 487, 2015.
62. Litter, M.I., Quici, N., Photochemical advanced oxidation processes for water and wastewater treatment. *Recent Patents on Engineering*, 4, 217, 2010.
63. Deepa, N.,.Meghna, P., Kandasamy, S., Experimental studies on decolorisation of malachite dye using continuous photocatalytic reactor. *Int. Res. J. Environ. Sci.*, 3(3), 14, 2014.
64. Rutala, W.A., Weber, D.J., Uses of inorganic hypochlorite (bleach) in healthcare facilities. *Clin. Microbiol. Rev.*, 10, 597, 1997.
65. Quader, A.K.M.A., Treatment of textile wastewater with chlorine: An effective method. *Chem. Eng. Res. Bullet.*, 14(1), 59, 2010.
66. Fukuzaki, S., Mechanisms of actions of sodium hypochlorite in cleaning and disinfection processes. *Biocontrol. Sci.*, 11, 147, 2006.
67. Ciblak, A., Mao, X., Padilla, I., Vesper, D., Alshawabkeh, I., Alshawabkeh, A.N., Electrode effects on temporal changes in electrolyte pH and redox potential for water treatment. *J. Environ. Sci. Health, Part A*, 47(5), 718, 2012.
68. Martínez-Huitle, C.A., Brillas, E., Decontamination of wastewaters containing synthetic organic dyes by electrochemical methods: a general review. *Appl. Catal. B.*, 87(3–4), 105–145, 2009.
69. Mook, W.T., Aroua, M.K. Issabayeva, G., Prospective applications of renewable energy based electrochemical systems in wastewater treatment: a review. *Renew. Sust. Energ. Rev.*, 38, 36, 2014.

70. Ghanbari, F., Moradi, M., A comparative study of electrocoagulation, electrochemical Fenton, electro-Fenton and peroxi-coagulation for decolorization of real textile wastewater: electrical energy consumption and biodegradability improvement. *J. Environ. Chem. Eng.*, 3(1), 499, 2015.
71. Guieysse, B., Norvill, Z.N., Sequential chemical–biological processes for the treatment of industrial wastewaters: review of recent progresses and critical assessment. *J. Hazard. Mater.*, 267, 142, 2014.
72. Anjaneyulu, Y., Chary, N.S., Raj, D.S.S., Decolourization of industrial effluents-available methods and emerging technologies. *Rev. Environ. Sci. Biotechnol.*, 4, 245, 2005.
73. Kiran, S., Ali, S., Asgher, M., Anwar, F., Comparative study on decolorization of reactive dye 222 by white rot fungi Pleurotusostreatus IBL-02 and Phanerochaetechrysosporium IBL-03. *Afr. J. Microbiol. Res.*, 6(15), 3639, 2013.
74. Naseer, A., Nosheen, S., Kiran, S., Kamal, S., Javaid, M.A., Mustafa, M., Tahir, A., Degradation and detoxification of Navy Blue CBF dye by native bacterial communities: an environmental bioremedial approach. *Desalin. Water Treat.*, 57(50), 24070, 2016.
75. Mahmoud, M.S., Decolorization of certain reactive dye from aqueous solution using Baker's Yeast (*Saccharomyces cerevisiae*) strain. *HBRC J.*, 12(1), 88, 2016.
76. Przystas, W., Zablocka-Godlewska, E., Grabinska-Sota, E., Efficacy of fungal decolorization of a mixture of dyes belonging to different classes. *Braz. J. Microbiol.*, 2, 415, 2015.
77. Ivana, C., Lepretti, M., Martucciello, S., Esposito, C., Enzymatic strategies to detoxify Glutein: Implications fot Celiac disease: review article. *Enz. Res.*, Article ID.17435, 2010.
78. Kalyani, D.C., Patil, P.S., Jadhav, J.P., Govindwar, S.P., Biodegradation of reactive textile dye Red BLI by an isolated bacterium *Pseudomonas Sp.* SUKI. *Biores. Technol.*, 99, 4635, 2008.
79. Sawant, S.S., Salunke, B.K., Taylor, L.E., Kim, B.S., Enhanced agarose and xylan degradation for production of polyhydroxyalkanoates by co-culture of marine bacterium, *Saccharophagusdegradans* and its contaminant, *Bacillus cereus. Appl. Sci.*, 7(3), 225, 2017.
80. Gupta, V.K., Suhas, A., Application of low cost adsorbents for dye removal, a review. *J. Environ. Manag.*, 90(8), 2313, 2009.
81. Robinson, T., Chandran, B.,. Nigam, P., Studies on the production of enzymes by white rot fungi for the decolorization of textile dyes. *J. Enz. Microb. Technol.*, 29, 575, 2011.
82. Rekha, R., Chauhan, P., Gangopadhyay, U.K., Zero effluent process by using membrane type solute separation systems for wet process house. *Asian Text. J.*, 18, 65, 2009.
83. Laszlo, J.A., Regeneration of azo-dye saturated cellulosic anion exchange resin by *Burkholderiacepacia* anaerobic dye reduction. *Environ. Sci. Technol.*, 34(1), 167, 2000.

84. Slokar, Y.M., Marechal, A.M., Methods of decoloration of textile wastewaters. *Dyes Pigm.*, 37, 335, 1997.
85. Ong, S.T., Keng, P.S., Lee, W.N., Ha, S.T., Hung, Y.T., Dyes waste water treatment-review. *Water*, 3, 157, 2011.
86. Rodriguez, E., Ruiz-Duenas, F.J., Kooistra, R., Ram, A., Martinez, A.T., Martinez, M.J., Isolation of two laccase genes from the white rot fungus *Pleurotuseryngii* and heterologous expression of the PcI3 encode protein. *J. Biotechnol.*, 134, 9, 2008.
87. Reshma, S.V., Spandana, S., Sowmya, M., Bioremediation technologies. World Congress of Biotechnology, India, 2011.
88. Cripps, C., Bumpus, J.A., Aust, D.A., Biodagradation of azo and heterocyclic dyes by *Phanerochaetechrysosporium*. *Appl. Environ. Microbiol.*, 56, 1114, 1993.
89. Chen, B.Y., Jo-Sh, C., Shan-Yu, C., Bacterial species diversity and dye decolorization of a two species mixed consortium. *Environ. Engng. Sci.*, 20, 337, 2003.
90. Patil, N.N., Shukla, S.R., Elucidation of degradation mechanism of reactive blue 171 dye by ceric ammonium nitrate. *Environ. Prog. Sustain. Energy*, 35(5), 1254, 2016.
91. Asad, S., Amoozegar, M.A., Pourbabaee, A.A., Sarbolouki, M.N., Dastgheib, S.M.M., Decolorization of textile azo dyes by newly isolated halophilic and halotolerant bacteria. *Biores. Technol.*, 98, 2082, 2007.
92. Kiran, S., Ali, S., Asgher, M., Anwar, F., Comparative study on decolorization of reactive dye 222 by white rot fungi *Pleurotusostreatus* IBL-02 and *Phanerochaetechrysosporium* IBL-03. *Afri. J. Microbiol. Res.*, 6(15), 3639, 2012.
93. Anku, W., Oppong, S.O.B., Shukla, S.K., Govender, P.P., Comparative photocatalytic degradation of monoazo and diazo dyes under simulated visible light using Fe^{3+}/C/S doped-TiO_2 nanoparticles. *ActaChimicaSlovenica*, 63(2), 380, 2016.
94. Mbarek, W.B., Azabou, M., Pineda, E., Fiol, N., Escoda, L., Suñol, J.J., Khitouni, M., Rapid degradation of azo-dye using Mn–Al powders produced by ball-milling. *RSC Adv.*, 7(21), 12620, 2017.

3

Polyaniline as an Inceptive Dye Adsorbent from Effluent

Raminder Kaur[1]* and Monika Duhan[2]

[1]*Department of Polymer Science and Chemical Technology, Delhi Technological University, India*

[2]*Department of Applied Chemistry and Polymer Technology, Delhi Technological University, Delhi, India*

Abstract

The wastewater is an "essential by-product" of any processing industry, and it is one of the main culprits which causes water pollution. The addition of effluent water to various water bodies results in serious pollution problems and causes negative effects on the eco-system and human life. Effluent from the textile industry contributes to a huge amount of dyes which are not only highly toxic but are also a matter of concern due to aesthetic point of view. They reduce the sunlight penetration and photosynthesis, thereby increasing the biological oxygen demand and causing lack of dissolved oxygen that is vital to sustain aquatic life. A variety of synthetic dyestuff released by the textile industry has been posing a threat to the safety of the environment. Moreover, some dyes are considered toxic or even carcinogenic for humans. Considering environmental aspects and legal restrictions, the removal of dye waste from the effluent is now one of the priorities of the textile industries. Among all the effluent treatment processes, adsorption has been found to be a process of high competence because of its low cost, effectiveness, insensitivity to toxic substances, feasibility, simplicity of design and operation. Operational difficulties and the need for effective regeneration of conventional adsorbents led the researchers to explore many novel adsorbents with characteristics such as low cost, good environmental stability, high surface area, easy to synthesize and regenerative, that is, natural clays, acid-activated red mud, fly-ash, polymeric hydrogels, chitosan, and various agriculture wastes, etc. Polyaniline with its unique physicochemical nature exhibits most of these desired properties. Studies reveal that, it has been used in the recent past for lowering the concentration of dyes from

**Corresponding author*: rkwalia14@gmail.com

Shahid-ul-Islam (ed.) Advanced Materials for Wastewater Treatment, (51–100)

their aqueous solutions. In this brief overview, a study has been conducted on the adsorption capability, parametric study and effective removal of the pollutants by polyaniline in its different forms,that is, acidic and basic forms, polyaniline composites and coatings, polyaniline salts, and grafted polyaniline (on solid substrates) and doped polyaniline. Main emphasis is given on the removal of dyes from the industrial effluents.

Keywords: Industrial effluent, wastewater treatment, adsorption, dyes, polyaniline

3.1 Introduction

When the quantity of any substance in the water increases to such an extent that it becomes harmful for the human beings and surrounding environment, then it is regarded as a pollutant. Pollutants at the low levels are not affective, as all the water bodies existing on the earth have ability to clean up assertive amount of pollution naturally by diluting and dispersing these pollution causing substances. But as the concentration or quantity of foreign substances gets increased, they start affecting the quality of water [1]. The plants, animals, and even the humans that consume such contaminated water also get affected. It is indeed the quantity of the pollutant or the volume of the water body that determines the extent of pollution. In addition to this surface water pollution (i.e., contamination of rivers, lakes, oceans, etc.), pollution may also affect underground aquifers, which broadly termed as ground water [2]. The ultimate consequence of pollution is the deterioration of the environment and the harm to the human beings. The main contributors of water pollution are: domestic wastes including sewage; chemical wastes from industries including harmful chemicals, dyes, soaps, and other chemicals; agriculture runoff including pesticides, insecticides, unused fertilizers; oil and petroleum products from refineries, oil wells or spell from ships, radio-active wastes from nuclear power plants and other types of wastes from hospitals, process industries, etc. [3]. In addition to these, discharge of huge amount of hot water from thermal power plants and commercial operations in water bodies also affect the environment and water animals and plants with thermal shock [1, 2].

Human health gets severely affected by the water pollution directly or indirectly. Consumption of polluted water can cause many fatal diseases. Water pollution poses a threat to the environment and human beings by various means such as it cause death of aquatic animals, disrupt food chains, spread diseases, and destroy ecosystems. In addition, the presence of organic compounds in the water body consumes a large sum of dissolved oxygen for degradation, which in turn lowers the amount of dissolved oxygen [4] and exhibits biochemical oxygen demand (BOD). Health risks from water pollution vary from area to area depending upon the discharge of effluent from different sources that vary in pollution-causing agents. According to reported work, presence of more than 500 mg/L total dissolved solids (TDS) in any water body is the indication of the water pollution. Higher amount of TDS in water may disturb many on-going physiological and biological processes in human body [5]. Toxic pollutants such as heavy metals and different types of chemicals may cause many toxicological effects on human health and the aquatic ecosystem.

Environmental pollution is becoming very problematic because of steadily increasing world population, industrialisation, and global & environmental variations. The major reasons for this are the deficit of knowledge of environmental importance for healthy life style, mistreatment of natural resources, and production and discharge of different types of pollutants in the environment [6]. Any type of pollution which perturbs the environment and human health is considered as the culprit for the environmental pollution. There is no denying of the fact that environment has to be protected and conserved to make the future life possible. Among different types of pollution, water pollution plays a major role and seeks immediate attention [7]. Throughout the last decades, awareness about the water pollution has grown worldwide and several new remedial measures are proposed and implemented for the prevention and conservation of available water resources [8, 9]. Environmental awareness is evidently one of the major concerns of today in all developed and developing countries. The different factors responsible for the water pollution are needed to deal wisely. The research should be more focussed on the environmental friendly products and technologies. The need of the hour is to conduct technological and social investigation to conserve and improve the environment.

3.1.1 Effluent from the Industries

Hazardous wastes released by a number of industries are the main focus of concern, as, if not remediated correctly, it can cause threat to the environment and the mankind. Since the industrial revolution had entered in the man's life, many things had revolutionary changed. But it affected the nature and the environment in different ways [10]. Along with many benefits from industries, many negative effects were also observed. Industries became one of the major contributors of water pollution as they discharge numerous pollutants which are very toxic in nature and affect the surrounding environment and human life adversely. Some industries use fresh water to bear away industrial wastes which includes: mercury, nitrates, sulfur, lead, asbestos, petrochemicals, and oils. All these remains ultimately get mixed with water bodies. During the earlier time, only small factories were build-up to meet the requirements of the people, which generated a limited waste, within the permissible limit of self-cleaning capacity of the water bodies. But as the time passed, these factories started to get converted into large industrial set-ups, which, knowingly or unknowingly started polluting the water bodies, with their huge discharge of industrial effluents [11].

Industries pollute the environment in many fronts. They contaminate sources of drinking water with their untreated or partially treated wastewater, poison air with the releases of harmful toxins and spoil the soil with burial of wasted chemicals and unnecessary by-products. In addition, noise pollution is the major threat in this era of industrialization. Moreover, sometimes, industrial mishaps may also lead to environmental disasters. The major causes of industrial pollution which are responsible for the environmental degradation are: unplanned industrial growth, use of out-dated technologies, inefficient waste disposal, lack of policies to control pollution, and low replenish rate of resources leached from the natural world. The main categories of water pollutants which significantly affect the aquatic component of our biosphere are: (1) suspended solids and sediments, (2) inorganic pollutants, (3) organic pollutants, (4) pathogens, (5) radioactive pollutants, (6) nutrients and agriculture runoff, (7) thermal pollutants, and (8) oil [5, 11]. In details:

- ***Suspended solids and sediments***: This class of pollutants consists of eroded minerals from land, slit, and sand particles. These pollutants can reduce the sunlight penetration in the water body and also can destruct the normal aquatic life by their deposition in the water. In addition, if these deposited solids are organic in nature and lead to the anaerobic conditions after their decomposition. Gills of fishes in water can be injured by these solid suspended pollutants.
- ***Inorganic pollutants***: Some nonbiodegradable compounds also contribute the water pollution from the industries. These pollutants comprise inorganic salts, mineral acids, metals, trace elements, metal complexes with organic compounds, sulfates and cyanides, etc. These pollutants are dangerous if present in the appropriate amount in the water and pose a bad impact on the existing organisms in the environment.

- ***Organic pollutants***: In addition to inorganic compounds coming from the industries, some organic compounds also enter the water resources in the form of pollutants and affect the quality of water. These organic pollutants are classified as (1) *Oxygen Demanding Wastes*– These pollutants are responsible for the depletion of the dissolved oxygen (DO) in the water, because they need oxygen for their decomposition by aerobic oxidation. (2)*Synthetic Organic Compounds*–such as pharmaceuticals, food additives, plastics, solvents, synthetic pesticides, insecticides, soaps & detergents, paints, synthetic fibres, and many volatile organic compounds in trace amount – These pollutants interfere with the microbial degradation during self-purification process of water body and also the quality of water.
- ***Pathogens***: The main sources of these pollutants are the sewage, poultries, and slaughterhouses. These pollutants include different types of bacteria and viruses, which can cause some fatal diseases.
- ***Radioactive pollutants***: The main sources of radioactive pollutants are nuclear power plants and nuclear reactors, testing of nuclear weapons, mining and processing of radioactive ores, industrial activities, medical institutes and research and development activities. They can cause serious problem after their deposition in the teeth and bones.
- ***Nutrients***: Some nutrients such as nitrates and phosphorous come from the effluent of the fertilizer industries and agricultural runoff. These nutrients promote "Eutrophication" and stimulate the growth of the algae and weeds in the water, which, in turn decreases the amount of dissolved oxygen in water, thus termed as pollutant. In addition to the aquatic life, human beings also affected by these pollutants, if polluted water is consumed by them directly.
- ***Thermal pollution***: The addition of large quantities of hot water in any water body is the cause of thermal pollution. The main culprits and contributors of thermal pollution are commercial Industries, nuclear power plants, and thermal power plants. Thermal pollution affects the aquatic life in different ways such as increase in the bacterial activity, reduction in DO level and cause thermal stratification in the water body.
- ***Oil***: This water pollutant is present in nature itself. It is degradable in the presence of bacterial action. The major sources of

this pollutant are leakage of oil pipes, oil spills and waste from the refineries and industries. It separates the contact of water body with the air and results into contraction of dissolved oxygen (DO) in the water [12]. In addition, photosynthesis activity of the aquatic plants is also affected by this pollutant, because it covers all the surface water and reduces the sunlight transmission through the surface to the aquatic flora.

3.2 Pollution Due to Dyes

Textile sector, also being a point of attraction from last two decades, is one of the major culprits for water pollution. This is due to the employment of a large volume of water for different processes of dyeing for good quality fabric production [13, 14]. A noticeable change has been seen in the things with the discovery of synthetic dyes such as more colors fast, production at low cost, brighter and easy application to the fabric, etc. However, these new changes made the world more advance and colorful but at a hefty cost. The toxic chemicals used to produce synthetic dyes may cause cancer or even death. Moreover, these are responsible for the unwanted color of the wastewater because of its recalcitrance behavior and resistance to sunlight penetration in water body [15]. Dyes are one of the major group of water-polluting chemicals among all water pollutants [16]. Water is used in different forms in the textile industries like 90–94% as process water and 6–10% as cooling water. Effluent from the textile industries contains a large range of dyes, derivatives, surfactants, solvents, salts, suspended particles, acids or bases, and additives, which makes it so complex to treat before its disposal to water bodies [17]. As per the reported work, in addition to different types of dyes (reactive, anionic, cationic, nonionic and basic dyes) used in the textile industries, several other chemicals that are used as wetting agents, antifelting, softening, sizing agents, carriers, surfactants, etc., are highly toxic, poorly biodegradable and persistent in the nature for long period and cause threat to the aquatic life [13, 18, 19]

Annual production of about 7×10^5 tonnes/year has been reported for more than 100,000 different commercial dyes in the recent literature [20]. However exact data is unknown, but more than 10,000 tonnes/year dyes are consumed around the world and out of this amount, roughly 100 tonnes/year of dyes is lost in the textile effluent [21]. Synthetic dyes are used not only for the dying purpose in the textile industries [22] but also playing a role in different sectors of progressive technology such as leather tanning [23], cosmetic, food processing, rubber, plastics, dye-synthesizing industries, and printing [24–26], paper industries [27], ground-water tracking [28], sewage [29] to determine the specific surface area of activated carbon [30], treatment of wastewater [31] and many more sectors. Different dyes employed in the textile industries are broadly categorized as: (1) Anionic dyes that include direct dyes, reactive dyes and acidic dyes. (2) Cationic dyes that include basic dyes. (3) Nonionic dyes including dispersed dyes. Reactive and basic dyes are easily water soluble, cheaper to make, bright colored and easy to apply on fabric. So, these dyes are generally preferred in the textile industries [32]. Type of the dye and dying process used to produce a colored fabric in the textile industries are some factors on which the concentration of unused dye in the wastewater is dependent. If the dying process used is improper and ineffective, usually about 25% of the total amount of dye gets lost in the wastewater and that contributes as water pollutant. It has been observed that the textile industries contributed a major part to the water pollution worldwide, in last two decades. Effluent coming from the textile industries consists of high amount of COD and BOD, mineral oils, unused dyes, and suspended solids, etc. According to literature survey, a significant amount of dyes are lost during dyeing process and a roughly 2–20% of dyes are directly flushed to the industrial effluent [33]. During this modern time, in addition to textile industries, several other industries such as paper, printing, carpet, food, cosmetic, and plastic industries also use a number of dyes to color their outputs. This also aided in the amount of colored effluent to water bodies and leads to the water pollution [34–37].

3.2.1 Lethal Effects of Dyes

With increasing demands for more colorful textile products, dyes are becoming an essential part of the modern life. But it is not safe to overuse these chemical compounds, because dyes can cause adverse effects on the environment and mankind. Different dyes are reported to have different lethal effects. Anthraquinon-containing dyes are difficult to degrade and have their impact in water body remains for a long period of time [38]. Complex dyes are known for their carcinogenic nature, as a majority of these are chromium based [39–42]. Even a very low concentration of the basic dye can affect the visibility in the water because of their high intensity of colors [39, 41–47]. Absorption and reflection of sunlight by these dyes hinder the penetration of the sunlight to the water body and ultimately affect the aquatic life. Also, the presence of some metals and aromatics in the effluent may influence the aquatic life [39–43, 48]. Toxicity of azo dyes is due to the existence of harmful amines in the textile effluent [38]. Industrial workers mainly working in the dye production may also get affected by dyes, which may harm their kidney, liver, or brain and may affect their central nervous system or reproductive system [49]. Due to water soluble nature of the reactive dyes, a small part of these may washed away during their use, as effluent and pollute the environment [50]. Moreover these dyes are quite stable chemically and resist breakdown and may have adverse impact on the environment. Reactive dyes, as reported in literature, are very effective in dying purpose due to their formation of covalent bonds with woollen, cellulose and PA fibres, but may also affect the living organisms by forming the bonds with – NH_2 and – SH group of proteins. The industry workers who deal directly with reactive dyes are reported to suffer from respiratory diseases [41, 42]. Dyes being persistent in nature due to their high thermal and photo stability affect the environment for long period. Even, after breakdown, that took a long span to decompose. Moreover, their breakdown products are highly toxic, mutagenic, and carcinogenic to life [49, 51]. Worst part is that, most of the dyes employed in the textile industries are known by their commercial/trade names only, and common workers are not aware of their chemical nature and different biological hazards associated with them. Though, it is possible to predict the toxicity of dyes [52], awareness among the workers and users is the major requisite. Keeping these lethal effects of dyes in view, necessary treatment of industrial effluent, especially that of textile industry prior to its discharge in water body is very important.

3.3 Methods Used for the Dye Removal

The removal of the coloring materials and dyes from the effluent of textile industries is an extremely demanding task because pigments and dyes confront biodegradation and remain in the environment for long span of time [53, 54]. These synthetic dyes and coloring matter are not only highly toxic but also a matter of concern due to aesthetic point of view. As the effluent from textile industries and other provenances is loaded with huge amount of dye, it needs necessary treatment prior to the discharge in water bodies. Scientists have reported the use of many physical and chemical treatment processes to eradicate the dye content from the industrial effluent such as coagulation, membrane separation, adsorption and advanced oxidation, etc. [55]. Characteristics of textile wastewater and the skill & ability of the operator are some factors which help to select the ideal treatment system. Literature provides the necessary information on the currently available water treatment techniques and also recommends many effective and economic alternatives for dye removal and decolorization of effluent [56, 57].

The major treatment methods used in the industries for dye removal are categorized into following classes:

(i) Physical methods:
- Coagulation /flocculation
- Adsorption
- Ion exchange
- Complexometric techniques
- Membrane separation

(ii) Chemical methods
- Chemical oxidation
- Irradiation

(iii) Biological methods
- Aerobic treatments
- Anaerobic techniques

(iv) Electrochemical methods

Every treatment process has few advantages and disadvantages, but in general, environmentally safe and low-cost treatment techniques are usually preferred [58, 59]. For example, as reported in the some previous study, azo dyes could be easily transformed into aromatic amine using direct anaerobic processes because most of the azo dyes are not degraded efficiently by biological treatments [60]. However these treatment methods are profitable and effective only when the concentration of the solute in the wastewater is relatively high [61–63]. Another study recommended adsorption as an effective and low cost alternate for removal of dyes and decolorisation [38]. As evident from the literature, the extensive removal of pollutants from industrial effluent may not be possibly achieved via a single technique, as a single treatment applied is capable to remove only specific type of pollutant. Different physical and chemical treatment methods [64] employed to remove dyes from the industrial effluent are reported in Table 3.1. The advantages and drawbacks of these available and commonly practiced treatment processes have also been summarized.

3.3.1 Removal of Dyes by Adsorption

Among all the available treatment processes, adsorption has been found to be a high calibre because of low cost, availability, and effectiveness of the adsorbent; insensitivity to toxic substances; feasibility and simplicity of design and operation [65–68]. Adsorption is a process in which the dissolved adsorbate molecules (in this case dyes or coloring agents) are

Table 3.1 Different effluent treatment techniques and their advantages and disadvantages [71].

Treatment technique	Advantages	Disadvantages
Adsorption (commonly used adsorbent: activated carbon)	Effective removal of various kinds of dyes	Costly, difficult to regenerate
Membrane separation method	Useful for almost all types of dyes	Production of concentrated sludge
Ion exchange method	Easier adsorbent regeneration	Applicable to limited dyes
Irradiation	Laboratory level effective oxidation	High amount of dissolved O_2 required for treatment
Electro kinetic Coagulation method	Low cost	More production of sludge
Oxidation method	Easy to apply	Requires activation of H_2O
Ozonation method	No increase in volume of effluent (ozone in gaseous form)	Short half-life (20 min)
Fenton's reagent	Effective decolorisation of soluble and insoluble dyes both	Sludge formation and its disposal
Electrochemical degradation	No production of sludge, no use of chemicals	Less removal of dyes due to high flow rate
Photochemical method	Less foul smell, no sludge production	High by-products production
Sod. Hypochlorite (NaOCl)	Azo-bond cleavage	Formation of aromatic amines

By white-rot fungi	Use of enzymes for degradation	Unreliable enzyme production
Microbial culture of mixed bacteria	Achieved decolorisation almost in a day	Azo dyes not degraded under aerobic conditions
Adsorption by biomass (living or dead)	Effective for dyes, affinity for microbial system	Useful only for some classes of dyes
Textile–dye bioremediation systems	Decolorization of azo and other water soluble dyes	Production of methane and hydrogen sulphide gas

attracted to the adsorbent surface by physical force or chemical bonding. Factors such as contact time, temperature, pH, particle size, and surface area are the basis for the adsorption of the dye present as pollutant [69]. Attachment of ionic and molecular species bearing different functional groups is dependent on the nature of the interaction between adsorbent and adsorbate molecules. It may be electrostatic interactions between oppositely charged sites at the adsorbent surface or it may be physiosorption due to weak van der waals forces. Or it can be chemisorption, involving strong adsorbate-adsorbent bonding or might be simple attachment of adsorbate molecule at a definite site on adsorbent surface [70]. When the adsorbent is not so costly and does not require any prior treatment, adsorption process seems to be quite effective [72]. In addition, adsorption process is also fruitful when biological treatment methods are not effective for remediation of the industrial effluent [72, 73].

Classically, the most preferred adsorbent used for the removal of the dyes is activated carbon, due to its powerful adsorption capacity. As reported in literature, the activated carbon is capable to adsorb both low molecular weight and larger molecules because of its porous structure and large surface area [70]. Although, activated carbon has some advantages in terms of more surface area and highly porous structure, its boundless use is narrowed because of its high cost and complicated preparation [74]. In addition, activated carbon is nonselective in nature and it is quite difficult to abstract the activated carbon from the solution and its dumping may lead to the secondary pollution [70]. Many researchers are working out to develop a broad range of alternative adsorbents for the treatment of the wastewater to substitute commercially available costly activated carbon. The main emphasis of the researchers is to develop new and novel alternative adsorbents, which have preferable performance characteristics, such as larger surface area, modified mechanical properties, low cost, high adsorption capacity, simple preparation, etc. In the recent past, many studies have been conducted and researchers have developed a variety of adsorbents from different precursors like agriculture by-products, natural materials, solid wastes of industries and have tested their possibility for the lowering the dye concentration from wastewater [75]. But many of these adsorbents have not been explored extensively at the industrial level due to some operational instability. Different adsorbents used for the dye removal from industrial effluent are given in Table 3.2.

3.3.1.1 *Factors Affecting Adsorption*

Adsorption is the aggregation of liquid or gaseous substance on the surface of a highly porous solid [72]. It is a surface phenomenon and is the outcome

Table 3.2 Different adsorbents used for the dye removal from industrial effluent.

Adsorbent	Dye used	Adsorption capacity (mg/g)	Ref.
1. Agriculture and industrial waste as adsorbent			
Rice husk	Indigo carmine	65.90	[76]
Activated carbon	Crystal violet	64.80	[77]
Sugarcane bagasse	Methylene blue	34.2	[78]
Sugarcane dust	Basic green-4	3.99	[79]
Cotton waste	Basic blue	277	[80]
Fly ash	Brilliant green	65.9	[81]
Sludge ash	Reactive blue-2	250	[82]
Sludge-based activated carbon	Acid brown-283	20.50	[83]
Activated carbon (chemviron GW)	Basic red-46	106	[83]
Sugar industry mud	Basic red-22	519	[84]
Activated carbon (jute fiber)	Reactive red	200	[85]
Metal hydroxide sludge	Reactive red-120	45.87	[86]
Charfines	Direct brown	6.40	[87]
Waste Fe(III)/Cr(III) hydroxide	Congo red	44	[88]
Activated carbon (filtrasorb 400)	Ramazol reactive black	434	[89]
Wheat straw	Remazol red	2.50	[90]
Corn-cob shreds	Remazol red	0.60	[90]
Barley husk	Congo red	171	[91]
Slag	Basic blue-9	9.95	[92]
Core Pith	Acid violet	1.60	[93]
Bamboo Dust Carbon	Methylene blue	143.20	[94]

(Continued)

Table 3.2 Cont.

Adsorbent	Dye used	Adsorption capacity (mg/g)	Ref.
2. Fruits and plants waste as adsorbent			
Orange peel	Rhodamine B	3.22	[95]
Yellow passion fruit	Methylene blue	16	[96]
Mandarin peel	Methylene blue	15.20	[96]
Banana pith	Direct red-28	18.20	[97]
Garlic peel	Methylene blue	142.86	[98]
Raw date pits	Methylene blue	80.29	[99]
Bagasse pith	Acid blue-25	21.70	[100]
Palm fruit bunch	Basic yellow-21	327	[101]
Jack fruit peel	Basic blue-9	285.71	[102]
Activated desert plant	Methylene blue	23	[103]
Guava leaf powder	Methylene blue	95.10	[104]
Tea waste	Methylene blue	85.16	[105]
Mango seed kernel	Methylene blue	142.90	[106]
Jute stick powder	Rhodamine B	87.70	[107]
Tree fern	Basic red-13	408	[108]
Saw dust-oak	Acid blue-25	27.85	[109]
Wood	Basic blue 69	77	[110]
Eucalyptus bark	Remazol BB	34.10	[111]
Wood chips	Remazol red	2.80	[112]
Beech Sawdust (original)	Red basic-22	20.20	[113]
Saw dust	Methylene blue	62.40	[114]
3. Natural inorganic minerals as adsorbent			
Clay	Methylene blue	6.30	[115]
Sepiolite	Reactive red-239	108.80	[116]
Red mud	Congo red	4.05	[117]

(*Continued*)

Table 3.2 Cont.

Adsorbent	**Dye used**	**Adsorption capacity (mg/g)**	**Ref.**
Zeolite	Reactive black-5	60.50	[116]
Ca–Montmorillonite	Basic green-5	156.30	[118]
Ti–Montmorillonite	Basic green-5	170.50	[118]
Raw kaolin	Methylene blue	13.99	[119]
Calcined alunite	Reactive blue-114	170.70	[120]
Halloysite nanotubes	Methylene blue	84.32	[121]
Dolomite	Brilliant red	900	[122]
Diatomaceous earth	Methylene blue	198	[123]
Anion clay hydrotalcite	Acid blue-29	34	[124]
Bentonite–Na	Acid blue-193	67.10	[125]
Crude clay	Basic red-46	54	[126]
Fly ash	Reactive blue-21	106.71	[127]
Sepiolite	Reactive blue-21	66.67	[127]
Natural zeolite	Basic red-46	8.56	[128]
Bentonite	Methylene blue	33	[129]
Silica	Basic blue-3	11.16	[130]
4. Bioadsorbents for dye removal			
Chitin and chitosan	Acid green-25	645.1	[131]
Chitosan bead (crab)	Reactive red-222	1106	[132]
Peat	Acid blue-25	12.7	[110]
Chitosan flake	Reactive red-222	339	[132]
White rot fungi	Remazol black B	60	[133]
Fungus (*Aspergillus niger*)	Congo red	14.72	[134]
Dried seagrape	Astrazon golden yellow	35.5	[135]
Non-cross-linked chitosan beads	Reactive red-189	1189	[136]

of the various liquid–solid intermolecular forces. Two types of adsorption are commonly observed, that is, physical adsorption and chemical adsorption. Physical adsorption is the result of weak van der Waal forces of attraction between adsorbate and adsorbent molecules. In addition to van der Waal forces, dipole–dipole π–π interaction, polarity and hydrogen bonding also play significant role [137]. It is reversible in nature. Unlike physical adsorption, when very strong chemical interactions exist between the molecules or ions of the adsobate and surface of the adsorbent, chemical adsorption takes place, which, is irreversible in nature [73]. Most of the dye adsorption process, as evident from literature, involves the physical adsorption. Main factors that affect the adsorption process are: initial dye concentration, pH of the solution, adsorbent surface area, particle size of adsorbent, temperature, contact time and adsorbent to adsorbate ratio [34, 72, 73]. Optimized values of these parameters help to remove the dyes or other harmful toxins from the water at industrial level. Various parameters that have significant effect on adsorption of dyes are discussed briefly in the subsequent section.

Effect of Initial Concentration of Dye: Initial dye concentration is one of the significant, which affects the adsorption of dye on any adsorbent. Generally, percentage of dye removal is observed to be decreased with an increment in the initial concentration of the dye. This happens because, as the dye concentration is increased, the number of available active sites, present initially, gets saturated due to the occupancy of more adsorbate moieties. After some time, when very little active sites remain available for subsequent adsorption, the adsorption percentage of the dye apparently decreases. In contrast to it, the adsorption capacity of the adsorbent is reported to be increased with increase in initial dye concentration because of the existence of high driving forces as a result of the concentration gradient which, in turn increases the rate mass transfer [138]. According to a recent literature study, when polyaniline-coated sawdust nanocomposites was used for the adsorption of acid violet 49 dye from effluent, it was observed that the adsorption capacity of the dye evidently increased from 48.42 to 173.41 mg/g as the initial dye concentration was increased from 25 to 100 mg/L [139]. Analogous to it, a study on activated bleaching earth for the removal of acid dye reported the similar trend in results. In an another study, methyl orange (MO) was removed from the water by using chitosan/alumina interface and the results showed the decrease in the percentage removal of the dye from 99.53% to 83.55% with increased MO concentration from 20 to 400 mg/L [140] A decrease in adsorption percentage of methylene blue dye was also reported with increasing its initial concentration from 10 to 90 mg/L for the pine leaves as adsorbent [21]. Table 3.3 reports the effect of initial dye concentration on the adsorption using different adsorbents.

Effect of Solution pH: pH value of the solution to be treated plays a significant role in the adsorption process. The degree of ionisation of the adsorbate, that is, dyes and the surface properties of the adsorbent depends upon the pH of the solution. In most of the adsorption studies, the effect of pH on the dye adsorption has been studied extensively [151]. A study on the effect of the pH for the adsorptive removal of Methyl Orange dye (MO) from its aqueous solution on polyaniline/saw dust (PANI/SD) as adsorbent, reported that the maximum adsorption percentage occurred at the pH values from 4 to 8 [152]. According to some another research, a decrease in the adsorption of the Acid Green 25 (AG25) on to SD and

Table 3.3 Effect of initial dye concentration on the adsorption using different adsorbents.

Dye used	Adsorbent	Initial dye conc. (mg/L)	Removal percentage (%) observed	Ref.
Direct brown 2	Sawdust	600–850	99.6–99.1	[141]
Basic blue 9	Sugarcane bagasse	250–500	94–55.5	[142]
Rhodamine B	Sugarcane bagasse	100–500	99.1–87.1	[142]
Astrazone black	Apricoc seed	50–500	91–62	[143]
Methylene blue	Raw mango seed	50–250	99.1–92.5	[144]
Methylene blue	Modified mango seed	50–250	99.9–96.9	[144]
Eriochrome black T	Activated carbon	30–150	45–10	[145]
Methylene blue	Modified sawdust	25–500	91.2–66.3	[146]
Methylene blue	Pine leaves	10–90	96–41	[21]
Methylene blue	Kaoline	10–40	90–62	[147]
Acid violet	Red mud	10–40	26.2–12.52	[148]
Malachite green	Rice husk	10–30	82.5–71	[149]
Congo red	Fly ash	5–30	99–84	[150]

PAni/SD was observed with increase in the pH value of the solution. It was reported that the more H^+ ions were present in the reaction system at lower pH, which facilitate the adsorption of anionic acid green dye on the positively charged adsorbent. On the other hand, at higher pH value (basic medium), more OH^- ions existed in the solution that competed with the negatively charged dye molecules for the available sites on the surface of the adsorbent. This resulted into less adsorption of dye at high pH value [153]. Figure 3.1 showing the effect of pH on adsorption of tartrazine dye onto polyaniline-coated sawdust (PANI/SD) and PANI/Al_2O_3 Composite. More removal has been observed at the lower pH, due to the higher concentration of the H^+ ions and as expected, less removal was observed at the higher pH value [154, 188].

The pH of the dye solution determines the adsorption of a particular dye on a particular adsorbent, depending upon the charges present on both. Depending upon the functional groups, ions or charges in the system, adsorption is observed to be different for different combination of dyes and adsorbents. Another study revealed the adsorption of Basic Green 4 dye by using Ananas comosus leaf powder. Results indicated that maximum adsorption occurred at pH 10 [155]. The removal of Congo red dye was also studied in literature. The pine cone was used as adsorbent and a maximum adsorption at pH 3.5 was reported [156]. RB4 dye was removed by modified barley straw as an adsorbent. 100% removal was claimed at the pH value of 3 that was reported reduced to the 50% with rise in pH value [157]. According to the most of the data available in literature, higher

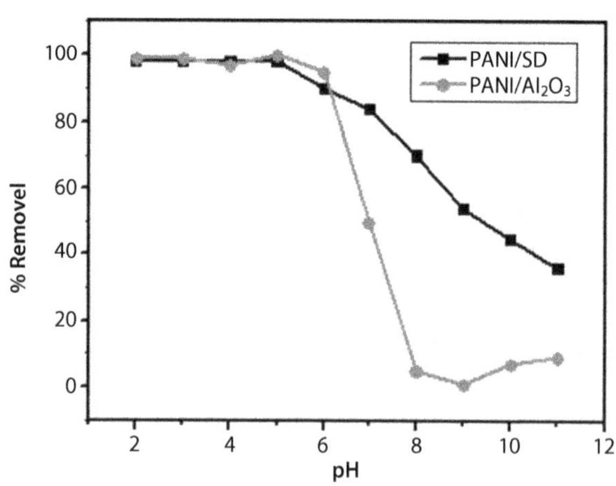

Figure 3.1 Effect of pH on adsorption of tartrazine dye onto polyaniline-coated sawdust (PANI/SD) [154] and PANI/Al_2O_3 composite [194].

adsorption was observed at the pH value less than 8 and above this pH value, the adsorption is dropped [158]. The effect of pH on the adsorption of dyes on different adsorbents is compiled and reported in Table 3.4.

Effect of Adsorbent Dosage: In addition to other adsorption parameters, amount of the adsorbent used during the process is also important. With increased amount of adsorbent, the more active sites are available for adsorption which favors the adsorption of more dye molecules. The same explanation holds true when the surface area of the adsorbent is increased or particle size of the adsorbent is reduced. The results of this parametric study are generally used to find out the minimum amount of adsorbent required for the effective removal of the dye from the aqueous solution [71]. During a study, congo red dye (CR) was removed by using a composite of polyaniline and chitosan (PANI/Ch) with a dosage range of 0.1–0.5 g/L, it has been found that, adsorption capacity of the adsorbent was increased up to 0.2 g/L and furthers no much increase was observed [166]. Table 3.5 reports the percentage adsorption of different dyes with the varying adsorbent dosage.

Effect of Temperature: Reaction temperature is another physiochemical factor, which affects the adsorption capacity of any adsorbent [171].

Table 3.4 Effect of pH on the adsorption of dyes on different adsorbents.

Dye used	Adsorbent	pH range	Removal percentage (%) observed	Ref.
Acid red 27	Fe_2O_3	1.5–10.5	98–27	[159]
Acid yellow 36	Activated-rice husk	2–9	80–45	[160]
Crystal violet	Kaolin	2–7	65–95	[161]
Methylene blue	Pine cone	3.47–7.28	63.83–94.82	[20]
Congo red	Pine cone	3.55–10.95	60.5–5.75	[156]
Methylene blue	Activated clay	2–9	60–95	[162]
Methylene blue	Tobacco stem ash	2.08–7.93	60–81	[163]
Methylene blue	Fly ash	2–8	36–45	[164]
Methylene blue	Pine leaves	2–11	20–80	[21]
Crystal violet	Modified-alumina	2.6–10.8	20–80	[165]

Table 3.5 Percentage adsorption of different dyes with the varying adsorbent dosage.

Dye used	Adsorbent	Amount of adsorbent	Percentage removal (%) observed	Ref.
Methylene blue	Modified mango seed	0.1–1.2 g	99.8–79	[144]
Methylene blue	Raw mango seed	0.1–1.2 g	99.5–68	[144]
Methyl orange	Chitosan/alumina composite	1–12 g/L	92.84–99.23	[140]
Acid orange 7	Tea waste	2–20 g/L	90–99	[167]
Crystal violet	Kaolin	0.25–4 g/L	75–97	[161]
Brilliant green	Treated sawdust	1–30 g/L	61–99.9	[168]
Crystal violet	Modified alumina	5–20 g/L	58–99	[165]
Methylene blue	Fly ash	8–20 g	45.16–96	[164]
Methylene blue	Modified sawdust	1.5–5 g	34.4–96.6	[146]
Reactive orange 16	Rice hull	0.02–0.08 g	21.7–56.2	[37]
Basic yellow 2	Tea waste	2–20 g/L	19–60	[167]
Malachite green	Treated sawdust	0.2–1.0 g/ 100 mL	18.6–86.9	[169]
Acid violet 17	Orange peel	50–600 mg/ 50 mL	15–98	[170]
Congo red	Pine cone	0.01–0.03 mg	13.45–18.96	[156]

As evident from literature, in the endothermic reactions, with increase in temperature, the available number of active sites on the surface of the adsorbent and the mobility of the dye molecules get increased. Both these factors favor the adsorption. On the other hand, in exothermic reactions, the adsorption was reported to be decreased with the rise in temperature due to the decrease in adsorptive forces between dye molecules and adsorbent [71]. An increase in the adsorption capacity, that is, from 91.96% to 97.32% was observed for the removal of Acid Violet 49 dye (AV49) by polyaniline-coated sawdust (PAC) with elevation in temperature from 30° to 45° [139].

3.4 Adsorption Kinetics

While designing any adsorption system, the determination of the rate of the reaction is the most important step. The rate expression of any reaction is derived from the kinetic study of the reaction system. Analysis of experimental data for an adsorption process can be carried out by using different available rate models such as (1) Pseudo-first order model, (2) Pseudo-second order model, (3) Intraparticle diffusion model to depict the experimental data. In detail:

(1). Pseudo-first-order model:
The linearized form of the Lagergrane pseudo-first-order model [151] is:

$$\log(q_e - q_t) = \log q_e - \frac{k_1 t}{2.303} \tag{3.1}$$

where,
q_t = adsorption capacity at time, t (mg/g)
q_e = adsorption capacity at equilibrium (mg/g)
k_1 = the rate constant of adsorption (min^{-1}) and
t = contact time (min).

A plot of $\log(q_e - q_t)$ versus t gives a straight line with slope = k_1 and intercept = q_e.

(2). Pseudo-second-order model:
The linearized form of the Lagergren pseudo-second-order model [151] is:

$$\frac{t}{q_t} = \frac{1}{k_2 q_e^2} + \frac{t}{q_e} \tag{3.2}$$

where,
q_t = adsorption capacity at time, t (mg/g)
q_e = adsorption capacity at equilibrium (mg/g)
k_2 = second order adsorption rate constant (g/mg/min)
$k_2 q_e^2 = h$, that is, the initial rate constant (cat $t \to 0$) (mg/g/min) and
t = contact time (min).

A plot of t/q_t versus t gives a straight line with slope = q_e and intercept = $k_2 q_e^2$.

(3). Intraparticle diffusion model:
According to this model, the adsorption takes place in many steps such as:

1. Transport of adsorbate molecules from the bulk aqueous phase to the surface of the adsorbent.
2. Transfer of adsorbate molecules from the surface of the solid adsorbent into the interior of the pores.

As reported in literature, the adsorption process that follows intraparticle diffusion, the adsorption occurs in proportional with $t^{1/2}$.The intraparticle diffusion model is expressed by the following equation [172]:

$$q_t = k_{id} t^{0.5} \tag{3.3}$$

where,
q_t = adsorption capacity at time t (mg/g)
$t^{0.5}$ = Half-life period/time (min) and,
k_{id} = rate constant of intra-particle diffusion (mg/g $min^{0.5}$).

A plot of q_t versus $t^{0.5}$ gives a straight line with slope = k_{id}.

To determine the rate of adsorption, the experimental data obtained is tested for all available models. The best fit rate equation is determined on the basis of the linear regression correlation coefficient values, that is, R^2 value. The higher value of R^2 is the indication of good fit. It has been noticed from extensive literature studies that most of the dye adsorption studies support pseudo-second-order rate model.

3.4.1 Adsorption Isotherms

Various isotherm models were proposed in past to explain the adsorption behavior of different adsorbents [173]. The mechanism of the adsorption can be studied by fitting the experimental data in different types of adsorption isotherms. Interactions between adsorbing species and adsorbent can be well described by the adsorption isotherms. Depending upon the interactions between surface of the adsorbent and the adsorbate particles and nature of the adsorption, that is, multilyered or monolayer adsorption, the results can be compared with these already established isotherms and proposed mechanism of adsorption can be predicted. In addition, removal capacity of any adsorbent can be measured to some extent by adsorption isotherms. Among all these models, Langmuir's and Freundlich's isotherm models are extensively used to explain the adsorption behavior, especially for monolayer adsorption.

Langmuir's Isotherm Model: The Langmuir's isotherm model is found to be best fitted when monolayer adsorption has been taken place. It is based on the assumption that, the adsorption takes place at specific homogeneous sites within the adsorbent, forming a single layer over the adsorbent surface. The Langmuir's isotherm is governed by the following equation [174]:

$$\frac{C_e}{q_e} = \frac{1}{bq_m} + \frac{1}{q_m} C_e \qquad (3.4)$$

where,

C_e = Concentration of dye adsorbed at equilibrium (mol/L)
q_e = Amount of dye adsorbed at equilibrium (mg/g)
q_m = Langmuir monolayer adsorption of dye (mg/g) and
b = Langmuir's constant.

On plotting C_e/q_e versus C_e, a straight line is obtained, with slope = q_m and intercept = b.

The separation factor, R_L (dimensionless) can be determined by following expression:

$$R_L = \frac{1}{1 + bC_0} \qquad (3.5)$$

where,

C_0 = Initial dye concentration.

The recommended value of R_L, for the favorable adsorption lies between 0 and 1.

Freundlich's Isotherm Model: Similar to Langmuir's model, the Freundlich's Isotherm model is also based upon the assumption of monolayer adsorption. It is best fitted for an adsorption on heterogeneous surface having unequal active sites and different energies of adsorption. The linearized equitation for the Ereundlich's isotherm model [175] is given as:

$$\log q_e = \log K_f + n_h \log C_e \qquad (3.6)$$

where,

q_e = Amount of dye adsorbed at equilibrium (mg/g)
C_e = Concentration of dye adsorbed at equilibrium (mol/L)
K_f and n_h are the Freundlich coefficients.

A plot of log q_e versus log C_e gives a straight line with slope = n_h and intercept = K_f.

3.5 Polyaniline: An Emerging Adsorbent

A variety of adsorbents have been used in the past for the eradication of dyes and toxins from the wastewater that include activated carbon, natural clay [176, 177], polymeric hydrogels [45, 46], agriculture wastes and by-products [18], industrial chemical remains [178, 179], modified clays, chitin and chitosan [72, 38], starch and its derivatives [73, 156], carbonized waste of wool [157], etc. The more details are provided in the Table 3.1. In last one decade, polyaniline (PANI) that is originally recognized and used as a conducting polymer [180, 181], has attracted the attentions of many researchers as an emerging adsorbent. Polyaniline, in addition to its unique and attractive chemical stability [182], is also supported by the low cost of the aniline (raw material) and high yield of polymerisation reaction [182, 183].

It is quite easy to oxidize or reduce the PANI because of its rare electronic behavior with its high electron affinity and low ionisation energy. With its improved properties, such as larger surface area, modified mechanical properties, low cost, high adsorption capacity, and simple preparation, polyaniline is now emerging as a novel adsorbent [193]. Polyaniline in its different forms: acidic (ES) and basic forms; PANI composites, coating of PANI on different solid substrates, salt form of the PANI and doped PANI is reported to had used as adsorbent for dyes and other organic compounds. Figure 3.2 represents the adsorption capacity of different forms of the Polyaniline for the methylene blue dye [193, 184–186]. Work done by different researchers on polyaniline as adsorbent has been summed up in Table 3.6.

3.5.1 Polyaniline in Dye Removal

The main driving force for the dye eradication by polyaniline is the interaction of positive fragment of PANI backbone with dye molecule (negatively charged part) and basic site of PANI with positively charged dye [208]. A successful attempt of the use of polyaniline as adsorbent, as reported in literature, was achieved by the 100% removal of Eriochrome Black T (EBT) dye from its aqueous as well as from the methanolic solution by acidic polyaniline and emeraldine base at lower concentration of dye in the solutions. Adsorption data was reported to follow the Langmuir isotherm.

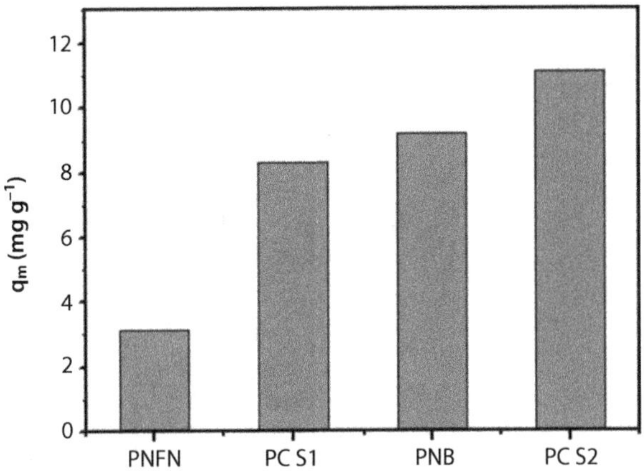

Figure 3.2 Adsorption capacity of different adsorbents for methylene blue dye: polyaniline nickel ferrite nanocomposite (PNFN) [184], Polyaniline-coated sawdust of wood (PCSD1) [185], Polyaniline nanotube base (PNB) [186], Polyaniline-coated sawdust of walnut (PCSD2) [193].

Adsorption of EBT was found to be endothermic and entropy dominated. Reaction was observed to have followed the first order. Different parameters such as particle size, adsorbent amount, dye concentration and contact time are reported to have a significant effect on the adsorption of dye. Rate constant was evaluated and its value was found to be comparable with the recorded data for adsorption of Chrome dye onto mixed adsorbent. Even the adsorption of the said polyaniline-based adsorbent was found to be greater than adsorption of Astrazone Blue on peat and Malachite green onto activated slag and carbon. Adsorption from methanolic solution was reported as more endothermic than aqueous solution. It was also observed that the conductivity of PANI changed with adsorption [209]. In an another study, an adsorption of sulfonated (anionic) dyes on Emeraldine salt (a form of the PANI) had been found to be influenced by chemical interaction between the adsorbate and adsorbent molecules. Figure 3.3 represents the variation of adsorption capacity of Congo red dye by HCl and p-toluene sulphonic acid (PTSA)-doped polyaniline (HCl/PTSA-doped PANI) and polyaniline/Chitosan composite (Pn/Ch composite) [166, 210–212]. As per the reported literature, thistype of study for the removal of anionic (sulfonated) dyes is quite new of its kind. A number of dyes had been used as adsorbate such as Methylene Blue, Coomassie Brilliant blue R-250, Orange G, Alizarin cyanine green, Remazol brilliant blue R and Rhodamine B for clearer prospect. According to the reported work, cationic or nonsulfonated dyes, that is, methylene blue and rhodamine B

Table 3.6 Work done by different researchers on polyaniline as adsorbent.

Adsorbent	Dye used	Adsorption capacity (q_m)	Ref.
Removal of dyes by polyaniline			
PANI (emeraldine salt)	Indigo carmine	99 mg/g	[182]
Polyaniline-coated onto wood sawdust (PANI/SD)	Methyl orange (MO)	99 mg/g	[187]
Polyaniline-coated sawdust	Acid violet 49	192.31 mg/g	[139]
PANI microspheres	methyl orange (MO)	154.56 mg/g	[188]
p-toluene sulfonic acid (PTSA) and camphor sulfonic acid (CSA)-doped Polyaniline.	Orange G(OG)	0.22 mg/g	[158]
Polyaniline-coated sawdust composite (PANI/SD)	Direct green 6 reactive orange 16	DG6–217.39 mg/g OG16–100 mg/g	[189]
Starch/polyaniline nanocomposite	Reactive black Reactive violet	RB-0.818 mmol/g RV-0.786 mmol/g	[190]
Polyaniline-Fe_2O_3 magnetic nanocomposite	Acid violet 19	0.995 mg/g	[191]
Polyaniline/Fe_3O_4 nanocomposite	Titan yellow	50 mg/g	[192]
Polyanilline (PANI) film – coated Sawdust	Methylene blue	11.2 mg/g	[193]
Pani/Al_2O_3 composite	Tartrazine	6.25 mg/g	[194]
Polyaniline nanotubes salt/silica composite	Acid green 25 (AG)	6.896 mg/g	[195]

Polyaniline-montmorrillonite composite	Congo red	2.091 mg/g	[196]
Nano-polyaniline	Amido black 10B	102.04 mg/g	[197]
PANI-mushroom Composite	Eosin yellow	429.18 mg/g	[198]
Removal of metal ions by polyaniline			
Polyaniline/coconut shell–activated carbon (PANI/GAC) composite	Cu(II)	38.97 mg/g	[199]
Polyaniline/humic acid composite	Hg(II) Cr(VI)	671 mg/g-Hg(II) 150 mg/g-Cr(VI)	[200]
Polyaniline/polystyrene composite	Hg(II)	0.671 mol/g × 106	[201]
Polyaniline (PANI)/rice husk nanocomposite	Zn(II)	24.3 mg/g	[202]
Polyaniline synthesized on jute fiber	Cr(VI)	62.8 mg/g	[203]
PANI/wheat bran	Cr(VI)	33.44 mg/g	[204]
Polyaniline/polyvinyl alcohol (PANI/PVA) composite	Cr(VI)	111.23 mg/0.1 g	[205]
PANI/saw dust PANI/rice husk	Cd (II)	PANI/ S.D-4.739 mg/g PANI/R.H-5.128 mg/g	[206]
Sulfuric acid-doped PANI	Cr(VI)	95.79 mg/g	[207]

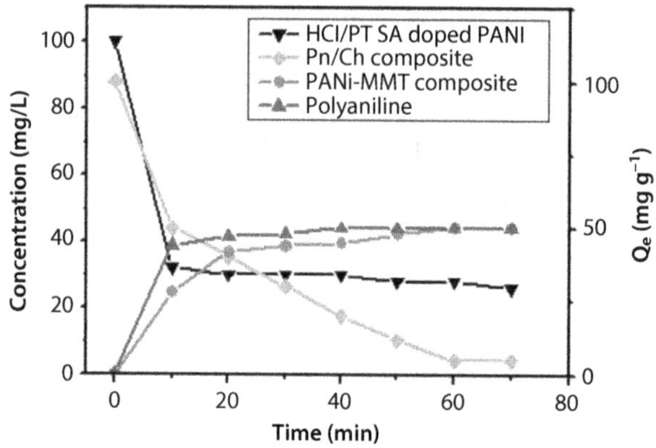

Figure 3.3 Variation in adsorption of congo red dye by HCl and PTSA-doped polyaniline (HCl/PTSA-doped PANI) [210], Polyaniline/chitosan composite (Pn/Ch Composite) [167], polyaniline [211] polyaniline momtmorrillonite composite (PANi-MMT composite) [212].

were not adsorbed by the polyaniline salt, which, evidently indicated that the interaction between dye and PANI were chemical interaction, that is, interaction between polar groups of sulfonated dyes and positively charged backbone of the PANI. Dye-loaded PANI has been confirmed by UV-VIS spectra, in which a blue shift in the peaks had been noticed. Less adsorption had been observed in the case of emeraldine base for sulfonated dyes. The data has been found to be supported by Langmuir isotherm and the adsorption was reported to have followed a second order reaction kinetics. Considerable decrease in the conductivity of PANI has been observed after the exposure to the dye solution [213]. In the study of removal of Indigo carmine dye from its aqueous solution using PANI-ES as an adsorbent, data had been reported to be fitted well in Langmuir's isotherm. According to the results of this work, with increased doping of polyaniline (emeraldine salt) by increasing the concentration of acid, the extent of the positive charge on the PANI backbone is more which, in turn increased the dye's adsorption capacity. Adsorption data of the concerned dye was observed to have followed the pseudo-second-order reaction kinetic [182].

Some workers doped the PANI by using cheaper and environment friendly Potash Alum and used it for the adsorption of heteropolyaromatic (methylene blue), azo (mordant yellow-10, orange-II) and xanthane (rhodamine-B) dyes. Significant data was obtained for MY-10 and O-II), that is, anionic dyes. However both MY and O-II dyes followed almost similar adsorption behavior initially, but with progress of reaction, more adsorption was observed in case of O-II than MY, as O-II dye was reported to attain a planar structure by intramolecular hydrogen bonding between –OH and –N=N– group. It had also been reported that 100 ppm initial dye concentration could be absorbed by using a comparatively lower amount of adsorbent, that is, 80 mg of doped PANI. Adsorption process of MY and O-II dyes on Potash Alum-doped PANI had been observed to be endothermic. Adsorption data of both the dyes followed the pseudo-second-order model and supported Langmuir's isotherm model. Subsequent desorption study indicated that percentage of desorption of MY and O-II were 92% and 94%, respectively at pH > 11, which supported the excellent regenerative capacity of PANI as an adsorbent [214].

As per reported literature, the effect of dopant was also studied for the adsorption of dyes on p-toulene sulfonic acid (PTSA)-doped PANI and camphor sulfonic acid (CSA)-doped PANI. A number of dyes were explored such as Methylene blue, Orange G, Alizarin cyanin green, Rhodamine B, Remazol brilliant blue R, Malachite green, and Coomassie brilliant blue R-250. Greater adsorption was seen in the case of CSA-doped PANI than PTSA-doped PANI because of the presence of more counter ions in case of CSA-doped polyaniline than PTSA-doped PANI, that increased the distance between chains and aided easier diffusion of anionic sulfonated dyes to the positively charged PANI backbone. A change in the conductivity (decrease) was also observed with adsorption in both the cases. Pseudo-second-order model described the adsorption behavior of the dye and Langmuir model's fitted well to the adsorption data [216].

Some researchers worked to remove Acid green 25 (AG 25) dye from textile industries wastewater by using polyaniline-coated onto wood sawdust (PANI/SD). Parameters such as, contact time, initial dye concentration, temperature, pH of the test solution, amount of adsorbent dose etc. had been studied and concluded that (PANI/SD) is an efficient adsorbent to remove the AG 25 from the wastewater in both batch and column systems. Moreover, the use of SD was discouraged over PANI being non profitable adsorbent for water treatment purpose. Adsorption process was spontaneous and endothermic at room temperature [137]. As per the literature, few researchers worked on the adsorption of Methyl orange (MO) dye from its aqueous solution by using polyaniline coated onto the sawdust (PANI/SD) as adsorbent. Adsorption efficiency of PANI/SD was compared with saw dust (without coating; SD) and activated carbon (granulated; GAC). Batch operations were carried out. It had been observed that the highest percentage of adsorption of MO could be achieved at the pH values from 4 to 8. H-bonding existed between the MO and PANI which results into higher adsorption at PANI/SD than SD and GAC at pH = 4–8. Also, the presence of strong intermolecular interactions between Methyl Orange dye (diazo groups and imine groups) and Polyaniline (imine and amino groups) reportedly enhanced the percentage removal of MO. Pseudo-second-order kinetics model was found to be best fitted to the adsorption data for all PANI/SD, GAC, SD. Significant result of this study were that more than 80% of MO dye was recovered by using 0.1 M HCl solution in case of PANI/SD, nearly 90% of MO was recovered in case of SD by using 0.05M solution of NaOH, but in case of GAC, a very little amount, that is, about <5% of dye could be recovered [187]. In another study, Eosin Y (EY) dye was removed by using two conducting polymers, that is, Polypyrrole (PPy) and Polyaniline (PANI), synthesized onto wood sawdust surface. Both column and batch systems were used to perform the adsorption experiments at room temperature. The adsorption was not found pH dependent for a wide pH range, that is, 2–10, but a decrease in the adsorption has been noticed at the higher pH values (pH ≥ 12). Efficient removal dye, that is, 96% was reported from PPy/SD column by using 0.05M NaOH solution. The higher adsorption of EY (anionic dye) in acidic conditions was reported to be due to more concentration of H^+ ions, which developed positive charge on the surface of adsorbents PPy/SD and PANI/SD that resulted into more attraction between adsorbate (EY) and adsorbents. But in case of alkaline solution OH – ions compete with EY and hence a lower adsorption was observed in this case. It has been found that, in addition to electrostatic interaction and H-bonding, ion-exchange reactions also played significant role to remove EY dye from its aqueous solution [158]. Figure 3.4 shows

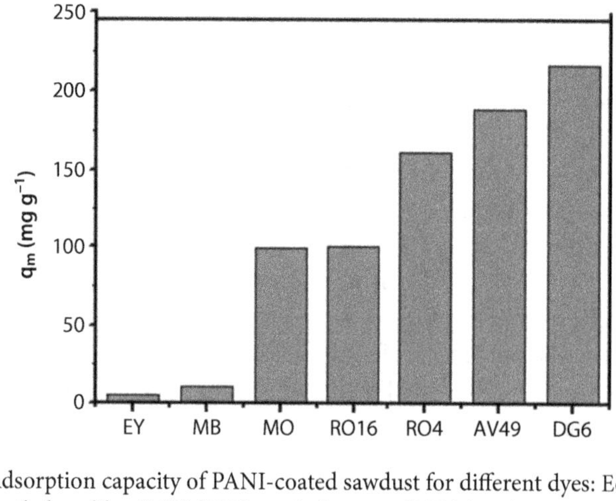

Figure 3.4 Adsorption capacity of PANI-coated sawdust for different dyes: Eosin Y (EY) [158], methylene blue (MB) [193], methyl orange (MO) [187], reactive orange 16 (RO16) [189], reactive orange 4 (RO4) [215], acid violet 49 (AV49) [139], direct green 6 (DG6) [189].

the adsorption of different dyes from aqueous solution using polyaniline-coated sawdust as an adsorbent [139, 158, 187, 189, 193, 215]. Another study conducted on the removal of anionic dye (Acid green 25) by using different adsorbents such as Polyaniline NT (nano tube) salt, conventional Polyaniline salt/silica composite and conventional Polyaniline salt. Among these adsorbents, Polyaniline NT salt/silica composite was reported as the best adsorbent and adsorbs comparatively more amount of dye per unit weight. The proposed order of the adsorption capacity of different adsorbents was: Polyaniline NT salt/silica composite > Polyaniline NT salt > Conventional Polyaniline NT salt/silica composite > Conventional Polyaniline salt. Higher adsorption on Polyaniline NT salt/silica composite was observed as compared to other concerned adsorbents, which was reported due to its higher surface area and availability of more binding sites. The adsorption was reported to be supported by intraparticle diffusion and surface adsorption simultaneously [217].

3.5.2 Polyaniline in Metal Ions Removal

As per the literature studies, polyaniline is also been found to be capable enough to remove harmful metal ions from the industrial effluent. A composite of PANI/GAC was synthesized by chemical oxidative polymerisation and was further used efficiently for the extraction of Cu (II) from the industrial effluent [199]. Also, polyaniline and humic acid (PANI/HA) composite was reported to be used for the removal of Hg (II) and Cr (VI) ions from the effluent [200]. Polyaniline and polystyrene were used to form a composite to separate out the Hg (II) ions from industrial wastewater. It had been found that removal of these harmful metallic ions from the wastewater increased with rise in concentration and temperature of solution under test. [201]. Nanocomposite of polyaniline with rice husk ash (RHA) was reviewed as a good adsorbent for the removal of Hg (II) from its aqueous solution [218]. In an another study, polyaniline/rice husk (RH) nanostructured composite was prepared by chemical oxidative polymerization and further used to eliminate the Zn from wastewater. This composite was found to be a potential adsorbent for the Zn removal [202]. Some researchers had prepared nanotubes of aniline, aniline sulfate and aniline hydrochloride by template or acids free oxidation polymerisation method. The capacities of the different forms of polyaniline to adsorb Hg were found to have followed the given order, that is, polyaniline (treated by alkali) > polyaniline sulfate > polyaniline hydrochloride. Different metal ions in their individual or mixture forms were used to study the removal capacity of these synthesized adsorbents and following order of competition between the metal ions was observed: Zinc(II), < Copper(II), < Cadmium(II), < Mercury(II), <Lead(II). This order is quite similar to that of the order of the respectively their atomic radius, atomic weight and electronegativity of these metal ions [219]. In addition, natural products such as jute fibers were used as a substrate to synthesize polyaniline and used it to investigate the absorptive removal of hexavalent chromium Cr (VI) from its aqueous solution in batch experiments [203]. In addition, Cr (VI) can be separated from the wastewater by using composites of polyaniline with maize bran, wheat bran and rice bran [204]. Removal of the chromium from the effluent was also carried out by using polyaniline and its composites. The adsorption capacities of these synthesized composites were also compared with some other adsorbents such as amberjet, anthracite and purolite-302. PANI/PVA composite was found to be a potential adsorbent for chromium removal. An effective removal of hexavalent chromium was observed by this composite of polyaniline and polyvinyl alcohol (PANI/PAV), consisted of aniline (0.2 mol/L) and PVA (2% w/v), generated by in situ polymerisation [205]. But the removal percentage is lower in case of PANI/anthracite composite in comparison to PANI [220]. A colorimetric determination of the adsorption was carried out in a study, in which chemically synthesized sawdust-polyaniline nanocomposite was used to clear away the Cr (VI) from the environment. It has been found that, all the examined composites reportedly achieved more than 80% adsorption during the first 30 minutes and the rate of adsorption depended upon the time [221]. Adsorption of Cd (II) ion by polyaniline/saw dust (PANI/SD) and polyaniline/rice husk (PANI/RH) was also reported in the literature. Results showed that these composites are suitable for the eviction of Cd (II) ions from the effluent [206]. According to a research, it is possible to expel the harmful Hg (II) ion from the industrial waste by using chemically synthesized polyaniline (PANI). The preferential adsorption method used during this research was batch experiment and the adsorbent was found to be efficient for the removal of Hg (II) ions [222]. Treatment of Cr (VI) polluted water was efficiently carried out by using sulfuric acid-doped PANI as an adsorbent [207]. Nanocomposite of polyaniline and rice husk (PANI-RH) was employed for the treatment of industrial effluent containing lead and cadmium ions. The reported results clearly indicated the higher adsorption capacity of the adsorbent for Pb (II) ions as compare to the Cd (II) ions [223]. In another study, a nanocomposite of polyaniline and humic acid (PANI/HA) was prepared and used for the separation of the Cr (VI) ions from the industrial effluent. Major role of humic acid (HA) in this composite was reported to stabilize the PANI by preventing its agglomeration [224].

3.5.3 Polyaniline in Phenols Removal

In addition to other water polluting moieties, phenolic compounds also play a considerable role in water pollution and lead to several water borne diseases. It is thus becomes mandatory to remove these pollutants from effluent before their discharge into fresh water. Several studies were conducted in this regard. Adsorption on polyaniline also reported good results for the phenol removal from the industrial effluents. As per the reported literature, mesoporous carbon material (CMK-1) and its converted form (CMK-1/Polyaniline) had been used to remove the p-cresol, phenol and resorcinol by batch adsorption. It had been noticed that CMK-1/Polyaniline form exhibited higher sorption of the phenolic moieties and could be used as a significant adsorbent for this significant water pollutants [225]. According to some other study, silica gel modified with polyaniline was found capable to treat the phenol polluted water in batch wise adsorption process. Results of that study showed that polyaniline-coated silica could be used as a potential adsorbent for phenols [226]. Moreover, a nanocomposite fabricated from PANI, that is, K10 polyaniline-montmorillonite (PANI/K10) was used for the eviction of phenolic impurities from aqueous solution in combination with gas chromatography and mass spectrometry. This composite was reported to be an efficient adsorbent because of its larger surface area and homogeneous dispersion [227]. Removal of 3-nitrophenol and phenol was conductedd using polyaniline electrode, for the polymerisation of these two phenols from an aqueous solution of NaCl, and resulted into the formation of composites of polyaniline/polyphenol and polyaniline/poly (3-nitrophenol). This study evidently concluded that the PANI films have goods capacity to remove phenols efficiently from wastewater [228].

3.5.4 Polyaniline in Acid Removal

Acids polluted industrial effluent can also be effectively treated with polyaniline. Polycarboxylic acids isomers such as trimellitic acid, pyromellitic and hemimellitic acids were separated by chemically synthesized polyaniline-polypyrrole, from their aqueous solutions. Removal of these isomers occurs in the following order: pyromellitic acid > hemimellitic acid > trimellitic acid [229]. Moreover, these three carboxylic acids isomers were also removed from the water by using a hemically synthesized polyaniline film grafted on red ceramic brick. The results indicated that this polyaniline-based adsorbent was a potential, cheaper, environment-friendly and reusable adsorbent [230]. Removal of the tannic acid from the effluent was studied by some researcher in batch adsorption experiment usingpolyaniline (PANI) synthesized by oxidative polymerisation.This study concluded that PANI is an efficient adsorbent for eviction of acids from effluent [231]. As per the reported work, Fe_3O_4 functionalized SiO_2-polyaniline nanoparticles (magnetically separable PANI) were employed to remove the humic acid (HA) from the wastewater. Higher adsorption was observed in this study, because of the electrostatic interaction between negative charge on HA and nitrogen of PANI [232].

3.6 Conclusion

Polyaniline has found to be used by different researchers for the absorptive removal of dyes and other chemicals from waste water. Originally regarded as a conducting polymer, polyaniline due to its unique chemical stability, electronic behavior and improved mechanical properties, is emerging as one of the inceptive adsorbent especially for the dyes. Being derived from less costly raw materials, simple preparation and high adsorption and regeneration capacity, polyaniline has shown its potential to be used as an adsorbent on commercial scale. Scientists are trying to explore the newly observed adsorption capacity of polyanile in its different forms, that is, acidic (ES) and basic forms of PANI; PANI composites, coating of PANI on different solid substrates, salt form of the PANI, doped PANI, etc. In addition to its high surface area, that is modified to be used as adsorbent, its ability of being easy to oxidize or reduce make it more attractive for the selective removal of different charged moieties from the effluent. In the nutshell, polyanile (PANI) and its derivatives have the calibre replace the commercially available adsorbent with its unique properties and excellent adsorptive and regeneration capacity.

References

1. Compagne, N.V.L., Pollution abatement: an industry's point of view. *Aquatic Pollutants*, 169–174, 1978.
2. Zouboulis, A.I., Moussas, P.A., Groundwater and soil pollution: bioremediation. *Earth Syst. Environ. Sci.*, 1037–1044, 2011.
3. Cho, C.W., Park, J.S., Zhao, Y., Yun, Y.S., Quantitative analysis of adsorptive interactions of ionic and neutral pharmaceuticals and other chemicals with the surface of *Escherichia coil* cells in aquatic environment. *Environ. Pollut.*, 227, 8–14, 2017.
4. Shuval, H.I., Health aspects of water recycling practices. *Aquatic Pollut.*, 395–403, 1978.

5. Vanesch, G.J., Aquatic pollutants and their potential biological effects. *Aquatic Pollut.*, 1-12, 1978.
6. Kunz,A., Zamora, P.P., Moraes, S.G., Durán, N., New tendencies on the textile effluent treatment. *Quim. Nova.*, 25, 78–82, 2002.
7. Carneiro, P.A., Osugi, M.E., Fugivara, C.S., Boralle, N., Furlan, M., Zanoni, M.V.B., Evaluation of different electrochemical methods on the oxidation and degradation of Reactive Blue 4 in aqueous solution. *Chemosphere*, 59, 431–439, 2005.
8. Mohan, S.V., Bhaskar, Y.V., Karthikenyan, J., Biological decolourization of simulated azo dye in aqueous phase by algae Spirogyra species. *Int. J. Environ. Pollut.*, 21, 211–222, 2004.
9. Vaghela, S.S., Jethva, A.D., Mehta, B.B., Dave, S.P., Adimurthy, S., Ramachandraiah, G., Laboratory studies of electrochemical treatment of industry azo dye effluent. *Environ. Sci .Technol.*, 39, 2848–2855, 2005.
10. Cardoso, J.C., Lizier, T.M., Zanoni, M.V.B., Highly ordered TiO2 nanotube arrays and photoelectrocatalytic oxidation of aromatic amine. *Appl. Catal .B.-Environ.*, 99, 96–102, 2010.
11. Ekhaise, F.O., Anyansi, C.C., Influence of breweries effluent discharge on the microbiological and physiochemical quality of Ikpoba River, Nigeria. *Af. J. Biotech.*, 4, 1062–1065, 2005.
12. Vinod, K., Chopra, A.K., Monitoring of physico-chemical and microbiological characteristics of municipal wastewater at treatment plant, Haridwar City (Uttarakhand), India, 2012.
13. Syeda, S.R., Ferdousi, S.A., Ahmmed, K.M.T., De-colorization of textile wastewater by adsorption in a fluidized bed of locally available activated carbon. *J. Environ. Sci. Heal. A.*, 47, 210–220, 2012.
14. O'Neill, C., Hawkes, F.R., Esteves, S.R.R., Hawkes, D.L., Wilcox, S.J., Anaerobic and aerobic treatment of a simulated textile effluent. *J. Chem. Technol. Biot.*, 74, 993–999, 1999.
15. Wong, Y.C., Szeto, Y.S., Cheung, W.H., McKay, G., Adsorption of acid dyes on chitosan—equilibriumisotherm analyses. *Process Biochem.*, 39, 695–704, 2004.
16. Attia, A.A., Girgis, B.S., Fathy, N.A., Removal of methylene blue by carbons derived from peach stones by H_3PO_4 activation: batch and column studies. *Dyes Pigm.*, 76, 282–289, 2008.
17. Pereira, R.S., Identificação e caracterização das fontes de poluição em sistemas hídricos. Revista Eletrônica de Recursos Hídricos. *IPH-UFRGS*, 1, 20–36, 2004.
18. Zaharia, C., Suteu, D., Coal fly ash as adsorptive material for treatment of a real textile effluent: operating parameters and treatment efficiency. *Environ. Sci. Pollut. R.*, 20, 2226–2235, 2013.
19. Gupta, V.K., Gupta, B., Rastogi, A., Agarwal, S., Nayak, A., A comparative investigation on adsorption performances of mesoporous activated carbon prepared from waste rubber tire and activated carbon for a hazardous azo dye—Acid Blue 113. *J. Hazard. Mater.*, 186, 891–901, 2011.

20. Sen, T.K., Afroze, S., Ang, H., Equilibrium, kinetics and mechanism of removal of methylene blue from aqueous solution by adsorption onto pine cone biomass of Pinus radiata. *Water Air Soil Pollut.*, 218, 499–515, 2011.
21. Yagub, M.T., Sen, T.K., Ang. H., Equilibrium, kinetics, and thermodynamics of methylene blue adsorption by pine tree leaves. *Water Air Soil Pollut.*, 223, 5267–5282, 2012.
22. Sokolowska, G., Synthetic dyes based on environmental considerations. *Dyes Pigm.*, 30, 1–20, 1996.
23. Kabdaşli, I., Tünay, O., Orhon, D., Wastewater control and management in a leather tanning district. *Water Sci. Technol.*, 40, 261–267, 1996.
24. Bensalah, N., Alfaro, M., Martínez-Huitle, C., Electrochemical treatment of synthetic wastewaters containing Alphazurine A dye. *Chem. Eng. J.*, 149, 348–352, 2009.
25. Wróbel, D., Boguta, A., Ion, R.M., Mixtures of synthetic organic dyes in a photoelectrochemical cell. *J. Photochem. Photobiol. A Chem.*, 138, 7–22, 2001.
26. Dawood, S., Sen, T.K., Phan, C., Synthesis and characterisation of novel-activated carbon from waste biomass pine cone and its application in the removal of Congo red dye from aqueous solution by adsorption. *Water Air Soil Pollut.*, 225, 1–16, 2014.
27. Ivanov, K., Possibilities of using zeolite as filler and carrier for dyestuffs in paper. *Papier.*, 50, 456–459, 1996.
28. Field, M.S., Wilhelm, R.G., Quinlan J.F., Aley, T.J., An assessment of the potential adverse properties of fluorescent tracer dyes used for groundwater tracing. *Environ Monit Assess.*, 38, 75–96, 1995.
29. Morgan-Sagastume, J., Jimenez, B., Noyola, A., Tracer studies in a laboratory and pilot scale UASB reactor. *Environ Technol.*, 18, 817–825, 1997.
30. He, L.M., Tebo, B.M., Surface charge properties of and Cu (II) adsorption by spores of the marine Bacillus sp. strain SG-1. *Appl Environ Microbiol.*, 64, 1123–1129, 1998.
31. Hsu, T.C., Chiang, C.S., Activated sludge treatment of dispersed dye factory wastewater. *J Environ Sci Health A.*,32, 1921–1932, 1997.
32. Namasivayam, C., Kavitha, D., Removal of congo red from water by adsorption onto activated carbon prepared from coir pith, an agricultural solid waste. *Dyes Pigm.*, 54, 47–58, 2002.
33. Ahmad, A., Idris, A., Hameed, B., Organic dye adsorption on activated carbon derived from solid waste. *Desalin. Water Treat.*, 51, 2554–2563, 2013.
34. Crini, G., Non-conventional low-cost adsorbents for dye removal: a review. *Biores. Technol.*, 97, 1061–1085, 2006.
35. Forgacs, E., Cserhati, T., Oros, G., Removal of synthetic dyes from wastewaters: a review. *Environ Int.*, 30, 953–971, 2004.
36. Muthukumar, M., Selvakumar, N., Studies on the effect of inorganic salts on decolouration of acid dye effluents by ozonation. *Dyes Pigm.*, 62, 218–221, 2004.

37. Ong, S., Lee C., Zainal, Z., Removal of basic and reactive dyes using ethylene-diamine modified rice hull. *Biores. Technol.*,98, 2792–2799, 2007.
38. Robinson, T., McMullan, G., Marchant, R., Nigam, P., Remediation of dyes in textile effluent: a critical review on current treatment technologies with a proposed alternative. *Biores. Technol.*, 77, 247–255, 2001.
39. Clarke, E., Anliker, R., Organic dyes and pigments. *Handb. Environ. Chem.*, 3, 181–215, 1980.
40. Mishra, G., Tripathy, M., A critical review of the treatments for decolouriza-tion of textile effluent. *Colourage*, 40, 35–38, 1993.
41. Banat, I.M., Nigam, P., Singh, D., Marchant, R., Microbial decolorization of textile-dye containing effluents: a review. *Biores. Technol.*, 58, 217–227, 1996.
42. Gupta, G., Prasad, G., Singh, V., Removal of chrome dye from aqueous solu-tions by mixed adsorbents: fly ash and coal. *Water Res.*, 24, 45–50, 1990.
43. Mittal, A.K., Gupta, S., Biosorption of cationic dyes by dead macro fungus Fomitopsis carnea: batch studies. *Water Sci. Technol.*, 34, 81–87, 1996.
44. Fu, Y., Viraraghavan, T., Fungal decolorization of dye wastewaters: a review. *Biores. Technol.*, 79, 251–262, 2001.
45. Chu, H., Chen, K., Reuse of activated sludge biomass: I. Removal of basic dyes from wastewater by biomass. *Process Biochem.*, 37, 595–600, 2002.
46. Fu, Y., Viraraghavan, T., Removal of congo red from an aqueous solution by fungus aspergillus niger. *Adv. Environ. Res.*, 7, 239–247, 2002.
47. Viraraghavan, F.Y., Removal of congo red from aqueous solution by fungus (*Aspergillus niger*). *Adv. Environ. Res.*, 7, 239–247, 2002.
48. Lazar, T., Color chemistry: synthesis, properties, and applications of organic dyes pigments. *Color Res. Appl.*, 30, 313–314, 2005.
49. Adirvelu, K., Kavipriya, M., Karthika, C., Radhika, M., Vennilamani, N., Pattabhi, S., Utilization of various agricultural wastes for activated carbon preparation and application for the removal of dyes and metal ions from aqueous solutions. *Biores. Technol.*, 87, 129–132, 2003.
50. Asgher, M., Bhatti, H.N., Evaluation of thermodynamics and effect of chemi-cal treatments on sorption potential of (Citrus) waste biomass for removal of anionic dyes from aqueous solutions. *Ecol. Eng.*, 38, 79–85, 2012.
51. Brown, D., Laboureur, P., The aerobic biodegrability of primary aromatic amines. *Chemosphere*, 12, 405–414, 1983.
52. Mathur, N., Bhatnagar, P., Sharma, P., Review of the mutagenicity of textile dye products. *Univers. J. Environ.*, 2, 1–18, 2012.
53. Hao, O.J., Kim, H., Chiang, P.C., Decolorization of wastewater. *Crit. Rev. Env. Sci. Tec.*, 30, 449–505, 2000.
54. Firmino, P.I.M., Silva, M.E.R., Cervantes, F.J., Santos, A.B., Colour removal of dyes from synthetic and real textile wastewaters in one – and two-stage anaerobic systems. *Biores. Technol.*, 101, 7773–7779, 2010.
55. Kumar, K.V., Ramamurthi, V., Sivanesan, S., Modeling the mechanism involved during the sorption of methylene blue onto fly ash. *J. Colloid Interf. Sci.*, 284, 14–21, 2005.

56. Robinson, T., McMullan, G., Marchant, R., Nigam, P., Remediation of dyes in textile effluent. *Biores. Technol.*, 77, 247–255, 2001.
57. Singh, K., Arora, S., Removal of synthetic textile dyes from wastewaters: a critical review on present treatment technologies. *Crit. Rev. Env. Sci. Technol.*, 41, 807–878, 2011.
58. De Souza, A.A.U., Brandao, H.L., Zamporlini, I.M., Soares, H.M., De Souza, S.M.A.G.U., Application of a fluidized bed bioreactor for cod reduction in textile industry effluent. *Res. Conserv. Recy.*, 52, 511–521, 2008.
59. Balamurugan, B., Thirumarimurugan, M., Kannadasan, T., Anaerobic degradation of textile dye bath effluent using *Halomonas* sp. *Biores. Technol.*, 102, 6365–6369, 2011.
60. Isik, M., Sponza, D.T., Anaerobic/aerobic treatment of a simulated textile wastewater. *Sep. Purif. Techanol.*, 60, 64–72, 2008.
61. Mohamme, M.A., Shitu, A., Ibrahim, A., Removal of methylene blue using low cost adsorbent: a review. *Res. J. Chem. Sci.*, 4, 91–102, 2014.
62. Panswad, T., Wongchaisuwan, S., Mechanisms of dye wastewater colour removal by magnesium carbonate-hydrated basic. *Water Sci. Tehanol.*, 18, 139–144, 1986.
63. Malik, P.K., Saha, S.K., Oxidation of direct dyes with hydrogen peroxide using ferrous ion as catalyst. *Sep. Purif. Techanol.*, 31, 241–250, 2003.
64. Panswad, T., Wongchaisuwan, S., Mechanisms of dye wastewater colour removal by magnesium carbonate-hydrated basic. *Sci. Technol. War*, 18, 139–144, 1986.
65. Namasivayam, C., Muniasamy, N., Gayatri, K., Rani, M., Ranganathan K, Removal of dyes from aqueous solutons by cellulosic waste orange peel. *Biores. Technol.*, 57, 37–43, 1996.
66. Mohammed, M.A., Shitu, A., Ibrahim, A., Removal of methylene blue using low cost adsorbent: a review. *Res. J. Chem. Sci.*, 4, 91–102, 2014.
67. Zhao, S., Zhou, F., Li, L., Cao, M., Zuo, D., Liu, H., Removal of anionic dyes from aqueous solutions by adsorption of chitosan-based semi-IPN hydrogel composites. *Compos. Part-B*, 43, 1570–1578, 2012.
68. Kant, R., Adsorption of dye eosin from an aqueous solution on two different samples of activated carbon by static batch method. *J .Water Res. Prot.*, 4, 93–98, 2012.
69. Patel, H., Vashi, R.T., Treatment of textile wastewater by adsorption and coagulation. *E. J. Chem*, 7, 1468–1476, 2010.
70. Sharma, P., Kaur, H., Sharma, M., Sahore, V., A review on applicability of naturally available adsorbents for the removal of hazardous dyes from aqueous waste. *Environ. Monit. Assess.*, 183, 151–195, 2011.
71. Salleh, M.A.M., Mahmoud, D.K., Karim, W.A.W.A., Irdis, A., Cationic and anionic dye adsorption by agricultural solid wastes: a comprehensive review. *Desalination*, 280, 1–13, 2011.
72. Dabrowski, A., Adsorption—from theory to practice. *Adv. Colloid Interface Sci.*, 93, 135, 2001.

73. Allen, S., Koumanova, B., Decolourisation of water/wastewater using adsorption. *J. Univ. Chem. Technol. Metall.*, 40, 175–92, 2005.
74. Li, X., Li, Y., Zhang, S., Ye, Z., Preparation and characterization of new foam adsorbents of poly(vinyl alcohol)/chitosan composites and their removal for dye and heavy metal from aqueous solution. *Chem. Eng. J.*, 183, 88–97, 2012.
75. Rehman, M.S.U., Kim, I., Han, J.I., Adsorption of methylene blue dye from aqueous solutionby sugar extracted spent rice biomass. *Carbohydr. Polym.*, 90, 1314–1322, 2012.
76. Lakshmi, U.R., Srivastva, V.C., Mall, I. D., Lataye, D. H., Rice husk ash as an effective adsorbent: Evaluation of adsorptive characteristics for indigo Carmine dye. *J. Environ. Manag.*, 90, 710–720, 2009.
77. Mohanty, K., Naidu, J.T., Meikap, B.C., Biswas, M.N. Removal of crystal violet from wastewater by activated carbons prepared from rice husk. *Ind. Eng. Chem. Res.*, 45, 5165–5171, 2006.
78. Filho, N.C., Venancio, E.C., Barriquello, M.F., Hechenleitner, A.A., Pineda, E. A. G., Methylene blue adsorption onto modified lignin from sugarcane baggasse. *Eclét. Quím.*, 32, 63–70, 2007.
79. Khattri, S., Singh, M.K., Colour removal from dye wastewater using sugarcane dust as an adsorbent. *Adsorpt. Sci. Technol.*, 17, 269–282, 1999.
80. McKay, G., Porter, J.F., Prasad, G.R., The removal of dye colors from aqueous solutions by adsorption on low-cost materials. *Water, Air, Soil Pollut.*, 114, 423–438, 1999.
81. Mane, V.S., Mall, I.D., Srivastava, V.C., Use of baggase fly ash as an adsorbent for the removal of brilliant green dye from aqueous solution. *Dyes Pigm.*, 73, 269–278, 2007b.
82. Aksu, Z., Biosorption of reactive dyes by dried activated sludge: equilibrium and kinetic modeling. *Biochem. Eng. J.*, 7, 79–84, 2001.
83. Martin, M.J., Artola, A., Balaguer, M.D., Rigola, M., Activated carbons developed from surplus sewage sludge for the removal of dyes from dilute aqueous solutions. *Chem. Eng. J.*, 94, 231–239, 2003.
84. Magdy, Y.H., Daifullah, A.A.M., Adsorption of a basic dye from aqueous solutions onto sugarindustry - mud in two modes of operations. *Waste Manage.*, 18, 219–226, 1998.
85. Senthilkumaar, S., Kalaamani, P., Porkodi, K., Varadarajan, P.R., Subburaam, C.V., Adsorption of dissolved reactive red dye from aqueous phase onto activated carbon prepared from agricultural waste. *Bioresour. Technol.*, 97, 1618–1625, 2006.
86. Netpradit, S., Thiravetyan, P., Towprayoon, S., Evaluation of metal hydroxide sludge for reactive dye adsorption in a fixed-bed column system. *Water Res.*, 38, 71–78, 2004b.
87. Mohan, S.V., Rao, N.C., Karthikeyan, J., Adsorptive removal of direct azo dye from aqueous phase onto coal based sorbents: A kinetic and mechanistic study. *J. Hazard. Mater. B.*, 90, 189–204, 2002b.

88. Namasivayam, C., Kadirvelu, K., Coirpith, an agricultural waste byproduct, for the treatment of dyeing wastewater. *Bioresour. Technol.*, 48, 79–81, 1994.
89. Al-Degs, Y., Khraisheh, M.A.M., Allen, S.J., Ahmad, M.N., Effect of carbon surface chemistry on the removal of reactive dyes from textile effluent. *Water Res.*, 34, 927–935, 2000.
90. Nigam, P., Armour, G., Banat, I. M., Singh, D., Marchant, R., Physical removal of textile dyes from e.uents and solid-state fermentation of dyeadsorbed agricultural residues. *Bioresour. Technol.*, 72, 219–226, 2000.
91. Chou, K., Tsai, J., Lo, C., The adsorption of congo red and vaccum pump oil by rice hull ash. *Bioresour.Technol.*, 78, 217–219, 2001.
92. Ramakrishna, K. R., Viraraghavan, T., Use of slag for dye removal. *Waste Manage.*, 17, 483–488, 1997.
93. Namasivayam, C., Kumar, M. D., Selvi, K., Begum, R. A., Vanathi, T., Yamuna, R. T., 'Waste' coir pith—A potential biomass for the treatment of dyeing wastewaters. *Biomass Bioenergy*, 21, 477–483, 2001b.
94. Kannan, N., Sundaram, M. M., Kinetics and mechanism of removal of methylene blue by adsorption on various carbons—A comparative study. *DyesPigm.*, *51*, 25–40, 2001.
95. Namasivayam, C., Muniasamy, N., Gayatri, K., Rani,M., Ranganathan, K. (1996). Removal of dyes from aqueous solutions by cellulosic waste orange peel. *Bioresour. Technol.*, 57, 37–43, 1996.
96. Pavan, F.A., Gushikem, Y., Mazzoczto, A.C., Dias, S.L.P., Lima, E.C., Statistical design of experiments as a tool for optimizing the batch conditions to methylene blue biosorption on yellow passion fruit and mandarin peels. *Dyes Pigm.*, 72, 256–266, 2007.
97. Annadurai, G., Juang, R., Lee, D., Use of cellulose based wastes for adsorption of dyes from aqueous solutions. *J. Hazard. Mater. B.*, 92, 263–274, 2002.
98. Hameed, B. H., Ahmad, A. A., Batch adsorption of methylene blue from aqueous solution by garlic peel, an agricultural waste biomass. *J. Hazard. Mater.*, 164, 870–875, 2009.
99. Banat, F., Al-Asheh, S. Al-Makhadmeh, L. Evaluation of the use of raw and activated date pits as potential adsorbents for dye containing waters. *Process Biochem.*, 39, 193–202, 2003.
100. McKay, G., Porter, J. F., Prasad, G. R. The removal of dye colors from aqueous solutions by adsorption on low-cost materials. *Water, Air, SoilPollut.*, 114, 423–438, 1999.
101. Nassar, M. M., Magdy, Y. H., Removal of different basic dyes from aqueous solutions by adsorption on palm-fruit bunch particles. *Chem. Eng.J.*, 66, 223–226, 1997.
102. Hameed, B.H., Removal of cationic dye from aqueous solution using jackfruit peel as a non-conventional and low-cost adsorbent. *J. Hazard. Mater.*, 162, 344, 2009.

103. Bestani, B., Benderdouche, N., Benstaali, B., Belhakem, M., Addou, A., Methylene blue and iodine adsorption onto an activated desert plant. *Bioresour. Technol.*, 99, 8441–8444.–350, 2008.
104. Ponnusami, V., Madhuram, R., Krithika, V., Srivastva, S.N., Effects of process variables on kinetics of methylene blue onto untreated guava powder: Statistical analysis. *Chem. Eng. J.*, 140, 609–613, 2008.
105. Uddin, M.T., Islam, M.A., Mahmud, S., Rukanuzzaman, M., Adsorptive removal of methylene blue by tea waste. *J. Hazard. Mater.*, 164, 53–60, 2008.
106. Kumar, K.V., Kumrana, A., Removal of methylene blue by mango seed kernel powder. *Biochem. Eng. J.*, 27, 83–93, 2005.
107. Panda, G.C., Das, S.K., Guha, A.K., Jute stick powder as a potential biomass for the removal of Congo red and rhodamine B from their aqueous solution. *J. Hazard. Mater.*, 164, 374–379, 2008.
108. Ho, Y., Chiang, T., Hsueh, Y., Removal of basic dye from aqueous solution using tree fern as a biosorbent. *Process Biochem.*, 40, 119–124, 2005b.
109. Ferrero, F., Dye removal by low cost adsorbent: hazelnut shells in comparison with wood sawdust. *J. Hazard. Mater.*, 142, 144–152, 2007.
110. Ho, Y.S., McKay, G., Kinetic models for the sorption of dye from aqueous solution by wood. *Trans I. Chem. E.*, 76, 183–191, 1998a.
111. Morais, L.C., Freitas, O.M., Gonçalves, E.P., Vasconcelos, L.T., Beça, C.G.G., Reactive dyes removal from wastewaters by adsorption on eucalyptus bark: variables that define the process. *Water Res.*, 33, 979–988, 1999.
112. Nigam, P., Armour, G., Banat, I.M., Singh, D., Marchant, R., Physical removal of textile dyes from e.uents and solid-state fermentation of dye adsorbed agricultural residues. *Biores. Technol.*, 72, 219–226, 2000.
113. Batzias, F.A., Sidiras, D.K., Dye adsorption by calcium chloride treated beech sawdust in batch and fixed-bed systems. *J. Hazard. Mater. B.*, 114, 167–174, 2004.
114. Garg, V.K., Amita, M., Kumar, R., Gupta, R., Basic dye (methylene blue) removal from simulated wastewater by adsorption using Indian Rosewood sawdust: a timber industry waste. *Dyes Pigm.*, 63, 243–250, 2004.
115. Gürses, A., Karaca, S., Dogar, C., Bayrak, R., Acikyildiz, M., Yalcin, M., Determination of adsorptive properties of clay/water system: methylene blue sorption. *J. Colloid Interface Sci.*, 269, 310–314, 2004.
116. Ozdemir, O., Arma˘gan, B., Turan, M., Celik, M. S., Comparison of the adsorption characteristics of azo-reactive dyes on mesoporous minerals. *Dyes Pigm.*, 62, 49–60, 2004.
117. Namasivayam, C., Arasi, D. J. S. E., Removal of Congo red from wastewater by adsorption onto waste red mud. *Chemosphere*, 34, 401–417, 1997.
118. Wang, C., Juang, L., Hsu, T., Lee, C., Lee, J., Huang, F., Adsorption of basic dyes onto montmorillonite. *J. Colloid Interf. Sci.*, 273, 80–86, 2004.
119. Ghosh, D., Bhattacharyya, K.G., Adsorption of methylene blue on kaolinite. *App. Clay Sci.*, 20, 295–300, 2002.

120. Özacar, M., Sengil, I.A., Adsorption of reactive dyes on calcined alunite from aqueous solutions. *J. Hazard. Mater. B.*, 98, 211–224, 2003.
121. Zhao, M., Liu, P., Adsorption behaviour of methylene blue on hallowsite nanotubes. *Microporous Mesoporous Mater.*, 112, 419–424, 2008.
122. Walker, G.M., Hansen, L., Hanna, J.A., Allen, S.J., Kinetics of a reactive dye adsorption onto dolomitic sorbents. *Water Res.*, 37, 2081–2089, 2003.
123. Al-Ghouti, M.A., Khraisheh, M.A.M., Allen, S.J., Ahmad, M.N., The removal of dyes from textile wastewater: a study of the physical characteristics and adsorption mechanisms of diatomaceous earth. *J. Environ. Manage.*, 69, 229–238, 2003.
124. Orthman, J., Zhu, H.Y., Lu, G.Q., Use of anion clay hydrotalcite to remove coloured organics from aqueous solutions. *Sep. Purif.Technol.*, 31, 53–59, 2003.
125. Özacar, M., Sengil, I.A., A two stage batch adsorber design for methylene blue removal to minimize contact time. *J. Environ. Manag.*, 80, 372–379, 2006.
126. Karim, A.B., Mounir, B., Hachkar, M., Bakasse, M., Yaacoubi, A., Removal of basic red 46 dye from aqueous solution by adsorption onto Moroccan clay. *J. Hazard. Mater.*, 168, 304–309, 2009.
127. Demirbas, E., Nas, M.Z., Batch kinetic and equilibrium studies of adsorption of reactive blue 21 by fly ash and sepiolite. *Desalination*, 243, 8–21, 2009.
128. Karadag, D., Akgul, E., Tok, S., Erturk, F., Kaya, M.A., Turan, M., Basic and reactive dye removal using natural and modification zeolites. *J. Chem. Eng. Data.*, 52, 2436–2441, 2007.
129. Al-Bastaki, N., Banat, F., Combining ultrafiltration and adsorption on bentonite in a one-step process for the treatment of colored waters. *Res. Conserv. Recy.*, 41, 103–113, 2004.
130. Ahmed, M.N., Ram, R.N., Removal of basic dye from waste-water using silica as adsorbent. *Environ. Pollut.*, 77, 79–86, 1992.
131. Wong, Y.C., Szeto, Y.S., Cheung, W.H., McKay, G., Adsorption of acid dyes on chitosan – equilibrium isotherm analyses. *Process Biochem.*, 39, 695–704, 2004.
132. Wu, F., Tseng, R., Juang, R., Comparative adsorption of metal and dye on flake – and bead – types of chitosans prepared from fishery wastes. *J. Hazard. Mater.*, 73, 63–75, 2000.
133. Swamy, J., Ramsay, J.A., The evaluation of white rot fungi in the decoloration of textile dyes. *Enzyme Microb. Technol.*, 24, 130–137, 1999.
134. Fu, Y., Viraraghavan, T., Removal of congo red from an aqueous solution by fungus *Aspergillusniger. Adv. Environ. Res.*, 7, 239–247, 2000b.
135. Punjongharn, P., Meevasana, K., Pavasant, P., Influence of particle size and salinity on adsorption of basic dyes by agricultural waste: Dried Seagrape (*Caulerpa lentillifera*). *J. Environ. Sci.*, 20, 760–768, 2008.
136. Chiou, M., Li, H., Equilibrium and kinetic modelling of adsorption of reactive dye on cross-linked chitosan beads. *J. Hazard. Mater. B.*, 93, 233–248, 2002.

137. Ali, H., Biodegradation of synthetic dyes—a review. *Water Air Soil Pollut.*, 213, 251–273, 2010.
138. Bulut, Y., Aydın, H., A kinetics and thermodynamics study of methylene blue adsorption on wheat shells. *Desalination*, 194, 259–267, 2006.
139. Baseri, J.R., Palanisamy, P.N., Sivakumar, P., Application of polyaniline nanocomposite for then adsorption of acid dye from aqueous solutions. *E-J. Chem.*, 9, 1266–1275, 2012.
140. Zhang, J., Zhou, Q., Ou, L., Kinetic, isotherm, and thermodynamic studies of the adsorption of methyl orange from aqueous solution by chitosan/alumina composite. *J. Chem. Eng. Data*, 67, 412–419, 2012.
141. Kannan, N., Sundaram, M.M., Kinetics and mechanism of removal of methylene blue by adsorption on various carbons—a comparative study. *Dye Pigment*, 51, 25–40, 2001.
142. Zhang, Z., O'Hara, I.M., Kent, G.A., Doherty, W.O.S., Comparative study on adsorption of two cationic dyes by milled sugarcane bagasse. *Ind. Crop. Prod.*, 42, 41–49, 2013.
143. Kahraman, S., Yalcin, P., Kahraman, H., The evaluation of low-cost biosorbents for removal of an azo dye from aqueous solution. *Water Environ J.*, 26, 399–404, 2012.
144. Kumar, P.S., Palaniyappan, M., Priyadharshini, M., Vignesh, A.M., Thanjiappan, A., Fernando, P.S.A., Ahmed, R.T., Srinath, R., Adsorption of basic dye onto raw and surface-modified agricultural waste. *Environ. Prog. Sust. Energ.*, 33, 87–98, 2013.
145. De Luna, M.D.G., Flores, E.D., Genuino, D.A.D., Wei Wan, C.M.F.M., Adsorption of Eriochrome Black T (EBT) dye using activatedcarbon prepared from waste rice hulls—optimization, isotherm and kinetic studies. *J. Taiwan Inst. Chem. Eng.*, 44, 646–653, 2013.
146. Zou, W., Bai, H., Gao, S., Li, K., Characterization of modified sawdust, kinetic and equilibrium study about methylene blue adsorption in batch mode. *Korean J. Chem. Eng.* 30, 111–122, 2013.
147. Tehrani-Bagha, A.R., Nikkar, H., Mahmoodi, N.M., Menger, F.M., The sorption of cationic dyes onto kaolin: kinetic, isotherm and thermodynamic studies. *Desalination*, 266, 274–280, 2011.
148. Namasivayam, C., Yamuna, R., Arasi, D., Removal of acid violet from wastewater by adsorption on waste red mud. *Environ. Geol.*, 41, 269–273, 2001.
149. Ramaraju, B., Kumar, M., Reddy, P., Subrahmanyam, C., Low cost adsorbents from agricultural waste for removal of dyes. *Environ. Prog. Sust. Energ.*, 2013, 1–9, 2013.
150. Mall, I.D., Srivastava, V.C., Agarwal, N.K., Mishra, I.M., Removal of congo red from aqueous solution by bagasse fly ash and activated carbon: kinetic study and equilibrium isotherm analyses. *Chemosphere*, 61, 492–501, 2005.
151. Nandi, B., Goswami, A., Purkait, M., Removal of cationic dyes from aqueous solutions by kaolin: kinetic and equilibrium studies. *Appl. Clay Sci.*, 42, 583–590, 2009.

152. Ansari, R., Mosayebzadeh, Z., Application of polyaniline as an efficient and novel adsorbent for azo dyes removal from textile wastewaters. *Chem. Papers.*, 65, 1–8, 2011.
153. Ansari, R., Alaie, S., Khah, A.M., Application of polyaniline for removal of Acid Green 25 from aqueous solutions. *J. Sci. Ind. Res.*, 70, 804–809, 2011.
154. Ansari, R., Keivani, M.B., Delaver, A.F., Application of polyaniline nanolayer composite for removal of tartrazine dye from aqueous solutions. *J. Polym. Res.*, 18, 1931–1939, 2011.
155. Chowdhury, S., Chakraborty, S., Saha, P., Biosorption of basic green 4 from aqueous solution by Ananas comosus (pineapple) leaf powder. *Colloids Surf. B Biointer.*, 84, 520–527, 2011
156. Dawood, S., Sen, T.K., Removal of anionic dye congo red from aqueous solution by raw pine and acid-treated pine cone powder as adsorbent: equilibrium, thermodynamic, kinetics, mechanism and process design. *Water Res.*, 46, 1933–1946, 2012.
157. Ibrahim, S., Adsorption of anionic dyes in aqueous solution using chemically modified barley straw. *Water Sci. Technol. J. Int. Assoc. Water Pollut. Res.*, 62, 1177, 2010.
158. Ansari, R., Mosayebzadeh, Z., Removal of Eosin Y, an Anionic Dye, from aqueous solution using conducting electroactive polymers. *Iran Polym. J.*, 19, 541–551, 2010.
159. Nassar, N.N., Kinetics, mechanistic, equilibrium, and thermodynamic studies on the adsorption of acid red dye from wastewater by γ-Fe2O3 nanoadsorbents. *Sep. Sci. Technol.*, 45, 1092–1103, 2010.
160. Imam, M.H., Hameed, B.H., Ahmad, A.L., Aqueous-phase adsorption of phenolic compounds on activated carbon. *Environ. Manag.*, 25, 569–574, 2003.
161. Nandi, B., Kinetic and equilibrium studies on the adsorption of crystal violet dye using kaolin as an adsorbent. *Sep. Sci. Technol.*, 43, 1382–1403, 2008
162. Weng, C.H., Pan, Y.F., Adsorption of a cationic dye (methylene blue) onto spent activated clay. *J. Hazard. Mater.*, 144, 355–362, 2007.
163. Ghosh, R.K., Reddy, D.D., Tobacco stem ash as an adsorbent for removal of methylene blue from aqueous solution: equilibrium, kinetics, and mechanism of adsorption. *Water Air Soil Pollut.*, 224, 1–12, 2013.
164. Kumar, K.V., Ramamurthi, V., Sivanesan, S., Modeling the mechanism involved during the sorption of methylene blue onto fly ash. *J. Colloid Interf. Sci.*, 284, 14–21, 2005.
165. Adak, A., Bandyopadhyay, M., Pal, A., Removal of crystal violet dye from wastewater by surfactant-modified alumina. *Sep. Purif. Technol.*, 44, 139–144, 2005.
166. Janaki, V., Oh, B.T., Shanthi, K., Lee, S.K., Ramasamy, A.K., Polyaniline/chitosan composite: an eco-friendly polymer for enhanced removal of dyes from aqueous solution. *Syn. Metals*, 162, 974– 980, 2012.
167. Khosla, E., Kaur, S., Dave, P.N., Tea waste as adsorbent for ionic dyes. *Desalin. Water Treat.*, 13, 1–10, 2013.

168. Mane, V.S., Babu, P., Studies on the adsorption of Brilliant Green dye from aqueous solution onto low-cost NaOH treated saw dust. *Desalination*, 273, 321–329, 2011.
169. Garg, V., Gupta, R., Bala Yadav, A., Kumar R., Dye removal from aqueous solution by adsorption on treated sawdust. *Biores. Technol.*, 89, 121–124, 2003.
170. Sivaraj, R., Namasivayam, C., Kadirvelu, K., Orange peel as an adsorbent in the removal of acid violet 17 (acid dye) from aqueous solutions. *Waste Manag.*, 21, 105–110, 2001.
171. Argun, M.E., Dursun, S., Karatas, M., Guru, M., Activation of pine cone using Fenton oxidation for Cd (II) and Pb (II) removal. *Biores. Technol.*, 99, 8691–8698, 2008.
172. Weber, W., Morris, J., Kinetics of adsorption on carbon from solution. *J. Sanit. Eng. Div. Am. Soc .Civ. Eng.*, 89, 31–60, 1963.
173. Srinivasan, A., Viraraghavan, T., Decolorization of dye wastewaters by biosorbents: a review. *J Environ Manag.*, 91, 1915–1929, 2010.
174. Langmuir, I., Adsorption of gases on glass, mica and platinum. *J. Am. Chem. Soc.* 40, 1361–1403, 1918.
175. Freundlich, H.Z., Over the adsorption in solution. *J. Phy. Chem.*, 57, 385–470, 1906.
176. Ramakrishna, K.R., Viraraghavan, T., Dye removal using low cost adsorbents. *Water Sci. Technol.*, 36, 189–196, 1997.
177. Babel, S., Kurniawan, T.A., Low-cost adsorbents for heavy metals uptake from contaminated water: a review. *J. Hazard. Mater.*, 97, 219–243, 2003.
178. Netpradit, S., Thiravetyan, P., Towprayoon, S., Application of 'waste' metal hydroxide sludge for adsorption of azo reactive dyes. *Water Res.*, 37, 763–772, 2003.
179. Acemioglu, B., Adsorption of congo red from aqueous solution onto calcium-rich fly ash. *J. Colloid Interf. Sci.*, 274, 371–379, 2004.
180. Ali, V., Kaur, R., Kamal, N., Singh, S., Jain, S.C., Kang, H.P.S., Zulfequer, M., Husain, M., Use of Cu^{+1} dopant and its doping effects on polyaniline conducting system in water and tetrahydrofuran. *J. Phys. Chem. Solids*, 67, 659–664, 2006.
181. Ali, V., Kaur, R., Lakshami, G.B.V.S., Kumar, A., Kumari, K., Kumar, S., Electrical conductivity and dielectric parameters of polyaniline doped with $CuClO_4$.4BN in aqueous DMSO solvent. *Adv. Polym. Techanol.*, 31, 374–379, 2012.
182. Yasar, M., Deligoz, H., Guclu, G., Removal of indigo carmine and Pb(II) Ion from aqueous solution by polyaniline. *Polym. Plast. Technol. Eng.*, 50, 882–892, 2011.
183. Zilberman, M., Siegmann, A., Narkis, M., Melt-processed electrically conductive polymer polyaniline blends. *J. Macromol. Sci. Phys-B.*, 37, 301–318, 1998.
184. Patil, M.R., Shrivastava, V.S., Adsorptive removal of methylene blue from aqueous solution by polyaniline-nickel ferrite nanocomposite: a kinetic approach. *Desalination Water Treat.*, 1–9, 2015.

185. Ansari, R., Mosayebzadeh, Z., Keivani, M. B., Khah, A.M., Adsorption of cationic dyes from aqueous solutions using polyaniline conducting polymer as a novel adsorbent. *J. Adv. Novel Ads.*, 2, 27–34, 2011.
186. Ayad, M.M., Abu El-Nasr, A., Adsorption of cationic dye (methylene blue) from water using polyaniline nanotubes base. *J. Phys. Chem. C.*, 114, 14377–14383, 2010.
187. Ansari, R., Mosayebzadeh, Z., Application of polyaniline as an efficient and novel adsorbent for azo dyes removal from textile wastewaters. *Chem. Papers.*, 65, 1–8, 2011.
188. Igberase, E., Osifo, P., Ofomaja, A., The adsorption of copper(II) ions by polyaniline graft chitosan beads from aqueous solution: equilibrium, Kinetic and desorption studies. *J. Environ. Chem. Eng.*, 2, 362–369, 2014.
189. Geetha, A., Palanisamy, P.N., Kinetics and equilibrium studies on the removal of anionic dyes using polyaniline coated sawdust composite. *Int. J. Chem. Tech. Res.*, 7, 2439–2447, 2014.
190. Janaki, V., Vijayaraghavan, K., Oh, B.T., Lee, K.J., Muthuchelian, K., Ramasamy, A.K., Kannan, S.K., Starch/polyanilie nanocomposite for enhanced removal of reactive dyes from synthetic effluent. *Carbohyd. Polym.*, 90, 1437–1444, 2012.
191. Patil, M.R., Shrivastava, V.S., Adsorption removal of carcinogenic acid violet19 dye from aqueous solution by polyaniline-Fe2O3 magnetic nanocomposite. *J. Mater. Environ. Sci.*, 6, 11–21, 2015.
192. Salem, M.A., Salem, I.A., Hanfy, M.G., Zaki, A.B., Removal of titan yellow dye from aqueous solution by polyaniline/Fe_3O_4 nanocomposite. *Eur. Chem. Bull.*, 5, 113–118, 2016.
193. Ansari, R., Mohammad, Z.M., Keivani, M.B., Khah, A.M., Adsorption of cationic dyes from aqueous solutions using polyaniline conducting polymer as a novel adsorbent. *J. Adv. Sci. Res.*, 2, 27–34, 2011.
194. Ahmed, Z., Gilani, S.R., Nazir, A., Naveed, M., Wajid, M.K., Ahmed, M., Naseer, Y., Application of pani/Al_2O_3 composite towards the removal of tartrazine dye from aqueous solution. *Sci. Int.(Lahore).*, 27, 319–323, 2015.
195. Ayad, M.M., El-Nasr, A.A., Anionic dye (acid green 25) adsorption from water by using polyaniline nanotubes salt/silica composite. *J. Nano. Chem.*, 3, 1–9, 2012.
196. Karthikaikumar, S., Karthikeyan, M., Kumar, K.K.S., Removal of congo red dye from aqueous solution by Polyaniline-Montomorrillonite composite. *Chem. Sci. Rev. Lett.*, 2, 606–614, 2014.
197. Tanzifi, M., Mansouri, M., Heidarzadeh, M., Kobra Gheibi, K., Study of the adsorption of amido black 10B dye from aqueous solution using polyaniline nano-adsorbent: kinetic and isotherm studies. *J. Water Environ. Nanotechnol.*, 1, 124–134, 2016.
198. Roy, S.B., Lohar, G., Adsorption isotherm, kinetic and equilibrium studies on the removal of basic dye eosin yellow from aqueous solution by the use of polyaniline and its composites. *Int. J. Chem. Eng. Res.*, 8, 13–27, 2016.

199. Fang, X., Xu, X., Wang, S., Wang, D., Adsorption kinetics and equilibrium of Cu(II) from aqueous solution by polyaniline/coconut shell-activated carbon composites. *J. Environ. Eng.*, 139, 1279–1284, 2013.
200. Ghorbani, M., Esfandian, H., Taghipour, N., Katal, R., Application of polyaniline and polypyrrole composites for paper mill wastewater treatment. *Desalination*, 263, 279–284, 2010.
201. Gupta, R.K., Singh, R.A., Dubey, S.S., Removal of mercury ions from aqueous solutions by composite of polyaniline with polystyrene. *Sep. Purif. Techanol.*, 38, 225–232, 2004.
202. Ghorbani, M., Eisazadeh, H., Ghoreyshi, A.A., Removal of zinc ions from aqueous solution using polyaniline nanocomposite coated on rice husk. *Iran J. Energy Environ.*, 3, 83–88, 2012.
203. Kumar, P.A., Chakaborty, S., Ray, M., Rmoval and recovery of chromium from wastewater using short chain polyaniline synthesized on jute fiber. *Chem. Eng. J.*, 141, 130–140, 2008.
204. Igberase, E., Osifo, P., Ofomaja, A., The adsorption of copper (II) ions by polyaniline graft chitosan beads from aqueous solution: equilibrium, kinetic and desorption studies. *J. Environ. Chem. Eng.*, 2, 362–369, 2014.
205. Rathinum, K., Sankaran, M., Adsorption study on removal of Cr(VI) ions by polyaniline composite. *Desalination Water Treat.*, 54, 1–11, 2014.
206. Kanwal, F., Rehman, R., Anwar, J., Mahmud, T., Adsorption studies of cadmium(II) using novel composites of polyaniline with rice husk and saws dust of *Eukalyptus camaldulensis*. *Electron. J. Environ. Agr. Food Chem.*, 10, 2972–2985, 2011.
207. Zhang, R., Ma, H., Wang, B., Removal of chromium (VI) from aqueous using polyaniline doped sulfuric acid. *Ind. Eng. Chem. Res.*, 49, 9998–10004, 2010.
208. Patra, B.N., Majhi, D., Removal of anionic dyes from water by potash alum doped polyaniline: investigation of kinetics and thermodynamic parameters of adsorption. *J. Phys. Chem. B.*, 119, 8154–8164, 2015.
209. Singla, M.L., Jain, D.V.S., Adsorption of eriochrome black T on polyaniline from aqueous and methanolic solutions. *Indian J. Chem.*, 39, 603–610, 2000.
210. Li, J., Wang, Q., Bai, Y., Jia, Y., Shang, P., Huang, H., Wang, F., Preparation of a novel acid doped polyaniline adsorbent for removal of anionic pollutant from wastewater. *J. Wuhan Uni. Technol. Mater. Sci. Ed.*, 1085–1091, 2015.
211. Chafai, H., Laabd, M., Elbariji, S., Bazzaoui, M., Albourine, A., Study of congo red adsorption on the polyaniline and polypyrrole. *J. Dispersion Sci. Technol.*, 1–19, 2016.
212. Karthikaikumar, S., Karthikeyan, M., Kumar, K.K.S., Removal of congo red dye from aqueous solution by polyaniline-montmorrillonite composite. *Chem. Sci. Rev. Lett.*, 2, 606–614, 2014.
213. Mahanta, D., Madras, G., Radhakrishnan S., Patil, S., Adsorption of dyes by polyaniline emeraldine salt and its kinetics. *J. Phys.Chem. B.*, 112, 10153–10157, 2008.

214. Bingol, D., Veli, S., Zor, S., Ozdemir, U., Analysis of adsorption of reactive azo dye onto CuCl2 doped polyaniline using Box–Behnken design approach. *Synth. Metals.*, 162, 1566–1571, 2012.
215. Baseri, J.R., Palanisamy, P.N., Sivakumar, P., Polyaniline nano composite for the adsorption of reactive dye from aqueous solutions: equilibrium and kinetic studies. *Asian J. Chem.*, 25, 4145–4149, 2013.
216. Mahanta, D., Madras, G., Radhakrishnan, S., Patil, S., Adsorption and desorption kinetics of anionic dyes on doped polyaniline, *J. Phys. Chem.*, 113, 2293–2299, 2009.
217. Ayad, M.M., Abu EI-Nasi, A., Anionic dye (acid green 25) adsorption from water by using polyaniline nanotubes salt/silica composite. *J. Nanostr. Chem.*, 3, 313–319, 2012.
218. Ghorbani, M., Lashkenari, M.S., Eisazadeh, H., Application of polyaniline nanocomposite coated on rice husk ash for removal of Hg(II) from aqueous media. *Synth. Met.*, 161, 1430–1433, 2011.
219. Morsi, R.E., Elsabee, M.Z., Polyaniline nanotubes: mercury and competative heavy metals uptake. *Am. J. Polym. Sci.*, 5, 10–17, 2015.
220. Eisazadeh, H., Removal of chromium from waste water using polyaniline. *J. Appl. Polym. Sci.*, 104, 1964–1967, 2007.
221. Phan, T.B., Do, N.Q., Mai, T.T.T., The adsorption ability of Cr(VI) on saw-dust-polyaniline nanocomposite. *Adv. Nat. Sci. Nano Sci. Nano Techanol.*, 1, 1–4, 2010.
222. Wang, J., Deng, B., Chen, H., Wang, X., Zheng, J., Removal of aqueous Hg(II) by polyaniline: sorption characterstics and mechanisms. *Environ. Sci. Techanol.*, 43, 5223–5228, 2009.
223. Pham, T.T., Mai, T.T.T., Bui, M.Q., Mai, T.X., Tran, H.Y., Phan, T.B., Nanostructured polyaniline rice husk composite as adsorption materials synthesized by different methods. *Adv. Nat. Sci. Nanosci. Nanotechanol.*, 5, 501–506, 2014.
224. Zhang, Y., Li, Q., Sun, L., Zhai, J., Batch adsorption and mechanism of Cr(VI) removal from aqueous solution by polyaniline/Humic acid nanocomposite. *J. Environ. Eng.*, 137, 1158–1164, 2011.
225. Anbia, M., Ghaffari, A., Adsorption of phenolic compounds from aqueous solutions using carbon nanoporous adsorbent coated with polymer. *Appl. Surf. Sci.*, 255, 9487–9492, 2009.
226. Belaib, F., Meniai, A.H., Lehocine, M.B., Elimination of phenol by adsorption onto mineral/polyaniline composite solid support. *Energy Proc.*, 18, 1254–1260, 2012.
227. Abolghasemi, M.M., Parastari, S., Yousefi, V., Microextraction of phenolic compounds using a fiber coated with a polyaniline-montmorillonite nanocomposite. *Microchimi. Acta.*, 182, 1323–1328, 2014.
228. Zhang, Y., Li, Q., Cui, H., Zhai, J., Removal of phenols from the aqueous solutions based on their electrochemical polymerization on the polyaniline electrode. *Electrochimi. Acta.*, 55, 7219–7224, 2010.

229. Laabd, M., Jaouhari, A.E., Bazzaoui, M., Albourine, A., Removal of polycarboxylic benzoic acids by adsorption onto polyaniline-polypyrrole from aqueous solution. *Int. J. Eng. Res. Techanol.*, 3, 224–231, 2014.
230. Laabd, M., Chafai, H., Aarab, N., Jaouhari, A.E., Bazzaoui, M., Kabli, H., Eljzouli, H., Albourine, A., Polyaniline films for efficient removal of aromatic acids from water. *Environ. Chem. Lett.*, 14, 395–400, 2016.
231. Sun, C., Xiong, B., Pan, Y., Cui, H., Adsorption removal of tannic acid from aqueous solution by polyaniline: analysis of operating parameters and mechanism. *J. Coll. Inter. Sci.*, 487, 175–181, 2017.
232. Wang, J., Lijuan, B., Ji, Y., Ma, H., Yin, X., Removal of humic acid from aqueous solution by magnetically separable polyaniline: adsorption behavior and mechanism. *J. Colloid Interface Sci.*, 430, 140–146, 2014.

4

Immobilized Microbial Biosorbents for Wastewater Remediation

Mohammad Asaduddin Laskar[1], Rajeev Kumar[2] and Mohamed A. Barakat[2,3*]

[1]Department of Chemistry, Faculty of Science, Jazan University, Jazan, Saudi Arabia

[2]Department of Environmental Sciences, Faculty of Meteorology, Environment and Arid Land Agriculture, King Abdulaziz University, Jeddah Saudi Arabia

[c]Central Metallurgical R & D Institute, Cairo, Egypt

Abstract

The increasing concern for the sustainability of environment and economy has drawn the attention toward the use of easily and abundantly available bioresources, including fungi, bacteria, and algae. These living/nonliving microorganisms possess abundant active sites (for binding with pollutants) by virtue of their constituents, namely, polysaccharides, protein, and nucleic acids. However, their low mechanical strength and low density restrict their applications in industrial scale. Hence, their immobilization into/onto an inert matrix would improve the physical and chemical resistance of the developed new biosorbent and enhance the latter's regenerability. In this chapter, different supports/matrices used for the immobilization of fungi, bacteria, and algae have been discussed. The application of these immobilized biosorbents in the removal of pollutants from wastewater has been discussed and the role of various operational parameters in the adsorption, such as contact time, solution pH, initial pollutant concentration, temperature, etc., has been discussed. This chapter also gives an insight of a detailed mechanism and future prospect for the adsorption of pollutants onto immobilized microbial sorbents for the removal of heavy metals form wastewaters.

**Corresponding author*: mabarakat@gmail.com

Shahid-ul-Islam (ed.) Advanced Materials for Wastewater Treatment, (101–128)

Keywords: Immobilized microbial biosorbents, wastewater purification, heavy etals, biosorption mechanism

4.1 Introduction

Pollutants, including heavy metals, pharmaceuticals, dyes, phenols, etc., are being continuously introduced into the environmental resources, especially water bodies. Various life forms, including human beings, are vulnerable to serious ill consequences due to the presence of pollutants [1–14]. Heavy metals, such as, arsenic, lead, chromium, mercury, copper, zinc, nickel, and cadmium, etc., are release into the environment from several source like industrial, mining, etc. As for instance, heavy metals are nonbiodegradable and tend to accumulate, within the environment as well as living organisms, upto a concentration level that may causes several health problems like DNA damage, renal abnormalities, organ failure, allergy, fertility reduction, and so forth [1–3, 5–8, 9–15]. Hence, it has been a great challenge to remove the heavy metals contaminants and reused the wastewater.

In contrast to adsorption method, various other methods, including chemical precipitation, evaporation, electrodeposition, ion-exchange and membrane separation, suffer from ineffectiveness and high expenditure in the treatment of wastewater having low concentration of heavy metals. In last few decades, a lot of research has been focused on the development of new adsorbents for the removal of the organic and inorganic contaminants from the wastewater, including nanocarbons (carbon nanotubes, graphene), metals oxides, polymers, clays and minerals, waste materials (industrial and agricultural) [1–15]. Hence, materials may have some merits and demerits such as cost, poor reservation, selectively, etc. Therefore, the increasing concern for the sustainability of environment and economy has drawn the attention toward the use of easily and abundantly available bioresources. As for instance, the remediation of wastewater may be achieved, in an eco-friendly manner, by using microbial biomass, in its natural form or after immobilization, as biosorbent for heavy metals scavenging.

The use of low-cost microbial organisms (living or dead) for the remediation of contaminated wastewater has emerged as an effective method of adsorption, popularly termed as biosorption [16]. The biosorption method offers numerous advantages which include reusability, low-operating cost, improved selectivity, minimal volume of diposable waste. The efficiency of such removal technique may depend on factors such as (1) size and/or charge of metal ions, (2) external parameters, namely, pH, temperature, ionic strength, contact time, and amount of biomass, and (3) species of microbes, including its physiological and metabolic nature [17]. Generally, the dead biomass adsorbs metal ions through physiochemical interactions that occur between the metal ions and the superficially available functional groups of the biomass adsorbent. These physical interactions may be classified as electrostatic interactions, ion exchange, chelation/complexation, reduction, and microprecipitation. The commonly available functional groups, on the surface of the dead biomass, may include hydroxyl, amine, carboxyl, and phosphoryl functional groups [18]. However, in case of living biomass, a dual mechanism, comprising of physicochemical interactions and intracellular accumulation, comes into play. The intracellular accumulation involves the transport of the metal ions into the cell, after crossing the cell membrane, and subsequently binding with the cellular organelle [17, 19].

However, the small units of microbial biomass have low density and exhibit low mechanical strength, which may lead to problematic swelling (of biomass) and separation (between solid and liquid) during operation [20]. Therefore, the use of stable matrices, such as alginate, polysulfone, polyacrylamide, polyurethane and silica, with high mechanical strength as biomass carrier can impart high resistivity (in harsh chemical environment), regenerability and easy separability (in contact solution) to the final microbial biocomposite. Hence, this chapter deals with the modified microbial biosorbent used for the removal of the heavy metals from wastewaters.

4.2 Immobilized Microbial Biosorbent

4.2.1 Algae Biosorbent

Algae are typically aquatic plant of diverse group of eukaryotic organisms, which may be unicellular or multicellular. $C_{106}H_{181}O_{45}N_{16}P$ is the stoichiometric formula for the most common elements in an average algae cell [21]. Cellulose is the major constituent of green algae and a high percent of cell walls are proteins which contains several types of functional groups such as hydroxyl, amino, carboxyl, sulphate, etc.[22]. These functional groups play a vital role in the binding of metal ions. A lot of work has been published on the removal of heavy metals by algal biomass. Now-a-day, researchers are exploring the adsorption properties of composite/immobilized algal biomass for the separation and removal of heavy metals. For the industrial scale-up, immobilized biosorbents could be ideal material because adequate size biosorbent particles can be prepared by immobilization, having better mechanical strength and higher regeneration properties [23].

Spent *Tricholoma lobayense* was employed for the preparation of a biocomposite by immobilizing it into sodium alginate through mixing in the presence of $CaCl_2$ [24]. In ternary system, Pb(II) could be effectively removed in the co-presence of Cd(II) and Cu(II), however, the presence of the former interfered with the removal of the co-existing Cd(II) or Cu(II) ions. The adsorption of Cu(II) and Cd(II) was unfavorable at higher temperature and initial metal ion concentration, while the recovery of Pb(II) was unfavorable at higher initial metal ion concentration. The retention of the metals by the biocomposite was mainly through chemisorptions. The immobilization of *C. vulgaris* onto Ca-alginate and agarose biopolymer resulted in the formation of two biocomposites [25]. The agarose-*C. vulgaris* biocomposites showed higher Cr(VI) adsorption, better kinetics, higher rate of adsorption than its counterpart. The *C. vulgaris* -Ca-alginate composite shows little improvement as compared to pure Ca-alginate. The rate of adsorption by *C. vulgaris*-Ca-alginate was restricted by the diffusion of the solute through the pores. The mixing of *Chlamydomonas reinhardtii* and *Anabena variabilis* with Na-alginate, in the presence of $CaCl_2$, rendered two biocomposites [26]. The *Chlamydomonas reinhardtii*–alginate composite showed higher adsorption capacity for Cd(II), Cu(II), Zn(II), and Ni(II) at pH 4–5. However, both the composites exhibited higher adsorption capacity for copper ion, while higher rate of adsorption was observed for Cd(II) and Cu(II). A mutual separation among the four metals was possible. Seaweed, namely *Ulva sp.*, was mixed with clay sepolite and subsequently subjecting the mixture to alternate drying and wetting steps [27]. The resultant biocomposite recovered the maximum amount of Uranium (VI), as UO_2^{2+} or $UO_2(OH)^+$, at pH 4.5. The adsorption of Uranium (VI) onto the composite was less favored at higher temperature, while de-sorption could be effected with HNO_3 (1 M). The immobilization of *Acinetobacter junii* onto coconut fiber was carried out by physical loading in a continuous flow reactor [28]. The biocomposite exhibited maximum sorption of Cr(VI) at pH 1–2, while the recovery decreased to 33% at pH 10. The oxyanions ($HCrO4^-$), formed at the lower pH, were probably responsible for the retention onto the positively charged adsorbent by virtue of electrostatic interaction. As compared to the pure/blank coconut fiber, the composite biosorbent showed lesser exhaustion time and shorter adsorption zone length. In another study [23], *Fucus vesiculosus* was immobilized into alginate xerogels. The composite biosorbent exhibited improved kinetic uptake and higher diffusion rates during the adsorption of Pb(II), Cu(II), and Cd(II). However, the composite showed decreased kinetic rate of adsorption. The heavy metals were retained by virtue of the presence of carboxyl group as well as through the displacement of calcium via "egg-box" model, whereby rendering a uniform and organized structure of the composite. This composite promises effective recovery of Pb(II) from dilute industrial and waste water. A marine alga, namely *Sargassum sp.*, was immobilized into polyvinylchloride-cryogel via freeze-thaw-freeze method [29]. The surface area of the composite was lower than that of the blank polyvinylchloride-cryogel, while the pore size increased significantly on immobilization. These characteristics may be the outcome of the evaporation of water from the water-laden entrapped biomass. Thus, the surface area and the pore size of the composite were highly dependent on the concentration of the alga contained in the polyvinylchloride-cryogel matrix. The composite exerted mass transfer resistance, during the uptake of Cu (II), and showed higher metal-sorption rate at higher biomass loading. The maximum adsorption capacity also gets enhanced at higher pH, up to pH 13. In another reported study [30] *Pediastrum boryanum* was immobilized, through entrapment, into alginate and alginate-gelatin. The endothermic sorption of Cr(VI), at pH 2.0, was exhibited by the composite by virtue of the functional groups, namely carboxyl, primary and secondary amine, and hydroxyl. The *Pediastrum boryanum*-entrapped composites showed improved maximum adsorption capacity as compared to the free algal cell, pure alginate and alginate-gelatin supports. A biocomposite sorbent, prepared by entrapping *Chlamydomonas reinhardtii* into carboxymethyl-cellulose, was investigated for the removal of Uranium from water [31]. It was observed that the composite showed lower sorption capacity for uranium than the free *Chlamydomonas reinhardtii.* The maximum adsorption capacity varied with the initial concentration of Uranium. The composite sorbent could be regenerated with 10 mM HNO_3. Peter et al., [32] reported the entrapment of *Spirulina platensis-maxima* into alginate and chitosan. The alganite-spirulina composite showed a lower adsorption capacity than the pure alginate sorbent. Again, chitosan-spirulina composite showed higher affinity for Cu(II) and lesser for Pb(II) and Cd(II) than the pure chitosan sorbent. The low adsorption capacity and reduced porosity was caused by the partial/or incomplete exposure of the active sites of the composite sorbents. The composite sorbent could be regenerated with mineral acids and water and used for at least 5 cycles of sorption studies. A freshwater algae, namely *Scenedesmus quadricauda*, was entrapped into Ca-alganite and then allowed to grow uniformly in an algal growth medium through incubation at an optimum environment [33]. At the optimum surface charge, rendered at pH 5.0, the maximum adsorption capacities for each metal increased with the increase in the initial metal ion concentration. The biosorbent composite exhibited selectivity for Cu(II) in the co-presence of Ni(II) and Zn(II). The immobilization of *Trametes versicolor* (white rot fungi) onto montmorillonite by mechanical mixing, through alternate wetting and drying stages, was carried out by Akar et al. [34]. The so developed sorbent could completely remove copper from wastewater/contaminated water within 45 minutes at pH 5.0. A chemical ion-exchange mechanism is reflected by the sorption energy within the range of 9.72–11.56 kJmol^{-1}. A collective contribution from the macro-porous structure and the functional groups, such as OH and Si-O-Si, lead to the retention of copper onto the sorbent. A *Sargassum* sp. was encased within a polymeric matrix of chitosan-epichilorohydrin [35]. The highest sorption capacity for nickel ions was obtained with the matrix containing the *Sargassum* sp. in a weight ratio of 3. The adsorption of nickel onto the adsorbent took place through ion exchange and complexation with the constituting functional groups, namely carboxyl, amino, alcoholic, and ether. The *Laminaria japonica* algae were immobilized on beads comprising of sodium alginate and glutin. The Ni (II) ions were retained onto the beads by virtue of the functional groups (OH,NH_2,CH_2,CH_3,C=O, NH,CH,CO, and C=O) constituting the algal cells [36]. The introduction of *Scenedesmus obliquus* CNW-N onto *L. cylindrica* (Loofa sponge) through column method, whereby the former get immobilized through weak interactions, was reported by Chen and coworkers [37]. The extent of immobilization was dependent on the flow rate and the frequency (and time) of replacement of the medium. The exposure of the system to optimum nutritional input enhanced the metal retention capacity of the resultant biosorbent. The size of 1 cm (in diameter) of the particles of Loofa sponge and a flow rate of 0.284 bed volume/min was found to give the best results for one cycle replacement of the medium. A *Cladophora sp.* was entrapped into a matrix constituting chitosan cross-linked with glutaraldehyde via coagulation [38]. This biomass composite exhibited higher affinity for Cd(II) and Zn(II) than the chitosan-glutaraldehyde-matrix devoid of the algae. However, there was no appreciable difference in the sorption capacity of Cr(III), Cu(II), and Ni(II).

4.2.2 Fungi Biosorbent

Fungi are single-celled organisms consist of yeasts, molds, and mushrooms, closely related to the animals rather than plant. Fungal cell walls are rich in protein, polysaccharides, and nucleic acids. Fungi have been widely explored for heavy metal remediation applications because they can accumulate and sequester, or bind metal ions as part of their metabolic processes or passive uptake in living form. While, binding of metals on nonliving is metabolism independent process and known as passive biosorption. The functional groups such as amine $-NH_2$, $-COOH$, $-OH$, $-PO_4^{3-}$, and $-SH$ are the main binding sites involved in adsorption of contaminants. However, direct use of fungi has some limitations like leaching and loss of biomass. Moreover, handling and storage of dead microbes are more easily compared to living microbes and can be easily immobilized with other materials to enhance the metal removal efficiency and easy separation from the aquatic environment.

Deng and Ting [15] prepared the *acrylic acid modified Penicillium chrysogenum* using the graft polymerization approach for removal of Cu(II) and Cd(II). The modified biomass shows the excellent adsorption capacity for the removal of both metal ions as shown in Figure 4.1 and the maximum monolayer adsorption capacities were 1.70 and 1.87 mmol/g for Cu(II) and Cd(II), respectively. Moreover, a rapid and high desorption of 97.2% Cu(II) and 98.6% Cd(II) was observed after 1 h treatment in 0.2M HCl.

The removal of U(VI) was investigated with several biocomposites prepared by immobilizing *Saccharomyces cerevisiae* onto Ca-alginate, Ca-alginate-graphene oxide, and polyvinyl alcohol-Ca-alginate-graphene oxide in $CaCl_2$-boric acid solution [39]. An ion-exchange mechanism played the major role in the adsorption of U(VI). Since hydrolyzed species are better adsorbed than the free hydrated ions, therefore at pH 2.6–5.0, the high degree of partitioning of $UO_2(OH)^+$, $(UO_2)_2(OH)_2^{2+}$, $(UO_2)_3(OH)^{5+}$, and $(4UO_3.H_2O)$ leads to higher adsorption of U(VI) onto

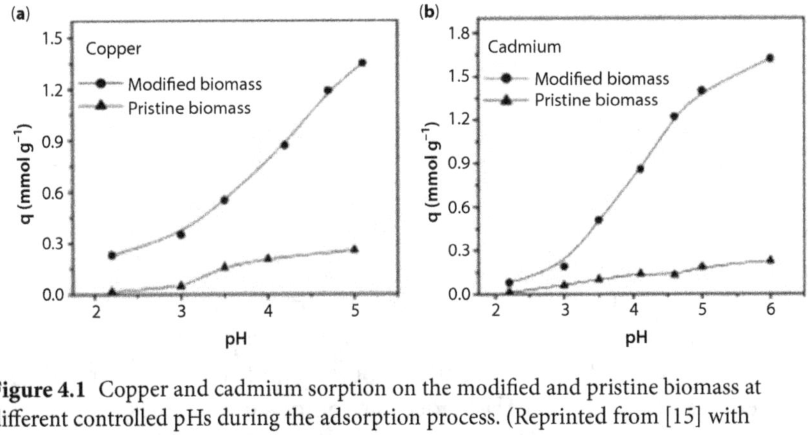

Figure 4.1 Copper and cadmium sorption on the modified and pristine biomass at different controlled pHs during the adsorption process. (Reprinted from [15] with permission, Copyright © 2005 American Chemical Society.)

the biocomposites. The loaded biocomposites could be recovered by eluting with 0.1 M HCl/HNO_3. In another work, dead *Agaricus bisporus* was immobilized on to Amberlite XAD-4 for obtaining a biocomposite for the removal of thorium [40]. The maximum adsorption of thorium onto the biocomposite occurred at pH 6 with saturation capacity of 18280 $\mu g\ g^{-1}$. Immobilization of *Trichoderma viride* on Ca-alginate rendered a biocomposite which was employed for the removal of Cr(VI), Ni(II), and Zn(II) from electroplating effluent [41]. With increasing flow-rate and initial metal ion concentration the breakthrough time and the time for saturation decreased. However, increase in the bed height (in the column) resulted in the increase of breakthrough time and sorption capacity. A macrofungi *Lentinus edodes* was supported onto polyvinyl-Na-alginate composite [42]. The developed biocomposite exhibited higher affinity for Pb^{2+} than Cd^{2+}. The free *Lentinus edodes* could reach equilibrium faster than that required by the biocomposite, which may be attributed to the difference in the influential factors, namely, mass transfer through diffusion, porosity and accessibility (of the analyte) to the active sites of the biomass. The sorption capacity for Uranium was improved by immobilizing *Trichoderma harzianum* onto Ca-alginate through ionotropic gelation [43]. The stability was demonstrated when 1.5 g of the biocomposite gave a breakthrough volume of 8.5 L of bacterial leach liquor (containing 58 mg/L of Uranium). The uranium-laden biocomposite could be regenerated with 0.1 N HCl. *Aspergillus species* was immobilized onto sodium montmorillonite in order to obtain a biocomposite sorbent that would remove Cr(VI) from acidic medium (pH 2.0) [44]. A breakthrough volume of 50 mL was obtained with 1.5 g of the biocomposite. The Cr(VI), as $HCrO_4^-$, interacted with the active sites, including protonated amine, carboxyl and hydroxyl groups of the biomass, as well as the $SiOH_2^+$, $AlOH_2^+$ groups, present on the surface of the clay support. Thus, the support (sodium montmorillonite) reinforced the interaction of Cr(VI) with the cell wall (of biomass), through the constituting functional groups, and enhanced the overall stability of the biocomposite. Hence, the biocomposite promises to be useful in the remediation of tannery and electroplating wastewaters containing Cr(VI). A bio-composite, in the form of membrane, was prepared by entrapping *Penicillium* into the cross-linked network of chitosan [45]. A plate column reactor (filled with the biocomposite) was successfully employed for the removal of Cu(II) from wastewater. The biosorbent could be used for 10 cycles of loading-elution. The batch biosorption capacity for Cu(II) was found to be 126.58 mg/g^{-1}. The shifting of the bands in Fourier transform infra-red spectra of the Cu(II)-loaded biocomposite indicates the participation of OH, NH, NH_2, CO, and CN functional groups. The removal of nickel was carried out with *Lactarius salmonicolor* immobilized silica gel [46]. The biocomposite had the isoelectric point at pH 2.05. It could attain a sorption capacity of 44.74 mg/g^{-1}, for Ni^{2+}, within 5 min of contact time at pH 6.5. The major rate-limiting mechanism for the retention of Ni^{2+} (onto the biocomposite) was chemisorption.

The entrapment of cross-linked Rhizopus arrhizus onto Ca-alginate was done for the removal of americium, europium and uranium from radioactive waste effluent [47]. The mere immobilization of the biomass resulted in a biocomposite having a soft nature and was prone to distortion in shape. Hence, the use of cross-linked biomass enhanced the mechanical strength and rigidity of the biocomposite. *Saccharomyces cerevisiae*, in association with histidine, was immobilized onto crosslinked chitosan beads [48]. At pH 6.0, the biocomposite exhibited an adsorption capacity of 104.2 mg/g for Ni(II). The retention of Ni(II) took place through its interaction with OH, NH_2, and COO groups, constituting the biocomposite.

The immobilization of surface engineered *Saccharomyces cerevisiae* cells onto cytopore rendered a Cd^{2+}-selective biocomposite [49]. The enhanced selectivity was achieved by exposing the Cd^{2+} binding peptide sequence, thereby increasing the covalent interaction between Cd^{2+} and the *Saccharomyces cerevisiae* cells. The biocomposite exhibited a detection limit of 1.1 mg/L, enrichment factor of 30 and a high tolerance toward co-existing metals. The dead biomass of *Pencillium fellutinum* was immobilized on Na-bentonite for the sorptive removal of Ni(II) and Zn(II) [50]. At pH 6 and 5, the endothermic adsorption process exhibited a sorption capacity for Ni(II) and Zn(II) as 161mg/g and 78.5 mg/g, respectively, which were reached within 30 min of contact time. The regeneration of the biocomposite was possible with 0.1 M NaOH for 7 cycles of loading elution process. A one-pot hydrothermal carbonization method was followed to immobilize *Geotrichum* sp. dwc-1 on attapulgite via covalent bonding [51]. The maximum adsorption capacity for U(VI) was 125 mg/g at pH 4.0. With the enhanced dispersive and stable nature, the biocomposite promises a successful application in the removal of U(VI) from the geological repositories and nuclear waste management. The removal of Ni(II) was established with three different biocomposites, which were prepared by entrapping *Candida lipolytica*, *Candida tropicalis*, and *Candida utilis*, separately onto Ca-alginate [52]. The chemical ion-exchange nature of adsorption of Ni(II) was established at pH 6.62 with equilibrium time of 3h. The *Candida tropicalis*-alginate biocomposite resulted in higher adsorption capacity than others. The biosorption process could be carried out continuously for five cycles with HCl as the eluent. A nanofibre composite comprising of chitosan and baker's yeast was prepared by the electrospinning method [53]. The biocomposite, with low crystalinity, showed a decending trend of the adsorption capacity, for U(VI) and Th(IV), in binary systems with co-metals in the order of Co(II) > Cd(II) > Cu(II) > Fe(II) > Ni(II) and Co(II) > Cu(II) > Ni(II) > Cd(II) > Fe(II), respectively. The biocomposite was regenerable for use for five cycles with Na_2CO_3 (0.1 M) and HNO_3 (0.1 M) as the eluting agents. In order to remove Cu(II), a biocomposite was prepared by immobilizing *Trichoderma asperellum* onto Ca-alginate [54]. In one case, the biocomposite was heated to inactivate the entrapped biomass, which resulted in the enhancement of the adsorption capacity (134.22 mg/g), which is higher than that of other biocomposite associated with viable biomass and plain alginate beads. Thus, the clogging and post-separation process, associated with the use of suspended biomass, may be avoided with this biocomposite. Since the biomass was isolated from polluted river water, therefore the entrapped biomass (supported on the Ca-alginate) will have better adaptability to applications in the treatment of industrial and other effluents. The immobilization of *Aspergillus niger* B77 onto polyurethane was carried out for the adsorption of Cu(II) [55]. The biomass laden biocomposite exhibited threefold increase in the uptake of Cu(II) than the free biomass by virtue of the increased number of theoretical plates. Again, it was also noted that the presence of glucose in the equilibrium mixture resulted in the further enhancement in the adsorption of Cu(II). The removal of Cu(II) and Pb(II), from their single and binary aqueous mediums, was undertaken with a biocomposite that was prepared by anchoring *Saccharomyces cerevisiae* onto a composite of Ca-alginate and activated carbon [56]. In the binary systems, the Pb(II) ions interfered significantly in the removal of Cu(II), while the later showed no appreciable interference during the recovery of the former. Hence, this antagonistic interference suggests that the biocomposite has higher affinity for Pb(II) at pH 5. Since, most of the metal-contaminated wastewater has an acidic pH, therefore this biocomposite promises efficient application in the remediation process. *Rhodotorula glutinis* was supported on the nanoparticles of iron oxide, whereby producing a magnetic biocomposite [57]. This biocomposite was applied in the removal of Uranium at pH 6.0. The commonly co-existing cations including K^+, Na^+, Ca^{2+}, and Mg^{2+} did not interfere with the removal of Uranium. The physiochemical uptake of Uranium was favorable at higher temperatures. At pH 5.0, a unit amount of the biocomposite of *Aspergillus fumigatus* supported Ca-alginate could remove U(VI) within 120 min [16]. Active sites for adsorption include the carboxyl, amino and phosphate groups which are spread over the surface of the biocomposite. However, at pH 6.0, the formation of the uranyl carbonate complexes restricts the availability of the binding sites (for uranyl ions) through competitive adsorption. Chitosan was used as a support for encapsulating *Zygosaccharomyces rouxii* to get a bicomposite for the removal of Cd(II) [58]. The composition of the ratio 4:10 by weight of *Zygosaccharomyces rouxii* to chitosan exhibited the highest recovery of Cd(II) at pH 3.0. The pH_{PZC} (pH at point zero charge) of the biocomposite was 2.52. The diminished leaching of the protein and nucleic acids from the regular spherical-shaped biocomposite contributes to the prevention of eutrophication. *Aspergillus* BRVR was entrapped into epichlorohydrin crosslinked cellulose matrix and the resultant biocomposite was used for the removal of Cr(VI) [59]. An increase in the bed-height resulted in the enhancement of the uptake capacity of the biocomposite, while the increase in the flow rate and Cr(VI) concentration lead to the reduction of its adsorption capacity. At pH 2.0, the functional groups including NH_2, OH, COOH (constituting the fungal cell wall) and those associated with cellulose get protonated whereby electrostatic interaction take place with $HCrO_4^-$. The biocomposite could be regenerated with 1.5 M NaOH.

4.2.3 Bacteria Biosorbent

Bacteria are unicellular microorganisms neither plants nor animals and have a number of shapes. Bacteria live in colony and approximately 5×10^{30} bacteria on earth [60]. The applications of bacterial cell as biosorbent have been widely investigated for the removal of a number of the contaminants. Immobilized cell systems for large-scale or column operations or batch method have gain a wide attention for the removal of pollutants from wastewaters.

The immobilization of *Pseudomonas* strain was accomplished during the radiation-induced formation of the polyacrylamide matrix [61]. In the porous polymeric matrix, the bacterial cells got entrapped onto the pore walls. At pH 5.0, a unit quantity of the biocomposite could retain 202 mg of uranium. The used biocomposite could be fully recovered with sodium-bicarbonate upto 4 loading-elution cycles. The entrapment of dead *Pseudomonas putida* cells into chitosan beads rendered a biocomposite sorbent whose adsorption capacity for U (VI) increased by 1.27 times as compared to the blank chitosan (devoid of the microorganism) [62]. The functional groups, namely, NH^{+3}, NH_2, OH, and COOH, were majorly involved in the retention of U(VI) species, such as $(UO_2)_3(OH)_2^{+2}$ and $(UO_2)_4(OH)_7^+$, at pH 5 as shown in Figure 4.2.

A biofilm of *Bacillus arsenicus* MTCC 4380 was coated on the surface of composite Neem leaves/$MnFe_2O_4$, prepared by the co-precipitation method [63]. Both As(III) and As(V) were retained on the bacterial cells.

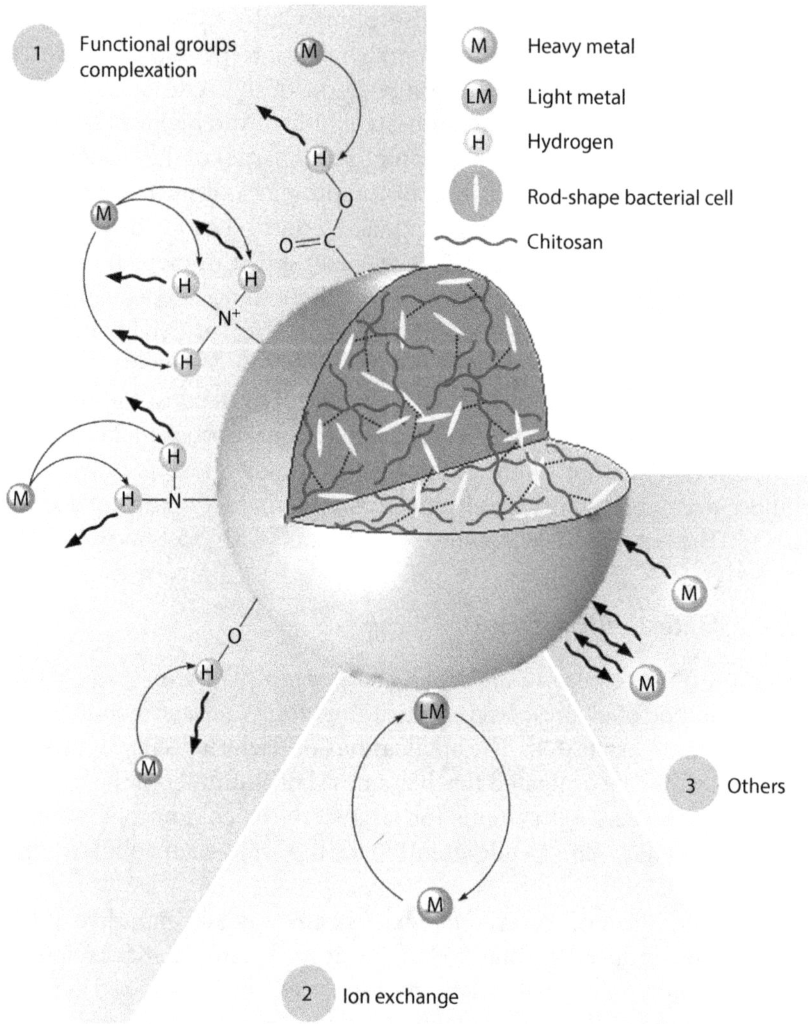

Figure 4.2 Diagram of U(VI) biosorption mechanisms onto the *Pseudomonas putida* @ chitosan bead (PICB). (Reprinted from [62] with permission from Elsevier.)

The contact time to reach the equilibrium increased with increasing initial concentration of the metals, but however, there was no effect of temperature. The mechanisms of ion-exchange and film diffusion were predicted to be the major routes for sorption. A composite comprising of *Escherichia coli* and polysulfone, prepared through spinning in *N,N*-dimethylformamide, was coated with polyethylenimine [64]. The sorbent was found to be highly stable in acetic acid and thus suitable for the recovery of ruthenium from the acetic acid effluent discharged during the Cativa process. It exhibited a breakthrough volume of 120 beds for flow-through column system. Again, the same sorbent [65] could recover Palladium from acidic aqueous effluent. A high sorption capacity for Pd was observed in 0.1–1.0 M solution of HCl. *Burkholderia vietnamiensis* C09V was entrapped into the composite comprising of polyvinyl alcohol, sodium alginate and kaolin [66]. The biocomposite could recover 89.4% crystal violet, 16.5% Cr(VI) and 64.1% Cu(II), when exposed for 42h. The Cr(VI) got reduced to Cr(III) when adsorbed by the entrapped bacterial cells. The adsorption of Zn^{2+} and Cd^{2+} was studied with several bio-composites by Piresa et al. [67]. These biocomposites were prepared by the capsule immobilization of inoculums of bacterial strains, namely, Cupriavidus (1C2), Sphingobacterium (1ZP4), and Alcaligenes (EC30), in Ca-alginate, Ca-pectate, and diethylamino-ethyl-methacrylate-based polymer. The bacterial strains, 1C2 and EC30, showed the maximum affinity for Zn(II) and Cd(II), respectively. In binary mixture, the removal of Cd (II) was higher, by 40%, than Zn(II) with the biocomposites. On comparing all the sorbents, the best result was obtained with 1C2 and EC30 immobilized on the synthetic polymer. The *Lysinibacillus sp.* NOSK was attached onto an electrospun polysulfone nanofibrous web [68] and used for the removal of Reactive Black 5 and Cr(VI) and maximum removal was 99.7% and 98.2%, respectively. Reduction of Cr(VI) to Cr(III) and adsorption/degradation of the dye by the bacterial cells were also studied. In another work, Cr(VI) was removed with *Ochrobactrum sp.* Cr-B4 immobilized polyvinyl-alcohol-alginate composite by virtue of adsorption on solid–liquid interface and enzyme catalyzed chromate reduction [69]. A granular-activated carbon-$MnFe_2O_4$ composite was coated with a biofilm of *Bacillus arsenicus* MTCC 4380 for the in-situ removal of As(III) and As(V) from groundwater [70]. An ion-exchange type, accompanied by film diffusion, mechanism, was followed for adsorption and the maximum adsorption capacity was attained within 180 min at 30°C at pH = 7. The *Pseudomonas putida* CZ1 entrapped goethite composite showed preferential sorption of Zn(II) over Cu(II) [71]. The toxicity of Cu(II) was higher than Zn(II) for the bacterial cells. The higher adsorption of the composite may be attributed to the variety of reactive sites rendered by functional groups, namely carboxylic, phosphoryl, and amino and hydroxyl, constituting the bacterial cells, in addition to the hydroxyl group of geothite. A Fe(III) selective biocomposite was prepared by anchoring pyoverdin (as secreted by pseudomonas fluorescens strain), through glycidoxypropyl linker, into the pores of mesoporous micelle-templated silicas [72]. The large surface area and the low density of the competitive silanols, within the pores, allowed higher retention of Fe(III). The larger pores facilitated the diffusion and thereby easy access to the complexation sites. A biofilm of *Corynebacterium glutamicum* MTCC 2745 was supported on a granular activated carbon/$MnFe_2O_4$ composite [73]. Since the growth of the bacteria cells is unfavorable at highly acidic and basic mediums, therefore, the maximum sorption of As(III) and As(V), existing as nonionic species H_3AsO_3, was observed at pH 7.0. *Pseudomonas aeruginosa* was supported on multiwalled carbon nanotubes. An adsorption capacity of 5.0496 mg/g^{-1} for Cu(II) was reached within 30 min [74]. Vermiculite clay was used as a support for anchoring *Streptococcus equisimilis* [75]. This biocomposite was used for treating wastewater, containing Ni(II), Cd(II), and diethylketone, discharged by metal refining and paint manufacturing industries. The increased adsorption capacity for Ni(II) and Cd(II) as well as diethylketone may be attributed to the presence of functional groups on the surface of the biofilm, which also promotes the degradation of diethylketone. The volcano rocks, comprising of porous aluminosilicate mineral, was coated with genetically modified *Pseudomonas putida* cells having protruding cyanobacterial metallothioneins [76]. The maximum adsorption capacity of 22.23 mg/g was attained for Cu(II) at pH 6 after a contact time of 60 min. The biocomposite exhibited improved mechanical strength and a regenerability of five continuous cycles of loading-elution. The removal of Cr(VI), Cd(II) and Co(II) was carried out with *Chryseomonas luteola* TEM 05 entrapped carrageenan and chitosan coated carrageenan gels [77]. The use of gelling agent KCl-chitosan, instead of KCl, resulted in the improvement of mechanical strength and thermal stability of the gel. The growth of bacterial cells is favorable on the Carrageenan gels, which also has a high diffusion coefficient. At the optimum condition of pH 6 (25 °C), the sorption of Co(II) and Cd(II), in the presence of all the metals, was found to decrease by 3-fold and 2-fold, respectively. Calcium alginate supported *Bacillus subtilis* was employed for the removal of Cd(II) at pH 5.92 [78]. An adsorption capacity of 251.91 mg/g (45 °C) and 188 mg/g (25 °C) was obtained for the initial cadmium concentration of 496.23 mg/L. The functional groups like hydroxyl, carboxyl, amines, amides and alcohol played the major role in the retention of Cd(II) onto the biocomposite sorbent.

4.3 Biosorption Mechanism

4.3.1 Algae-Based Biocomposite

The uptake of heavy metals by living microalgae involves both nonmetabolic and metabolic pathways [17]. Kumar et al. [17] represented the several mechanisms involved the translocation, sequestration, and uptake in living as well as, nonliving microalgae (Figure 4.3). In its reversible nonmetabolic route, the functional groups, constituting the cell surface, undergoes electrostatic interaction with the heavy metals and thus retain the latter through physical adsorption, ion exchange, chemisorptions, coordination, chelation, precipitation, entrapment (into the polymeric network of polysaccharides), and diffusion (into cell membrane). The ion-exchange mechanism is involved in the replacement of Ca^{2+}, Mg^{2+}, Na^{+}, and K^{+} with heavy metals [79]. The constituents of the cell wall, namely, amino acids (comprising protein) and polysaccharides, interact with the heavy metals through its amino, carboxyl, and sulfate groups. The formation of complexes through coordination bonds with heavy metals is observed for alginate and sulfated polysaccharides. The histidine and the peptide bond provide imidazole and

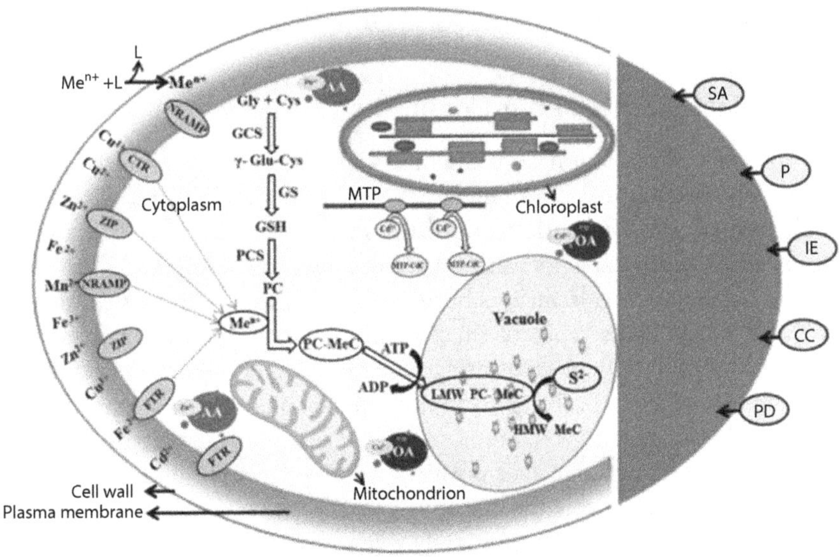

Figure 4.3 Schematic representation of several mechanisms of heavy metal translocation, sequestration, and uptake in living (Left), as well as, nonliving (Right, brown shaded) microalgae; including Me^{n+}-Metal ion, L-liquid(Me^{n+} + L represents metal ion in liquid); Metal-ion transporters (such as NRAMP, CTR, ZIP, and FTR); Phytochelatinbio-synthesis pathway, PC complexes, and enzymes involved in the PC synthesis (GCS-glutamyl–cysteinyl synthase, GS-Glutathione synthase, PCS-phytochelatin synthase); AA-Amino Acids; OA-Organic Acids; LMWPC-MeC-Low Molecular Weight Phytochelatin Metal Ion Complexes; HMWPC-MeC-High Molecular Weight Phytochelatin MetalIon Complexes; MTP-Metallothionein Protein; SA-surface adsorption; P-Precipitation; IE-Ion Exchange; CC-Complexation and Chelation and PD-Passive diffusion. (Reprinted from [17] with permission from Elsevier.)

nitrogen/oxygen atoms, respectively, to form coordination bonds with the heavy metals [19]. Besides these forces and factors, the biosorption of the metal ions is also controlled by the experimental conditions. In Table 4.1, a comparison in the adsorption capacity of various algae-based biosorbents and the experimental conditions have been summarized.

In the irreversible metabolic pathway [17], which occurs within the cell, the heavy metals get transported across the cell membrane and ultimately accumulate inside the cell through interaction with the intracellular organelle. The heavy metals are carried into the cytoplasm by membrane transport protein, including the Group A transporter's members, such as Natural Resistance-Associated Macrophage Proteins, ZRT-IRT-like Protein, Fe–Transporter, and Cu–Transporter [80]. The dead algae adsorb heavy metals through the reversible non-metabolic mechanism [79]. In dead algae, the fibrillar cellulose and the embedding matrices of alginic acid/alginate and sulfated polysaccharides play the major role in the retention of heavy metals onto the cell wall [79].

4.3.2 Bacteria-Based Bio-Composite

The sorption of metal ion onto the bacterial biocomposite would depend on the physiochemical parameters, such as pH, temperature, density of binding sites, and the initial concentration of the metal ions. The maximum adsorption capacity and experimental condition used for the removal of heavy metals various bacteria-based biosorbent are summaries in Table 4.2. Various types of mechanisms, namely micro-precipitation, surface-accumulation, electrostatic interaction, chelation or ion exchange, may be involved in the bio-sorption of metal ions [81]. As reported by Podder and Majumder [82], at lower pH, the pro-binding sites on the surface (of the biocomposite) get positively charged, which undergoes electrostatic interaction with the negatively charged As(III) and As(V). The superficial positive charge was mainly due to the $MnFe_2O_4$ portion of the biocomposite. In presence of water, the surfaces of both *B. arsenicus* MTCC 4380 and GAC/$MnFe_2O_4$ biocomposite undergoes hydration whereby forming coordination shells with the OH groups [83]. As the pH of the medium changes, these OH groups may bind/release H^+ ions so that the surface remains positively charged. These superficial OH groups may get replaced by the approaching AsO_3^{-3}/$H_2AsO_3^-$/AsO_4^{3-}/$H_2AsO_4^-$, whereby the latter get retained onto the biocomposite. Additionally, the positively charged surface may exert electrostatic/columbic forces on the negatively charged forms of As(III)/As(V) ions [73]. As reported by Qu et al. [84], the common functional groups constituting the surface of *Leucobacter* sp.

Table 4.1 The maximum adsorption capacity and experimental condition used for the removal of heavy metals various algae-based biosorbent.

Biosorbent	Metal	Adsorption capacity (mg/g)	Experimental conditions						Ref.
			pH	Conc. (mg/L)	Volume (mL)	Temp. (°C)	Time (h)	Dose (g)	
A. junii -coconut fiber	Cr(VI)	17.4	2	25–200	50	30	8	0.5	[28]
P. boryanum cells	Cr(VI)	17.3	2	20–400	25	25	-	0.1	[30]
Alginate /*P. boryanum* cells	Cr(VI)	23.8	2	20–400	25	25	-	0.1	[30]
Alginate–gelatin/*P. boryanum* cells	Cr(VI)	29.6	2	20–400	25	25	-	0.1	[30]
C. reinhardtii cells/carboxymethyl cellulose	U(VI)	232.6	4.5	50–1500	50	-	2	0.125	[31]
Ca-alginate/*S. quadricauda*	Cu(II),	75.6,	5	25–600	25	25	3	-	[33]
Ca-alginate/*S. quadricauda*	Zn(II)	55.2	5	25–600	25	25	3	-	[33]
Ca-alginate/*S. quadricauda*	Ni(II)	30.4	5	25–600	25	25	3	-	[33]

Table 4.2 The maximum adsorption capacity and experimental condition used for the removal of heavy metals various bacteria-based biosorbent.

Biosorbent	Metal	Adsorption capacity (mg/g)	Experimental conditions						Ref.
			pH	Conc. (mg/L)	Volume (ml)	Temp. (°C)	Time (h)	Dose (g)	
Volcanic rock/*P. putida*	Cu(II)	22.23	6	10–40	100	30	0.50	25	[76]
C. glutamicum /granular-activated carbon/$MnFe_2O_4$	As(III)	2584.66	7	50–2000	100	30	3.66	0.9	[73]
C. glutamicum /granular-activated carbon/$MnFe_2O_4$	As(V)	2651.67	7	50–2000	100	30	3.66	0.9	[73]
Polyethylenimine-polysulfone/*E. coli*	Pb(II)	216.9	–	0–500	30	25	24	0.2	[65]
Polyethylenimine-polysulfone/*E. coli*	Ru	121.28 ± 13.15	–	76.4–713.6	30	25	24	1	[64]
Pseudomonas putida @ chitosan	U(VI)	625	5	100–1000	100	30	6	0.4	[62]

N-4 were hydroxyl, carbonyl, sulfonate, amide, imidazole, phosphate, and phosphodiester groups. The retention of Ni(II) ions onto *Leucobacter* sp. N-4 may have been caused by the chemical interaction of the former with the hydrogen atoms constituting the carboxyl, hydroxyl, and amine groups on the latter's surface.

4.3.3 Fungi-Based Biocomposite

A living fungal cell adsorbs metal ions through metabolic and non-metabolic processes, while a dead fungal would take the nonmetabolic pathway for adsorption [85]. Hence, the retention of metal ions by fungi may involve various processes, such as outer surface intake, intracellular accumulation, and extra cellular precipitation. The outer surface intake is accomplished with irreversible nonmetabolic physiochemical interactions, namely ion exchange, surface adsorption, and complex/chelate formation, between the constituent functional groups of fungi and the metal ions. Fungal cell wall contains polysaccharides, proteins, lipids, polyphosphates, and some inorganic ions, by virtue of which carboxyl, hydroxyl, sulphate, phosphate, and amino groups are available for binding with metal ions [86]. As reported by Sathvika et al. [59], at pH 2.0–3.0, the surface of aspergillus-cellulose biocomposite becomes positively charged, by virtue of the protonated amines, carboxyl and hydroxyl groups, and thus retains the negatively charged $HCrO_4^-$ through electrostatic interaction. On the other hand, at pH 6.5–7.5, the positively charged metal hydroxides, such as $Cr(OH)^{2+}$ and $Cr(OH)_2^+$, get adsorbed onto the negatively charged surface of the biocomposite through electrostatic interaction. The metabolic intracellular accumulation of metal ions involves the cell membrane of viable fungal cells. Fungal species may form widespread network of mycelium and thus accumulate pollutants within this network [87]. The metabolic extracellular precipitation occurs when the intracellular secretion interacts with metal ions to facilitate precipitation. However, the nonmetabolic precipitation may be induced by certain chemical interactions between the metal ions and the surface constituents. Some fungi may exhibit metal selectivity [88]. Again, a consortium of different fungal species may have 100% efficiency in the removal of metal ions [89]. As Akra et al. [46] reported the adsorption of Ni(II) onto immobilized *L. salmonicolor* fungal may involve a collective method involving ion-exchange mechanism and interaction with hydroxyl, amino and carboxyl groups. Kulal el. al. [90] proposed a complexation mechanism in Figure 4.4 for the adsorption of U(VI) from *Penicillium chrysogenum* immobilized on silica gel. Besides the functional groups, the experimental conditions and immobilizing material also play a

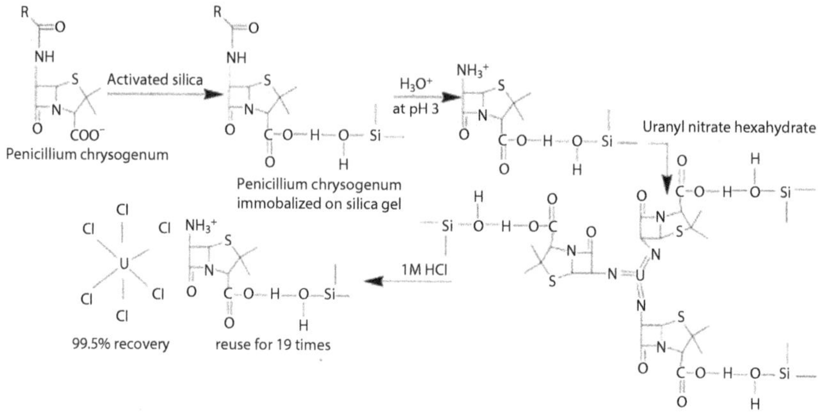

Figure 4.4 Probable mechanism for sorption and desorption of uranium(VI) with *Penicillium chrysogenum* on activated silica.

vital role in the adsorption process. As shown in Table 4.3, the adsorption capacity highly depends on the type of fungi and immobilizing material metal. The adsorption capacity of the biosorbent changes with the change in the experimental conditions like solution pH, solution volume, biosorbent dose, reaction temperature, and initial metal concentrations.

4.4 Conclusion

The scavenging efficiency of microbes is influenced by several biotic factors (such as strain, tolerance, size, and life stages) and environmental factors (such as pH, temperature, concentration of pollutants, etc.). The use of freely suspended microbes is losing charm due to the latter's small particle size, low density, poor mechanical strength, less rigidity, which causes difficulties in column processes, due to biomass swelling, clogging and poor regenerability. The technique of immobilizing microbes, onto a stable support matrix, has offered an alternative eco-friendly method for the remediation of wastewater. The immobilized microbes enhance the performance of the overall modified sorbent by virtue of their special features, namely: (1) capability to detect and take up pollutants at low concentrations, (2) free of toxic sludge, (3) easily culturable and maintainable, and (4) high specific surface area. Apart from being of low cost and high abundance, such microbes also help in carbon dioxide mitigation.

Although, physical methods of modification are simple and inexpensive, however, the chemical modification of the surface of microbes, by

Table 4.3 The maximum adsorption capacity and experimental condition used for the removal of heavy metals various fungi-based biosorbent.

Biosorbent	Metal	Adsorption capacity (mg/g)	Experimental conditions						Ref.
			pH	Conc. (mg/L)	Volume (ml)	Temp. (°C)	Time (h)	Dose (g)	
Magnetic *R. glutinis,*	U(VI)	226 ± 27	6	10–320	-	50	0.5	0.00515	[57]
Polyurethane/*A. niger,*	Cu(II)	110.6	5	–	100	30	–	1	[55]
Ca-alginate- activated carbon –*S. cerevisiae*	Cu(II)	64.90	5	10–400	100	30	4	2	[56]
Ca-alginate-activated carbon –*S. cerevisiae*	Pb(II)	166.31	5	10–400	100	30	4	2	[56]
Alginate/viable *T. asperellum*	Cu(II)	117.65	5	50–600	20	28 ± 2	8	0.4	[54]
Alginate/non-viable *T. asperellum*	Cu(II)	140.85	5	50–600	20	28 ± 2	8	0.4	[54]
Chitosan/baker's yeast nanofibre.	U(VI)	219	6	50–500	50	25	3	0.05	[53]
Chitosan/baker's yeast nanofibre.	Th (IV)	131.9	6	50–500	50	25	3	0.05	[53]
Ca-alginate/*C. lipolytica*	Ni(II)	123.43	6.62	25–460	50	45	3	0.05	[52]
Ca-alginate/*C. tropicalis*	Ni(II)	197.68	6.62	25–460	50	45	3	0.05	[52]
Ca-alginate/*C. utilis*	Ni(II)	178.06	6.62	25–460	50	45	3	0.05	[52]

introducing new binding groups and/or modifying the existing functional groups, may make the latter more effective. The mechanical strength of the microbes may also be enhanced by grafting long polymer chains onto the microbial surface. The genetic manipulation of the microbes, so as to optimize culture growth, can improve its sorption capability.

The lack of complete success in their commercial use may be because of our incomplete knowledge about the mechanisms, kinetics and thermodynamics. The diversity of the functional groups and their elusive extracellular availability (in microbes) have collectively made it difficult to get conclusive information. A collaborative work of chemist, engineers and biologists may be able to develop microbial biosorbents which could address issues, namely, (1) recovery of precious metals, proteins, antibodies, and peptides, (2) increased resistance to heat so that autoclaving and purification of pollutants may be effectively carried out, (3) enrichment of microelements, biological food supplements and fertilizers, and (4) treatment of real industrial effluents with multiple pollutants.

References

1. Kumar, R., Oves, M., Almeelbi, T., Al-Makishah, N.H., Barakat, M.A., Hybrid chitosan/polyaniline-polypyrrole biomaterial for enhanced adsorption and antimicrobial activity. *J. Colloid Interface Sci.*, 490, 488, 2017.
2. Barakat, M.A., Al-Ansari, A.M., Kumar, R., Synthesis and characterization of Fe–Al binary oxyhydroxides/MWCNTs nanocomposite for the removal of Cr (VI) from aqueous solution. *J. Taiwan Inst. Chem. Eng.*, 63, 303, 2016.
3. Barakat, M.A., Al-Ansari, A.M., Kumar, R., Synthesis and characterization of Fe– Al binary oxyhydroxides/MWCNTs nanocomposite for the removal of Cr (VI) from aqueous solution. *J. Taiwan Inst. Chem. Eng.*, 63, 303, 2016.
4. Ahmad, A., Mohd-Setapar, S.H., Chuong, C.S., Khatoon, A., Wani, W.A., Kumar, R., Rafatullah, M., Recent advances in new generation dye removal technologies: novel search for approaches to reprocess wastewater. *RSC Adv.*, 5, 30801, 2015.
5. Ahmad, A., Siddique, J.A., Laskar, M.A., Kumar, R., Mohd-Setapar, S.H., Khatoon, A., Shiekh, R.A., New generation Amberlite XAD resin for the removal of metal ions: a review. *J. Environ. Sci.*, 31, 104, 2015.
6. Barakat, M.A., Kumar, R., Synthesis and characterization of porous magnetic silica composite for the removal of heavy metals from aqueous solution. *J. Ind. Eng. Chem.*, 23, 93, 2015.
7. Kumar, R., Ehsan, M., Barakat, M.A., Synthesis and characterization of carbon/AlOOH composite for adsorption of chromium (VI) from synthetic wastewater. *J. Ind. Eng. Chem.*, 20, 4202, 2014.

8. Kumar, R., Khan, M.A., Haq, N., 2014. Application of carbon nanotubes in heavy metals remediation. *Critical Reviews Environ. Sci. Technol.*, 44, 1000, 2014.
9. Gupta, V.K., Kumar, R., Nayak, A., Saleh, T.A., Barakat, M.A., 2013. Adsorptive removal of dyes from aqueous solution onto carbon nanotubes: a review. *Adv. Colloid Interf. Sci.*, 193, 24, 2013.
10. Kumar, R., Ansari, M.O., Barakat, M.A., DBSA doped polyaniline/multi-walled carbon nanotubes composite for high efficiency removal of Cr (VI) from aqueous solution. *Chem. Eng. J.*, 228, 748, 2013.
11. Kumar, R., Barakat, M.A., Daza, Y.A., Woodcock, H.L., Kuhn, J.N., EDTA functionalized silica for removal of Cu (II), Zn (II) and Ni (II) from aqueous solution. *J. Colloid Interface Sci.*, 408, 200, 2013.
12. Ahmad, R., Kumar, R., Laskar, M.A., Adsorptive removal of Pb^{2+} form aqueous solution by macrocyclic calix [4] naphthalene: kinetic, thermodynamic, and isotherm analysis. *Environ. Sci. Poll. Res.*, 20, 219, 2013.
13. Kumar, R., Kumar, M., Ahmad, R., Barakat, M.A., l-Methionine modified Dowex-50 ion-exchanger of reduced size for the separation and removal of Cu (II) and Ni (II) from aqueous solution. *Chem. Eng. J.*, 218, 32, 2013.
14. Ahmad, R., Kumar, R., Haseeb, S., Adsorption of Cu $^{2+}$ from aqueous solution onto iron oxide coated eggshell powder: Evaluation of equilibrium, isotherms, kinetics, and regeneration capacity. *Arabian J. Chem.*, 5, 353. 2012.
15. S. Deng, Y. P. Ting. Fungal biomass with grafted poly(acrylic acid) for enhancement of Cu(II) and Cd(II) biosorption. *Langmuir*, 21, 5940, 2005
16. Bai, J., Yao, H., Fan, F., Lin, M., Zhang, L., Ding, H., Lei, F., Wu, X., Li, X., Guo, J., Qin, Z., Biosorption of uranium by chemically modified Rhodotorula glutinis. *J. Environ. Radioact*, 101, 969, 2010.
17. Kumar, K.S., Dahms, H.U., Won, E.J., Lee, J.S., Shin, K.H., Microalgae–a promising tool for heavy metal remediation. *Ecotoxicol. Environ. Safety*, 113, 329, 2015.
18. Won, S.W., Kotte, P., Wei, W., Lim, A., Yun, Y.S., Biosorbents for recovery of precious metals. *Bioresour. Technol.*, 160, 203–212, 2014.
19. Monteiro, C.M., Castro, P.M., Malcata, F.X., Metal uptake by microalgae: underlying mechanisms and practical applications. *Biotechnol. Prog.*, 28, 299, 2012.
20. Zawierucha, I., Kozlowski, C., Malina, G., Immobilized materials for removal of toxic metal ions from surface/groundwaters and aqueous waste streams. *Environ. Sci.: Processes Impacts*, 18, 429, 2016.
21. Stumm, W., Morgan, J.J., *Aquatic chemistry: an introduction emphasizing chemical equilibria in natural waters*. Wiley, New York, 1981.
22. Dönmez, G.Ç., Aksu, Z., Öztürk, A., Kutsal, T., A comparative study on heavy metal biosorption characteristics of some algae. *Pro. Biochem.*, 34, 885, 1999.
23. Mata, Y.N., Blázquez, M.L., Ballester, A., González, F., Munoz, J.A., Biosorption of cadmium, lead and copper with calcium alginate xerogels and immobilized Fucus vesiculosus. *J. Hazard. Mater.*, 163, 555, 2009.

24. Cao, Y.R., Liu, Z., Cheng, G.L., Jing, X.B., Xu, H., Exploring single and multimetal biosorption by immobilized spent Tricholoma lobayense using multistep response surface methodology. *Chem. Eng. J.*, 164, 183, 2010.
25. Aksu, Z., Egretli, G., Kutsal, T., A comparative study for the biosorption characteristics of chromium (VI) on ca-alginate, agarose and immobilized c vulgaris in a continuous packed bed column. *J. Environ. Sci. Health Part A*, 34, 295, 1999.
26. Kondo, K., Hirayama, K., Matsumoto, M., Adsorption of metal ions from aqueous solution onto microalga entrapped into Ca-alginate gel bead. *Des. Wat. Treat.*, 51, 4675, 2013.
27. Donat, R., Esen, K., Cetisli, H., Aytas, S., Adsorption of uranium (VI) onto Ulva sp.-sepiolite composite. *J. Radioanalyt. Nucl. Chem.*, 279, 253, 2008.
28. Pulimi, M., Samuel, J., Natarajan, C., Mukherjee, A., Adsorptive removal of Cr (VI) by acinetobacter junii VITSUKMW3 immobilized on coconut fiber in batch and continuous flow reactor. *Asian J. Chem.*, 26, 2649, 2014.
29. Sheng, P.X., Wee, K.H., Ting, Y.P., Chen, J.P., Biosorption of copper by immobilized marine algal biomass. *Chem. Eng. J.*, 136, 156, 2008.
30. Ozer, T.B., Erkaya, I.A., Udoh, A.U., Duygu, D.Y., Akbulut, A., Bayramoglu, G., Arica, M.Y., Biosorption of Cr (VI) by free and immobilized Pediastrum boryanum biomass: equilibrium, kinetic, and thermodynamic studies. *Environ. Sci. Poll. Res.*, 19, 2983, 2012.
31. Erkaya, I.A., Arica, M.Y., Akbulut, A., Bayramoglu, G., Biosorption of uranium (VI) by free and entrapped Chlamydomonas reinhardtii: kinetic, equilibrium and thermodynamic studies. *J. Radioanalyt. Nucl. Chem.*, 299, 1993, 2014.
32. Kőnig-Péter, A., Csudai, C., Felinger, A., Kilár, F., Pernyeszi, T., Column studies of heavy metal biosorption by immobilized Spirulina platensis-maxima cells. *Des. Wat. Treat.*, 57, 28340, 2016.
33. Bayramoğlu, G., Arıca, M.Y., Construction a hybrid biosorbent using Scenedesmus quadricauda and Ca-alginate for biosorption of Cu (II), Zn (II) and Ni (II): kinetics and equilibrium studies. *Bioresour. Technol.*, 100, 186, 2009.
34. Akar, S.T., Akar, T., Kaynak, Z., Anilan, B., Cabuk, A., Tabak, Ö., Demir, T.A., Gedikbey, T., Removal of copper (II) ions from synthetic solution and real wastewater by the combined action of dried Trametes versicolor cells and montmorillonite. *Hydrometallurgy*, 97, 98, 2009.
35. Yang, F., Liu, H., Qu, J., Chen, J.P., Preparation and characterization of chitosan encapsulated Sargassum sp. biosorbent for nickel ions sorption. *Bioresour. Technol.*, 102, 2821, 2011.
36. Fan, W.H., Xu, Z.Z., Li, Q., Kinetics and mechanism of nickel (II) ion biosorption by immobilized brown Laminaria japonica algae. *Ads. Sci. Technol.*, 28, 499, 2010.
37. Chen, B.Y., Chen, C.Y., Guo, W.Q., Chang, H.W., Chen, W.M., Lee, D.J., Huang, C.C., Ren, N.Q., Chang, J.S., Fixed-bed biosorption of cadmium

using immobilized Scenedesmus obliquus CNW-N cells on loofa (Luffa cylindrica) sponge. *Bioresour. Technol.*, 160, 175, 2014.
38. Sargın, İ., Arslan, G., Kaya, M., Efficiency of chitosan–algal biomass composite microbeads at heavy metal removal. *React. Functional Polymers*, 98, 38, 2016.
39. Chen, C., Wang, J., Uranium removal by novel graphene oxide-immobilized Saccharomyces cerevisiae gel beads. *J. Environ. Radioact.*, 162, 134, 2016.
40. Ozdemir, S., Okumuş, V., Dündar, A., Kılınç, E., The use of fungal biomass Agaricus bisporus immobilized on amberlite XAD-4 resin for the solid-phase preconcentration of thorium. *Bioremediat. J.*, 18, 38, 2014.
41. Kumar, R., Bhatia, D., Singh, R., Rani, S., Bishnoi, N.R., Sorption of heavy metals from electroplating effluent using immobilized biomass Trichoderma viride in a continuous packed-bed column. *International Biodeteriorat. Biodegrad.*, 65, 1133, 2011.
42. Zhang, D., Zeng, X., Li, W., He, H., Ma, P., Falandysz, J., Selection of optimum formulation for biosorbing lead and cadmium from aquatic solution by using PVA-SA's immobilizing Lentinus edodes residue. *Desalin. Wat. Treat.*, 31(1–3), 107, 2011.
43. Akhtar, K., Khalid, A.M., Akhtar, M.W., Ghauri, M.A., Removal and recovery of uranium from aqueous solutions by Ca-alginate immobilized Trichoderma harzianum. *Bioresour. Technol.*, 100, 4551, 2009.
44. Sathvika, T., Rajesh, V., Rajesh, N., Prospective application of Aspergillus species immobilized in sodium montmorillonite to remove toxic hexavalent chromium from wastewater. *RSC Adv.*, 5, 107031, 2015.
45. Xiao, G., Zhang, X., Su, H., Tan, T., Plate column biosorption of Cu (II) on membrane-type biosorbent (MBS) of Penicillium biomass: optimization using statistical design methods. *Bioresour. Technol.* 143, 490, 2013.
46. Akar, T., Celik, S., Gorgulu Ari, A., Tunali Akar, S., Nickel removal characteristics of an immobilized macro fungus: equilibrium, kinetic and mechanism analysis of the biosorption. *J. Chem. Technol. Biotechnol.*, 88, 680, 2013.
47. Tripathi, S.C., Kannan, R., Dhami, P.S., Naik, P.W., Munshi, S.K., Dey, P.K., Salvi, N.A., Chattopadhyay, S., Modified Rhizopus arrhizus biomass for sorption of 241Am and other radionuclides. *J. Radioanalyt. Nuclear Chem.*, 287, 691–695, 2011.
48. Nguyen, M.L., Juang, R.S., Modification of crosslinked chitosan beads with histidine and Saccharomyces cerevisiae for enhanced Ni (II) biosorption. *J. Taiwan Inst. Chem. Eng.*, 56, 96, 2015.
49. Yang, T., Zhang, X.X., Chen, M.L., Wang, J.H., Highly selective preconcentration of ultra-trace cadmium by yeast surface engineering. *Analyst*, 137, 4193, 2012.
50. Rashid, A., Bhatti, H.N., Iqbal, M., Noreen, S., Fungal biomass composite with bentonite efficiency for nickel and zinc adsorption: a mechanistic study. *Ecol. Eng.*, 91, 459, 2016.

51. Cheng, W., Ding, C., Sun, Y., Wang, X., Fabrication of fungus/attapulgite composites and their removal of U (VI) from aqueous solution. *Chem Eng. J.*, 269, 1, 2015.
52. Haydar, S., Ahmad, M.F., Hussain, G., Evaluation of new biosorbents prepared from immobilized biomass of Candida sp. for the removal of nickel ions. *Desalin. Wat. Treat.*, 57, 5601, 2016.
53. Hosseini, M., Keshtkar, A.R., Moosavian, M.A., Electrospun chitosan/baker's yeast nanofibre adsorbent: preparation, characterization and application in heavy metal adsorption. *Bull.Mat. Sci.*, 39, 1091, 2016.
54. Tan, W.S., Ting, A.S.Y., Efficacy and reusability of alginate-immobilized live and heat-inactivated Trichoderma asperellum cells for Cu (II) removal from aqueous solution. *Bioresour. Technol.*, 123, 290, 2012.
55. Tsekova, K., Marinov, P., Ilieva, S., Kaimaktchiev, A., Copper adsorption by free and immobilized on polyurethane foam cells of Aspergillus niger. *Biotechnol. Biotechnol. Equip.*, 15, 93, 2001.
56. Lu, M., Liu, Y.G., Hu, X.J., Ben, Y., Zeng, X.X., Li, T.T., Wang, H., Competitive adsorption of Cu (II) and Pb (II) ions from aqueous solutions by Ca-alginate immobilized activated carbon and Saccharomyces cerevisiae. *J. Cent South Uni.*, 20, 2478, 2013.
57. Bai, J., Wu, X., Fan, F., Tian, W., Yin, X., Zhao, L., Fan, F., Li, Z., Tian, L., Qin, Z., Guo, J., Biosorption of uranium by magnetically modified Rhodotorula glutinis. *Enzyme Microb. Technol.*, 51, 382, 2012.
58. Jiang, W., Xu, Y., Li, C., Lv, X., Wang, D., Biosorption of cadmium (II) from aqueous solution by chitosan encapsulated Zygosaccharomyces rouxii. *Environ. Progress Sustain. Energy*, 32, 1101, 2013.
59. Sathvika, T., Rajesh, V., Rajesh, N., Adsorption of chromium supported with various column modelling studies through the synergistic influence of Aspergillus and cellulose. *J. Environ. Chem. Eng.*, 4, 3193, 2016.
60. Whitman, W.B., Coleman, D.C., Wiebe, W.J., Prokaryotes: the unseen majority. *Proceed. National Acad. Sci.*, 95, 6578, 1998.
61. D'Souza, S.F., Sar, P., Kazy, S.K., Kubal, B.S., Uranium sorption by Pseudomonas biomass immobilized in radiation polymerized polyacrylamide bio-beads. *J. Environ. Sci. Health Part A*, 41, 487, 2006.
62. Sohbatzadeh, H., Keshtkar, A.R., Safdari, J., Fatemi, F., U (VI) biosorption by bi-functionalized Pseudomonas putida@ chitosan bead: Modeling and optimization using RSM. *Int. J. Bio. Macromol.*, 89, 647, 2016.
63. Podder, M.S., Majumder, C.B., Study of the kinetics of arsenic removal from wastewater using Bacillus arsenicus biofilms supported on a Neem leaves/ $MnFe_2O_4$ composite. *Ecol. Eng.*, 88, 195, 2016.
64. Kim, S., Choi, Y.E., Yun, Y.S., Ruthenium recovery from acetic acid industrial effluent using chemically stable and high-performance polyethylenimine-coated polysulfone-Escherichia coli biomass composite fibers. *J. Hazard. mater*, 313, 29, 2016.

65. Cho, C.W., Kang, S.B., Kim, S., Yun, Y.S., Won, S.W., Reusable polyethylenimine-coated polysulfone/bacterial biomass composite fiber biosorbent for recovery of Pd (II) from acidic solutions. *Chem. Eng, J.*, 302, 545, 2016.
66. Cheng, Y., Zhou, F., Li, S., Chen, Z., Removal of mixed contaminants, crystal violet, and heavy metal ions by using immobilized stains as the functional biomaterial. *RSC Adv.*, 6, 67858, 2016.
67. Pires, C., Marques, A.P., Guerreiro, A., Magan, N., Castro, P.M., Removal of heavy metals using different polymer matrixes as support for bacterial immobilisation. *J. Hazard. Mat.*, 191, 277, 2011.
68. San Keskin, N.O., Celebioglu, A., Sarioglu, O.F., Ozkan, A.D., Uyar, T., Tekinay, T., Removal of a reactive dye and hexavalent chromium by a reusable bacteria attached electrospun nanofibrous web. *RSC Adv.*, 5, 86867, 2015.
69. Hora, A., Shetty, V.K., Kinetics of bioreduction of hexavalent chromium by poly vinyl alcohol-alginate immobilized cells of Ochrobactrum sp. Cr-B4 and comparison with free cells. *Desalin. Wat. Treat*, 57, 8981, 2016.
70. Podder, M.S., Majumder, C.B., Kinetic, mechanistic and thermodynamic studies of removal of arsenic using Bacillus arsenicus MTCC 4380 immobilized on surface of granular activated carbon/$MnFe_2O_4$ composite. *Groundwat. Sus. Devel.*, 2, 53, 2016.
71. Chen, X., Chen, L., Shi, J., Wu, W., Chen, Y., Immobilization of heavy metals by Pseudomonas putida CZ1/goethite composites from solution. *Colloids Surf. B: Biointer.*, 61, 170, 2008.
72. Renard, G., Mureseanu, M., Galarneau, A., Lerner, D.A., Brunel, D., Immobilisation of a biological chelate in porous mesostructured silica for selective metal removal from wastewater and its recovery. *New J. Chem.*, 29, 912, 2005.
73. Podder, M.S., Majumder, C.B., Corynebacterium glutamicum MTCC 2745 immobilized on granular activated carbon/MnFe 2 O 4 composite: A novel biosorbent for removal of As (III) and As (V) ions. *Spectrochimica Acta Part A: Mol. Biomolecul. Spectro.*, 168, 159, 2016.
74. Liu, X., Wei, W., Zeng, X., Tang, B., Liu, X., Xiang, H., Copper adsorption kinetics onto pseudomonas aeruginosa immobilized multiwalled carbon nanotubes in an aqueous solution. *Analyt. Let.*, 42, 425, 2009.
75. Costa, F., Tavares, T., Biosorption of nickel and cadmium in the presence of diethylketone by a Streptococcus equisimilis biofilm supported on vermiculite. *Int. Biodeterior. Biodegrad.*, 115, 119, 2016.
76. Ni, H., Xiong, Z., Ye, T., Zhang, Z., Ma, X., Li, L., Biosorption of copper (II) from aqueous solutions using volcanic rock matrix-immobilized Pseudomonas putida cells with surface-displayed cyanobacterial metallothioneins. *Chem. Engineer. J.*, 204, 264, 2012.
77. Baysal, Ş.H., Önal, S., Özdemir, G., Biosorption of chromium, cadmium, and cobalt from aqueous solution by immobilized living cells of Chryseomonas luteola TEM 05. *Preparative Biochem. Biotechnol.*, 39, 419, 2009.

78. Ahmad, M.F., Haydar, S., Bhatti, A.A., Bari, A.J., Application of artificial neural network for the prediction of biosorption capacity of immobilized Bacillus subtilis for the removal of cadmium ions from aqueous solution. *Biochem. Engineer, J.*, 84, 83, 2014.
79. He, J., Chen, J.P., A comprehensive review on biosorption of heavy metals by algal biomass: materials, performances, chemistry, and modeling simulation tools. *Bioresour. Technol.*, 160, 67, 2014.
80. Blaby-Haas, C.E., Merchant, S.S., The ins and outs of algal metal transport. *Biochimica et Biophysica Acta (BBA)-Mol. Cell Res.*, 1823, 1531, 2012.
81. Mosbah, R., Sahmoune, M.N., Biosorption of heavy metals by Streptomyces species—an overview. *Cent. Eur. J. Chem.*, 11, 1412, 2013.
82. Podder, M.S., Majumder, C.B., Removal of arsenic by a Bacillus arsenicus biofilm supported on GAC/$MnFe_2O_4$ composite. *Groundwa. Sus. Develop*, 1, 105, 2015.
83. Giri, A.K., Patel, R.K., Mahapatra, S.S., Mishra, P.C., Biosorption of arsenic (III) from aqueous solution by living cells of Bacillus cereus. *Environ. Sci. Pollut. Res.*, 20, 1281, 2013.
84. Qu, Y., Li, H., Li, A., Ma, F., Zhou, J., Identification and characterization of Leucobacter sp. N-4 for Ni (II) biosorption by response surface methodology. *J. Hazard. Mat.*, 190, 869, 2011.
85. Shakya, M., Sharma, P., Meryem, S.S., Mahmood, Q., Kumar, A., Heavy metal removal from industrial wastewater using fungi: uptake mechanism and biochemical aspects. *J. Environ. Eng.*, 142, C6015001, 2015.
86. Verma, A., Singh, A., Bishnoi, N.R., Gupta, A., Biosorption of Cu (II) using free and immobilized biomass of Penicillium citrinum. *Ecol. Eng.*, 61, 486, 2013.
87. Harms, H., Schlosser, D., Wick, L.Y.,. Untapped potential: exploiting fungi in bioremediation of hazardous chemicals. *Nat. Rev. Microbiol.*, 9, 177, 2011.
88. Iskandar, N.L., Zainudin, N.A.I.M., Tan, S.G., Tolerance and biosorption of copper (Cu) and lead (Pb) by filamentous fungi isolated from a freshwater ecosystem. *J. Environ. Sci.*, 23, 824, 2011.
89. Mishra, A., Malik, A., Novel fungal consortium for bioremediation of metals and dyes from mixed waste stream. *Bioresour. Technol.*, 171, 217, 2014.
90. Kulal, D. K., Pansare, A. V., Tetgure, S. R., Karve, M., Patil, V. R., Determination of uranium(VI) using *Penicillium chrysogenum* immobilized on silica gel and spectrophotometer. *J. Radioanal. Nucl. Chem.*, 307, 1253, 2016.

5

Remediation of Cr (VI) Using Clay Minerals, Biomasses and Industrial Wastes as Adsorbents

Rashmi Acharya*, Satyabadi Martha and K. M. Parida*

Centre for Nano Science and Nano Technology, Siksha 'O' Anusandhan University, Bhubaneswar, India

Abstract

Trivalent chromium [Cr (III)] and hexavalent chromium [Cr (VI)] are two major forms of chromium that exist in the aqueous environment. The later form is acutely toxic, mutagenic, teratogenic, and carcinogenic in biological systems. It is released to the aqueous environment as industrial effluents in form of different oxy anionic species like $HCrO_4^-$, $Cr_2O_7^{2-}$, CrO_4^{2-}, etc., speciation of which depends on pH and concentration of solution. In recent years, many studies have been carried out with different adsorbents like clay minerals, biomasses and industrial wastes for remediation of Cr (VI) from contaminated water as these are chief and abundantly available. This chapter reports on the evaluation of adsorption capacities of these adsorbents with brief discussion of mechanism of adsorption. In this chapter, isotherm models, thermodynamics and kinetics of adsorption and pH of solution were briefly discussed. Maximum adsorption capacity obtained by the adsorbents with optimum adsorption conditions such as pH, temperature, contact time, initial Cr (VI) concentration, adsorbent dose along with fitted isotherm, kinetic model, values of thermodynamic parameters like $\Delta H°$, $\Delta G°$, $\Delta S°$, and mechanism of adsorption were presented in tables.

Keywords: Adsorption capacity, clay minerals, biomass, industrial wastes, Cr (VI)

**Corresponding authors*: rashmiacharya@soauniversity.ac.in, kulamaniparida@soauniversity.ac.in

Shahid-ul-Islam (ed.) Advanced Materials for Wastewater Treatment, (129–170)

5.1 Introduction

In recent years, rapid growth of industrialization and urbanization has resulted in increased environmental pollution, which has been a serious cause of public concern [1]. In particular, water pollution is one of the imortant global environmental issues. Discharge of organic and inorganic wastages, sediments, radioactive materials, effluents, sewages and heavy metals in to the water bodies cause pollution of water. Among these pollutants, heavy metals are responsible to a considerable extent because of their toxicity, bioaccumulation, persistence and nonbiodegradable nature [2]. Since the worldwide discharge of chromium containing wastewater is approximately 170,000 tons per annum, chromium has been considered as a major pollutant among the toxic heavy metals [3].

Chromium is the twenty-fourth most abundant element in Earth's crust [4]. It exhibits a wide range of possible oxidation states ranging from +II to +VI with the predominance of trivalent [Cr (III)] and hexavalent [Cr (VI)] species in aquatic environments.This is evident from Eh-pH diagram given in Figure 5.1. Cr (III) is the most thermodynamically stable oxidation state under reducing conditions and predominates at pH < 3.0 [5]. The appreciabe less toxic property of Cr (III) at low concentrations makes it an essential nutrient and a component of the glucose tolerance factor [3]. It exists as a cation in the aqueous environment. Being a cation, it is less mobile in the soil environment and can easily be precipitated out from aqueous solution

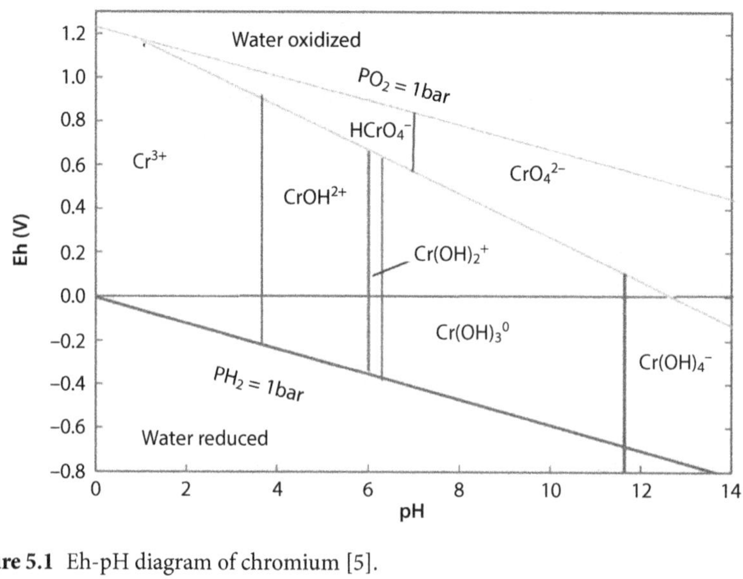

Figure 5.1 Eh-pH diagram of chromium [5].

as Cr $(OH)_3$ [6]. In consrast, Cr (VI) exists in aqueous medium as different oxy anionic species such as chromate (CrO_4^{2-}), dichromate ($Cr_2O_7^{2-}$), etc., speciation of which depends on its concentration and pH of the solution. Figure 5.2 shows that H_2CrO_4 is the dominant species at pH below 1.0 where as $HCrO_4^-$ and $Cr_2O_7^{2-}$ predominate between the pH 1.0 and 6.0. The stable Cr (VI) species above pH 6.0 is CrO_4^{2-} [7]. Further, there is a possibility of transformation of $Cr_2O_7^{2-}$ to $HCrO_4^-$ below $1.26–1.74 \times 10^{-2}$ mol/L of total Cr(VI) concentration in acidic medium [8].

Cr(VI) is acutely toxic, mutagenic [9], teratogenic [10], and carcinogenic [11] in biological systems. It is nonbiodegradable and bioaccumulative in living tissues. Hence, different Cr (VI) species move through the cell membrane and get reduced to Cr (III) by reactive oxygen species (ROS) such as superoxide, hydroxyl radicals, and hydrogen peroxide resulting in intracellular damage. Cr (III) ions so formed make stable coordination complexes with nucleic acids and proteins. Therefore, reduction of Cr (VI) inside the cell is responsible for genotoxic damage as well as other forms of toxicity [12].

As Cr (VI) compounds are toxic and corrosion inhibitors, these are widely used in various industrial processes such as leather tanning, electroplating, metal finishing, refractories, pigment, dye synthesis, alloy, and steel manufacturing [13–15]. The wastewaters released from these industries usually contain Cr (VI) of concentration up to 500 ppm [3]. Contamination of Cr (VI) also occurs through accidental leakage and improper disposal [12]. In addition, Cr (VI) can be produced naturally in the environment due to oxidation of the Cr (III) present in ultramafic derived soils and ophiolitic rocks [16]. So the total amount of chromium that is discharged into the

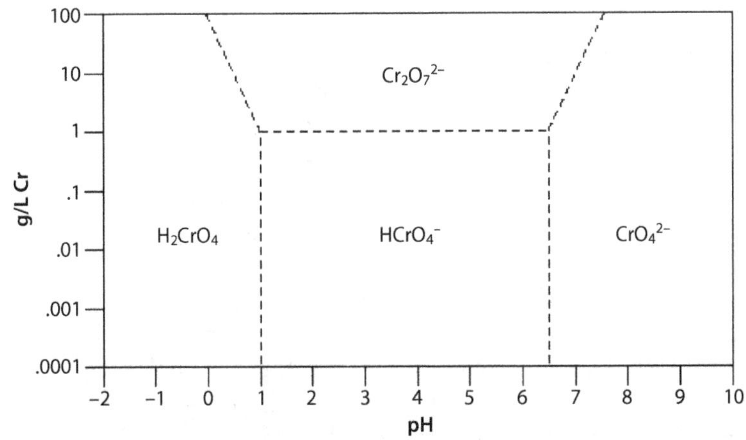

Figure 5.2 Speciation diagram of Cr (VI) [7].

environment is mainly in form of Cr (VI). The negatively charged Cr(VI) species are poorly adsorbed by the negatively charged soil particles due to repulsive electrostatic interactions and hence are transferred freely into the aqueous environment [17]. Moreovere, these species are highly mobile in natural water ecosystems and cause threat not only to aquatic life but also to human beings. Due to the detrimental effect of Cr (VI), stringent regulations have been imposed for the presence of Cr (VI) in water. The US EPA has fixed 0.05 mg/L and 0.1 mg/L as the maximum permissible limit of Cr (VI) in drinking water and in surface water, respectively [18, 19]. The concentration of Cr (VI) in water must be reduced to permissible level before being discharged into the aqueous environment to minimize its effects on aquatic life and downstream users. Therefore, treatment of Cr (VI) contaminated water is an important and necessary environmental issue. Several techniques like ion exchange, ultra filtration, electro dialysis, reverse osmosis, chemical precipitation, electrochemical reduction, sulphide precipitation, cementation, ion exchange, solvent extraction, membrane filtration, evaporation, etc., have been developed for remediation of Cr (VI) in contaminated water [20–22]. High operational cost, low feasibility of scale-up for industrial applications and ineffectiveness at high concentrations are some of the limitations for most of these techniques that restrict their applications [23]. Photocatalytic reduction technique is recently developed for this purpose [24–26]. However, adsorption is one of the effective and promising techniques, commonly used for remediation of Cr (VI) [27–29]. It is because this method has a large number of prominent advantages such as flexibility and simplicity of design, ease of operation, insensitivity of toxic pollutants and avoidance of the formation of secondary pollutants [30].

A large number of adsorbents, including polymeric resins, clays, biopolymers, biomass, activated carbon, and grapheme oxides are used for remediation of Cr (VI) and are compared with respect to their removal efficiency, adsorption capacity, regeneration and other features [31]. Activated carbon is widely used as an excellent adsorbent for the treatment of industrial effluents contaminated with Cr (VI) due to its microporous structure, high surface area and large number of positive charges on the surface [32, 33]. Mohan et al. reviewed adsorption efficiency of activated carbons for removal of hexavalent chromium from water/wastewater and compared their efficiency with that of low-cost adsorbents such as agricultural byproducts, seeds, zeolites, lignite, peat, chars, coals, clay minerals, etc. It was shown that some of these materials have equal or more adsorption capacity than activated carbons [34]. Moreover, the high cost and limited selectivity of activated carbon restricted its large scale application

[35–37]. Therefore, a large number of studies have been reported in recent years, on remediation of Cr (VI) by adsorbents like clay minerals, biomasses, and industrial wastes as these are chief and abundantly available. The adsorption reactions for remediation of Cr (VI) are characterized by isotherm models, kinetics and thermodynamics of adsorption. The adsorption capacity is influenced by various adsorption conditions such as pH, temperature, contact time, initial Cr (VI) concentration and adsorbent dose. Isotherm models, thermodynamics, kinetics of adsorption and pH of solution were briefly discussed in this chapter. An overview was made on evaluation of adsorption capacities with brief discussion of mechanism of adsorption. Maximum adsorption capacity obtained by the adsorbents with optimum adsorption conditions such as pH, temperature, contact time, initial Cr(VI) concentration, adsorbent dose along with fitted isotherm, kinetic model, values of thermodynamic parameters like $\Delta H°$, $\Delta G°$, $\Delta S°$, and mechanism of adsorption were presented in tables.

5.2 Isotherm Models

Adsorption isotherm correlates the adsorbate concentration in bulk solution with the amount of adsorbate adsorbed at the interface. It also provides information about a favorable adsorption process, nature of adsorption surface and type of layer of adsorption. Hence several isotherm models have been used in various adsorption processes. Langmuir, Freundlich and Dubnin–Radushkevich isotherm models have been used for the adsorption process involving uptake of Cr (VI) from aqueous solution by different adsorbents. These isotherm models are discussed in this section.

5.2.1 Langmuir Isotherm Model

This model was developed by Irving Langmuir in 1916. According to it, adsorption of metal ion occurs on a homogeneous surface and follows mono layer adsorption mechanism without any interaction between adsorbed ions. This model assumes that once a metal ion occupies a site, no further adsorption occurs at that site and a dynamic equilibrium is attained between the adsorbed phase and the liquid phase. It also assumes that all adsorption sites are identical and energetically equivalent. The isotherm is represented by the following equation [38]:

$$C_e/q_e = 1/Q_0 b + C_e/Q_0 \tag{5.1}$$

where

C_e is the equilibrium concentration (mg/L).

q_e is the amount adsorbed on unit mass of adsorbent (mg/g) at equilibrium time.

Q_0 is the maximum quantity of adsorbate required to form a single monolayer on unit mass of adsorbent (mg/g) and is a measure of adsorption capacity.

b is a constant related to the affinity of binding sites with the metal ions.

An adsorption process will follow Langmuir isotherm model if the plot of (C_e/q_e) versus C_e yields a straight line or the correlation coefficient should possess highest value. The value of Q_0 and that of b can be calculated from the slope and intercept of this linear plot. Further, feasibility of adsorption process can be predicted by the magnitude of dimensionless equilibrium parameter or separation factor (R_L), which is obtained from Langmuir isotherm by the equation given below [39]:

$$R_L = 1/(1 + bC_0) \tag{5.2}$$

where, C_0 is the initial concentration of Cr(VI) in mg/L.

Adsorption of Cr (VI) on to different adsorbents becomes feasible when the value of separation factor (R_L) lies in between 0 and 1 (i.e., $0 < R_L < 1$). If it is greater than one, the adsorption process becomes unfavorable. Linear nature of adsorption process is indicated when R_L is equal to one. If R_L is equal to zero, the adsorption process is said to be irreversible.

5.2.2 Freundlich Isotherm Model

This model was developed by Freundlich in 1907. It assumes that uptake of adsorbates occurs on a heterogeneous surface by multilayer adsorption and that adsorbed amount increases with an increase in concentration. Freundlich isotherm equation can be written as follows [40]:

$$\log q_e = \log K_f + (1/n) \log C_e \tag{5.3}$$

where

q_e is the amount of adsorbate adsorbed at equilibrium (mg/g).

C_e is the equilibrium concentration of adsorbate in solution (mg/L).

K_f and n are the constants incorporating all factors affecting the adsorption process (adsorption capacity and intensity).

Both these parameters can be obtained from the intercept and slope of the linear plot of log q_e versus log C_e. Adsorption of Cr (VI) by an adsorbent follows Freundlich isotherm model if the plot of log q_e versus log C_e will be a straight line. The values of K_f and n can be calculated from the intercept and slope of this plot. Larger the values of K_f and n higher is the adsorption capacity.

5.2.3 Dubnin–Radushkevich Isotherm Model

This model was used to evaluate the maximal adsorption capacity and to provide the mechanism responsible for metal uptake. The isotherm equation is given as [41]:

$$\log q_e = \log q_D - 2B_D R^2 T^2 \log (1 + 1/C_e) \tag{5.4}$$

where

q_e is the amount adsorbed at equilibrium.

q_D is maximal adsorption capacity and its value gives an idea about adsorption capacity of adsorbent.

C_e is the equilibrium concentration.

T is the absolute temperature.

R is the universal gas constant.

B_D is a constant related to the adsorption energy (E) and is expressed in mol^2/J^2.

The adsorption energy (E) and B_D are related as [42]:

$$E = 1/\sqrt{2B_D} \tag{5.5}$$

Mechanism of adsorption can be predicted from the value of adsorption energy (E) [43].When its value lies between 8 and 16 kJ/mol, the adsorption process is said to follow ion exchange path, while physical adsorption prevails for its value less than 8 kJ/mol [44]. An adsorption process follows this model, when the plot of log q_e versus log $(1 + 1/C_e)$ gives a straight line. Values of q_D and B_D are calculated from the slope of this plot.

5.3 Thermodynamics of Adsorption

Adsorption of Cr (VI) on different adsorbents depends on various thermodynamic parameters like standard free energy change ($\Delta G°$), standard enthalpy change $\Delta H°$, and standard entropy change ($\Delta S°$). Free energy change of an adsorption process gives an idea about feasibility of the process. An adsorption process is said to be feasible or spontaneous when change in standard free energy is negative ($\Delta G° < 0$). Moreover, it suggests the mechanism of adsorption. Change in standard free energy value between −20 and 0 kJ/mol is consistent with physical adsorption while that in the range from −80 to −400 kJ/mol indicates chemisorptions [43]. It is calculated by Eqn. 5.6 [45]:

$$\Delta G° = -RT \ln (K) \tag{5.6}$$

The value of change in standard enthalpy ($\Delta H°$) suggests exothermic or endothermic nature of adsorption process. The adsorption process is said to be exothermic if the value of $\Delta H°$ is negative while positive value of $\Delta H°$ is an indicative of endothermic adsorption process [46].

The spontaneity of adsorption can be evident from the positive value of change in standard entropy ($\Delta S° > 0$). Further, it reflects about the increased randomness at the interface between adsorbent and solution during the adsorption process [46]. The values of $\Delta H°$ and $\Delta S°$ can be obtained according to Van't Hoff equation (Eqn. 5.7):

$$\ln(K) = (\Delta S°/R) - (\Delta H°/RT) \tag{5.7}$$

The values of $\Delta H°$ and $\Delta S°$ are calculated from the slope and intercept of van't Hoff plot of $\ln(K)$ versus $1/T$.

5.4 Kinetics of Adsorption

Different kinetic models were used to explain mechanism of adsorption. Moreover, experimental adsorption capacity of an adsorbent can be compared with that of calculated value obtained from the slope of a linear plot. Pseudo-first-order and pseudo-second-order kinetic models have been followed for adsorption of Cr (VI) onto surfaces of different adsorbents. Therefore, these two kinetic models have been discussed in this section.

5.4.1 Pseudo-First-Order Kinetics

Lagergren derived pseudo-first-order model, which finds wide application [47]. The rate equation of this model is given by:

$$dq_t/dt = k_{ad} (q_e - q_t) \tag{5.8}$$

where

q_e is the amount of adsorption at equilibrium time.
q_t is the amount of adsorption at time t.
k_{ad} is the rate constant of the pseudo-first-order adsorption process.

The linear form of above equation can be written as:

$$\ln(q_e - q_t) = \ln q_e - k_{ad}t \quad (5.9)$$

If the plot of log $(q_e - q_t)$ versus t gives a straight line, the adsorption process is said to follow first order kinetics and rate constant (k_{ad}) is calculated from the plot. Again first order kinetics is followed when the value of correlation coefficient (R^2) is equal to or greater than 0.96.

5.4.2 Pseudo-Second–Order Kinetics

Pseudo-second-order model was given by HO and Mckay [48]. Equation 5.10 represents the rate equation for pseudo second order model.

$$q_t = kq_e^2t/(1 + k\,q_e t) \text{ or, } t/q_t = 1/(kq_e^2) + 1/q_e \quad (5.10)$$

where

k is second order rate constant.
q_e is the amount adsorbed at equilibrium.
q_t is the amount adsorbed at time t.

If the plot of t/q_t versus t gives a linear relationship, the adsorption process follows pseudo second order kinetics. Again the adsorption process will follow pseudo second order kinetics, when the regression coefficient (R^2) is equal to or greater than 0.96.

5.5 Solution pH

The adsorption of metal ion is strongly influenced by the pH of the medium [49]. Generally, pH of the solution provides a favorable adsorbent surface charge for adsorption to occur and hence affects the extent of Cr (VI) adsorption on to various adsorbents [50]. Surfaces of adsorbent are protonated at acidic solution pH (pH < 7.0) and there appears development of enough positive surface charge which attract oxyanionic species of Cr (VI) through electrostatic attraction resulting in effective adsorption

(Figure 5.3 (a)). Therefore, most of the studies for adsorption of Cr (VI) have been carried out at pH < 7.0. When pH of the solution rises, concentration of OH^- ions increases gradually. As a result, there occurs competition between negatively charged Cr (VI) species and large number of OH^- ions for the adsorption sites leading to decrease in uptake of Cr (VI) as shown in Figure 5.3 (b). Further, adsorption energy of different Cr (VI) species may be responsible for decrease in adsorption at higher pH. The dominant species of Cr (VI) at pH > 7.0 is CrO_4^{2-}, which has higher value of adsorption energy as compared to $HCrO_4^-$, the dominant species at pH < 7.0. This indicates $HCrO_4^-$ ions with smaller adsorption energy possess greater affinity for adsorption and hence adsorption occurs to a greater extent at acidic pH. On the other hand, $HCrO_4^-$ ions are gradually converted to CrO_4^{2-} ions with gradual rise of pH and CrO_4^{2-} ions with higher value of adsorption energy, get adsorbed to a small extent resulting in decrease in adsorption at pH > 7.0 [51].

Point of zero charge (pH_{ZPC}) of the adsorbent is another factor that determines extent of adsorption. When pH of the solution becomes less than pH_{ZPC}, surface of the adsorbent acquires positive charges and hence,

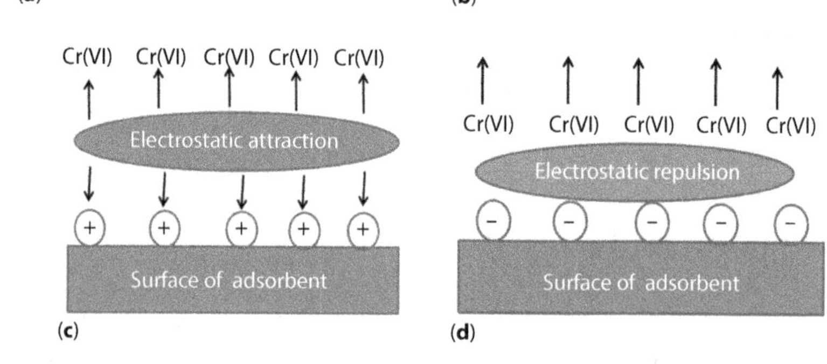

Figure 5.3 (a) Development of positive charges on the surface of adsorbent at pH < 7.0 and electrostatic attraction of negatively charged Cr (VI) species. (b) Competition between CrO_4^{2-} species and OH^- ions at pH > 7.0. (c) Development of surface positive charges at pH < pH_{ZPC} of the adsorbent causing electrostatic attraction of Cr (VI) species for the surface of adsorbent. (d) Development of negative surface charges at pH > pH_{ZPC} causing electrostatic repulsion of Cr (VI) species from the surface of adsorbent.

negatively charged Cr (VI) species get effectively adsorbed on the adsorbent surface through strong electrostatic attraction [52]. This was shown in Figure 5.3 (c). On the other hand, adsorption surface becomes negatively charged if pH of the solution is greater than pH_{ZPC}. This causes electrostatic repulsion between negatively charged Cr (VI) species and negatively charged surface of adsorbent, leading to decrease in adsorption (Figure 5.3 (d)).

5.6 Clay Minerals

Clays were used as adsorbents for removal of Cr (VI) due to their high specific surface area, low cost and ubiquitous presence in most soils [53–55]. However, natural clay minerals exhibit poor adsorption capacity for removal of Cr (VI) because of their negatively charged surfaces which repel anionic Cr (VI) species [56]. Efforts were made to improve adsorption capacity of natural clay minerals either by adding reducing agents or by modifying their surfaces. The present section reviews the adsorption behavior of natural clays and their modified forms. Optimum adsorption conditions, maximum adsorption capacity, mechanism of adsorption, kinetic model, isotherm model, and thermodynamic parameters were listed in Table 5.1.

5.6.1 Natural Clay Minerals

Adsorption of Cr (VI) onto natural clay minerals usually occurs through electrostatic attraction between positively charged surfaces and negatively charged Cr (VI) species as shown in Figure 5.4. Pumice and scoria volcanic rocks exhibited an adsorption capacity of 0.046 and 0.0448 mg/g, respectively, for a solution of initial Cr (VI) concentration 0.5–10.0 mg/L with an adsorbent dose of 100 g/L at pH 2.0 [57]. Khan et al. reported that

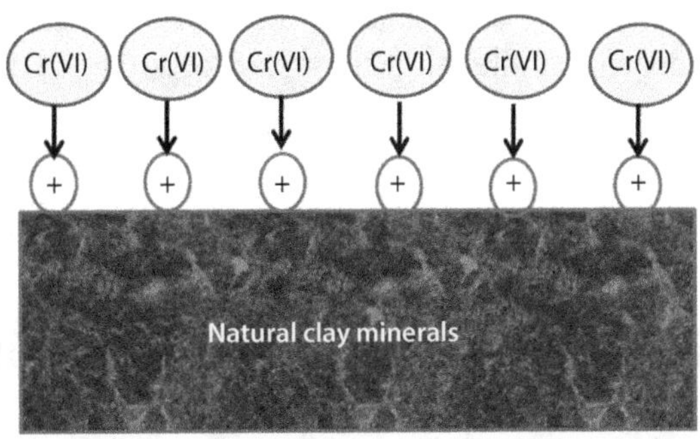

Figure 5.4 Adsorption of Cr (VI) onto natural clay minerals.

natural bentonite exhibited maximum adsorption capacity of 0.57 mg/g at pH 2.0 [46]. Maximum adsorption capacity obtained by Wollastonite was 0.826 mg/g [58]. According to Akar et al., maximum adsorption capacity obtained by natural Turkish montmorillonite clay was 3.16 mg/g [59]. Natural Akadama clay was used for removal of Cr (VI) from contaminated water [60]. The surface of the clay acquired high degree of protonation as shown in scheme 5.1. This facilitated adsorption of negatively charged Cr (VI) species onto the protonated clay surface through electrostatic attraction with a maximum adsorption capacity of 4.29 mg/g.

5.6.2 Natural Clay Minerals Along with Reducing Agents

In order to improve the adsorption capacity some researchers employed the concept of reduction of Cr (VI) to Cr (III) by an organic or inorganic reducing agent followed by subsequent adsorption of the latter [61–63]. Clay minerals rich in Fe (II) could remove Cr (VI) considerably by reduction/adsorption mechanism [56, 64]. Effect of humic acid on adsorption of Cr (VI) on to kaolin was investigated by Li et al. They observed that kaolin showed an enhanced adsorption capacity in presence of humic acid added after addition of the adsorbent and Cr (VI) solution. The increase in adsorption capacity was attributed to the high reducing power and strong complexing capacity of humic acid. Humic acid reduced negatively charged Cr (VI) species to Cr (III) which was adsorbed on to Kaolin. Maximum adsorption capacity calculated from Langmuir isotherm model was found to be 10.4 mg/g at pH 6.0. Addition of inorganic salts (Na^+, Ca^{2+} and Mg^{2+}) reduced adsorption of Cr (VI) because of their high coagulation power [65].

5.6.3 Modified Clay Minerals

Usually surfaces of clay minerals are modified either by metallic hydroxy polycations or by cationic surfactants. Modification of clay surfaces provides enough adsorption sites to facilitate adsorption of Cr (VI) [66, 67]

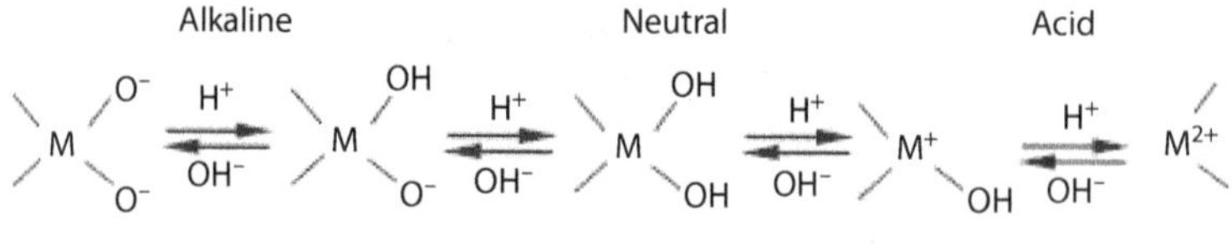

Scheme 5.1 Protonation on the surface of natural akadama clay [60]. (M represents Al, Fe, etc.)

and hence an enhanced adsorption capacity is obtained as compared to unmodified ones.

Fe-functionalized akadama clay has shown higher removal efficiency for Cr(VI) as compared to Mg, Mn, Ca, and Al functionalized akadama clay samples and pristine akadama clay. It is because of its higher value of zeta potential which contributes higher degree of electrostatic repulsion between adjacent and similarly charged particles in a dispersion and greater interaction with negatively charged Cr (VI) species resulting in better adsorption. Moreover, iron modification provides more number of Fe-functional groups by combination force of the acidification and iron precipitation, with formation of large number of available adsorption sites for Cr (VI). The adsorption process involves protonation of surface hydroxy groups followed by adsorption of oxy-anions of Cr (VI) through coulombic interactions and replacement of hydroxyl ions in the iron crystal lattices by Cr (VI) oxyanions through ligand exchange. In addition to this, there occurs partial reduction of Cr (VI) species by some organic matter such as humus, accumulated during the formation and transformation of volcanic soils [68]. Adsorption of Cr (VI) on to Fe/Zr pillared montmorillonite was carried out by batch mode through electrostatic interaction as well as ion exchange mechanism. Ion exchange between Cr (VI) and –OH groups of the adsorbent could be explained by following equations.

$$\text{M-OH} + \text{HCrO}_4^- \rightarrow \text{M-HCrO}_4 + \text{OH}^- \quad (5.11)$$

$$\text{M-OH}_2^+ + \text{HCrO}_4^- \rightarrow \text{M-HCrO}_4 + \text{H}_2\text{O} \quad (5.12)$$

M represents Fe, Zr, or mixed Fe/Zr.

Maximum uptake of Cr (VI) by $Fe/Zr_{4:1}$ pillared montmorillonite calculated from Langmuir isotherm equation was found to be 22.35 mg/g^{-1}. Presence of phosphate, sulfate and tartarate inhibited markedly the adsorption process due to their competition with Cr (VI) species for the active surface sites where as nitrate, chlorate and acetate did not have any significant effect bacause of their little affinity toward active surface sites [69]. Mg–Al hydrotalcite modified kaolin clay adsorbed Cr (VI) species both by repalcement of surface or interlayer NO_3^- ions and surface complexation of positively charged adsorbent surface with the anionic Cr (VI) species as shown in Eqns. 5.13 and 5.14, respectively, with maximum adsorption capacity 15.7 mg/g at pH 7.5 [70].

$$\text{MKC} - \text{OH}_2^+/\text{NO}_3^- + \text{A} \rightarrow \text{MKC} - \text{OH}_2^+/\text{A} + \text{NO}_3^- \quad (5.13)$$

$$MKC - OH_2^+ + A \rightarrow MKC - OH_2^+/A \tag{5.14}$$

where MKC and A represent modified kaolin clay and Cr (VI) species, respectively.

Moroccan stevensites were used for removal of Cr (VI) from aqueous solutions. Three different stevensites, namely natural stevensites, Al- stevensite (moroccan stevensite pillared by Keggin aluminium hydroxypolycation) and CTA- stevensite (cationic surfactant cetyltrimethylammoniumbromide) were considered as adsorbents. The adsorption capacities evaluted by Dubinin–Kaganer–Radushkevich (DKR) isotherm model were 0.712, 3.92 and 10.17 mg/g for natural stevensites, Al- stevensite and CTA- stevensite respectively [71]. Similar results were obtained for adsorption of Cr (VI) onto Al-montmorillonite and CTA- montmorillonite [66]. The increase in adsorption capacity of Al- stevensite over natural stevensites might be attributed to the increase in Al- OH_2^+ functional groups which facilitated the formation of Al- OH_2^+ - $HCrO_4^-$ surface complexes at low pH values [71]. Further, the enhanced affinity of CTA- stevensite for Cr (VI) with respect to Al- stevensite was probably due to presence of large number of positive charges resulting from formation of surfactant bilayer on the external surface [72]. The adsorbed CTA^+ bilayer on the clay minerals resulted in positively charged surfaces which were compensated by adsorption of Br^-. There occured exchange of Br^- ions by $HCrO_4^-$. The anion exchange reaction of $HCrO_4^-$ for Br^- can be written as:

$$(\text{CTA-clay-Br}) + HCrO_4^- \rightarrow (\text{CTA-clay–}HCrO_4) + Br^- \tag{5.15}$$

Adsorption energy values obtained by DKR isotherm model for adsorption of Cr (VI) onto Al– stevensite and CTA- stevensite were found to be 1.44 and 1.34 kJ/mol, respectivly, which suggested that ion exchange is the major mechanism that governed the Cr (VI) adsorption [73]. Hu and Luo synthesized inorganic – organic pillared montmorillonites by modifying with hydroxyaluminum and cetyltrimethylammonium bromide (CTMAB) in order to compare their adsorption behavior for removal of Cr (VI) [74]. Three adsorbents, namely CAM, ACM, and ACMCO were prepared by intercalation of CTMAB into montmorillonite before hydroxyaluminum, after hydroxyaluminum and simultaneous intercalation of CTMAB and hydroxyaluminum, respectively. Adsorption capacities of ACM, ACMCO,and CAM calculated from Langmuir isotherm model were found to be 11.97, 9.09, and 6.54 mg/g, respectively. It could be due to the significant adsorption of CTMAB on the surface of montmorillonite after its modification with hydroxyaluminum. CTMAB formed an admicelle surfactant configuration and reversed the clay surface charge to positive. This causes surface complexation reaction between anionic Cr (VI) species with $CTMA^+$ and amphoteric surface hydroxyl groups at the edge of the clay according to following equations [75, 76].

$$CTMA^+ + HCrO_4^- \rightarrow CTMAHCrO_4 \tag{5.16}$$

$$2CTMA^+ + CrO_4^{2-} \rightarrow (CTMA)_2CrO_4 \tag{5.17}$$

$$2CTMA^+ + Cr_2O_7^{2-} \rightarrow (CTMA)_2Cr_2O_7 \tag{5.18}$$

$$MOH_2^+ + HCrO_4^- \rightarrow MOHHCrO_4^- + H^+ \tag{5.19}$$

$$MOH_2^+ + CrO_4^{2-} \rightarrow MOHCrO_4^{2-} + H^+ \tag{5.20}$$

$$MOH_2^+ + Cr_2O_7^{2-} \rightarrow MOHCr_2O_7^{2-} + H^+ \tag{5.21}$$

MOH is the amphoteric surface hydroxyl groups (Si–OH or Al–OH).

Three different organic modified rectorites were prepared by modifying natural rectorite with three different surfactant namely dodecyl benzyl dimethyl ammonium chloride, hexadecyl trimethyl ammonium bromide and octadecyl trimethyl ammonium bromide to study their adsorption behavior for removal of Cr (VI). Maximum adsorption capacity calculated from Langmuir isotherm model were 0.97, 2.39, and 3.57 mg/g for rectorite modified with dodecyl benzyl dimethyl ammonium chloride, hexadecyl trimethyl ammonium bromide and octadecyl trimethyl ammonium bromide, respectively. Higher adsorption capacity of later two adsorbents might be attributed to their high surface area which was evident from larger d- spacing [77]. Gładysz-Płaska et al. reported that Cr(VI) ions and phenol could be simultaneously adsorbed onto natural red clay modified with hexadecyltrimethylammonium (HDTMA) bromide. Formation of a bilayer and presence of $HDTMA^+$ cation micelles on the mineral surface was probably responsible for the strong affinity of Cr(VI) anions to the surface of red clay. Adsorption of Cr (VI) was decreased by increasing the ionic strength due to competition of negatively charged Cr(VI) species with Cl^- and NO_3^- for the positively charged surface [78]. Ramos et al. reported that maximum adsorption capacity of 5.19 mg/g was achieved by zeolite modified with hexadecyltrimethylammonium (HDTMA) bromide at pH 6.0 and at temperature 25 °C [79]. Benzyl dimethyl octadecyl ammonium bentonite or stearalkonium bentonite exhibited adsorption capacity of 13 mg/g [80]. Sarkar et al. synthesized two organo clay adsorbents having surfactant– clay ratio (w/w) 2.38:1 and 4.75:1, respectively, by modifying Queensland bentonite with di (hydrogenated tallow) dimethyl ammonium chloride for remediation of Cr (VI). Modification of bentonite caused development of enough positive surface charges, which attracted negatively charged Cr (VI) species resulting in adsorption of latter through electrostatic attraction. In addition to this, there occurred ion exchange between weakly held counter ions of surfactants and more strongly held counter anions of Cr (VI) leading to enhancement of adsorption capacity. Maximum adsorption capacities were found to be 8.51 and 14.64 mg/g for the adsorbents with surfactant clay ratio 2.38:1(modified Queensland bentonite 1) and 4.75:1 (modified Queensland bentonite 2), respectively. Higher adsorption capacity of the later was due to formation of large number of cationic adsorption sites by higher surfactant loading [81]. Montmorillonites modified with cetylpyridinium bromide were used as adsorbents for removal of Cr (VI) from water. The equilibrium data were well described by Langmuir isotherm, which estimated the maximum adsorption capacity of 16.55 and 18.05 mg/g for the adsorbent modified by contact with 7.5 m mol/L (modified Montmorillonite 1) and 10.0 m mol/L (modified Montmorillonite 2) of cetylpyridinium bromide respectively. High adsorption capacity at relatively high concentration of cetylpyridinium bromide was due to presence of large positively charged surface [82]. Bentonite modified with cetylpyridinium bromide was used for removal of Cr (VI) [30]. The organo- bentonite removed 36.4 mg/g of Cr (VI) at 27 °C. Tannin-immobilized activated clay was used as a promising adsorbent for removal of Cr (VI) [83]. This adsorbent showed maximum adsorption capacity of 24.09 mg/g where as activated clay exhibited maximum adsorption capacity of 0.47 mg/g under identical experimental conditions. High adsorption capacity of tannin-immobilized activated clay was explained by the following three step mechanism.

1. A few amount of Cr (VI) got adsorbed on the porous surface of tannin-immobilized activated clay.
2. A part of Cr (VI) anionic species were adsorbed through acid catalysed esterification reaction of phenolic groups present in tannin to form oxoanionic species as shown in Scheme 5.2 (A).
3. Cr (VI) species were reduced to Cr (III) by H^+ ions where as phenolic groups of tannin were oxidized to carboxylic groups. This caused ion exchange of Cr (III) ions with hydroxyl or/

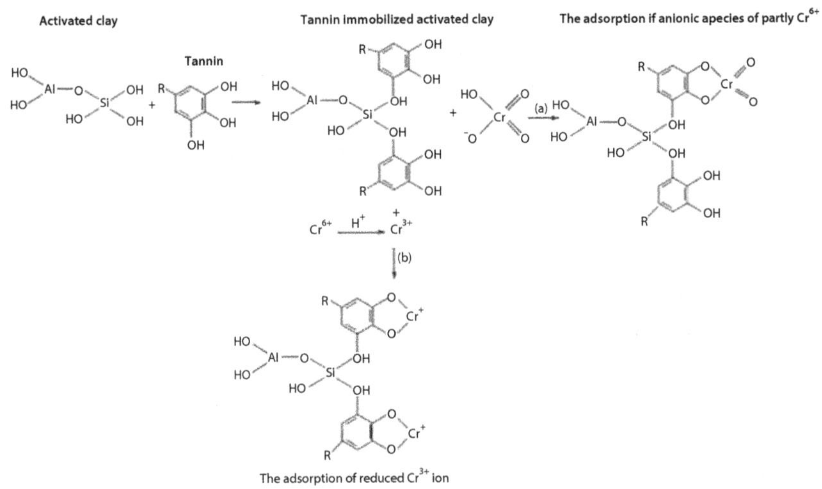

Scheme 5.2 (a) Adsorption of Cr (VI) and (b) Reduction of Cr (VI) to Cr (III) followed by adsorption of the latter onto tannin-immobilized activated clay. (Reprinted with permission from [83], Order No. 4065850957543).

and carboxyl groups of tannin. In other words, Cr(III) ions so formed by the reduction reaction got adsorbed by ion exchange mechanism. It was shown in the Scheme 5.2 (B).

Adsorption behavior of natural sepiolite, acid-activated sepiolite, functionalized natural sepiolite and functionalized acid-activated sepiolite were investigated by Marjanovic and his coworkers [17]. The mercaptosilane functionalized sepiolites exhibited greater adsorption capacity as comapared to unfunctionlized ones due to strong electrostatic attraction between protonated surface mercapto groups (–SH) and anionic Cr (VI) species at pH 2.0. Protonation of surface mercapto groups (–SH) was shown in Eqn. 5.22.

$$SiOSi(OCH_3)_2-(CH_2)_3-SH \xrightleftharpoons{H^+} SiOSi(OCH_3)_2-(CH_2)_3-SH_2^+ \tag{5.22}$$

Adsorption capacity of functionalized natural sepiolite and functionalized acid-activated sepiolite were found to be 2.69 and 5.90 mg/g, respectively,at an initial pH 2.0. The increase in adsorption capacity of later was due to presence of large number of mercapto groups which is a consequence of better functionalisation of mercaptosilane on acid-activated sepiolites. The adsorption capacity was further increased to 8.03 mg/g by riseing the initial pH to 3.0. It might be due to reduction of Cr (VI) species to Cr (III) by mercapto groups (-SH) which itself got oxidized to sulphonic groups ($-SO_3H$) at this pH as per Eqn. 5.23. Sulphonic groups ($-SO_3H$) so obtained were then ionized at the same pH to form $-SO_2O^-$ groups, which held Cr (III) ions through electrostatic attraction, leading to higher adsorption.

$$2\,HCrO_4^- + (-SH) + 8H^+ \rightarrow 2Cr^{3+} + (-SO_3H) + 5H_2O \qquad (5.23)$$

Further increase in initial pH to 4.5, adsorption capacity was decreased to 5.65 mg/g due to availability of less number of positive charges on its surface. The adsorption capacities of natural sepiolites and acid-activated sepiolites functionalized with amine–silane groups were approximately found to be 37.0 and 60.0 mg/g, respectivly, which were about ten times greater than that of mercapto silane functionalized sepiolites under identical experimental conditions. These results indicated that adsorption capacities of functionalized sepiolites depended both on the type of groups covalently grafted to the sepiolite surface and the number of surface silanol groups that are available for covalent grafting. At a final pH 2.0, large number of protonated amino groups ($-NH_2^+$ and $-NH_3^+$) were formed on the surface of amine - silane functionalized sepiolites. This facilitated adsorption of large number of predominately available $HCrO_4^-$ ions by electrostatic attraction, leading to high adsorption capacity. Simultaneously, this study ruled out the possibility of reduction of Cr (VI) to Cr (III) since the later could be adsorbed neither through columbic interaction because of unavailability of negatively charged groups nor through ion exchange mechanism due to low cation exchange capacity of amine - silane functionalized sepiolites. Moreover, Cr (III) could not be precipitated out as $Cr(OH)_3$ in the strongly acidic conditions. Therefore, maximum adsorption capacity of amine - silane functionalized sepiolites at pH 2.0 might be explained by remarkable protonation of amine groups and subsequent adsorption of anionic Cr (VI) species through electrostatic attraction. Further, the high adsorption capacity of amine - silane functionalized acid-activated sepiolites might be attributed to their better amine functionalisation as compared to that of natural sepiolites [84].

5.7 Biomasses

Biomasses are biological materials derived from living organisms. These materials were used effectively for biosorption of heavy metals either by metabolically mediated or by physico-chemical pathways of uptake from water and wastewater [85–87]. Being inexpensive and abudantly available,

Table 5.1 Maximum adsorption capacity, optimum adsorption conditions like pH, temperature, initial Cr(VI) concentration, adsorbent dose, fitted isotherm, kinetic model, thermodynamic parameters and mechanism of adsorption of clay minerals.

Name of the adsorbent	Maximum adsorption capacity (mg g^{-1})	Optimum pH	Temperature (°C)	Initial Cr(VI) concentration (mg/L)	Adsorbent dose (g/L)	Contact time (minutes)	Fitted isotherm	Kinetic model	Thermodynamic parameters	Mechanism of adsorption	Ref.
Pumice volcanic rocks	0.046	2.0	22.4	0.5–10	100	3960	Langmuir and Freundlich	pseudo-second-order	–	Electrostatic attraction and surface complexation	[57]
Scoria volcanic rocks	0.0448	2.0	22.4	0.5–10	100	1800	Langmuir and Freundlich	pseudo-second-order	–	Electrostatic attraction and surface complexation	[57]
Natural bentonite	0.57	2.0	40	–	50	–	–	–	–	–	[46]
Wollastonite	0.826	2.5	50	5.2	20	100	Langmuir	first order	$\Delta H° = +ve$ $\Delta G° = +ve$ $\Delta S° = +ve$	–	[58]
Natural Turkish montmorillonite clay	3.16	1.0	20	250	10	420	Langmuir and D-R	pseudo-second-order	–	Ion exchange	[59]
Natural Akadama clay	4.29	2.0	25	50	5	180	Freundlich	pseudo-second-order	–	Electrostatic attraction	[60]
Fe-functionalized akadama clay	5.34	9.2	–	28.24	5	240	–	–	–	Electrostatic attraction, ligand exchange, Reduction of Cr(VI) to Cr(III)	[68]

(Continued)

Table 5.1 Cont.

Name of the adsorbent	Maximum adsorption capacity (mg g^{-1})	Optimum pH	Temperature (°C)	Initial Cr(VI) concentration (mg/L)	Adsorbent dose (g/L)	Contact time (minutes)	Fitted isotherm	Kinetic model	Thermodynamic parameters	Mechanism of adsorption	Ref.
Kaolin with humic acid	10.4	6.0	30	12	0.2	1440	Langmuir	–	ΔH^o = +ve ΔG^o = –ve ΔS^o = +ve	Reduction of Cr(VI) to Cr(III) followed by electrostatic attraction	[65]
rectorite modified with dodecyl benzyl dimethyl ammonium chloride	0.97	6.0	26	100	1	40	Langmuir	–	ΔH^o = -ve ΔG^o = -ve ΔS^o = -ve	Electrostatic attraction	[77]
Hexadecyl trimethyl ammonium bromide	2.39	6.0	26	100	1	40	Langmuir	–	ΔH^o = -ve ΔG^o = -ve ΔS^o = -ve	Electrostatic attraction	[77]
Octadecyl trimethyl ammonium bromide	3.57	6.0	26	100	1	40	Langmuir	–	ΔH = -ve ΔG = -ve ΔS = -ve	Electrostatic attraction	[77]
Natural red clay modified with hexadecyltrimethylammonium bromide	4.47	5.8	20	5.2	4	300	Langmuir and Freundlich	first order	ΔH^o = -ve ΔG^o = +ve	Electrostatic attraction	[78]
Zeolite modified with hexadecyltrimethylammonium	5.19	6.0	25	10–150	25	7200	Langmuir	–	ΔH = -ve	Electrostatic attraction	[79]
Natural sepiolite	<0.01	3.0-5.0	25	10	20	1440	–	–	–	Weak electrostatic attraction	[17]

Acid-activated sepiolite	0.1	3.0-5.0	25	10	20	1440	–	–	–	Weak electrostatic attraction	[17]
Functionalized natural sepiolite	2.69	2.0	25	5–100	4	1440	D-R	–	ΔG^{o} = -ve	Electrostatic attraction with Cr(VI) species, Reduction of Cr(VI) to Cr(III) followed by electrostatic attraction of Cr(III)	[17]
Functionalized acid-activated sepiolite	8.03	3.0	25	5–100	4	1440	Dubinin-Radushkevick	–	ΔG^{o} = -ve	Electrostatic attraction with Cr (VI) species, Reduction of Cr(VI) to Cr(III) followed by electrostatic attraction of Cr(III)	[17]
Natural stevensite	0.712	3.0	Room temperature	26–520	5	120	DKR	–	–	–	[71]
Al- stevensite	3.92	3.0	Room temperature	26–520	5	120	DKR	pseudo-second-order	–	Ion exchange	[71]
CTA- stevensite	10.17	3.0	Room temperature	26–520	5	120	DKR	pseudo-second-order	-	Ion exchange	[71]
HDTMA– Kaolinite	1.56	≤ 1.0	-	2340–23400	10	10–60	-	-	-	Electrostatic attraction	[66]

(*Continued*)

Table 5.1 Cont.

Name of the adsorbent	Maximum adsorption capacity (mg g^{-1})	Optimum pH	Temperature (°C)	Initial Cr(VI) concentration (mg/L)	Adsorbent dose (g/L)	Contact time (minutes)	Fitted isotherm	Kinetic model	Thermodynamic parameters	Mechanism of adsorption	Ref.
HDTMA– American bentonite	41.34	≤ 1.0	-	23400	10	30	Langmuir	-	-	Electrostatic attraction	[66]
HDTMA–Indian bentonite	41.34	≤ 1.0	-	23400	10	30	Langmuir	-	-	Electrostatic attraction	[66]
ACM	11.97	4.0	25	50	10	60	Langmuir	pseudo-second-order	-	Electrostatic attraction	[74]
ACMCO	9.09	4.0	25	50	10	60	Langmuir	pseudo-second-order	-	Electrostatic attraction	[74]
CAM	6.54	4.0	25	50	10	60	Langmuir	pseudo-second-order	-	Electrostatic attraction	[74]
Benzyldi methyloctadecyl ammonium bentonite or stearalkonium bentonite	13	7.0–7.1	23	5.2–52	2	240	Langmuir -Freundlich	first order	-	Electrostatic attraction	[80]
modified Queensland bentonite 1	8.51	4.7	37	10 - 80	5	180	Langmuir	pseudo-second-order	-	Electrostatic attraction and ion-exchange	[81]
modified Queensland bentonite 2	14.64	4.7	37	10 - 80	5	180	Langmuir	pseudo-second-order	-	Electrostatic attraction and ion-exchange	[81]
Mg–Al hydrotalcite modf. kaolin clay	15.7	7.5	25	49.92	8.0	480	Langmuir	pseudo-second-order	-	Ion-exchange and surface complexation	[70]

modified Montmorillonite 1	16.55	4.5	–	30	10	240	Langmuir	-	-	Electrostatic attraction	[82]
modified Montmorillonite 2	18.05	4.5	–	30	10	240	Langmuir	-	-	Electrostatic attraction	[82]
Fe/Zr pillared montmorillonite	22.35	3.0	25	25	2	120	Langmuir	pseudo-second-order	ΔH = -ve ΔG = -ve ΔS = +ve	Electrostatic attraction and ion-exchange	[69]
Tannin-immobilized activated clay	24.09	2.5	57	60–300	6	180	Freundlich	pseudo-second-order	-	Electrostatic attraction with Cr (VI) species, Reduction of Cr(VI) to Cr(III) followed by ion exchange	[83]
Activated clay	0.47	2.5	57	60-300	6	180	-	-	-	-	[83]
organo- bentonite	36.4	6.0-6.3	27	52	10	120	Langmuir	pseudo-second-order	ΔH^o = +ve ΔG^o = -ve ΔS^o = +ve	Ion exchange	[30]
Natural sepiolites functionalized with amine –silane	37. 0	2.0	25	10–250	2	1440	Freundlich	pseudo-second-order	ΔG^o = -ve	Electrostatic attraction	[84]
Acid-activated sepiolites functionalized with amine –silane	60.0	2.0	25	10-250	2	1440	Freundlich	pseudo-second-order	ΔG^o = -ve	Electrostatic attraction	[84]

a large number of biomasses have been used for remediation of Cr (VI) from aqueous solution. The present section reviews biosorption of Cr (VI) by different biomasses. Optimum adsorption conditions, maximum adsorption capacity, mechanism of adsorption, kinetic model, isotherm model and thermodynamic parameters were listed in Table 5.2.

Jain et al. prepared two adsorbents from sunflower stem waste and studied their efficiency for removal of Cr (VI) from aqueous solutions under different process conditions. Maximum adsorption capacity for boiled sun flower stem and formaldehyde treated sunflower stem was found to be 4.9 and 3.6 mg/g, respectively [40]. Treated Waste News paper contains functional polar groups such as alcohols and ethers along with phosphates impregnated into the cellulose matrix during treatment process. These groups are protonated at low pH and adsorb negatively charged Cr (VI) species effectively through electrostatic attraction with maximum uptake capacity of 59.88 mg/g [88]. Chemically treated waste tamarind hull exhibited maximum adsorption capacity of 81mg/g at low pH by the reduction of Cr (VI) to Cr (III) as per following equation [89].

$$Cr_2O_7^{2-} + 14H^+ + 6e^- \rightarrow 2Cr^{3+} + 7H_2O \tag{5.24}$$

Moussavi and Barikbin used pistachio hull waste biomass for removal of Cr (VI) from industrial wastewater and obtained an uptake capacity of 116.3 mg/g [90]. Adsorption process involved electrostatic attraction between protonated –OH, –CH, –C=O, –C–O and –N–H groups present on the surface of the adsorbent and anionic Cr (VI) species.

Treated mangifera indica saw dust adsorbed Cr (VI) from aqueous solutions by electrostatic attraction with maximum adsorption capacity of 12.78 mg/g [91]. Adsorption of Cr (VI) onto saw dust was investigated by Gupta and Babu. Surfaces of saw dust contained oxo groups (C_xO and C_xO_2), which got hydrolysed in presence of water to produce positively charged species ($C_xOH_2^{2+}$) according to Eqn. 5.25.

$$2H_2O + C_xO_2 \Leftrightarrow C_xOH_2^{2+} + 2OH^- \tag{5.25}$$

These species on the surfaces of sawdust adsorbed negatively charged Cr (VI) species through electrostatic attraction as shown in the Eqn. 5.26.

$$C_xOH_2^{2+} + HCrO_4^- \Leftrightarrow C_xO_2H_3CrO_3^- \tag{5.26}$$

As a result, the adsorbent exhibited maximum adsorption capacity of 41.52 mg/g. In addition to this, large specific surface area and less bulk

Table 5.2 Maximum adsorption capacity, optimum adsorption conditions like pH, temperature, initial Cr(VI) concentration, adsorbent dose, fitted isotherm, kinetic model, thermodynamic parameters and mechanism of adsorption of biomass.

Name of the adsorbent	Maximum adsorption capacity (mg/g)	Optimum pH	Temperature (°C)	Initial Cr(VI) concentration (mg/L)	Adsorbent dose (g/L)	Contact time (minutes)	Fitted isotherm	Kinetic model	Thermodynamic parameters	Mechanism of adsorption	Ref.
Formaldehyde-treated sunflower stem	3.6	2.0	26	50	4.0	120	Freundlich, Dubinin–Radushkevich	–	–	Electrostatic attraction	[40]
Boiled sun flower stem	4.9	2.0	26	50	4.0	120	Langmuir, Dubinin–Radushkevich	–	–	Electrostatic attraction	[40]
Rice husk	11.398	1.5	30	100	10	360	Langmuir	Pseudo-second-order	ΔH^o = +ve, endothermic ΔG^o = -ve, spontaneous ΔS^o = +ve	Electrostatic attraction	[94]
Rice straw	12.172	2.0	30	100	10	180	Freundlich	Pseudo-second-order	ΔH^o = +ve, endothermic ΔG^o = -ve, spontaneous ΔS^o = +ve	Electrostatic attraction	[94]
Rice bran	12.341	2.0	30	100	10	300	Freundlich	Pseudo-second-order	ΔH^o = +ve, endothermic ΔG^o = -ve, spontaneous ΔS^o = +ve	Electrostatic attraction	[94]
Treated mangifera indica sawdust	12.78	2.6	30	50	4.0	60	Langmuir, Freundlich	Pseudo-first-order	ΔH^o = -ve, endothermic ΔG^o = -ve, spontaneous ΔS^o = -ve	Electrostatic attraction	[91]
Hyacinth roots	15.281	2.0	30	100	10	240	Freundlich	Pseudo-second-order	ΔH^o = +ve, endothermic ΔG^o = -ve, spontaneous ΔS^o = +ve	Electrostatic attraction	[94]

(*Continued*)

Table 5.2 Cont.

Name of the adsorbent	Maximum adsorption capacity (mg/g)	Optimum pH	Temperature (°C)	Initial Cr(VI) concentration (mg/L)	Adsorbent dose (g/L)	Contact time (minutes)	Fitted isotherm	Kinetic model	Thermodynamic parameters	Mechanism of adsorption	Ref.
Neem leaves	15.954	2.0	30	100	10	240	Langmuir	Pseudo-second-order	ΔH^o = +ve, endothermic ΔG^o = -ve, spontaneous ΔS^o = +ve	Electrostatic attraction	[94]
Coconut shell	18.695	2.0	30	100	10	240	Langmuir	Pseudo-second-order	ΔH^o = +ve, endothermic ΔG^o = -ve, spontaneous ΔS^o = +ve	Electrostatic attraction	[94]
green coconut shell	22.96	2.5	28	30	5	40	Langmuir	Pseudo-second-order	–	–	[96]
Chitosan coated fly ash	33.27	5.0	15	100	0.66	50	Langmuir	–	–	Electrostatic attraction and reduction of Cr(VI) to Cr(III)	[103]
Saw dust	41.52	1.0	30	500	10	1050	Langmuir	Second order	–	Electrostatic attraction	[92]
seeds of Artimisia absinthium	46.99	2.0	30	60	10.0	1440	Langmuir	Pseudo-second-order	ΔH^o = +ve, endothermic ΔG^o = -ve, spontaneous ΔS^o = +ve	Electrostatic attraction	[97]
Coir pith treated with 50% H_2O_2	51.06	2.0	–	1,171	–	–	–	–	–	–	[86]

Treated Waste Newspaper	59.88	3.0	45	5-70	4.0	60	Langmuir	Pseudo-second-order	ΔH^o = +ve, ΔG^o = -ve, ΔS^o = +ve	Electrostatic attraction	[88]
Activated powdered neem leaves	62.9	2.0	30	700	10.0	4020	Langmuir isotherm	Second-order kinetics	-------------------------	Electrostatic attraction	[98]
Coconut coir	70.4	3.0	25	500	2	7200	–	First order	–	Electrostatic attraction followed by reduction of Cr(VI) to Cr(III	[100]
Puresorbe	76.92	2.0	–	40 - 100	2.0	230	Redlich-Peterson	second order	ΔH^o = +ve, endothermic ΔG^o = -ve, spontaneous ΔS^o = +ve	Electrostatic attraction	[52]
Chemically treated waste tamarind hull	81	2.0	50	100	1.0	1200	Freundlich, Redlich–Peterson and Fritz–Schlunder	Pseudo-first-order	ΔH^o = +ve, endothermic ΔG^o = -ve, spontaneous ΔS^o = +ve	Reduction of Cr(VI) to Cr(III)	[89]
Cross-linked chitosan resin	86.81	3.0	30	30	2.0	120	Langmuir	Pseudo-second-order	ΔH^o = +ve, ΔG^o = +ve, ΔS^o = +ve	Electrostatic attraction and ion exchange	[105]
Pistachio hull waste biomass	116.3	2.0	25	50-500	2.5	360	Langmuir	Pseudo-second-order	-----------------	Electrostatic attraction	[90]
2 M acrylamide-grafted coir pith	128.29	2.0	–	1,171	–	–	–	–	–	–	[102]

(*Continued*)

Table 5.2 Cont.

Name of the adsorbent	Maximum adsorption capacity (mg/g)	Optimum pH	Temperature (°C)	Initial Cr(VI) concentration (mg/L)	Adsorbent dose (g/L)	Contact time (minutes)	Fitted isotherm	Kinetic model	Thermodynamic parameters	Mechanism of adsorption	Ref.
Coir pith	165	2.0	30	1,171	13	1320	Langmuir	–	–	–	[86]
2M acrylic acid grafted coir pith	196	2.0	30	1,171	13	1320	Langmuir	–	ΔH^{o} = +ve, ΔG^{o} = -ve, ΔS^{o} = +ve	Electrostatic attraction and reduction of Cr(VI) to Cr(III	[86]
Acid treated banana skins	–	4.0	25	100	7.5	600	–	–	–	Reduction of Cr(VI) to Cr(III) followed by electrostatic attraction	[95]
mango leave dust	250.23	2.0	40	200	20.0	120	Langmuir	Pseudo-first-order	ΔH^{o} = +ve, ΔG^{o} = -ve, ΔS^{o} = +ve	electrostatic attraction	[93]
Amine-cross linked and carboxyl- chelated wheat straw	227.3	4.2	–	100	2.0	360	Langmuir	–	–	–	[99]

density of saw dust were responsible for high adsorption capacity [92]. Hydroxyl, carboxylic, carbonyl, amino and nitro groups present in different compounds of mango leave dust get protonated and adsorb negatively charged Cr (VI) species effectively through electrostatic attraction resulting in high adsorption capacity of 250.23 mg/g at pH 2.0 and at 40 °C [93].

Singha and Das used six different adsorbents namely rice straw, rice bran, rice husk, hyacinth roots, neem leaves and coconut shell to study their adsorption behavior for removal of Cr (VI) from aqueous solution [94]. Maximum adsorption capacity was found to be 12.172, 12.341, 11.398, 15.281, 15.954, and 18.695 mg/g for rice straw, rice bran, rice husk, hyacinth roots, neem leaves and coconut shell respectively. Coconut shell exhibited highest removal efficiency (94.90%) for Cr (VI) from wastewater of an electroplating unit and rice husk showed minimum efficiency (53.11%) among all the adsorbents studied. Cr (VI) was removed from aqueous solutions by acid treated banana skins [95]. Banana skin mainly contains hydroxyl and carboxyl groups. The hydroxyl groups present in banana skins are oxidized by Cr (VI) species in acidic medium to form carbonyl groups which on further oxidation produce carboxyl groups. As a result, Cr (III) species were formed as the reduction product and were subsequently adsorbed by the carboxyl groups present in banana skins and obtained as an oxidation product. Maximum adsorption capacity for green coconut shell was found to be 22.96 mg/g at pH 2.5 [96]. Seeds of Artimisia absinthium were used for adsorption of Cr (VI). The adsorbent exhibited maximum adsorption capacity of 46.99 mg/g at pH 2.0 due to ionic interaction between negatively charged Cr (VI) species and the positively charged surface of the adsorbent [97]. Babu and Gupta prepared activated powdered neem leaves in order to study its adsorption behavior for removal of Cr (VI) from aqueous solutions. The maximum adsorption capacity was recorded 62.9 mg/g. High adsorption capacity of activated powdered neem leaves was mainly due to its large specific surface area and development of acidic surface oxides [98]. Zhong et al. prepared amine-cross linked and carboxyl- chelated wheat straw amphoteric adsorbent for adsorption of Cr (VI) and Cu (II) in single and binary systems. Maximum adsorption capacity calculated from Langmuir isotherm model was found to be 227.3 mg/g in single system [99].

Uptake capacity of coconut coir for Cr (VI) was found to be higher than that of its chars due to decrease in surface functional groups in the later case by the process of pyrolysis. The authors have suggested that Cr (VI) species were adsorbed by the functional groups of coconut coir through electrostatic attraction and then were reduced completely to Cr (III) by the hydroxyl and carbonyl groups present on the surface of the adsorbent. Cr (III) ions so formed were adsorbed by the carbonyl and carboxyl groups.

Maximum removal capacity of 70.4 mg/g was achieved [100]. Obtained adsorption capacity was higher than that for coconut coir reported by Gonzalez et al. [85]. Puresorbe, a coir based adsorbent, was used for remediation of Cr (VI). Superphosphate treated puresorbe contained protonated active sites such as carboxylates, super phosphates, phenolic and aliphatic hydroxyl groups which adsorbed negatively charged Cr (VI) species through electrostatic attraction as per the Eqn. 5.27, 5.28 and 5.29.

$$\mathrm{RCOOH + 2H_2O + HCrO_4^- \rightarrow RCO_3H_4O_3Cr^+ + 2OH^-} \quad (5.27)$$

$$\mathrm{ROH + 2H_2O + HCrO_4^- \rightarrow RO_2H_4O_3Cr^+ + 2OH^-} \quad (5.28)$$

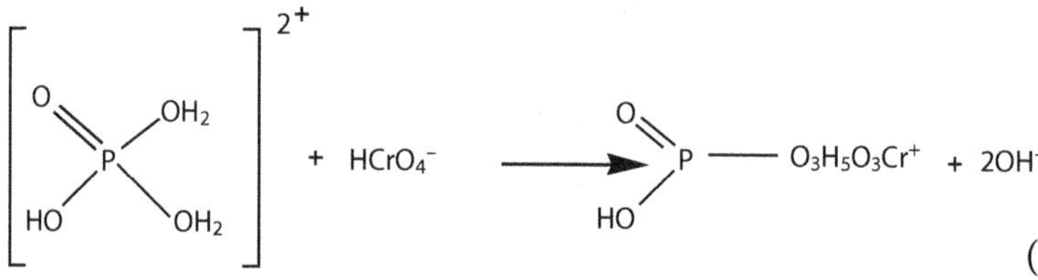

(5.29)

The adsorption capacity of puresorbe was found to be 76.92 mg/g [52]. It was observed that coir pith grafted with 2M acrylic acid showed improved uptake capacity of 196 mg/g. The observed uptake capacity was found to be higher than that obtained by coir pith treated with 50% H_2O_2, 2 M acrylamide-grafted coir pith and unmodified coir pith. High adsorption behavior of acrylic acid grafted coir pith was due to presence of increased number of surface carbonyl groups. Mechanism of adsorption involves electrostatic attractions between positively charged surfaces of adsorbent and anionic Cr (VI) species followed by reduction of Cr (VI) species to less toxic Cr (III) by hydroxyl and carbonyl groups present on the surfaces of adsorbent. Cr (III) ions formed coordinate covalent bond with the carbonyl groups (CO) and methoxy groups (–O–CH_3) of acrylic acid grafted coir pith and hence got adsorbed onto it [74]. Reduction of Cr (VI) to Cr (III) during biosorption was also reported [101, 102].

Chitosan coated fly ash containing 30 wt. % chitosan exhibited maximum adsorption capacity of 33.27 mg/g. The adsorption process involves electrostatic attraction between the active functional groups like Si-OH, NH_3^+ and Al^{3+} of the adsorbent and anionic Cr (VI) species as well as reduction of Cr (VI) to Cr (III) by the -OH and $-NH_2$ groups present in the chitosan-coated fly ash [103]. In addition to this, coating of chitosan over fly ash could form an exfoliated structure which enhances adsorption of Cr (VI) [104]. A cross-linked chitosan resin was synthesized by the cross- link of Epichlorohydrin through an improved dropsphere-forming method [105].

The adsorbent so formed exhibited high adsorption capacity of 86.81 mg g^{-1} for uptake of Cr (VI). Adsorption of Cr (VI) onto cross-linked chitosan resin takes place through ionic interaction and ion exchange mechanism.

5.8 Industrial Wastes

In recent years, large amount of industrial wastes are produced due to rapid industrialization. These wastes create disposal management problems. Attempts have been made to use these materials in remediation of Cr (VI). In this section, we have reviewed adsorption of Cr (VI) by industrial wastes. Name of these adsorbents, their adsorption capacity, adsorption parameters, isotherm, kinetics, thermodynamics and mechanism of adsorption were given in Table 5.3.

Spent activated clay contains silanol (Si–O^-) and aluminol (Al–O^-) groups as the active sites, which acquired positive charges at pH less than pH_{zpc} (3.8). As a result, negatively charged Cr (VI) species get adsorbed onto the surface of the clay through coulombic interaction as shown in Eqn. 5.30.

$$S\text{–}OH_2^+ + HCrO_4^- \rightarrow S\text{–}HCrO_4 + H_2O \quad (5.30)$$

where S represents Si or Al.

Neutral surface hydroxyl functional groups (S–OH°) might be the active sites at pH greater than pH_{zpc}. There appears competition between OH^- ions and oxyanionic species of Cr (VI) for the active sites leading to decrease in adsorption [49]. Asl and his coworkers prepared zeolite from raw fly ash obtained from the combustion of powdered coal in thermal power plant to study its adsorption behavior for removal of Cr (VI). The adsorbent exhibited maximum adsorption capacity of 37.4 mg/g. Further, the adsorbent was used to remove Cr (VI) from industrial wastewater containing 85.3 mg/L and the removal efficiency was found to be 82.15% [106]. Maximum uptake capacity of bagasse fly ash obtained from a sugar industry was found to be 1.8 mg/g [107]. The waste produced by the incomplete combustion of fuel in fertilizer industry is called carbon slurry which was chemically treated and thermally activated in order to study adsorption of Cr (VI) from aqueous solution. The adsorbent exhibited maximum adsorption capacity of 15.244 mg/g [108]. Mallick et al. used manganese nodule leach residue, a waste obtained during the pilot plant testing of NH_3-SO_2 leaching of Indian Ocean nodules for removal of Cr (VI) from aqueous solutions [28]. They reported that water washed residue possessed higher surface area, surface oxygen and surface hydroxyl groups due to

Table 5.3 Maximum adsorption capacity, optimum adsorption conditions like pH, temperature, initial Cr(VI) concentration, adsorbent dose, fitted isotherm, kinetic model, thermodynamic parameters and mechanism of adsorption of industrial wastes.

Name of the adsorbent	Maximum adsorption capacity (mg/g)	Optimum pH	Temperature (°C)	Initial Cr(VI) concentration (mg/L)	Adsorbent dose (g /L)	Contact time (minutes)	Fitted Isotherm	Kinetic model	Thermodynamic parameters	Mechanism of adsorption	Ref.
Spent activated clay	1.422	2.0	40	0.2-10.0	1.0	120	Langmuir	Pseudo-first-order	ΔH^o = +ve, endothermic ΔG^o = −ve, spontaneous ΔS^o = +ve	coulombic interaction	[50]
Bagasse fly ash	1.8	5.0	30	20.0	10.0	40	Langmuir and Freundlich	–	ΔH^o = +ve, endothermic ΔG^o = −ve, spontaneous ΔS^o = +ve	–	[107]
Distillery sludge	5.7	3.0	–	10.0	7.0	105	Langmuir and Freundlich	First order	-	Electrostatic attraction	[109]
Carbon slurry	15.244	2.0	30	100.0	4.0	70	Langmuir and Freundlich	Pseudo-second-order	ΔH^o = −ve, endothermic ΔG^o = −ve, spontaneous ΔS^o = +ve	Electrostatic attraction, Reduction of Cr(VI) to Cr(III).	[108]
Fly ash	37.4	2.0	40	50.0	4.0	240	Langmuir	Pseudo-second-order	ΔH^o = +ve, endothermic ΔG^o = −ve, spontaneous ΔS^o = +ve	–	[106]
Water washed manganese nodule leach residue	888.68	3.0	30	10.0	2.0	180	Langmuir	Second order	ΔH^o = +ve, endothermic ΔG^o = −ve, spontaneous ΔS^o = +ve	Electrostatic attraction	[28]
Heat activated red mud	0.3	2.0	30	0.08	5	120	Langmuir	–	ΔH^o = +ve, endothermic ΔG^o=+ve, nonspontaneous ΔS^o = −ve	Electrostatic attraction and ligand exchange	[111]

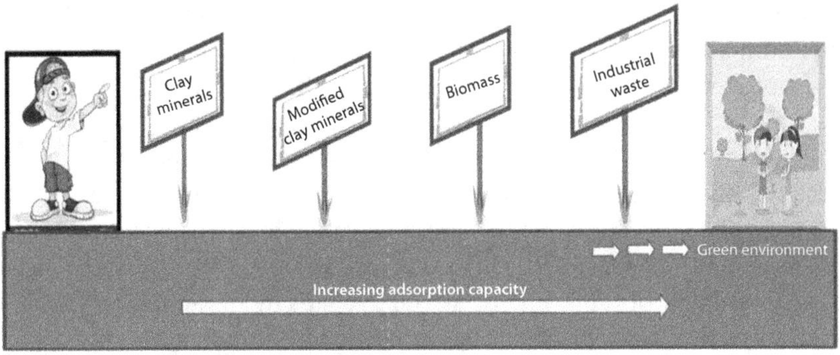

Figure 5.5 Road map showing increasing order of adsorption capcpities of clay minerals, modified clay minerals, biomass and industrial wastes for remediation of Cr (VI).

which it exhibited higher adsorption capacity as compared to unwashed residue. Maximum adsorption capacity of distillery sludge was found to be 5.7 mg/g for removal Cr (VI). It was observed that about 91% of Cr (VI) could be removed from electroplating effluent by using 15 g/L of distillery sludge [109]. Red mud, an aluminum industry waste was used for remediation of Cr (VI) [110, 111]. At low pH, the surface charge on activated red mud is neutralized and hence diffusion of dichromate ions is facilitated resulting in higher adsorption of Cr (VI). At higher pH, presence of large amount of OH^- ions hinder the diffusion of Cr (VI) species leading to decrease in adsorption [110]. Heat-acid treated red mud adsorbed Cr (VI) species through electrostatic attraction and ligand exchange at low pH (2.0). Enough positive charges are developed on the surface of the adsorbent onto which $HCrO_4^-$ ions get adsorbed as follows [111].

$$\equiv S\text{–}OH^{+} + H^{+} \rightarrow \equiv S\text{–}OH_2^{+} \quad (5.31)$$

$$\equiv S\text{–}OH_2^{+} + HCrO_4^{-} \rightarrow \equiv S\text{–}OH_2^{+}\text{–}HCrO_4^{-} \quad (5.32)$$

$$\equiv S\text{–}OH_2^{+} + HCrO_4^{-} \rightarrow S\text{–}HCrO_4 + H_2O \quad (5.33)$$

Figure 5.5 shows a road map for the discussed adsorbrnts with increasing order of adsorption capacity for remediation of Cr (VI).

5.9 Conclusion

Natural clay minerals exhibited poor adsorption capacity for removal of Cr (VI) because of electrostatic repulsion between negatively charged clay surfaces and oxyanionic Cr (VI) species. Natural Akadama clay exhibited maximum adsorption capacity of 4.29 mg/g. However, adsorption capacity was improved in presence of reducing agents like humic acid. An adsorption capacity of maximum 10.4 mg/g was obtained for kaolin clay in presence of humic acid. The surfaces of clays were also modifyed with metallic hydroxy polycations and cationic surfactants in order to increase adsorption capacity. Protonated -OH groups in metallic hydroxy polycations adsorbed Cr (VI) species through electrostatic attractions. There might be the possibility of replacement of -OH groups by oxyanions of Cr (VI). Large number of adsorption sites were formed in surfactant modified clays as compared to that modified by metallic hydroxy polycations. Therefore, the former exhibited higher uptake capacity. Further, concentration of surfactant loading enhanced adsorption sites on surfaces of clay minerals resulting in high adsorption capacities. Moreover, nature of surfactant group played a key role in improving adsorption capacity. Acid activated sepiolites functionalized with amine-silane groups exhibited highest adsorption capacity of 60.0 mg/g among the modified clay minerals probably due to better amine functionalisation and remarkable protonation of amine groups.

Biomasses usually contained hydroxy, carboxylic, carbonyl, amino and nitro groups as surface functional groups. Positive charges were developed on these groups by chemical treatment. Moreover, C=O groups were hydrolysed in presence of water to form positive charged species. As a result, Cr (VI) species got adsorbed onto biomasses through electrostatic attraction. In addition to this, Cr (VI) species were reduced by hydroxy and carbonyl groups to Cr (III). Further, chemical modification of surfaces of these materials enhanced number of surface functional groups as well as facilitated reduction of Cr (VI).

The maximum adsorption capacity of 250.23 mg/g was obtained by mango leaf dust among the biomasses.

Electrostatic attraction was the major mechanism for adsorption of Cr (VI) on to industrial wastes. Water washed manganese nodule leach residue exhibited highest adsorption capacity of 888.68 mg/g. Though adsorbents differ in adsorption capacities, yet comparison in their performance is difficult as adsorption capacities were achieved at different adsorption conditions like pH, temperature, initial Cr (VI) concentration, adsorbent dose and contact time, etc. More over, most of the studies have been carried out in laboratory scale and performances of adsorbents were not tested either in pilot plant scale or in industrial scale. So it is still in the infant stage and various issues need to be solved concerning the development of more facile processes to obtain the composite adsorbents. Thus, much work is necessary for better understanding of adsorption phenomenon and demonstration of possible useful technology.

References

1. Sun, Z., Yao, G., Zhang, X., Zheng, S., Frost, R.L., Enhanced visible-light photocatalytic activity of kaolinite/g-C3N4 composite synthesized via mechanochemical treatment. *Appl. Clay Sci.*, 129, 7, 2016.
2. Dubey, R., Bajpai, J., Bajpai, A.K., Green synthesis of grapheme sand composite (GSC) as novel adsorbent for efficient removal of Cr (VI) ions from aqueous solution. *J. Wat. Proces. Eng.*, 5, 83, 2015.
3. Lytras, G., Lytras, C., Argyropoulou, D., Dimopoulos, N., Malavetas, G., Lyberatos, G., A novel two-phase bioreactor for microbial hexavalent chromium removal from wastewater. *J. Hazard. Mater.*, 336, 41, 2017.
4. Emsley, J., *Nature's building blocks: an A-Z guide to the elements*, pp. 495–498, Oxford University Press, 2001.
5. Natural attenuation of hexavalent chromium in ground water and soils, EPA 540 S-94-505, 1994.
6. Eary, L.E., Dhanpat, R., Chromate removal from aqueous wastes by reduction with ferrous ion. *Environ. Sci. Technol.*, 22, 972, 1988.
7. Determination of Cr (VI) in water, wastewater and solid waste extracts, 26 LPN 34398-011M 7-96, 1996.
8. Bhowal, A., Datta, S., Studies on transport mechanism of Cr (VI) extraction from an acidic solution using liquid surfactant membranes. *J. Memb. Sci.*, 188, 1, 2001.
9. Maleki, A., Hayati, B., Naghizadeh, M., Joo, S.W., Adsorption of hexavalent chromium by metal organic frameworks from aqueous solution. *J. Indust. Eng. Chem.*, 28, 211, 2015.
10. Abbasi, S.A., Soni, R., Teratogenic effects of chromium (VI) in environment as evidenced by the impact on larvae of amphibian Rana Tigrina: implications in the environmental management of chromium. *Int. J. Environ. Studies*, 23, 131, 1984.
11. Hamdan, S.S., El-Naas, M.H., Characterization of the removal of Chromium (VI) from groundwater by electrocoagulation. *J. Indust. Eng. Chem.*, 20, 2775, 2014.
12. Gorny, J., Billon, G., Noiriel, C., Dumoulin, D., Lesven, L., Madé, B., Chromium behavior in aquatic environments: a review. *Environ. Rev.*, 24, 503, 2016.
13. Wang, G., Wang, S., Sun, W., Sun, Z., Zheng, S., Synthesis of a novel illite@ carbon nanocomposite adsorbent for removal of Cr (VI) from wastewater. *J. Environ. Sci.*, in press, 2016.
14. D., Zhao, X., Gao, C., Wu, R., Xie, S., Feng, C., Chen, Facile preparation of amino functionalized graphene oxide decorated with Fe_3O_4 nanoparticles for the adsorption of Cr (VI). *Appl.Surf. Sci.*, 384, 1, 2016.

15. Mishra, S., Verma, N., Carbon bead-supported hollow carbon nanofibers synthesized via templating method for the removal of hexavalent chromium. *J. Indust. Eng. Chem.*, 36, 346, 2016.
16. Kazakis, N., Kantiranis, N., Voudouris, K.S., Mitrakas, M., Kaprara, E., Pavlou, A., Geogenic Cr oxidation on the surface of mafic minerals and the hydrogeological conditions influencing hexavalent chromium concentrations in groundwater. *Sci. Total Environ.*, 514, 224, 2015.
17. Marjanovic, V., Lazarevic, S., Jankovic-Castvan, I., Potkonjak, B., Janackovic, D., Petrovic, R., Chromium (VI) removal from aqueous solutions using mercaptosilane functionalized sepiolites. *Chem. Engg. J.*, 166, 198, 2011.
18. Du, Y., Wang, L., Wang, J., Zheng, G., Wu, J., Dai, H., Flower- wire and sheet like MnO_2-deposited diatomites: highly efficient absorbents for the removal of Cr(VI). *J. Environ. Sci.*, 29, 71, 2015.
19. Tian, L., Xie, G., Li, R., Yu, X., Hou, Y., Removal of Cr (VI) from aqueous solution using MCM-41. *Desalin. Water Treat.*, 36, 334, 2011.
20. X., Cheng, Li, N., Zhu, M., Zhang, L., Deng, Y., Deng, C., Positively charged microporous ceramic membrane for the removal of Titan Yellow through electrostatic adsorption. *J. Environ. Sci.*, 44, 204, 2016.
21. Namasivayam, C., Yamuna, R.T., Adsorption of chromium (VI) by a low-cost adsorbent: biogas residual slurry. *Chemosphere*, 30, 561, 1995.
22. Iqbal, M., Saed, A., Akhtar, N., Petiolar felt-sheath of palm: a new biosorbent for the removal of heavy metals from contaminated water. *Bioresour. Technol.*,81, 151, 2002.
23. D., Kang, X., Yu, M., Ge, F., Xiao, H., Xu, Novel Al-doped carbon nanotubes with adsorption and coagulation promotion for organic pollutant removal. *J. Environ. Sci.*, 39, 1, 2016.
24. Mohapatra P., Samantaray, S.K., Parida, K.M., Photocatalytic reduction of hexavalent chromium in aqueous solutionover sulphate modified titania. *J. Photochem. Photobiol. A: Chem.*, 170, 189, 2005.
25. Das, D.P., Parida, K.M., De, B.R., Photocatalytic reduction of hexavalent chromium in aqueous solution over titania pillard zirconium phosphate and titanium phosphate under solar radiation. *J. Mol. Catal. A: Chem.*, 245, 217, 2006.
26. Parida, K.M., Sahu, N., Visible light induced photocatalytic activity of rare earth titania nanocomposites. *J. Mol. Catal. A: Chem.*, 287, 151, 2008.
27. Das, J., Das, D., Dash, G.P., Das, D.P., Parida, K., Studies on Mg/Fe Hydrotalcite – like – compound (Htlc): removal of chromium (VI) from aqueous solution. *Int. J. Environ. Studies*, 61, 605, 2004.
28. Mallick, S., Dash, S.S., Parida, K.M., Adsorption of hexavalent chromium on manganese nodule leached residue obtained from NH3–SO2 leaching. *J. Colloid Interf. Sci.*, 297, 419, 2006.
29. Parida, K.M., Mishra, K.G., Dash, S.K., Adsorption of toxic metal ion Cr (VI) from aqueous state by TiO2- MCM-41: equilibrium and kinetic studies. *J. Hazard. Mater.*, 241–242, 395, 2012.

30. Yang, J., Yu, M., Chen, W., Adsorption of hexavalent chromium from aqueous solution by activated carbon prepared from longan seed: Kinetics, equilibrium and thermodynamics. *J. Indust. Eng. Chem.*, 21, 414, 2015.
31. Kalidhasan, S., Kumar, A.S.K., Rajesh, V., Rajesh, N., The journey traversed in the remediation of hexavalent chromium and the road ahead toward greener alternatives—a perspective. *Coord. Chem. Rev.*, 317, 157, 2016.
32. Shawabkeh, R.A., Adsorption of chromium ions from aqueous solution by using activated carbo-aluminosilicate material from oil shale. *J. Colloid Interf. Sci.*, 299, 530, 2006.
33. Sharma, A., Bhattacharya, K.G., Adsorption of chromium (VI) on Azadirachta Indica (Neem) leaf powder. *Adsorption*, 10, 327, 2004.
34. Mohan, D., Pittman Jr., C.U., Activated carbon and low cost adsorbents for remediation of tri- and hexavalent chromium from water. *J. Hazard. Matter.*, 137, 762, 2006.
35. Bansal, M., Singh, D., Garg, V.K., A comparative study for the removal of hexavalent chromium from aqueous solution by agriculture wastes' carbons, *J. Hazard. Mater.*, 171, 83, 2009.
36. Wahab, M.A., Jellali, S., Jedidi, N., Ammonium biosorption onto sawdust: FTIR analysis, kinetics and adsorption isotherms modeling. *Bioresour. Technol.*, 101, 5070, 2010.
37. Wang, X.S., Chen, L.F., Li, F.Y., Chen, K.L., Wan, W.Y., Tang, Y.J., Removal of Cr (VI) with wheat-residue derived black carbon: reaction mechanism and adsorption performance. *J. Hazard. Mater.*, 175, 816, 2010.
38. Langmuir, I., The adsorption of gases on plane surfaces of glass, mica and platinum. *J. Am. Chem. Soc.*, 40, 136, 1918.
39. Kadirvelu, K., Thamaraiselvi, K., Namasivayam, C., Adsorption of nickel (II) from aqueous solution onto activated carbon prepared from coirpith. *Sep. Purif. Technol.*, 24, 497, 2001.
40. Jain, M., Garg, V.K., Kadirvelu, K., Chromium (VI) removal from aqueous system using Helianthus annuus (sunflower) stem waste. *J. Hazard. Mater.*, 162, 365, 2009.
41. Hall, K.E., Eagleton, L.C., Acrivos, A., Vermeulen, T., Pore- and solid-diffusion kinetics in fixed-bed adsorption under constant-pattern conditions. *Ind. Eng. Chem. Fund.*, 5, 212, 1966.
42. Hasany, S.M., Chaudhary, M.H., Sorption potential of Haro river sand for the removal of antimony from acidic aqueous solution. *Appl. Radiat. Isotopes*, 47, 467, 1996.
43. Ozcan, A., Oncu, E.M., Ozcan, A.S., Kinetics, isotherm and thermodynamic studies of adsorption of Acid Blue 193 from aqueous solutions onto natural sepiolite. *Colloid Surf. A*, 277, 90, 2006.
44. Onyango, M.S., Kojima, Y., Aoyi, O., Bernado, E.C., Matsuda, H., Adsorption equilibrium modeling and solution chemistry dependence of fluoride removal from water by trivalent-cation-exchanged zeolite F-9. *J. Colloid Interf. Sci.*, 279, 341, 2004.

45. Salvestrini, S., Leone, V., Iovino, P., Canzano, S., Capasso, S., Considerations about the correct evaluation of sorption thermodynamic parameters from equilibrium isotherms. *J. Chem. Thermodynamics,* 68, 310–316, 2014.
46. Khan, S.A., Rehman, R., Khan, M.A., Adsorption of chromium (III), chromium (VI) and silver (I) on bentonite, *Waste Manage.,* 15, 271, 1995.
47. S. Lagergren, Zur theorie der sogenannten adsorption geloster stoffe, kunglingasvenska vetenskapsakademiens. *Handlingar,* 24, 1, 1898.
48. Ho, Y.S., Porter, J.F., McKay, G.A., Equilibrium isotherm studies for the sorption of divalent metal ions onto peat: copper, nickel and lead single component systems. *Water Air Soil Pollut.,* 141, 1, 2002.
49. Weng, C.H., Sharma, Y.C., Chu, S.H., Adsorption of Cr (VI) from aqueous solutions by spent activated clay. *J. Hazard. Matter.* 155, 65, 2008.
50. Hamadi, N.K., Chen, X.D., Farid, M.M., Lu, M.G.Q., Adsorption kinetics for the removal of chromium(VI) from aqueous solution by adsorbents derived from used tyres and sawdust. *Chem. Eng. J.,* 84, 95, 2001.
51. Hu, J., Chen, G., Lo, I.M.C., Removal and recovery of Cr(VI) from wastewater by maghemite nanoparticles. *Water Res.,* 39, 4528, 2005.
52. Nityanandi, D., Subbhuraam, C.V., Kinetics and thermodynamic of adsorption of chromium (VI) from aqueous solution using puresorbe. *J. Hazard. Mater.,* 170, 876, 2009.
53. Sharma, Y.C., Cr (VI) removal from industrial effluents by adsorption on an indigenous low cost material. *Colloid Surf. A: Physicochem. Eng. Asp.,* 215, 155, 2003.
54. Fritzen, M.B., Souza, A.L., Silva, T.A.G., Souza, L., Nome, R.A., Fiedler, H.D., Nome, F., Distribution of hexavalent Cr species across the clay mineral surface-water interface. *J. Colloid Interf. Sci.,* 296, 465, 2006.
55. Guerra, L., Oliveira, H.C.P., Corrêa da Costa, P.C., Viana, R.R., Airoldi, C., Adsorption of chromium (VI) ions on Brazilian smectite: effect of contact time, pH, concentration and calorimetric investigation. *Catena,* 82, 35, 2010.
56. Taylor, R.W., Shen, S., Bleam, W.F., Tu, S., Chromate removal by dithionite - reduced clays: Evidence from direct X-Ray adsorption near edge spectroscopy (XANES) of chromate reduction at clay surfaces. *Clays Clay Miner.,* 48, 648, 2000.
57. Alemayehu, E., Thiele-Bruhn, S., Lennartz, B., Adsorption behaviour of Cr (VI) onto macro and micro-vesicular volcanic rocks from water. *Sep. Purif. Technol.,* 78, 55, 2011.
58. Sharma, Y.C., Effect of temperature on interfacial adsorption of Cr (VI) on wolastonite. *J. Colloid Interf. Sci.,* 233, 265, 2001.
59. Akar, S.T., Yetimoglu, Y., Gedikbey, T., Removal of Cr (VI) ions from aqueous solution by using Turkish montmorillonite clay: effect of activation and modification. *Desalination,* 244, 97, 2009.
60. Zhao, Y., Yang, S., Ding, D., Chen, J., Yang, Y., Lei, Z., Feng, C., Zhang, Z., Effective adsorption of Cr (VI) from aqueous solution using natural Akadama clay. *J. Colloid Interf. Sci.,* 395, 198, 2013.

61. Eary, L.E., Rai, D., Kinetics of chromate reduction by ferrous ions derived from hematite and biotite at 25°C. *Am. J. Sci.*, 289, 180, 1989.
62. Brigatti, M.F., Lugli, C., Cibin, G., Marcelli, A., Giuli, G., Paris, E., Mottana, A., Wu, Z., Reduction and Sorption of Chromium by Fe(II)-Bearing Phyllosilicates: Chemical treatments and x-ray absorption spectroscopy (xas) studies. *Clays Clay Miner.*, 48, 272, 2000.
63. Bolan, N.S., Adriano, D.C., Natesan, R., Koo, B.J., Effects of organic amendments on the reduction and phytoavailability of chromate in mineral soil. *J. Environ. Qual.*, 32, 120, 2003.
64. Parthasarathy, G., Choudary, B.M., Sreedhar, B., Kunwar, A.C., Srinivasan, R., Ferrous saponite from the Deccan Trap, India and its application in adsorption and reduction of hexavalent chromium. *Am. Miner.*, 88, 1983, 2003.
65. Li, Y., Yue, Q., Gao, B., Effect of humic acid on the Cr (VI) adsorption onto Kaolin. *Appl. Clay Sci.*, 48, 481, 2010.
66. Krishna, B.S., Murty, D.S.R., Jai Prakash, B.S., Surfactant- modified clay as adsorbent for chromate. *Appl. Clay Sci.*, 20, 65, 2001.
67. Li, Z., Alessi, D.S., Zhang, P., Bowman, R.S., Organo-illite as a low permeability sorbent to retard migration of anionic contaminants. *J. Environ. Eng.*, 128, 583, 2002.
68. Ji, M., Su, X., Zhao, Y., Qi, W., Wang, Y., Chen, G., Zhang, Z., Effective adsorption of Cr (VI) on mesoporous Fe functionalized Akadama clay: optimization, selectivity and mechanism. *Appl. Surface Sci.*, 344, 128, 2015.
69. Zhoua, J., Wu, P., Dang, Z., Zhu, N., Li, P., Wu, J., Wang, X., Polymeric Fe/Zr pillared montmorillonite for the removal of Cr(VI) from aqueous solutions. *Chem. Eng. J.*, 162, 1035, 2010.
70. Deng, L., Shi, Z., Li, B., Yang, L., Luo, L., Yang, X., Adsorption of Cr (VI) and phosphate on Mg–Al hydrotalcite supported kaolin clay prepared by ultrasound-assisted coprecipitation method using batch and fixed-bed systems. *Ind. Eng. Chem. Res.*, 53, 7746, 2014.
71. Benhammou, A., Yaacoubi, A., Nibou, L., Tanouti, B., Chromium (VI) adsorption from aqueous solution onto Moroccan Al- pillared and cationic surfactant stevensite. *J. Hazard. Mater.*, 140, 104, 2007.
72. Li, Z., Willms, C.A., Kniola, K., Removal of anionic contaminants using surfactant–modified palygorskite and sepiolite. *Clays Clay Miner.*, 51, 445, 2003.
73. Krishna, B.S., Murty, D.S.R., Jai Prakash, B.S., Thermodynamics of chromium (VI) anionic species sorption onto surfactant–modified montmorillonite. *J. Colloid Interf. Sci.*, 229, 230, 2000.
74. Hu, B., Luo, H., Adsorption of hexavalent chromium onto montmorillonite modified with hydroxyaluminum and cetyltrimethylammonium bromide. *Appl. Surf. Sci.*, 257, 769, 2010.
75. Benhammou, A., Yaacoubi, A., Nibou, L., Tanouti, B., Study of the removal of mercury (II) and chromium (VI) from aqueous solutions by Moroccan stevensite. *J. Hazard. Mater.*, 117, 243, 2005.

76. Majdan, M., Maryuk, O., Pikus, S., Olszewska, E., Kwiatkowski, R., Skrzypek, H., Equilibrium, FTIR, scanning electron microscopy and small wide angle X-ray scattering studies of chromates adsorption on modified bentonite. *J. Mol. Struct.*, 740, 203, 2005.
77. Huang, Y., Ma, X., Liang, G., Yan, Y., Wang, S., Adsorption behavior of Cr (VI) on organic- modified rectorite. *Chem. Eng. J.*, 138, 187, 2008.
78. Gładysz-Płaska, A., Majdan, M., Pikus, S., Sternik, D., Simultaneous adsorption of chromium (VI) and phenol on natural red clay modified by HDTMA. *Chem. Eng. J.*, 179, 140, 2012.
79. Leyva-Ramos, R., Jacobo-Azuara, A., Diaz-Flores, P.E., Guerrero-Coronado, R.M., Mendoza-Barron, J., Berber-Mendoza, M.S., Adsorption of chromium (VI) from an aqueous solution on a surfactant-modified zeolite. *Colloids Surf. A: Physicochem. Eng. Asp.*, 330, 35, 2008.
80. Maryuk, O., Pikus, S., Olszewska, E., Majdan, M., Skrzypek, H., Zieba, E., Benzyldimethyloctadecylammonium bentonite in chromates adsorption. *Mater. Let.*, 59, 2015, 2005.
81. Sarkar, B., Xi,Y., Megharaj, M., Krishnamurti, G. S.R., Rajarathnam, D., Naidu, R., Remediation of hexavalent chromium through adsorption by bentonite based Arquad® 2HT-75 organoclays. *J. Hazard. Mater.*, 183, 87, 2010.
82. Brum, M.C., Capitaneo, J.L., Oliveira, J.F., Removal of hexavalent chromium from water by adsorption onto surfactant modified montmorillonite. *Miner. Eng.*, 23, 270, 2010.
83. Li, W., Tang, Y., Zeng, Y., Tong, Z., Liang, D., Cui, W., Adsorption behavior of Cr(VI) ions on tannin-immobilized activated clay. *Chem. Eng. J.*, 193–194, 88, 2012.
84. Marjanović, V., Lazarević, S., Janković-Častvan, I., Jokić, B., Janaćković, Dj., Petrović, R., Adsorption of chromium (VI) from aqueous solutions onto amine-functionalized natural and acid-activated sepiolites. *Appl. Clay Sci.*, 80–81, 210, 2013.
85. Gonzalez, M.H., Araujo, G.C.L., Pelizaro, C.B., Menezes, E.A., Lemos, S.G., de Sousa, G.B., Nogueira, A.R.A., Coconut coir as biosorbent for Cr(VI) removal from laboratory wastewater. *J. Hazard. Mater.*, 159, 252, 2008.
86. Suksabye, P., Thiravetyan, P., Cr (VI) adsorption from electroplating wastewater by chemically modified coir pith. *J. Environ. Manage.*, 102, 1, 2012.
87. Wang, X.S., Tang, Y.P., Sheng, R.T., Kinetics, equilibrium and thermo dynamics study on removal of Cr (VI) from aqueous solutions using low cost adsorbent Alligator weed. *Chem. Eng. J.*, 148, 217, 2008.
88. Dehghani, M.H., Sanaei, D., Ali, I., Bhatnagar, A., Removal of chromium (VI) from aqueous solution using treated waste newspaper as a low-cost adsorbent: Kinetic modeling and isotherm studies. *J. Molecular Liq.*, 215, 671, 2016.
89. Verma, A., Chakraborty, S., Basu, J.K., Adsorption study of hexavalent chromium using tamarind hull-based adsorbents. *Sep. Purif. Technol.*, 50, 336, 2006.

90. Moussavi, G., Barikbin, B., Biosorption of chromium (VI) from industrial waste water onto pistachio hull waste biomass. *Chem. Eng. J.*, 162, 893, 2010.
91. Kapur, M., Mondal, M.K., Mass transfer and related phenomena for Cr (VI) adsorption from aqueous solutions onto Mangifera indica sawdust. *Chem. Eng. J.*, 218, 138, 2013.
92. Gupta, S., Babu, B.V., Removal of toxic metal Cr (VI) from aqueous solutions using sawdust as adsorbent: Equilibrium, kinetics and regeneration studies. *Chem. Eng. J.*, 150, 352, 2009.
93. Saha, R., Saha, B., Removal of hexavalent chromium from contaminated water by adsorption using mango leaves (Mangifera indica). *Desalin. Water Treat.*, 52, 1928, 2014.
94. Singha, B., Das, S.K., Biosorption of Cr (VI) ions from aqueous solutions: Kinetics, equilibrium, thermodynamics and desorption studies. *Colloids Surf. B: Biointerfaces*, 84, 221, 2011.
95. Lopez-Garcia, M., Lodeiro, P., Herreroa, R., Barriada, J.L., Rey-Castro, C., David, C., de Vicente, M.E.S. de Vicente, Experimental evidences for a new model in the description of the adsorption-coupled reduction of Cr(VI) by protonated banana skin. *Bioresour. Technol.*, 139, 181, 2013.
96. Kumar, S., Meikap, B.C., Removal of chromium (VI) from waste water by using adsorbent prepared from green coconut shell, *Desalin. Water Treat.*, 52, 3122, 2014.
97. Rao, R.A.K., Ikram, S., Uddin, M.K., Removal of Cr (VI) from aqueous solution on seeds of Artimisia absinthium (novel plant material). *Desalin. Water Treat.*, 54, 3358, 2015.
98. Babu, B.V., Gupta, S., Adsorption of Cr (VI) using activated neem leaves: Kinetic studies. *Adsorption*, 14, 85, 2008.
99. Zhong, Q.Q., Yue, Q.Y., Gao, B.Y., Li, Q., Xu, X., A novel amphoteric adsorbent derived from biomass materials: synthesis and adsorption for Cu (II)/Cr (VI) in single and binary systems. *Chem. Eng. J.*, 229, 90, 2013.
100. Shen, Y.S., Wang, S.L., Tzou, Y.M., Yan, Y.Y., Kuan, W.H., Removal of hexavalent Cr by coconut coir and derived chars – the effect of surface functionality. *Bioresour. Technol.*, 104, 165, 2012.
101. Escudero, C., Fiol, N., Villaescusa, I., Bollinger, J.C., Effect of chromium speciation on its sorption mechanism onto grape stalks entrapped into alginate beads. *Arab. J. Chem.*, 10, S1293, 2013.
102. Sillerova, H., Chrastny, V., Kovacs, E., Komarek, M., Isotope fractionation and spectroscopic analysis as an evidence of Cr (VI) reduction during biosorption. *Chemosphere*, 95, 402, 2014.
103. Wen, Y., Tang, Z., Chen, Y., Gu, Y., Adsorption of Cr (VI) from aqueous solutions using chitosan-coated fly ash composite as biosorbent. *Chem. Eng. J.*, 175, 110, 2011.
104. Copello, G.J., Varela, F., Vivot, R.M., Díaz, L.E., Immobilized chitosan as biosorbent for the removal of Cd (II), Cr (III) and Cr (VI) from aqueous solutions. *Bioresour. Technol.*, 99, 6538, 2008.

105. Wu, Z., Li, S., Wan, J., Wang, Y., Cr (VI) adsorption on an improved synthesised cross-linked chitosan resin. *J. Molec. Liq.*, 170, 25, 2012.
106. Asl, S.H., Ahmadi, M., Ghiasvand, M., Tardast, A., Katal, R., Artificial neural network (ANN) approach for modeling of Cr(VI) adsorption from aqueous solution by zeolite prepared from raw fly ash (ZFA). *J. Indust. Eng. Chem.*, 19, 1044, 2013.
107. Gupta, V.K., Ali, I., Removal of lead and chromium from wastewater using bagasse fly ash– a sugar industry waste, *J. Colloid Interf. Sci.*, 271, 321, 2004.
108. Gupta, V.K., Rastogi, A., Nayak, A., Adsorption studies on the removal of hexavalent chromium from aqueous solution using a low cost fertilizer industry waste material. *J. Colloid Interf. Sci.*, 342, 135, 2010.
109. Selvaraj, K., Manonmani, S., Pattabhi, S., Removal of hexavalent chromium using distillery sludge. *Bioresour. Technol.*, 89, 207, 2003.
110. Gupta, V.K., Gupta, M., Sharma, S., Process development for the removal of lead and chromium from aqueous solutions using red mud an aluminum industry waste. *Wat. Res.*, 35, 1125, 2001.
111. Ma, M., Lu, Y., Chen, R., Ma, L., Wang, Y., Hexavalent chromium removal from water using heat-acid activated red mud. *Open J. Appl. Sci.*, 4, 275, 2014.

6

Microbial Diversity as a Tool for Wastewater Treatment

Sadia Ilyas and Haq Nawaz Bhatti*

Enviromental Chemistry laboratory, Department of Chemistry, University of Agriculture, Faisalabad, Pakistan

Abstract
Diversity of microorganisms in wastewater performs different roles and has different operational conditions for their optimal activity and growth. The comprehensive understanding of adequate and active microbial population and their biological role is essential to degrade organic wastes with insignificant dissolved oxygen demand and for limited growth of photosynthetic organisms. The diversity of microbial community, the role they perform in wastewater treatment, their optimal growth condition and substrate utilization by aerobic and anaerobic mode are described in this chapter. Furthermore, the futuristic prospects and change in community of every treatment are also highlighted.

Keywords: Water pollution, activated sludge, microbial diversity, aerobic degradation, anaerobic degradation

6.1 Overview of Wastewater; Sources, Pollutants, and Characteristics

Water pollution may be known as the use of water as a resource by impairmention of human activities. Water pollution due to industrial, agricultural and municipal waste has a serious problem for human being and it is increased due to explodation of population. The main sources of effluent water are electroplating industry, release of textile industry, miscellaneous, and paper and pulp industry [1].

**Corresponding author*: hnbhatti2005@yahoo.com, haq_nawaz@uaf.edu.pk

Shahid-ul-Islam (ed.) Advanced Materials for Wastewater Treatment, (171–222)

The wastes of textile industry relates with process of textile manufacturing and the use of fibre for textile. Bisulfite, chlorine, peroxides, hypochlorites, soda ash, sodium hydroxide, sodium silicate, sodium softeners, starch, polyvinyl alcohol, fats, waxes, greases, vegetable oil, resins, phenolics, chromium, organic soaps and alkalis are major components of textile industry waste [2]. Pulp and paper industry effluent contains different chemicals like NaOH, Na_2CO_3, MgSO3, Ca(HSO3)2, $A1_2(SO_4)_3$, $CaOCl_2$, $BaSO_4$, Na_2S, $BaSO_4$, $C1O_2$, TiO_2, $CaCO_3$. The waste of electroplating contains sulfuric acid, hydrochloric acid, nitric acid, along different metals like silver, cobalt, arsenic, selenium, molybdenum, iron, lead, tin, chromium, cadmimium, zinc, gold, nickel, and copper [1].

There are five major classes of pollutants that play a major role in water pollution, for example, inorganic pollutants, suspended solids, radioactive pollutants, nutrients, thermal pollutants, sediments, and pathogen. Organic pollutants are divided into: sewage, oxygen-demanding wastes, agricultural runoff, oil pollutant, disease-causing wastes and synthetic organic compounds [3, 4]. Domestic waste, biodegradable organic compounds, meat-packing plants, tanneries, pulp and paper mills, industrial wastes from food processing plants, slaughter houses, animal sewage, and agriculture runoff are oxygen demanding effluents [4]. In the presence of dissolved oxygen these waste are breakdown due to activities of microorganisms. Due to this process the dissolved oxygen is decreased rapidly in water and that is very dangerous to water body organisms 4–6 ppm is the optimum value of dissolved oxygen in water that favors the aquatic life. The reduction in dissolved oxygen of water leads to water pollution caused by oxygen demanding wastes that are already mentioned. At low level of dissolved oxygen in water the water bodies cannot retain in water [3].

Pathogenic microorganisms that cause diseases present in water with sewage and effluent water that are very dangerous to human health. These pathogens are different types of bacteria and viruses that are very harmful and cause severe water-borne diseases like polio, dysentery, hepatitis, cholera, and typhoid. Thus, purification and sanitization are main process to remove water pollution [1, 2].

Plasticizers, synthetic fibers, detergents, pesticides, insecticides, food additives, solvents, plastics, industrial chemicals, and pharmaceuticals are synthetic organic compounds that are made by man. All of these act as water effluent and enter into water table due to sewage, leakage during use it may be accidently or intentionally from industrialist [3].

Many of these chemical act as effluent are very hazardous to humans, animals and plants. The trace amount of bio-refractory organics just like aromatic chlorinated hydrocarbons produces change in odor, color, and taste of water and make it unacceptable for use. The chemicals like alkyl benzene sulfonate from synthetic detergents do not easily degrade and become persistant. Alcohols, gasoline, ethers, and aldehydes are volatile substances and causes explosion in sewers [5].

The growth of aquatic weeds and algae becomes stimulated by the sewage and run-off of water supply from agricultural lands. The excess growth of plants destroys water bodies, required for recreation and other uses. Due to eutrophication the dissolved oxygen of water is lost and thus, become a dead pool [4].

The likage from oil pipe lines, oil spill production from oil tankers, losses during off shore exploration, oil spills due to oil tankers on the sea or oil tanker accidents like in the Gulf War between Iraq and U.S.-led allied forces in the year 1991 and crossing waterways causes oil pollution [3].

The photosynthesis of marine plants is affected most, due to oil pollution light cannot be transmitted through the surface. The water bodies' plants, animals, and water birds become endangered due to oil pollution. Thus, oil pollution causes to bring marine life and water plants to delete rate. Due to increased oil-based technologies, international oil slicks during international hostilities, accidental oil spillage oil pollution has been increased in seas in recent years [1].

All of the metalic compounds, sulfates, cyanides, nitrates, trace elements, inorganic salts, mineral acids and complexes of metals with organics, organometallic compounds are the inorganic pollutants in natural source of water. Metal involve with some organic compounds like fulvic acids and EDTA synthetic organic species [6].

Metal having interaction with organic compounds are influenced due to acid-base reaction, redox reaction, formation of colloid and reaction by microorganisms in water. Due to metal and organic compound intraction the toxicity of metals and growth of algae are also increased in water [1]. Due to human activities many of the metal and metallic compounds are released into water natural resources. But in biological process trace of metals plays a vital role but it lead to toxicity in high concentration. Heavy metals like Cd, Pb Hg, Sb, Se, As, and some of metalloid are considered the most toxic and dangerous to human health. In most enzymes –SH bonds are attacked by most of the heavy metals and unable to move because heavy metals have strong affinity for sulfur. Some of macromolecules like carboxlic acid, amino group, and protein are also interacted with heavy metals [7].

Across the cell wall heavy metals are associated with the cell membrane and interfere with ETC (electron transport chain). Heavy metals also lead to enchance the breakdown and reduce the phosphate biocompounds. Due to street dust, industrial waste and domestic sewage heavy metals are also diluted in water resources. The most common water pollutants are algal nutrients and phosphates of detergents [3, 4].

In natural process of soil erosion, mining and development in agricultural science tends to increase sediments. Due to motion of water some of the solid particles are remain in water that act as solid suspend or colloids. To check the water quality they act as indicator [4].

They are abbreviate as SS but don't be confuse with it because sometime settleable solids also refer as SS. The growth of disease-causing microorganisms also enhances due to suspended solid particles that also cause unpleasant taste and smell. Over a period of time at the bottom of aquatic system some of solid suspended particles are placed. Some large size particles like sand, gravel, etc., are also settle down where little or no flow of water take place. Thus, this type of settling leads to clarity of water from larger objects. But the remaining non-settleable particles are known as colloidal. They are too large or too small that not be able to settle down on the bottom. While settleable-suspended solids are known as bedload or bedded sediments [4, 7]. These sediments are classified from small particle of clay and slit to larger particles of gravel and sand on the basis of rate of flow of water. However some of these sediments can move without reunion of solid suspended particles. By a strong flow of water some of settleable solids are move with bottom water body and it is known as bedload transport [7].

Effluent from industries such as slaughter houses, and drainage system are main source through which disease-causing microorganisms pathogens entered into water. Some fetal water-borne disease like dysentery, polio, cholera, hepatitis, and typhoid are caused by bacteria and viruses. Nitrogen and phosphrous that are substantial amount of nutrients and comes from wastewater of fertilizer industry, sewage, and agricultural runoff [8].

All of these nutrients from wastewater are provide to plants that stimulate the algal growth and other aquatic plants. Thus the level of water body is low. For a long time, water body decreased the dissolved oxygen that leads to eutrophication and results as deed pool of water. In this dead pool of water some people are swimming that leads to eye infections, skin allergy, vomiting, gastroenteritis due to growth of blue-green algae. There are high potential risks to children especially under six months baby due to high concentration of nitrogen in water. Due to increase amoun of nitrogen in water leads to methaemoglobin that involves decrease in oxygen carrying capacity of blood (known as blue baby disease) because NO^{-}_{3} rapidly oxidized Fe^{2+} in blood [1].

Due to discharge of hot water from nuclear power plants, thermal power plants and industries where water act as coolant are main cause of thermal pollution. This is due to Decrease the dissolved oxygen in water by increasing the temperature of water bodies and having bad effect on water bodies. The organisms can adjust their self at dissolve oxygen level and temperature. If organic matter is also present in water along high temperature the bacterial activities increase and lead to decrease the dissolved oxygen level. Stratification in water body is due to hot water because hot water remain at the surface [8].

Due to processing of ores from mines for industrial activites, agriculture, research and medical radioactive matter produced like C^{14}, Ca^{45}, P^{32}, S^{35}, Co^{60}, I^{131}, etc. From nuclear reactors and nuclear plants the main sources of radioactive discharge are Uranium-235, Uranium-238, Cesium Cs^{137}, Plutonium Pu^{248} Sr^{90} and used in nuclear weapons. In water if amount of pollutant is not less than the accepted value then essential step should be taken to reduce and remove these pollutants from water with the help of some sophisticated treatments like chemical precipitation, sedimentation, biological oxidation, adsorption by activated carbon, and filteration [7].

On the basis of various chemical, biological and physical parameters the pollutant of wastewater are characterized. Temperature, radioactivity, color, smell, dissolved oxygen, corrosive properties and insoluble substance include in physical parameters. While chemical oxygen demand, TDS (total dissolved substance), pH (alkanity and acidity), hardness, chemicals demand like chlorine, Some of inorganic and organic components like hydrocarbon oils, phenols, surfactants and greases included in chemical parameters. In biological parameters presence of disease-causing bacteria (pathogens), hazardous to human, plants and water bodies, biological oxygen demand (BOD) are included [1].

6.1.1 Biodiversity of Wastewater Plants

The biodiversity of wastewater treatment plants as a prokaryotes or eukaryotes is classified depending upon the structure and functioning of its cell wall. In eukaryotes membrane bounded cell organelles and well defined nucleus is present which is absent in prokaryotes (Figure 6.1) [8–15]. Figure 6.1 shows the difference between two cells. In prokaryotes (1) true bacteria or eubacteria present while (2) archaebacterial or ancient bacteria. In biological wastewater treatment plants both of these microbes eubacteria and archaebacterial are most important. Combinly both can be called as bacteria.

Figure 6.2 represents the life cycle of a bacterial cell having lag phase, log phase, stationary phase and dead phase.The log phase is bacially the logrithmic or exponential growth phase because growth of bacterial cell occur in this phase.The log phase is comprized of three basic portions that are: (1)

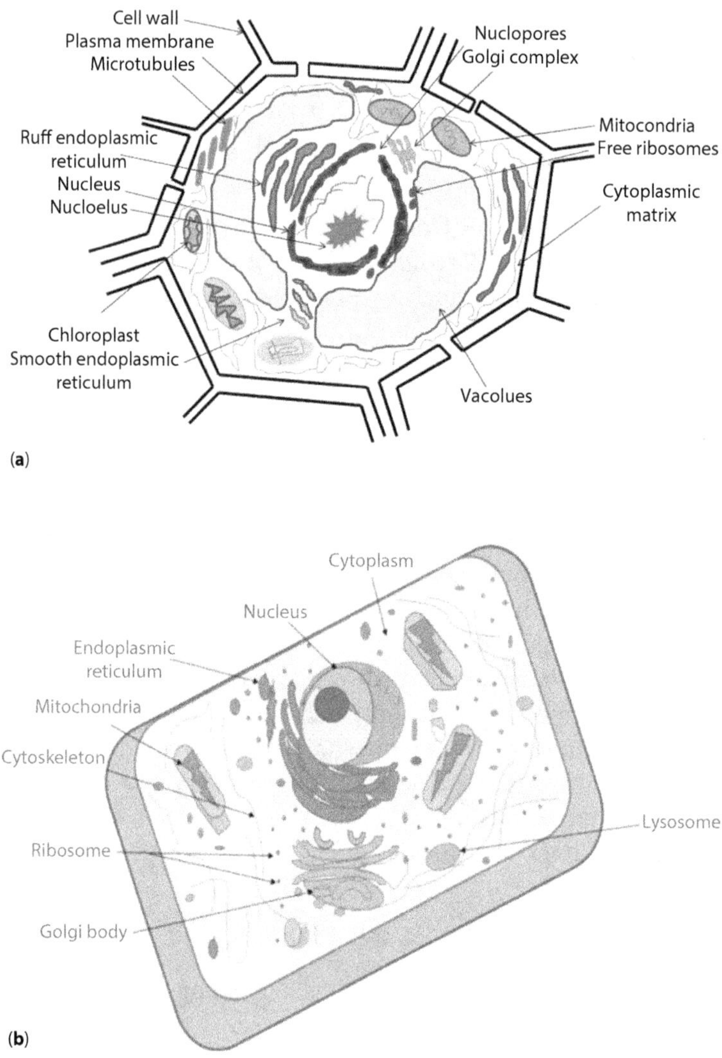

Figure 6.1 (a) prokaryotic cells and (b) Eukaryotic (E.M.Armstrong, 2001).

uptake of substrate, (2) cell synthesis and fast growth rate, and (3) cell synthesis and decline phase growth [16]. During first step due to uptake of substrate the bacterial cell size increases due to adsorption or absorption of substances. Rapid growth of bacterial cell is obtained during this phase of cycle. During third phase the bacterial cell use all the available substrate and decline phase starts. The gradual leveling off of the growth curve represents the decrease in growth and leads toward the stationary phase of

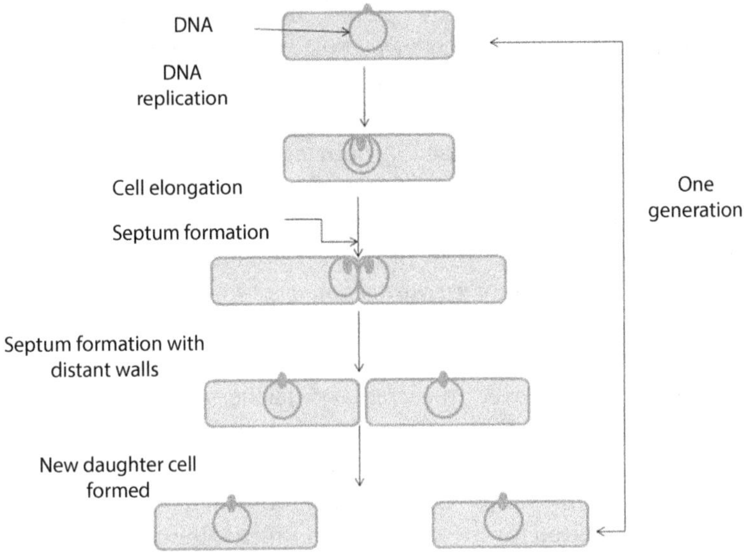

Figure 6.2 Binary fission and cell separation.

growth. The biological treatment unit has been reached to the stationary phase of growth which has carried the maximum cells of bacteria [17]. The growth of bacteria is affected due to (1) lack of ever incresing amount of substrate, (2) lack of free electron acceptors like oxygen or nitrates, and (3) metabolic waste production and accumulation. During the stationary phase the cell number is balanced by the same rate of growth and death.

The death rate of bacterial cell becomes increased as compare to growth during the decline phase of growth. The slope of decline phase of growth is more prominent toward the death rate as compare to growth in batch culture when compared with negative slope of continuous culture. Within the batch culture limited quantity of substrate is added as compared to continuous culture which always receives some quantity of substrate [18]. So, the death rate in batch phase is more as compare to continuous culture. But the amount of substrate received by continuous culture varies in the day and week. Thus the slope of decline phase changes according the mount of substrate added [17, 18]. The reproduction in bacteria occurs by the process of binary fission in which when a cell becomes double in size after growth known as mother cell divides into two cells known as daughter cells or offspring (Figure 6.2).

Due to the incomplete isolation of daughter cells of some bacteria specific arrangements are obtained like they exist as tetrads or filaments [17]. Mostly, the eukaryotic microbes like protozoa, nematodes, fungi, and rotifers are present in wastewater plants. These eukaryotic non-harmful microbes enter in the body of wastewater plants bodies either by infiltration or by inflow in the plant body [10, 15].

The saprophytic microorganism is mostly the fungi and is classified by their reproduction modes. These saprophytes obtain their food from dead bodies and most of them are free living just like mushrooms, yeast, and molds. Those fungi which are aerobic can tolerate low pH and nutrient environment. Mostly the growth of fungi is obtained in the range of pH 2–9 but at pH 5.6 maximum growth is obtained so, taken as the optimum pH and the requirement of nutrients to fungi is half than required by bacteria [19]. The filamentous fungi can attribute to settle ability and proliferate the problems occurred in secondary clarifiers, during the activated sludge process. The filamentous fungi having proliferation requires less nutrients and work at pH of less than 5.6 in activated sludge process the settle ability problems are solved by filamentous fungi. So a large population of fungi is required for the treatment of organic industrial wastewater. Fungi can grow in low pH, low nutrients amount, less moisture conditions and have the ability to degrade cellulose. *Saccharomyces* is the example of unicellular fungi. The reproduction in *Saccharomyces* occurs by budding in which many of the daughter cells can produce from one mother cell. Yeast produces carbon dioxide and water by degradation of organic compounds and some facultative anaerobes have ability to degrade sugars into ethanol in the absence of molecular oxygen.

Protozoa are the single cell organisms. Mostly fungi exist as single cell but some also exist in groups (colonies) they are mostly aerobic but some exceptions like flagellates and amoebae can tolerate an aerobic environment [10, 20]. Five-group system is most commonly used for arranging protozoa into different categories depending upon their mode of locomotion in activated sludge process like crawling silicates, stalked ciliates, free swimming ciliates, amoebae and flagellates [9, 15, 20]. But most widely used protozoa in activated sludge process are ciliated protozoa. These protozoa have small hair like cilia which move in one direction to force water back and also for moving food (bacteria) into the mouth. For betterment in settle ability ciliated protozoa purify the wastewater, add weight to floc particles and use wide spread cells from the top of floc and regenerate the nutrients like phosphorus and nitrogen [21].

Most of the multicellular organisms like nematodes and rotifers are also advantageous in activated sludge process. The metazoan are the protozoa which can penetrate in floc particles and help in degradation of substances present inside the floc particles by adding oxygen, nutrients and nitrates into the body. Bacteria obtain carbon and other energy sources from substrate for its metabolic activity but in some cases cBOD and nBOD are also required for proper functioning [9, 10, 12, 15, 20].

6.2 Role of Dominant Wastewater Treatment Communities in Biodegradation

6.2.1 Hydrolytic Microbial Community

In biological treatment section, a large number and various population of bacteria and its enzymes are important to degenerate the enormous amount and diversification of substrates due to various class of bacteria regenerate at various rates [22]. For suitable biological action a treatment portion like as solids retention time (SRT) or mean cell residence time (MCRT) should be set for growing the needed bacterial population and its suitable enzymes. The three steps involved in biological action such as in first step the degeneration of specific substrate, in second step floc generation and in last step elimination of particular pollutants like as phosphorus (P) [23] The mean cell residence time (MCRT) is also be able to set for avoid the degeneration of a particular substrate, for example, a activated sludge operation of a MCRT can be set to avoid nitrification and improve biological phosphorus elimination. For appropriate biological actions of microbes, the mean cell residence time or solids population time is essential and is effected by temperature. When increase the temperature as a result the bacterial activity or biomass in a biological treatment section is also increase so, little number of bacteria or amount of biomass is needed and it also enables to decrease in mean cell residence or solids retention time. When decrease the temperature as a result the biological activity or amount of biomass are also increase in biological treatment section so, greater number of bacteria or amount of biomass is needed and it permits a increase in mean cell residence time or solid retention time [21]. Bacteria in dispersed form like as a suspended growth anaerobic digester or flocculated form like as activated sludge method are present frequently in biological treatment section.

A substrate is adsorbed or adsorbed by bacteria in biological treatment sections. Absorbed substrate is define as which are simple in structure, rapidly enter bacterial cell and also soluble. For example, absorded substrates or soluble cBOD involve ethanol (CH_3CH_2 OH), glucose ($C_6H_{12}O_6$) acetone (CH_3COOH). These absorded substrate are readily degenerate by using endoenzymes therefore, the substrate present an electron carrier molecules like as free molecular nitrate or oxygen and a quickly needed for nutrients. Adsorbed substrates are define as which are complex in structure, prevent the entry in bacterial cells and weakly soluble or insoluble. For example, adsorbed substrate involve starches like as proteins, cellulose, lipids (oil and fats) and disaccharides like as lactose and maltose. To enter into bacterial cell, these substrate should be hydrolyzed into simple soluble molecules [24, 25]. Nondegenerateable pollutants as well as insoluble and weakly soluble substrates are adsorbed on the bacterial cell or floc particles surface. Indirectly elimination of pollutant and adsorbed substrate from waste streams by coating action of secretions of higher life form, ciliated protozoa and metazoan, specially free living nematodes and rotifers and directly through electrochemical method [26].

If pollutants and substrates have well suited charge, they connect to bacterial cells which containing negative charge fibrils as a result it expand into bulk solution. While pollutants and substrates have no well-suited charge for adsorption to fibrils of bacterial cells and by coating action of these higher life form its charges become compatible for adsorption. In biological treatment section hydrolytic bacteria formed exoenzymes by enough residence time occur. The substrates are hydrolyzed into soluble and little substrates so that they are absorbed by the nonhydrolytical and hydrolytic bacteria in the biological treatment section, when exoenzymes are moved and attach with the adsorbed substrates. When at the end it absorbed so the soluble and little substrates are degenerated by endoenzymes [27].

Hydrolytic bacteria contain of a association of gram-positive, facultative anaerobic bacteria, rod-shaped and anaerobic bacteria that disintegrate weakly soluble and insoluble complex lipids, proteins and carbohydrates into soluble and simple sugar, glycerine ($CH_2OHCHOHCH_2OH$ and fatty acids, and amino acids, correspondingly. By using a large number of bacteria for degeneration and absorption of these required soluble substrates. Hydrolysis process is define as addition of water (hydro) to complex molecules in the presence of bacteria to break down specific chemical bonds present in the complex molecules so allowing the formation and liberation of simple and soluble molecules [26]. The breakdown of specific chemical bonds and water addition is catalyzed in the presence of exoenzymes, for example, lipases (hydrolyze lipids), proteases (hydrolyze proteins) and cellulose (hydrolyze starches or carbohydrates). Hydrolysis process are slowly occurred. When the complex molecules is hydrolyzed slowly so, the highly soluble substrate are become existing to the biomass and an instant requirement for electron carrier molecules like as free molecular nitrate or oxygen is avoided and and nutrients [28].

In biological treatment section, there are two essential roles of hydrolysis occurred. In the first role, various sections obtained only complex substrate. While in second role, hydrolysis allows the cellular parts to degeneration and solubilization as compared to association and bacteria die in any biological section [29].

The capability of anyone biological treatment section to degenerate a substrate which influence on the existence of the following significant effects:

1. A sufficient number of enzymes
2. A substrate having molecular structure.
3. A various population of enzymes and bacteria.
4. An adequate performed condition involving mean cell residence time.

In first, bacteria degenerate the more simple and soluble substrate and after that the complex and insoluble substrate. In biological treatment, a substrate like as sugar is fastly degenerate due to the following conditions exists:

1. Rate of all biochemical reactions that are essential for total degradation of the substrate are optimum.
2. For degeneration of a substrate in the presence of all essential enzymes are firstly occurs in appropriate number.
3. A sufficient residence time is provided for substrate degeneration. Wax or chitin like as substrate is weakly degenerateable due to following situation exists in biological treatment section:

Rate of all biochemical reactions that are essential for total degeneration of the substrate are not optimum.

- For degeneration of a substrate in the presence of all essential enzymes are not firstly occurs in appropriate number.

6.2.2 Acetogenic, Coliforms, and Cyanobacterial Community

Member of Acetobacteracae family which produced acetate and is essential for the degeneration of soluble cBOD into methane (CH_4) in anaerobic digesters. It is a special group of fermentative bacteria which converts alcohols, ketones, and organic acids into carbon dioxide, hydrogen, and acetate, respectively. Meethanogen (methane producing bacteria)use acetate as a main substrate for producing methane. The essential acetogenic bacteria involved is *Acetobacter*, *Syntrobacter*, and *Syntrophomonas* [18].

Gram-negative rods which is a member of the coliform group of bacteria so that at 37 °C it ferment the sugar lactose and formation of gas. Complete coliform group involves these genera of the Enterobacteriacease family: *Enterobacter, Escherichia, Hafria, Serratia, Citrobacter, Yerinai and Klebisella.* The coliform bacteria existence is an indication of fecal contamination and mainly indicative genus in fecal contamination is *Escherichia* [30].

Photosynthetic bacteria are members of cyanobacteria. It was considered a blue-green algae before their prokaryotic cell structure was indicated. Cyanobacteria can be present as a chain of filament or cells (*Oscillatoria*) or a individual cells (*Chlorella*). In activated sludge method, the filament of cynobacteria exits and it may participate to settle ability troubles.

Cyanobacteria commonly are found on the surface of biofilm in trickling filters and enter the activated sludge process in the effluent of trickling filters when the filters are used upstream of the activated sludge process as "roughing" towers or to pretreat industrial wastewaters. Examples of cyanobacteria include *Anabaena, Chlorella, Euglena,* and *Oscillatoria* [18].

6.2.3 Denitrifying, Fecal Coliforms, and Fermentative Microbial Community

Denitrifying bacteria involve facultative anaerobic bacteria which using a nitrate without free molecular oxygen to degenerate into soluble cBOD. The results showed that the use of nitrate is converted to nitrogen and enters into atmosphere in the form of nitrous oxide (N_2O) and molecular nitrogen (N_2). For a wastewater treatment industry a denitrification is apply to satisfy a complete nitrogen discharge demand. During denitrification process it causes problems such as foaming in anaerobic digestor and clustering in secondary clarifiers. However, there are large number of genera that consists denitrifying bacteria and the three genera that consist the most species of denitrifying bacteria, for example, *Alcaligenes, Bacillus,* and *Pseudomonas* [31].

Fecal coliforms involve all coliforms that are entirely fecal in source and at 45 °C it can ferment lactose or formation of colonies. Fecal contamination is represented by fecal coliform and contains principally of species of the genus *Escherichia* [32].

Acidogenic (acid forming) or fermentative bacteria that convert fatty acids, sugars and amino acids into organic acids involving butyrate (CH_3CH_2COOH), formate ($HCOOH$), acetate (CH_3 $COOH$), propionate (CH_3CH_2COOH), and butyrate (CH_3CH_2COOH). Fermentative or acidogenic bacteria are essential in anaerobic digestors so they are converted complex substarte into simple substrates by using methane-producing bacteria.

In Biological phosphours removal sections fermentative bacteria are essential and it formed the required organic acids which are important to removal of phosphours by poly-P bacteria. There are a large number of genera of fermentative bacteria involve *Clostridium, Escherichia, Lactobacillus, Proteus, Bacteroides,* and *Bifidobacteria* [31]. In activated sludge method facultative anaerobic rods, gram-negative are composed nearly 80% of the bacteria and in the anaerobic digester a significant number of the bacteria. They are included with acetate formation, biological phosphorous elimination, degeneration of soluble cBOD, floc production, hydrolysis of cBOD, denitrification and acid formation. In bacterial genera that consist Gram negative, facultative anaerobic rods involve *Escherichia, Proteus, Klebsiella, Aeromonas, Salmonella and Klebsiella Escherichi* [32].

6.2.4 Floc-Forming and Gram-Negative Microbial Community

FLoc formation process is started by using a floc-forming bacteria in the activated sludge method. When the life time of sludge is increase, floc-forming bacteria formed cellular parts which require to attach each other or grouped. The amount of Floc-forming bacteriais small which involve Aerobactor, *Achromobacter, Flavobacterium, Citromonas, zoogloea,* and *Pseudomonas* [31]. In activated sludge method in which gram-positive, aerobic cocci, and rods are composed nearly 20% of bacteria. They are included in biological phosphours elimination, floc production, nitrification and degeneration of soluble cBOD. In bacterial genera which consists Gram negative, aerobic cocci and rods involve *Acinetobacter, Nitrobacter, Pseudomonas, Zoogloea, Alcaligens, Nitrosomonas, and Acetobacter* [33].

6.2.5 Nocardioforms and Methane-Forming Microbial Community

Spore-producing and dram-positive bacteria which are special group of actinomycetes or nocardioforms. These filamentous organisms are comparatively long and short branched. There are some growing characters present in nocardioforms like as real branching which occur in fungi.

Norcardia and it reslted genera which are most frequently realted with a foam formation in the activated slugde method involve *Arthrobacter, Actinomadura, Micromonospora,* and *Corynebacteriu. Norcardia* species are most frequently described as difficult involve *N. caviae, N. asteroids, N. pinesis, N. amarae, N. rhodochrus* and *Norcardia* is the most generally noted foam formation actinomycete [33].

Methanogens and methane-producing bacteria form methane (CH_4) from restricted amount of substrates in anaerobic digestors. There are two main routes exits for methane formation: (1) breakdown of acetate and (2) use of hydrogen and carbon dioxide. For example, genera of methane-producing bacteria involve *Methanococcus, Methanosarcinia, Methanomonas* and *Methanobacterium.*

6.2.6 Nitrifying Microbial Community

For the living organisms nitrogen is key element. Nitrogen is present in – 3 oxidation state in amino group ($-NH_2$) and used in amino acid like glycine (Figure 6.3). Amino acid is consider the building block of protein. Protein is essential for making of structural material, genetic material, and enzymes.Most of the macromolecules are considered as organic nitrogen compounds such as protein and amino acid. The waste of food and fecal contain these macromolecule compounds and release in municipal wastewater. Although some amino acids are water soluble due to their simple structure while protein are insoluble due to their complex structure. Proteins are suspended in water because they are colloids and having large surface area [26].

In sewerage system due to bacterial action amino acids undergo to deamination (Figure 6.4). Deamination is known as libration of ($-NH_2$) and formation of reduced nitrogen. Reduced nitrogen is present in two types one is ammonium ions or ammonia or ionized ammonia. Each type of reduced nitrogen is pH dependant for their quantity. NH_3 toxic and liberate in atmosphere from biological treatment unit during mixing action, aeration, turbulence, or sewage system [33]. But ionized ammonia is nontoxic and utilize by bacteria in the form of nitrogen as nutrient. Most of the reduced nitrogen is present in ionized ammonia at pH lower than 9 [24, 26].

Urea (H_2NCONH_2) is another remarkable organic nitrogen compound [34]. Urea is a main constituent of urine that are released into municipal

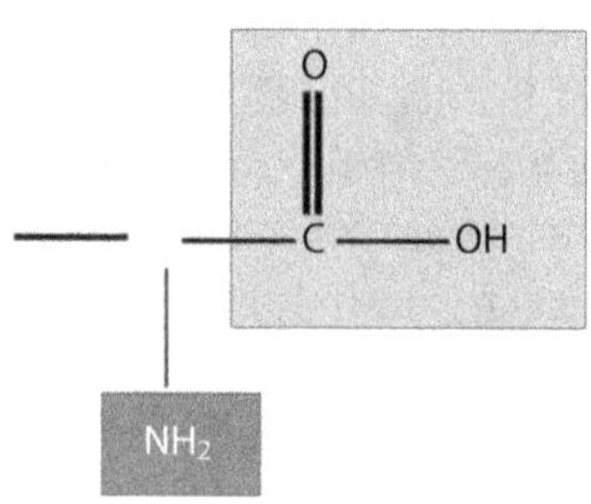

Figure 6.3 Structure of simple amino-acid (glycine).

wastewater. Due to bacterial action urea is hydrolyzed in sewer system. Hydrolysis is defined as due to addition of water in compound it breakdown into simpler units. Reduced nitrogen is produced due to hydrolysis of urea [35]. The out flow or influent of municipal wastewater having reduced nitrogen in the form of ionized ammonia from inorganic nitrogen as well as organic nitrogen [36]. In water treatment about 40% nitrogenous waste is consist of inorganic waste while 60% consist of organic nitrogen. In most of municipal wastewater treatment influent contains 25–30 mg ionized ammonia. In municipal wastewater treatment plant the additionally nitrogenous waste is polymers that mostly use in industrial and wastewater treatment plant. In most of the industrial waste not only ionized nitrogen are present but some of nitrite and nitrate ions are also present. In industrial discharge the presence of (NO_2^-) and (NO_3^-) ions act as indicator [35, 36]. In sewer system these ions are not formed. In activated sludge process the formation of nitrite and nitrate ions take during nitrification. It may be define as the biological oxidation of ionized ammonia to nitrite and nitrite to nitrate [34]. *Nitrosomonas* and *Nitrosospira* oxidize ionized NH_3 to (NO_2^-), *Nitrospira and Nitrobacter oxidized* (NO_2^-) *into* (NO_3^-). For cellular activity including reproduction energy is obtained due to nitrifying bacteria that help to oxidize the NH_3 to (NO_2^-). Sludge is increase in amount during activated sludge process due to this reproduction. From bicarbonate alkalinity carbon is produced used by nitrifying bacteria. From oxidation of ammonia and nitrite small amount of energy is obtained by nitrifying bacteria, and growth of bacteria or sludge formation is low. Each pound of ionized ammonia oxidized into nitrate by about 0.06 pounds of nitrifying bacteria [37].

Nitrifying bacteria are considered chemolithoautotrophs. They are defined as the bacteria that oxidized a mineral and obtain cellular energy and from inorganic carbon that don't have hydrogen obtain their carbon. Carbon dioxide is the major source of inorganic carbon. These bacteria

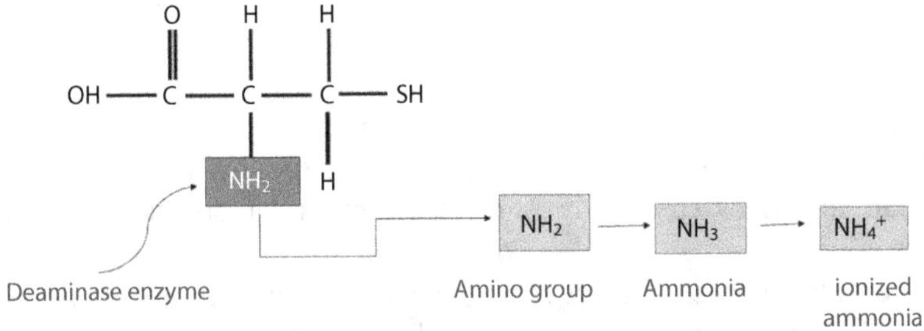

Figure 6.4 Schematic diagram of deamination of sulfur amino-acid (cysteine).

consumed carbon in the form of bicarbonate alkalinity, it is formed by dissociation pf carbonic acid that form due to addition of H_2O and CO_2 [35].

In oxidized ammonia and nitrite nitrifying bacteria are very useful and active due to their (1) cytomembranes, (2) having specified nitrifying enzyme system. In cellular membranes cyotemembranes are infolding that helps to increase the surface areas of membrane and provide space for nitrification process. They are found everywhere in plant, soil etc because they are free living. They become the part of wastewater treatment by infiltration and influent. They live in one to two inches in top of soil and considered strict aerobes. They reproduced in limited because they are not having enough from comes from oxidation of ionized ammonia and nitrite. About 8–10 hours are optimum time of generation gab. But it is increased from 2 to 3 days when the conditions are strong during activated sludge process. Thus, for efficient nitrification nitrifying bacteria needs relative high mean cell residence times (MCRT) to form population during activated sludge process. Nitrifying bacteria's generation time as well as activity is temperature dependant. They reproduce in high concentration as temperature increased. At 5 °C to 40 °C of temperature range they reproduce fastly because they are very active at this temperature. But the optimum temperature range of nitrifying bacteria for reproduction is 30 °C. At this temperature nitrifying bacteria have short generation time and have maximum reproduction, for example, *Nitrosomonas* [24].

The bacteria involve in nitrification process are poor floc-forming organisms. Due to opposite charge between nitrifying bacteria and floc particle they are attached with floc particle and they adsorb on floc particles through coating action of higher form of life. In activated sludge process there are many microorganisms like rotifers, free living nematodes and ciliated protozoa discharge some secretions on the surface of nitrifying bacteria and make it suitable for adsorption of floc particles.

The amount of energy substrate are responsible for the maximum activity and size of population of nitrifying bacteria but there are some other factors that are responsible of the size of population, activity of nitrifying bacteria and efficiency of activated sludge process for nitrification. These factors are dissolved oxygen, toxicity, pH and temperature and MCRT. Both temperature and MCRT are important factors. MCRT and temperature both are considered the most delicate factors that affects the nitrification process as well as efficiency of bacteria and their population size in activated sludge process. There is inverse relationship between temperature and MCRT or nitrifying bacteria. As we decrease the temperature MCRT is increased and bacterial population is decreased because at low temperature of wastewater treatment they are inactive and reduce efficiency of nitrification during activated sludge process.

There is no alkalinity in nitrification process. Through denitrification and deamination of organic nitrogen compounds the alkaline media transfer to activate sludge naturally. Ionized ammonia is produced due to deamination of amino acids. The increase in basicity indicates ionized ammonia. Alkalinity is produced due to denitrification in two ways (1) Indirectly: formation of bicarbonates ions due to withdrawing of carbon dioxide into wastewater and produce alkalinity and (2) Directly: formation of OH^- produced alkalinity [24].

6.2.7 Denitrifying Microbial Community

Denitrification is defined as breakdown of soluble cBOD by using nitrite and nitrate with the help of facultative bacteria. The industrial effluent contain nitrates and nitrites and release to activated sludge process. For processing of denitrification there is production of nitrite and nitrate during activated sludge process. Anoxic condition is required for denitrification process. In order to maintain anoxic condition there are some remarkable criteria that are summarized as presence of (NO_2^-) and (NO_3^-), absence of free O_2 presence of soluble cBOD and oxygen gradient, presence of denitrifying bacteria (facultative anaerobic bacteria).

In order to obtain energy and carbon for cellular growth facultative anaerobic bacteria are eligible to use nitrite, nitrate and free molecular oxygen for breakdown of soluble cBOD. But denitrifying bacteria are eligible only to use one molecule at a time. They prefer that molecule that is present at that time with high amount of energy and carbon for cellular growth and that is usually free oxygen molecule. Free oxygen molecule use for breakdown of the soluble CBOD is more reliable and preferable as compare to nitrate because free O_2 gives more energy for growth.

As soil and water organisms facultative anaerobic bacteria pass into fecal waste of activated sludge through inflow and infiltration. Most of the denitrifying bacteria are involved with floc due to opposite charge on the surface of floc particle that are usually secreted by free living nematodes, rotifers and ciliated protozoa. Almost 80% of denitrifying bacteria are dispersed in activated sludge process and incoprate with floc particle that are one billion of denitrifying bacteria per gram of floc particle. There are many genera of bacteria that include denitrifying bacteria. *Alcaligens*, *Bacillus*, and *Pseudomonas* are most commonly genera that have denitrifying bacteria. The reproduction phenomena of denitrifying bacteria is very fast. About 15–30 mins are generation time for denitrifying bacteria.

Under anoxic condition the production of enzymes that are essential for use of nitrate and nitrite is very fast. Nitrates that are produce during denitrification process are not enough and accumulated in activated sludge process during insufficient process of nitrification. Denitrification does not occur unless free oxygen molecule are present. So, for denitrification the presence of oxygen gradient and absence of free molecular oxygen is essential. When the dissolved oxygen concentration is about <1 mg outside a floc particle whose diameter is >150 μm then oxygen gradient is maintain. When denitrifying bacteria used dissolved oxygen then nitrates moves toward the core of floc particles. Dissolved oxygen is used for breakdown of soluble cBOD hence it is not available and denitrifying bacteria start to use nitrates. In the presence of measurable dissolved oxygen denitrification is premited by oxygen gradient.

In process of denitrification the concentration of substrate or soluble cBOD is one of main factor that effect the process. Higher the amount of substrate high the demand of electron acceptor that is nitrate and free oxygen molecule. Electrons are released from degradable substrate because soluble cBOD is breakdown inside the cell. With the help of electron acceptor from bacterial cell released electrons are withdrawed. Hence high the concentration of substrate higher the amount of electron acceptor needed. Dinitrification occurs fastly as oxygen is removed more fastly.

A huge variety of soluble organic compounds use as substrate by denitrifying bacteria usually found in domestic waste. Methanol (CH_3OH), acetate (CH_3COOH), glucose ($C_6H_{12}O_6$) and ethanol (CH_3CH_2OH), are some common used organic compound in denitrification process. Methanol is mostly used as organic compound because it is easily absorbed by cell and degenerate also simple form of soluble cBOD.

One of method of nitrate reduction is denitrification. During bacterial activity oxygen is removed from nitrate when it is reduced. When free molecular oxygen is not available denitrifying bacteria reduce nitrate to breakdown soluble cBOD. This is known is dissimilatory nitrate reduction or denitrification process because in cellular material the nitrogen of nitrate is not associated with it and escape from bacterial cell as N_2 and N_2O (Figure 6.5).

The nitrogen nutrient is provided to bacterial cell by assimilatory nitrate reduction and dissimilatory nitrate reduction is often used for respiration uses. For the preparation of cellular material nitrogen nutrient is provided from ionized ammonia source. The bacterial cell always uses nitrogen which is present in – 3 oxidation state. As the nitrogen present in ionized ammonia is in – 3 state it is mostly selected by bacterial cell. In the absence of ionized ammonia nitrate acts as a source of nitrogen. But in

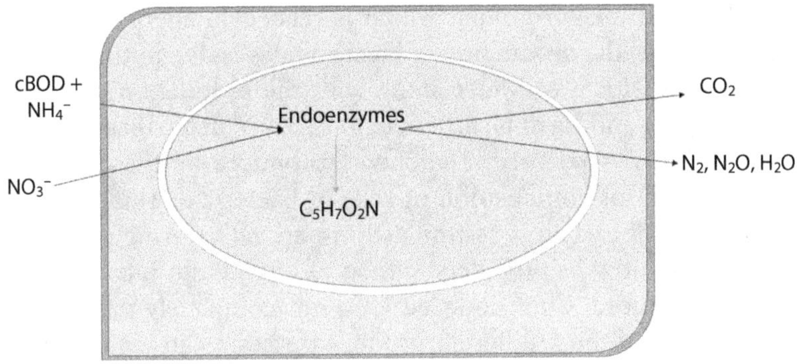

Figure 6.5 Dissimmilaterity nitrate.

nitrate +5 oxidation state of nitrogen exist. So when bacteria uses nitrate as an energy source it causes to add hydrogen in the cell along with the removal of oxygen from inside of cell. As the hydrogen is added along with removal of oxygen nitrogen is converted into – 3 oxidation state and again ionized ammonia appears. The nitrogen that is produced during assimilation got entered in the cell again. The nitrogen production by this process is called assimilated nitrate reduction. Within the activated sludge process both assimilation nitrogen reduction and dissimilatory nitrogen reduction occur spontaneously. When nitrate ions are available along with cBOD in the absence of ionized ammonia and molecular oxygen the degradation of cBOD occur by nitrate.

During denitrification molecular nitrogen is produced after gradual reduction of nitrate. During denitrification 4 biochemical steps and 5 nitrogenous molecules incorporate. Five of these molecules are molecular nitrogen, nitrous oxide, nitric oxide, nitrite, and nitrate. During these processes first substrate is always nitrate but sometimes nitrite also acts as first substrate or mostly act as intermediate. Nitrous and nitric oxide are intermediate products while molecular nitrogen is formed in final step.

The facultative anaerobic bacteria result in production of molecular nitrogen from nitrates. But in some denitrifying bacteria deficiency of enzymes occur which convert nitrate into molecular nitrogen. As in *Escherichia coli* conversion of nitrate to nitrite occurs but next step is not possible. However, due to adverse operation conditions, in facultative anaerobic bacteria production of intermediate compounds starts. In denitrification the process complete in 4 stages and only two generate energy. The two steps are the reduction of nitrates and nitrites. So the energy production is related to the whole process that occurs in nitrate and nitrite step conversion. The formation of bicarbonates and hydroxyl ions produce basic environment which is related to energy production in whole process. The production of bicarbonates is due to the mixing of carbon dioxide with wastewater along with the production of carbonic acid which splits and form bicarbonate ions. After denitrification half of the loss alkalinity is restored. The energy produced by the degradation of cBOD is used for composition of cells in bacteria or sludge [38, 39]. As low amount of carbon is assimilated in new cells during degradation of cBOD using nitrate so the degradation of cBOD into nitrate much of the CO_2 is produced. The produced CO_2 not completely participate in wastewater. Out of the 5 different produced gases 3 can escape in environment. But in some cases these gases become enclosed in floc particles and cause problem in secondary clarifiers before they enter in air. At the redox value of +100 mV and −100 mV denitrification process is started and it is effected by three remarkable factors that are nutrients, pH and temperature 7.0 to 7.5 is the optimum range of pH for denitrification but at <6.0 and >8.0 values of pH the depressed action of denitrifying bacteria are occurred. As the temperature of waste increases denitrification process increased. Wastewater has less affinity for dissolved oxygen as its temperature increased that leads to denitrification more fastly. At <5 °C denitrification process is prohibited [38, 39].

For growth of bacterial cells and degeneration of soluble cBOD phosphrous and nitrogen are essential nutrients. The huge amount of nitrogen and phosphrous along with free molecular oxygen are essential for the breakdown of cBOD and they refer as present in aeration tank as mix solution of effluent with minor quantity of both nutrients that is 0.5 mg/L $H_2PO_4^-$ or HPO_4^{2-} and 1.0 mg/L NH_4^+ or 3.0 mg/L NO_3^- for phosphrous and nitrogen. Less bacterial growth is produced than free molecular oxygen when the breakdown of cBOD with nitrates. In denitrifying tank of effluent filtrate they have huge concentration of phosphrous and nitrogen. When anoxic condition is maintained then denitrification is started. These conditions are originated intentionally or accidently. Intentionally generated conditions refer as i) Total nitrogen release from denitrifying tank and its use ii) To enhance the efficiency of treatment plant by anoxic period. Accidently conditions originate are not usually mention in secondary clarifiers.

6.2.8 Phosphorous Solubilizing Microbial Community

Phosphorus (P) is a major nutrient which is necessary for all living cells. It is the key element for the production of adenosine triphosphate or ATP, nucleic acids, DNA and RNA, phospholipids, teichoic acids, and teichuronic acids. ATP acts as a high-energy molecule and used in the transfer of energy within the cell. Phospholipids are essential components in the structure of cell membranes, while teichoic acids and teichuronic acids are essential components in the structure of cell wall of gram-positive bacteria. Phosphorus also stored in cells as intracellular volutin granules or polyphosphates. Phosphorus may be 1–3% of the dry weight of a bacterium. Although the contets of phosphorus is approximately one-fifth of the contents of nitrogen of the bacterium, the actual phosphous content may vary from one-seventh to one-third of the nitrogen content that depend on the environmental conditions. Phosphorus is a nutrient required for growth of aquatic plants. Often phosphorus is a limiting nutrient and the concentration of phosphorus in waters determines the quantity of vegetative growth. However, the introduction of trace amounts of phosphorus into receiving waters may have greater effect on the quality of receiving waters [24]. Undesired growth of algae is triggered at orthophosphate concentrations as low as 0.5 mg/L. Presence of algal blooms and phytoplankton results in a rapid and remarkable deterioration in the quality of water, Phosphorus pollution of natural waters is usually responsible for eutrophication and occurs mainly as a result of phosphorous-rich discharges from wastewater treatment plants. Many environmental problems are related with the sudden growth of marine plants. These problems cover clogging of receiving waters as well as color, odor, taste, and turbidity matter, if the receiving water used as drinkable water source. Moreover, and importantly, the die-off of large number of marine plants provide oxygen depletion and eutrophication. The oxygen is removed from receiving waters by using bacteria and other organisms as they can easily decompose the dead plants. Eutrophication or rapid aging of receiving waters arise as the non-decomposable part of dead plants assemble in receiving waters. Rapid growth of marine plants also lowers the value of wastewaters for fishing, industrial use, and recreational use.

Most often phosphorus is found in wastewater in greater quantity than those required for growth of marine plants. Thus, in order to inhibit or lowers the phosphorous-related water quality issues, state and federal regulatory agencies need phosphorus removal at wastewater treatment plants. Because the phosphorus reacts rapidly with minerals such as aluminum, calcium, and iron, little phosphorus leaches from soil. Besides, little phosphorus leaches from soil when it is applied to the soil as a fertilizer. Requirement for phosphorus removal is becoming more usual for municipal and industrial wastewater treatment plants. The discharge limits for total phosphorus in these plants are usually ≤2 mg/L. The phosphorus exists inorganic and inorganic forms. The inorganic forms of phosphorus involve orthophosphates and polyphosphates. The orthophosphates are available for biological metabolism without further breakdown and are considered to be easily available nutrients for phosphorus and for bacterial use in wastewater treatment plants and marine plants in natural waters. The orthophosphates include PO_4^{3-}, HPO_4^{2-}, $H_2PO_4^-$, and H_3PO_4. The commonly used forms of orthophosphate in wastewater treatment plants are HPO_4^{2-} and H_2PO^-. Relative quantity of each form is pH-dependent. Orthophosphate present is produced through dissociation. Within the pH operating range of most wastewater treatment plants; HPO_4^{2-} is dominant at pH values greater than 7, while $H_2PO_4^-$ is dominant at pH values greater than 7. The hydrolysis of polyphosphates is influenced by many factors including retention time and pH in an aeration tank. Principle form of orthophosphate acquire from hydrolysis is pH-dependent. Due to their stability in water, polyphosphates are easily sequestering minerals such as aluminum, calcium, and iron. The phosphate tied to organic compounds is referred as organic phosphorus. The organic phosphorous compounds are of least importance in domestic wastewater, but these compounds can be of significant interest in industrial wastewater and crude. The common organic forms of phosphorus involve inositol phosphates, nucleic acids, phospholipids, and phytin. The phytin is an organic acid found in vegetables such as corn and soybean and is difficult to destroy. Average concentration of total phosphorus in municipal wastewater is in range of 10–20 mg/L. The total phosphorus consists of inorganic phosphorus and organic phosphorus. The major sources of phosphorus released to municipal wastewater treatment plants consist of human waste, detergents, and industrial waste. The orthophosphate makes up approximately 50–70% of total phosphorus, while polyphosphates and organic phosphorus make up the remaining 30–50% of total phosphorus. When the orthophosphate, polyphosphate, and organic phosphorous compounds enter an activated sludge process these compounds undergo biological and chemical changes and experience several fates. Some organic phosphorous compounds are removed from wastewater when particulate, organic phosphorous compounds or phosphorous compounds are adsorbed to solids surface and settle out in the primary clarifier.

In activated sludge process, the phosphorous compounds undergo several fates. With sufficient hydraulic retention time (HRT), organic phosphorous compounds are destroyed through microbial activity, and the orthophosphates are released in aeration tank. With sufficient hydraulic retention time (HRT), polyphosphates are biologically and chemically hydrolyzed, and orthophosphate released in the aeration tank. The principal organisms that are responsible for mineralization or degradation of phosphorous compounds involve actinomycetes such as *Streptomyces,* bacteria such as *Arthrobacter* and *Bacillus*, and fungi such as *Aspergillus* and *Penicillium.* These organisms produce phosphatase, an enzyme that liberate orthophosphate from phosphorus-containing compounds.

The orthophosphate is readily accessible phosphorous nutrient for bacterial growth and energy transfer. As a readily accessible nutrient, phosphorus is removed from bulk solution from aeration tank and incorporated or assimilated into the cellular material as bacterial degrade substrate (soluble cBOD) and reproduce (sludge production). However, assimilated phosphorus makes up 1–3% of bacterial weight (mixed liquor volatile suspended solids).

If the deficiency for orthophosphate occurs in the activated sludge process, production of nutrient-deficient floc particles or sludge and undesired and excessive growth of nutrient-deficient filamentous organisms may occur. Nutrient-deficient floc particles and nutrient-deficient filamentous organisms adversely affect the solids settleability in the secondary clarifier and may responsible for foam production and accumulation [40, 41, 42].

Soluble cBOD is absorbed by bacterial cells in floc particles during nutrient deficiency for orthophosphate (<0.05 mg/liter). Therefore, the soluble cBOD cannot be destroyed due to lack of adequate phosphorus. However, the cBOD is converted into an insoluble polysaccharide (starch) and stored in floc particle, until orthophosphate are available for its degradation. Stored polysaccharide is less dense than the water, and its storage between bacterial cells results in the loss of floc particle density. Polysaccharides capture air and gas bubbles also. Captured air and gas bubbles contribute to the loss of floc particle density and formation of the foam. The foam produced from orthophosphate deficiency may be billowy white or greasy gray. The billowy white foam is related with young sludge age, while greasy gray foam is related with old sludge age. As bacteria age in floc particles, their secreted oils assemble in the floc particles and are transferred to the foam. The transfer of oils changes the texture and color of foam from billowy white to greasy gray [24].

There are several filamentous organisms that grow rapidly in a nutrient-deficient condition for phosphorus or nitrogen. These filamentous organisms outgrow floc bacteria in a nutrient deficient condition because (1) they require less nutrients than floc bacteria or (2) they may compete more efficiently for nutrients when nutrients are limited in quantity. The effective competition for nutrients is provided by greater surface area of the filamentous organisms that are exposed to the bulk solution which contains nutrients as compared to surface area of the floc bacteria. Of the filamentous organisms that grow rapidly in a nutrient deficient condition, the nocardioforms are foam producers. The Foam typically of nocardioforms is viscous and chocolate brown. In aeration tank, the orthophosphate is incorporated into floc particles as insoluble hydroxyapatite ($CaOH(PO_4)_3$). This occurs naturally without chemical addition. The pH of the aeration tank decreases if dissolved oxygen concentration of the aeration tank is decreased and much of the carbon dioxide released from the degradation of soluble cBOD remains in solution (i.e., it is not stripped to the atmosphere). The decrease of CO_2 occurs because carbon dioxide dissolves in the mixed liquor and carbonic acid (H_2CO_3) is formed. Under these conditions, orthophosphate remains in the solution as $H_2PO_4^-$ ion.

Though, if the dissolved oxygen concentration of the aeration tank is increases and much of the carbon dioxide in the aeration tank is stripped to the atmosphere, little carbonic acid is produced and the pH of the aeration tank increases. Under these conditions, orthophosphate is present as HPO_4^{2-} ion. If this happens in hard water (containing calcium as Ca^{2+}), orthophosphate is precipitated out from the solution as hydroxyapatite and incorporated into the floc particle.

The orthophosphate may remain in the solution in aeration tank in two forms. It may remain in solution in the ionic form as determined by pH, or it may be remained in the solution sequestered (bonded in solution) to an alkali metal. The effluent phosphorus from the activated sludge process contains approximately 90% of orthophosphate. Orthophosphate may be as soluble ions or sequestered orthophosphate. To limit the concentration of effluent phosphorus from activated sludge process, an advanced wastewater treatment measure is needed. The advanced wastewater treatment consists of biological, chemical, and physical measures that remove the inorganic and organic suspended solids, the phosphorus-containing and nitrogen-containing compounds that contribute to eutrophication, slowly degradable or nondegradable organic compounds. The phosphorus can be removed in municipal wastewater treatment plants through biological and chemical treatments. Several of the measures are considered to be advanced wastewater treatment measures that include chemical precipitation of phosphorus, enhanced biological phosphorus removal (EBPR), and biological-mediated/chemical precipitation of phosphorus.

The enhanced biological phosphorus removal (EBPR) or "luxury uptake of phosphorus" occurs when the phosphorus uptake by bacteria is in excess of cellular requirements. Typically, the activated sludge phosphorus contents are approximately 1–3%, while the activated sludge phosphorus contents are approximately 6–7% when EBPR is used. The EBPR is relatively inexpensive and is capable of removing phosphorus to lower concentrations. The EBPR also limits the chemical costs and sludge disposal costs that is related to the chemical precipitation of phosphorus. The EBPR incorporates the use of two groups of bacteria, fermentative bacteria and poly-P bacteria. The Poly-P bacteria are also known as phosphorus accumulating organisms (PAO). At least the two treatment tanks, an aerobic (fermentative) tank and anaerobic tank, are used for EBPR. Fermentative bacteria are facultative anaerobes, while the poly-P bacteria with exception are strict aerobes. Fermentative bacteria and the poly-P bacteria enters the EBPR process as fecal bacteria and soil and water bacteria from inflow and infiltration. The key to EBPR is the exposure of poly-P bacteria to change anaerobic and aerobic conditions. In anaerobic tank, fatty acids (particularly acetate and propionate) are formed through anaerobic activity of the fermentative bacteria. These compounds serves as a substrate for proliferation of aerobic poly-P bacteria that is in the anaerobic tank. Though, the poly-P bacteria cannot use (degrade) the fermentative compounds in the anaerobic condition. Degradation of these compounds by poly-P bacteria occurs only in the presence of free molecular oxygen (aerobic tank) or nitrate.

The free molecular oxygen and nitrate ions should be absent in the anaerobic (fermentative) tank. A residual quantity of either free molecular oxygen or nitrate are quickly exhausted in anaerobic tank, but needs a longer retention time and more soluble cBOD in anaerobic tank in order to exhaust free molecular oxygen or nitrate ions. Presence of free molecular oxygen or nitrate in anaerobic tank interferes with the phosphorus removing ability of the EBPR process. The interference occurs as a result of an increase in redox potential and inability to form fatty acids that are essential for release of phosphorus. The fermentation is the microbial degradation of soluble organic compounds (cBOD) without use of free molecular oxygen or nitrate. The significant fermentative organic compounds or substrate formed in the anaerobic tank involves alcohols and variety of soluble fatty acids. In the anaerobic tank, poly-P bacteria quickly absorb the fatty acids and polymerize (store) the acids as an insoluble starch (poly-β-hydroxybutyrate or PHB). PHB in poly-P bacteria serves two important functions. Firstly, it helps the bacteria to grow and rebuild polyphosphates by taking up soluble phosphate. Secondly, PHB along with polyphosphates help aerobic poly-P bacteria to live in an anaerobic condition. Polymerization of the fatty acids needs an expenditure of cellular energy by the poly-P bacteria. This expenditure or energy is the breakdown or release of orthophosphate from the poly-P bacteria to the bulk solution of the anaerobic tank. As a result of the liberation of orthophosphate from the poly-P bacteria, the anaerobic tank contains two pools of phosphorus, namely, the phosphorus in the influent wastewater (feed phosphorus) and the released phosphorus by the poly-P bacteria. In the aerobic tank the poly-P bacteria use free molecular oxygen to destroy the stored PHB as a carbon and energy source. Simultaneously, the poly-P bacteria absorb ortho-phosphorus in order to store the energy released from the degraded PHB. Poly-P bacteria get so much energy from degraded PHB that they may absorb not only the released phosphorus but also large quantity of feed phosphorus. Absorbed phosphorus is assimilated into macromolecules and stored as polyphosphate granules or volutin. The phosphorus removal is attained when the bacteria (sludge) are wasted from the secondary clarifier. Sludge that is not wasted is returned to the anaerobic tank where EBPR process is repeated again. By exposing the poly-P bacteria to alternating anaerobic and aerobic conditions, the poly-P bacteria are stressed and take up phosphorus in excess amount of normal cellular requirements. There are numerous processes available for EBPR. Alternative exposure of poly-P bacteria to anaerobic and aerobic tanks may be accomplished in main biological treatment process. A mainstream process for biological phosphorus removal contains an anaerobic tank alongwith the main liquid process stream from influent to effluent. A side stream process contains an anaerobic tank which is aside of the main liquid process steam [40, 41].

All nutrients removal processes for phosphorus remove the excess phosphorus biologically, except the Phostrip process that include chemical precipitation of phosphorus. The removed phosphorus from these processes is found biologically in bacterial cells or chemically precipitates within the sludge. When biological phosphorus removal is combined with nitrification and denitrification for nitrogen removal, the removal of phosphorus and nitrogen is known as biological nutrient removal (BNR) or combined phosphorus/nitrogen removal. The nitrification is biological oxidation of ionized ammonia (NH_4^+) to nitrate (NO_3^-). The denitrification is biological used of nitrate to degrade soluble cBOD in the absence of free molecular oxygen. When the nitrate is used to degrade soluble cBOD, the nitrogen in the nitrate leaves the wastewater and returned to the atmosphere as molecular nitrogen (N_2) and nitrous oxide (N_2O). There are two EBPR processes that remove the phosphorus only. These processes are the A/O and the Phostrip. A/O process is a mainstream process. Phostrip process is a side-stream process and includes biological and chemical measures for the removal of phosphorus. Phostrip process has a stripper tank where anaerobic condition permits the release of phosphorus by poly-P bacteria from return activated sludge (RAS) as shown in fig.6.6 and 6.7.

The liberated phosphorus is removed from the stripper tank by elutriation water. The lime is introduced to the stripper tank overflow to precipitate the liberated phosphorus.

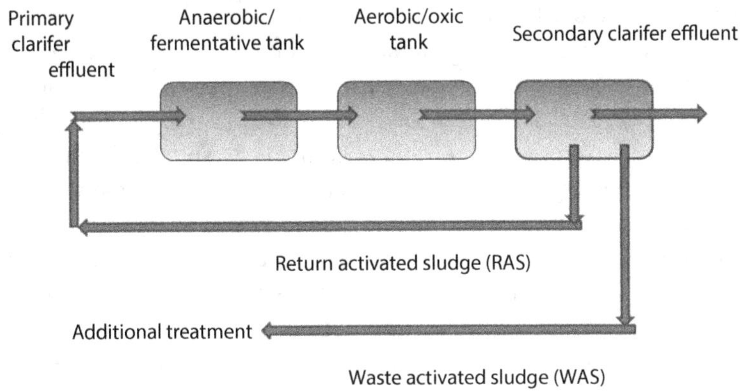

Figure 6.6 Flowsheet diagram of A/O process.

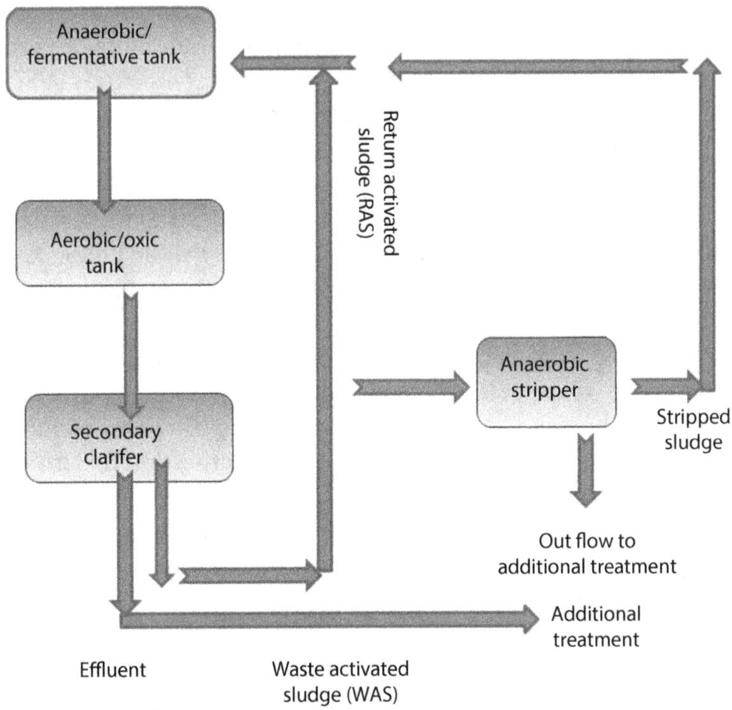

Figure 6.7 Flowsheet diagram of Phostrip process.

6.2.9 Sulfur Oxidizing and Reducing Microbial Community

The sulfur is found in various compounds of natural and pollutant origin and is very common in wastewater. The domestic wastewater contains approximately 3–6 mg/liter of the organic sulfur as proteinaceous wastes and approximately 4 mg/L of the organic sulfur as sulfonates obtained from detergents. The domestic wastewater may also contain approximately 20 mg/L of sulfate. In sludge, total sulfur may be 1–2% of the dry weight of the sludge.

The sulfur is excreted by aerobic organisms as sulfate. Though, upon the death of these organisms, sulfur is released from the organisms in the reduced state. Less sulfur or sulfide is produced from the oxidized forms of sulfur (sulfate, sulfite, thiosulfate, and elemental sulfur) during anaerobic degradation of substrate by anaerobic bacteria such as *Desulfovibrio desulphuricans.* Degradation of sulfur-containing compounds has an adverse impact on the wastewater treatment plants and water quality. The biological compounds which contain sulfur include amino acids and proteins. The sulfur is present in natural waters and wastewater and in many functional groups. Because a large variety of organic sulfur compounds exist, several sulfur products and biochemical reactions are related with the degradation of organic sulfur compounds. Degradation of sulfur-containing amino acids may result in the production of volatile sulfur compounds (VSC). In addition to hydrogen sulfide, major VSC related to the degradation of sulfur-containing organic compounds are methyl thiol (CH_3SH) and dimethyl disulfide (CH_3SSCH_3). The industrial wastewaters may also contain large quantities of sulfur-containing compounds. The industrial wastewaters are those that comprize sulfonated detergents and sulfite waste liquor.

A large variety of detergents that include alkyl benzene sulfonates (ABS) are used for commercial, domestic, and industrial purposes. When present in wastewater, ABS represent a source of sulfur and many other detergents. ABS also represents numerous additional operational concerns which include; the decreased rate of oxygen dissolution, proliferation of anaerobic bacteria, resistance to biological degradation and toxicity. There are five major groups of sulfur bacteria. These groups contain the sulfate-reducing bacteria (SRB), sulfur-oxidizing bacteria, colorless sulfur bacteria, sulfur-oxidizing photosynthetic-green bacteria, and sulfur-oxidizing photosynthetic purple bacteria. The sulfur bacteria, mainly SRB, grow with other bacteria, and their presence cannot easily be detected by microscopic examination. SRB usually are outnumbered by other groups of bacteria, except in special environments such as anaerobic digesters. Although, changes in the sulfur profile include sulfate and sulfides across a treatment unit may be the indicative of presence of specific sulfur bacteria. The operational concerns associated to the activity of SRB include malodor production, undesired growth of the filamentous organisms *Beggiatoa, Thiothrix,* and competition with methane-producing bacteria. If sulfates are present in anaerobic digester sludge, *Desulfovibrio desulfuricnas* proliferates. This

SRB limit the sulfate to hydrogen sulfide by using hydrogen. The methane-forming bacteria also use hydrogen to produce methane. Presence of sulfate in anaerobic digester sludge results in competition for hydrogen by SRB and methane-producing bacteria. Methane-producing bacteria appropriately use less substrate and produce less methane. In addition, the hydrogen sulfide produced by the SRB has an inhibitory effect on methane-producing bacteria. When SRB uses sulfate to degrade substrate such as ethanol, they produce acetate. However, SRB acts as a competitor for ethanol with methane-producing bacteria and substrate producer for methane-producing bacteria.

In the absence of sulfate, many SRB are able to adapt and continue to increase rapidly. Some grow with hydrogen-consuming, methane-producing bacteria. When this happens, the SRB produces substrates such as acetate and hydrogen that may use by methane-producing bacteria. In the absence of free molecular oxygen or nitrate or presence of an oxygen and nitrate gradient, SRB uses low molecular weight carbon sourses as substrate Either in the presence or absence of free molecular oxygen or nitrate gradient. These carbon sources are formed by the fermentation of carbohydrates, lipids, and proteins.

The sulfides discharged to an anaerobic digester or formed in an anaerobic digester combined with soluble metals such as cadmium, iron, and zinc to form highly insoluble salts. In Addition metals that combine with soluble sulfide include chromium, copper, lead, and nickel. For examples cadmium sulfide (CdS), ferrous sulfide (FeS), and zinc sulfide (ZnS). These metallic sulfides and immobilized heavy metals turn into sludge black. The hydrogen sulfide does not assemble in an anaerobic digester, until the metals are removed from the solution. The hydrogen sulfide may be chemically and biologically oxidized to sulfate. The hydrogen sulfide is an oxygen scavenger; when it is exposed to air or oxygenated water, it may be readily converted to sulfate. The hydrogen sulfide may be biologically oxidized to sulfate by sulfur-oxidizing bacteria. There are two commonly occurring sulfur-oxidizing bacteria that are important in malodor control of wastewater. These bacteria are *Chlorobium* and *Chromatium*. The photosynthetic bacteria oxidize sulfides to sulfur compounds that are not noxious.When it is present in highly concentrated masses, *Chromatium* gives red color to the wastewater in the presence of organic overloading and septicity.

In activated sludge process the sulfur filamentous organisms, *Beggiatoa,* and *Thiothrix*oxidize the sulfides to elemental sulfur, and *Thiobacillus* oxidizes the elemental sulfur to sulfate.

The sulfur-oxidizing bacteria oxidize inorganic sulfur in low oxidation states by adding oxygen to inorganic sulfur. The sulfur-oxidizing bacteria get energy from the oxidation of inorganic sulfur. Some of the sulfur-oxidizing bacteria are chemoautotrophs and use energy that they get from the oxidation of inorganic sulfur for the production of organic compounds from carbon dioxide (CO_2). The sulfur-oxidizing chemoautotrophs include species that are in the genera *Thiobacillus*, *Thiospirillopsis*, and *Thiovulum*. Because most reduced form of sulfur needed by these bacteria are formed by the metabolic activity of SRB, the sulfur-oxidizing chemoautotrophs commonly grow to greater numbers at the interface between anaerobic conditions (source of sulfide) and aerobic conditions (source of oxygen).

The sulfur-oxidizing bacteria like SRB are also known for operational issues related to their growth. They transform reduced form of sulfur to sulfuric acid (H_2SO_4). Concrete (a stone bonded by calcium carbonate ($CaCO_3$) corrodes or "rots" in the presence of sulfuric acid. Even with the development of sulfuric acid, sulfur-oxidizing bacteria continue to grow. For example, *Thiobacillus* is very tolerant of sulfuric acid and can grow at a pH values <1.

Growth of purple sulfur bacteria is enhanced by the presence of hydrogen sulfide. The purple sulfur bacteria are photosynthetic and involves the genera *Chromatium*, *Thiocapsa*, and *Thiopedia*. These bacteria oxidize the sulfides to elemental sulfur to get energy and deposit the sulfur intracellularly as granules. In the dense populations the purple sulfur bacteria can turn the wastewater red. The oxygen is toxic to the purple sulfur bacteria. However, the growth of these bacteria is prevented in the presence of free molecular oxygen [24].

6.3 Methods for the Treatment of Wastewater

6.3.1 Preliminary Treatments

The wastewater treatment is related to type of waste specifically present in it. The cleanity of water id related to amount of dissolved oxygen present in it. The amount of pollutants in wastewater is measured by calculating the amount of dissolved oxygen in it which is basic requirement of biochemical oxygen demand (BOD) test. For the aerobic treatments of industrial waste and sewage sludge and also for maintaining oxygen level in water amount of dissolved oxygen is calculated. An improved method known as Winkler's method use iodometric analysis for determining the amount of dissolved oxygen in the water.

Firstly, within the water sample 48% of manganese sulfate was added related to total volume. Then addition of 15% KI in 70% KOH result in the pinkish brown precipitates. In the basic medium Mn^{2+} was converted into Mn^{4+}.

$$2\ MnSO_4(s) + O_2(aq) \rightarrow 2\ MnO(OH)_2(s) \qquad (6.1)$$

The brown color precipitates of $MnO(OH)_2$ are formed. The formation of precipitates does not confirm that Mn exist as Mn^{3+} or Mn^{4+}. The hydrated MnO_2 is found brown in color but the precipitates of $Mn(OH)_3$ are also brown in color.

$$4\ Mn(OH)_2(s) + O_2(aq) + 2\ H_2O \rightarrow 4\ Mn(OH)_3(s) \qquad (6.2)$$

In the second step of Winkler test acidification occur. The dissolution of precipitates occur again. The acid help in formation of elemental iodine by dissolving the manganese containing iodine precipitates. In the acidic medium manganese sulfate causes to convert iodide ions into iodine and itself got reduced.

$$Mn(SO_4)_2 + 2\ I^-(aq) \rightarrow Mn^{2+}(aq) + I_2(aq) + 2\ SO_4^{2-}(aq) \qquad (6.3)$$

For the titration of iodine thiosulfate is used along with starch indicator.

$$2\ S_2O_3^{2-}(aq) + I_2 \rightarrow S_4O_6^{2-}(aq) + 2\ I^-(aq) \qquad (6.4)$$

The given data result in stoichiometric calculation

$$1 \text{ mole of } O_2 \rightarrow 2 \text{ moles of } MnO(OH)_2 \rightarrow 2 \text{ mole of } I_2 \rightarrow 4 \text{ mole of } S_2O_3^{2-} \qquad (6.5)$$

Thus the number of moles of oxygen in water sample can be calculated after the calculation of number of moles of iodine produced and amount of oxygen is expressed in mg/dm^3.

The amount of oxygen required by biological organisms at a particular time and temperature to dissociate the organic matter is known as biochemical oxygen demand (BOD). The amount of biochemical oxygen demand required for degradation of organic pollutant is calculated in mg/L of oxygen consumed in five days of incubation at 20°C. Commonly BOD is calculated using standard dilution method. The standard dilution method used for testing of wastewater and water is labeled by U.S.EPA having label method 5210B. To calculate biochemical oxygen demand the level of dissolved oxygen is calculated before and after incubation and is maintained by using dilution factor. The analysis of 300 mL of incubated bottles having buffer water is mixed with seed microorganisms and placed in a dark room at 20 °C for five days to stop the dissolved oxygen supply for photosynthesis was performed. Along with the dilution water control blanks this process of biochemical oxygen demand also requires seed control and glucose glutamic acid control. The quality of diluted water required for dilution of other samples is guaranteed by blank water dilution. This is essential because if any impurity present in dilution water it can change the results of samples. The type (quality) of seed is determined by GGA system which is standardized solution used to check the given quality which is 198 mg/L ± 30.5 mg/L of biochemical oxygen demand. To calculate the carbonaceous biochemical oxygen demand, after sample dilution inhibitor is also added. This inhibitor does not allow the oxidation of ammonia nitrogen which provides nitrogenous biochemical oxygen demand. For calculating the biochemical oxygen demand only carbonaceous Biochemical oxygen demand is calculated because measurement of nitrogenous biochemical oxygen demand does not reflect the oxygen required for decomposing organic matter. This is because cBOD is produced during break down of organic matter while nBOD is produced during break down of proteins.

$$\text{Unseeded BOD}_5 \text{ is calculated by: } (D_0 - D_5) \tag{6.6}$$

$$\text{Seeded BOD}_5 \text{ is calculated by: } (D_0 - D_5) - (B_0 - B_5)p/f \tag{6.7}$$

where

D_0 is the dissolved oxygen (DO) of the diluted solution after preparation (mg/L)

D_5 is the DO of the diluted solution after 5 day incubation (mg/l)

p is the decimal dilution factor

B_0 is the DO of diluted seed sample after preparation (mg/l)

B_5 is the DO of diluted seed sample after 5-day incubation (mg/L)

f is the ratio of seed volume in dilution solution to seed volume in BOD test on seed.

The amount of oxygen which can be oxidized by using $K_2Cr_2O_7$, equivalent to the amount of organic matter in waste is calculated by COD. To determine the pollutants strength in industrial waste, sewage and streams COD is one of the best methods. American society of testing and materials has set a reference point that chemical oxygen demand is the oxygen required for oxidation of organic and oxidizeable matter under controlled conditions for determining the amount of chlorides. The main focus of COD is to oxidize the organic matter present in wastewater samples by mixing with silver sulfate and mercury sulfate in the presence of excess amount of $K_2Cr_2O_7$ in 50 % sulfuric acid and got converted into carbon dioxide, ammonia and water. The amount of dichromate which left untreated is titrated against ferrous ammonium sulfate solution. The calculation of chemical oxygen demand is as follow:

$$\text{COD in mg/L} = (V_1 - V_2)\, N \times 8 \times 1000/x \tag{6.8}$$

In this V_1 and V_2 represent the amount of ferrous ammonium sulfate present in sample and blank respectively and X is the amount of sample taken.

As the COD test is performed for both biologically inert and oxidizable sample its value is always more as compare to BOD.

The removal of gross solids like suspended solid particles, oil and grease, grit and large floating particles is the first objective of preliminary treatment when they are in amount which is noticeable. The removal of large particles like canes, wood cloth and other floating rubbish are removed by using metal bars used to stain the wastewater which moves below them. The speed of wastewater is reduced by increasing the area of grit settling chamber as compare to first step. The gross solids are removed by moving water from moving or mixed screens. Bar screen, hand raked, drum screen, wire rope screen and bar screen are the different types of screen available for cleansing.

The different types of circular overhead fed vibratory units, self-cleansing, gravity type units and rotary are the advanced mechanical screens used for filtration. These modern screens are higher in cost but most effective in removal of suspended solids as compare to conventional screens. In some cases macerators or comminutors are used for cutting the gross solids into smaller size instead of using screens.

To avoid the settling in pipe lines the detritus is removed in the preliminary steps either by grit channels or tanks to protect the pumps and equipments against any damage or abrasion. As the weight of grit is more than other waste it can be easily separated by controlling the flow in grit tanks the grit produced in the system is continuously separated and manually removed and added to land fill, road making or sludge drying beds. If the amount of oil and grease is more in wastewater then it is better to remove that first to protect the whole plant. The removal of oil and grease is done in skimming tanks which causes the oil and grease to skim off from wastewater. The skimming is more effectively done by flotation, aeration or chlorination. The removal of oil and grease from emulsified form as present in wool - scouring waste is most difficult. For such purpose the removal should be done by using chemical reagents in primary sedimentation tanks.

6.3.2 Primary Treatments

One of the primary treatments is the sedimentation. The suspended solids are removed from wastewater after the removal of gritty material, oil, grease and solid particles. This suspension is done to reduce the caring capacity and move toward the secondary treatment. All of the suspension and sediment particles can be removed more effectively by sedimentation process. Sedimentation process is most successfully applied where waste contains higher concentration of settle able solids or it contains waste related to sewage. These sediment tanks causes separation of waste by using gravity separation method related to the size of particles. Mostly center feed circular clarifiers and horizontal flow sedimentation tanks are used for sedimentation. By using mechanical scrapping the settled sludge is removed from sedimentation tanks and moved in hoppers and pumped out. In a well-organized sedimentation tank 50% of the separation of suspended solids is obtained in about two hours. The most effective sedimentation system would have the ability to remove 90% of suspended solids and 40% of organic matter. The wastewater having higher concentration of industrial wastes requires large time to mix up the waste with heavy loads within the purification plant. By using simple sedimentation process the fine particles could not be removed more effectively. For that purpose chemical coagulation or mechanical flocculation is used.

Within the chemical coagulation or mechanical flocculation wastewater is passed from rotating paddles are operated at a fixed speed of 0.43 m/s for the fixed time of thirty minutes. Due to this continuous mixing the fine particles got coagulated and settle down. *Clarifocculator*is are the special type of equipment available in sedimentation tanks just like flocculating chamber of sediment tank. The floc is produced when wastewater is treated with chemical which causes absorption or suspension and form colloidal particles. Mostly alum $A1_2(SO_4)_3.18H_2O$, ferric chloride, chlorinated copper $FeSO_4.Cl$ which is mixture of chloride and ferric sulfate and hydrated lime are used as coagulants. For treatment of wastewater or water alum is most commonly used. For efficient separation, the chemicals are mixed in baffled channels for formation of coagulant as done in mechanical flocculation before sedimentation. For obtaining the efficient results of flocculation pre aeration is done for ten minutes for the removal of carbon dioxide and hydrogen sulfide gas.

For the removal of impurities most efficient process is the coagulation. Different types of chemicals known as coagulant aiding chemicals also used in little amount to enhance the floc settling and increasing the rate of coagulation. Mostly hydrolyzed higher molecular weight polymers of acrylamide or acrylonitrile having mass of 10^4–10^6, activated silica and poly electrolytes like polymer of methacrylic acid, acrylic and cyanamide are used as coagulants. For preperation of dilute solutions polyelectrolytes are used. Due to their specific properties specific poly electrolyte should be selected. The working of synthetic coagulants for aid depends upon two processes.

- The coagulant aids, having long chain molecular structure, are absorbed on two or more particles, thus drawing them together.
- By reducing the charge on the particles and thus reducing the repulsive power of the like charges on the particles (1) to compensate the alkalinity lime is added, (2) coagulant addition for quick mixing of four to six minutes, and (3) the addition of coagulant helps in proper mixing with less agitation for 40 minutes [1].

Within the primary treatment neutralization and equalization is done. Different industries form different waste having different concentration at different interval of time. To solve this issue all of the waste is mixed in large holding tanks for specific time period. To obtain homogenized and equally distributed effluent the waste of one tank is mixed with other. Mixing of waste effluent by paddle is most effective method to obtain homogenize mixer of waste along with the aeration process. The neutralization of most acidic or basic solution should be properly done before discharging out. The acidic waste should be neutralized by adding lime slurry or lime stone depending upon its type or quantity carbon dioxide, waste boiler or sulfuric acid should be used to neutralize the basicity if both type of waste like acidic and basic are produced and stored separately then mixing of both to neutralize each other is the most cheapest method.

6.3.3 Secondary/Biological Treatments

By using rectangular beds, one to three meter deep, using analytical graded source (broken stones, coke, PVC, coal, In secondary treatment, the colloidal and dissolved organic media in waste containing water is escaped by biological method included microorganism and bacteria. It may be anaerobic or aerobic [11]. In aerobic method, microorganisms and bacteria use organic media as a food. They have following changes:

- Flocculation and coagulation of colloidal media.
- Oxidation of soluble organic media to carbon dioxide
- Degeneration of nitrogenous organic media to NH_3, which is then change into nitrite and gradually into nitrate.

Content of BOD decreased by using secondary treatment. It also eliminates the quantity of phenol and oil. Therefore, maintenance and commissioning are more costly by using secondary treatment. The stream from 1° sedimentation boilers is the first object to system of aerobic oxidation, like as activated sludge elements, trickling filters, oxidation pools or oxidation channels. However, sludge got by aerobic method, both with that gained in 1° sedimentation reservoir, is exposed into anaerobic absorption system in sludge digesters.

All organic-containing matter, with some omissions, like as ethers and hydrocarbons, can be dissolved by biological aerobic method. Multifaceted protein materials and cell tissues are prepared by this method, which are swell and escape from the waste by using settling process impervious organics and germicidal, like as phenols and cyanides can be devastated by particular kinds of microorganisms after a long period of time.

Under anaerobic conditions (i.e., without broke down oxygen or vaporous oxygen), certain gatherings of micro-organism, for example, hydrolyte and methane framing life forms, can do the absorption of complex organic consumes. The hydrolyte life forms change over complex natural mixes to straightforward and low-sub-atomic weight natural acids and alcohols. These are then changed over by methane microorganisms to $C0_2$ and CH_4. Anaerobic treatment process can be completed top to bottom without the requirement for vast surface range. It can happen in blended or improved culture and can, along these lines, be kept up effortlessly on extensive scale. The process can be connected to most sorts of substrates aside from a couple like lignin and mineral oil. The procedure is more affordable, however, the last effluent is less attractive, when contrasted with that from high-impact treatment, as a result of the dim shading, smell and higher remaining BOD. Anaerobic treatment is for the most part utilized for the absorption of sludges. Be that as it may, natural fluid wastes from dairy, butcher house and so forth, were dealt with by this technique monetarily and successfully. The effectiveness of this procedure relies on pH, temperature, waste stacking, absence of oxygen and dangerous materials. Some of the usually utilized natural treatment procedures are portrayed underneath.

6.3.3.1 *Aerated Lagoons and Bioaugmentation*

These are vast holding tanks or lakes having a profundity of 3–5 m and are fixed with concrete, polythene or elastic. The effluents from essential treatment procedures are gathered in these tanks and are circulated air through with mechanical gadgets, for example, coasting aerators, for around 2–6 days. During this time, a solid flocculent muck is formed which realizes oxidation of the broke down natural matter. Bod removable to the degree of 90% could be accomplished with proficient operation. The operation and upkeep are generally basic. The real inconveniences are the bigger space prerequisites and the bacterial defilement of the lagoon effluent which necessitates further biological purification in maturation pond or by secondary sedimentation and sludge digestion.

Bioaugmentation or biomass upgrade is the expansion of financially arranged bacterial culture to a wastewater treatment framework to (1) increment the thickness of coveted microorganisms and their enzymes and (2) accomplish a particular operational objective—for instance, decrease slop generation or control rottenness creation. The expansion of bacterial societies expands the thickness of wanted microorganisms without essentially expanding the solids inventories and solids home circumstances of an enacted slime prepare or anaerobic digester. Adequate expansion of bioaugmentation items may empower an administrator to diminish MLVSS fixation and MCRT. Decrease in MLVSS and MCRT help to control the undesired development of filamentous living beings. Treatment proficiency, allow consistence, and operational expenses at a civil wastewater treatment plant are impacted incredibly by the enzymatic exercises and capacities of an extensive populace and assorted qualities of coli-aerogens. Coli-aerogens are microbes that possess the gastrointestinal tract of people and enter wastewater treatment plant in fecal waste [24].

Cases of critical exercises of coli-aerogens that are of essential to wastewater treatment plants include:

- Nutrient and disintegrated oxygen necessities
- Products acquired from the debasement of substrates
- Rates of degradation of the substrates
- Types of substrates that can be degraded

Cases of critical capacities of coli-aerogens that are of significance to wastewater treatment incorporate

- Adverse conditions that are endured
- Competition with different creatures
- Floc-shaping capacity
- pH development extend
- Temperature development extend

The exercises and capacities of coli-aerogens are upheld by a littler populace of a few imperative genera of saprophytic and nitrifying microscopic organisms that enter the treatment plant as soil and water living beings through inflow and penetration (I/I). The saprophytic microorganisms and their enzymes frameworks are more productive in corrupting a bigger assortment of substrates than the coli-aerogens. Also, numerous saprophytic microbes have a kind of capacities that empower them to survive and stay dynamic under unforgiving natural or operational conditions that are not endured well by the coli-aerogens. Saprophytic microorganisms are principally in charge of the debasement of natural mixes (substrates) in nature. Notwithstanding, saprophytic microbes don't enter wastewater treatment plants in noteworthy numbers and don't develop in substantial numbers in wastewater or ooze, because of the nearness of vast quantities of coliaerogens that enter wastewater treatment plants. Therefore, the productive enzymatic exercises and one of a kind capacities of the saprophytic microscopic organisms are "weakened" by the coli-aerogens. Because of the moderately modest number of saprophytic microscopic organisms in a wastewater treatment plant when contrasted with the huge number of coli-aerogens, a wastewater treatment plant may encounter challenges in treating particular substrates, enduring unfavorable conditions, or rectifying an operational problem [24].

The purpose of mixing the substances which are used for the biodegradation of contaminant is to make the working ability of plant better. There are certain things which decide the products and culture to be used, these are purpose of installing the treatment plant. There is addition of bacterial culture to the places where there is installation of wastewater treatment plant. The place where bacterial culture is added, decided according the requirement of treatment plant and set duration of bacteria. The mean of adjustment period is that period needs to bacteria for making their enzyme in the new environment. If adjustment period is longer than after longer time the bacteria will be added to treatment plant. There is limited position of addition of products which are used for the finishing the pollutant but there is no location of addition of bacterial culture, and bacteria travel throughout the treatment plant. There are two type of addition of bacterial culture to the treatment plant one is introductory dose and other is maintenance dose. In introductory dose, amount of dose used is less than 2ppm and it is applied for 2–4 weeks. The amount of dose in the maintenance dose is mostly 2ppm and its application is either daily or after week or according to the need. The reproduction place of augmented saprophytic bacteria is wastewater treatment plant; their number cannot cross the number of coli-aerogels addition of which is continuous. There are millions of colli-aerogens in one milliliter of liquor or can cross billion in one flock particle. The addition of saprophytic is to up to level where there is easiness in making the observation of their working and abilities. Bioaugmented bacteria do not produce the diseases. But the some products like preservative used to stop the growth of bacteria when bacteria are stored or when they are transported can cause allergy. There precautionary clothing should be used for protection against allergic reaction; therefore the goggle and long shirts should be used while dealing the bioaugmentation products. The washing of organ should be carried out which come in contact with the product. These bacteria do not produce diseases. However some chemical like preservative used for stopping their growth while stored or transporting can cause allergy. So the protective clothing should be used for the protection against the allergic reaction, so the long shirt with sleeves and Google should be used while dealing with the augmented products. The washing of part of body should be carried out which are in more chance to come in contact with these product. These bacteria which are required are obtained by the sampling of water and soil from their living places. After getting the bacteria screening is carried out to check the enzymatic actions of bacteria as well as their other activites.

The screening tell us the

- How fast they can cause breakdown of organic compound
- The condition to which they can bear and work
- The type of organic matter to which they can break

The large number of population of bacteria can be grown using the carbon source. The growth can be stopped using the different techniques like by making their suspension, by keeping them in freeze-drying, to carry bacteria liquids and dry form of chemical should be used. The various numbers of species and genera are used for specific and general purpose. Many species of saprophytic bacteria are used for producing the augmented products which are used for finishing the pollutant from the soil and water.

Beside saprophytic bacteria the other microorganism used for producing the augmented product are fungi and nitrifying bacteria. But the microorganism that can efficiently can produce augmented products are saprophytic bacteria. The degradation of cBOD which is difficult to degrade can easily be degrade using the saprophytic bacteria. The degradation of pollutant can be carried out using different enzyme of saprophytic bacteria and these enzyme are released outside the body and named as exoenzyme. The finishing of pollutant is such that enzyme of saprophytic bacteria diffuses through the particle and as result of diffusion interact with the different things like carbohydrate, lipid and protein, and as result of interaction dissolve them. For the breakdown of sugar, fatty acid and other different chemical beside saprophytic bacteria coli-aerogens are also used. Sometime there is competition between the saprophytic bacteria for nutrients and dissolved oxygen, on the other hand there is less competition between the coli-aerogen bacteria and filamentous organism. There is increase in performance of wastewater treatment plant as result of dissolving the cBOD and the dissolving and breakdown of cellulose. The starch and cellulose are made up of smaller subunit which is glucose, these glucose is not soluble this is due unique bonding present between the molecule glucose. The breakdown of cellulose which occur in plants wood cannot be done by human and coli-aerogen, so the cellulose is not digestible in the body and it is removed from the body through fecal. There is presence of cellulose in sludge which is either primary or secondary and cellulose is 8–15% part of sludge. The bacteria named as cellumonas have ability to breakdown the cellulose, this is due to reason that it produces the enzyme named as cellulase. This can cause the decomposition of bonding of cellulose and hence can dissolve cellulose.

However there is a problem that this bacteria which can cause breakdown of cellulose do not occur in natural population of bacteria which are present in sludge. These bacteria are augmented in it processing. By the mixing the augmented bacteria in the sludge the breakdown of strong bonding of glucose molecule in the cellulose and starch can be broken down and cellulose and starch can be make soluble. The breakdown of glucose molecule generates carbon dioxide, water and other give rise to bacteria. According to estimation if 0.6 pound of glucose breaks then it produces 0.4 pound of carbon dioxide and 0.2 pound of water. If a one pound of cellulose breaks then it becomes part of sludge and it makes 0.6 pound of sludge. If one pound of cellulose do not solubilize then it makes the 1 pound solid which is part of sludge but it is not activated sludge process. As result of addition of cellumonas the quantity of cellulose reduces up to 40% to produce the sludge. There is addition of cellumonas from activated sludge to either aerobic or anaerobic digester. The continuous breakdown of cellulose in aerobic and anaerobic digester. There is addition of bioaugmentation product to solve many problems of treatment efficiency. The use of bioaugmentation product in wastewater treatment plant is beneficial economically. Some specific bacterial culture is used for the breakdown of sulfur containing smell-producing compound. The breakdown of methyl sulfide, dimethyl sulfide and dimethyl disulfide can be can done using the the bacteria named as thiobacillus and hypomicrobium. The following thing should know about the bioaugmentation product for the selection and using the bacterial culture [17].

- The present situation of treatment plant
- The recognition of situation to be given with augmentation product
- Gathering and reading of data for consumption in the bio-augmentation product
- To chose the proper bioaugmentation product gathering and reading of data
- Sight the possible effect of augmentation product
- Learn the method to store and use
- See the time which is required to observe the effect of augmentation product
- See the condition of checking the product are working
- Check the price of augmentation product

6.3.3.2 Trickling Filter Process

The trickling filter contained the round resin which are synthetic size of which is 40 mm to 150 mm, water is sprayed evenly on the whole bed with the help of distributer which is moving and have two face and nozzles on it [30]. Hence there is flow of wastewater through the media. The setting of the filter is such that flow of air is opposite to the flow of effluent flow. There is layer of viscous material which contained microorganism which are aerobic and bacterial it is collectively known as zooglea is kept on the filter medium which trickle the nutrient from the wastewater [43]. There is attachment of organic matter on the surface of viscous material when ir passes through the film of viscous material, where the reoxidation take place which is done by bacteria and other microorganism which are present there. When the layer increases small part of it sloughed off, and flows with the effluent. The effluent coming from the trickling filter is allowed to settle down and then its separation is carried out. There is pumping of

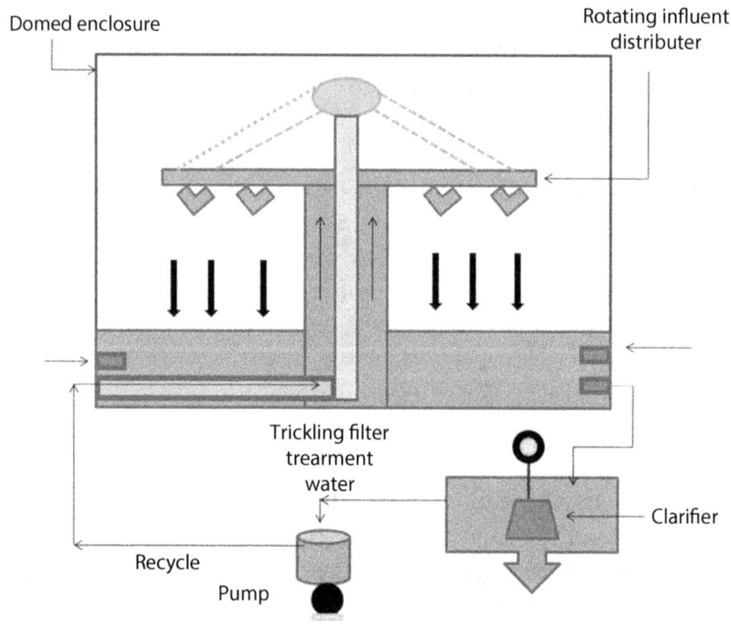

Figure 6.8 Trickling filter for wastewater treatment

ofsludge to the unite where the digestion take place. The diagram of trickling filter is shown in Figure 6.8 [44].

The process of trickling filter is through the aerobic so it is the facultative system. There is presence of species of bacteria named as desulfovivrio. It form spore and respire aerobically. The presence of micrococcus is also ensure in the trickling filter. Algae, fungi, ciliates, protozones also present feed of which is microorganism are also present in the trickling filter [29].

The material which is made of plastic is also used in filter and is useful for the industrial waste. The better result can be obtained using the smaller media as compared to larger media but it causes the restriction in the flow of air. The media which is not natural is useful for the material which comes out from the industries. The surface area of synthetic material is large but and they are also economical to bear the film of microorganism. The temperature is harmful for the layer of microorganisms. The rate of metabolism of microorganisms depends on the heat of water passing through the filter. Therefore the capacity of working the filter decreases in the winter season [29, 30].

The efficiency depend on the component of waste, size of waste, and evenness of waste, and spraying of water on the surface of filter. There is greater power of trickling filter to avoid the toxicity. However huge load should be avoided because it lowers efficiency, efficiency can decrease temporary or permanently [36]. The working of filter is very easy and BOD can be reduced up to 65–85% using the trickling filter paper which is function of filtration. Further the repeated attention to the manual is not needed in case of trickling filter paper. The trickling filter paper produces the effluent which are of same composition and further processing is not needed [29].

The only harm of using this process is that the it require higher cost for its installing and ventilation is also required for this process, this ventilation is needed for the drain system which is on lower side. The efficiency lower as the amount of wastewater is increased. To avoid this decrease in efficiency the water is mixed with effluent with the which is obtained from the previous process. The trickling filter is used for the treatment of wastewater from different sources like water from distillery, brewery, cannery, etc [27].

6.3.3.3 *Activated Sludge Process*

This is process for the treatment of wastewater the by oxidation using microorganism, this is for the water which contained the solid in dissolved form like colloids, coarse solid and organic matter. There is suspension of floc in the tank, and this tank is used for the treatment of industrial wastewater [45]. The activated sludge is termed as the aerobic bacteria which after growth remains suspended. The sludge and the effluent separation is carried out and then discharging is carried out. To produce perfect population small part of sludge is restored and is used for fresh treatment plant. The sludge obtained after the primary sedimentation as well extra sludge is send to the digester. Proper aeration is provided to the sludge for 6–24 hours, for industrial waste the sludge aeration time is 6–24 hours. The BOD can be removed up to the 90–95% [29].

The microbial floes formed in this process comprize of Zoogleal masses of living organisms, embedded with their food and slime material and act as active centres for biological oxidation and that is why it is called "Activated Sludge." A young light sludge is preferred to old heavy sludge, because the latter would be mineralized and become devoid of oxygen. For this process to be efficient, at least 0.5 ppm oxygen must be present all the time [45].

Oxygen is supplied either by mechanical aeration or by diffused aeration systems. The addition of some nutrient is also necessary for the microorganism, these nutrient are nitrogen and phosphorous, to provide these nutrient the urea and mono – or di – ammonium hydrogen us used, there is also requirement of other nutrient which are potassium, magnesium and calcium. The pH, temperature oxidation and reduction potential is also the key factor which affect the efficiency of activated sludge. The pH for this process is 6.5 to 9. When temperature is decreased then process of metabolism decrease while high temperature enhances the rate of metabolism, use of oxygen is fast and as a result the anaerobic respiration starts. There are some chemical which cannot broken down like alkyl benzene sulfate and polyethylene glycol, these are all synthetic compound. There is formation of foam due to these compound and as a result the use of antifoaming agent is essential [42].To check the activity of activated sludge process the indicators are used. There is presence of large number of population of microorganism. If there is presence of ciliates then it mean there is deficiency of nitrogen and potassium, and presence of filamentous indicated by the retriction in floc compaction and turbid effluent are produced [43].

The active sludge process produces a high quality effluent with relatively small areas. Activated sludge process is an efficient process that reduce 95% BOD with retention time of 13 to 14 days. The disadvantages include high running costs and maintenance, require special attention and impact sensitivity to loads of toxic and organic substances. The active sludge process has been successfully used for the treatment of effluents from food, sugar, textile processing, industrial antibiotic production, etc [45].

Various amendments of the conservative sludge-active process have come into fashion and differ mainly in the air supply method. These contain: conical aeration process, compressed air process, prolonged aeration process, aeration dispersion process, contact stabilization, simplex process aerobic high speed processing process, aeration rotor process and Swedish INK A process. The configuration represented Figure 6.9 activated sludge process [45].

6.3.3.4 *Oxidation Ditch and Oxidation Pond Process*

It is supposed that adjustment of predictable activated sludge method. Oxidation channel containing of an oval geometry proper channel and about one to two meter lined and deep with butyl rubber, tar and plastic [1]. Waste-containing water after the process of comminution or screening in the 1° treatment permits into the oxidation channel. The mixed liquor containing the sludge solids is removed in ditch by using of mechanical containing rotors. There is need to larger retention times (R_t).The commonly hydraulic retention twelve to twenty four hour and for solids, it is twenty to thirty days. Maximum formation of sludge is reused for succeeding treatment set. The extra amount of sludge can be dehydrated without using dehydrating bed [35].

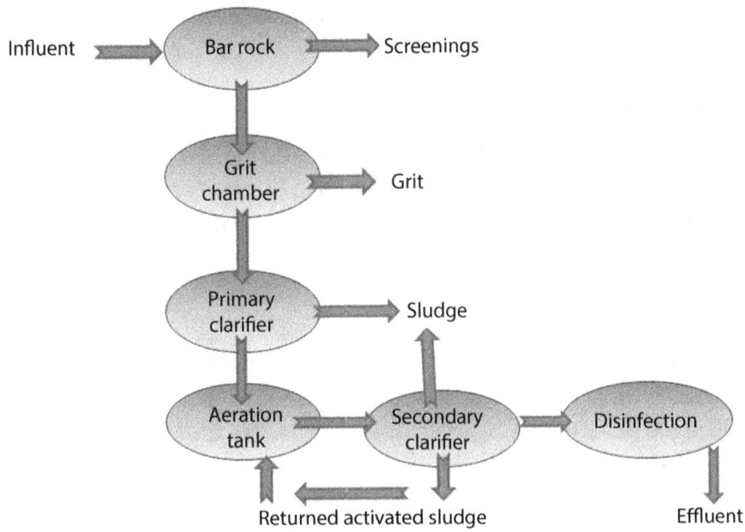

Figure 6.9 Activated sludge process flowsheet.

The main benefits of oxidation channel such as simple in handle, soft maintenance, low amount of construction, maintenance and operation, average flexibility and efficiency. These methods are commonly used for waste containing having less amount of BOD. Oxidation of ditch method is used more for the waste containing water treatment for manufacturing of beet sugar, it also involves the fruit and vegetable sugar cane industry, meat containing packing industry and all edible oil industries and refineries [1].

An oxidation pool is a largest shallow pool having one meter to 2 meter depth with prepare to measure the outflow and inflow, the waste containing water enter the pool at one end and stream is escaped at the other end point. Stability of organic media in waste-containing water bringing about by bacteria like as Flavo bacterium, Alcaligenes and Pseudmonas by using flagellated protozoa family [16]. The amount of oxygen for the process of metabolism is delivered by algae in the presence of the pond. The algae consume the carbon dioxide escaped by their photosynthesis [43].

Although the purification pond is generally regarded as an aerobic process, purification is performed by a combination of optional aerobic and anaerobic processes. Wastewater present at the top of the tank (which accounts for most of the residue) undergoes aerobic oxidation in CO_2 and H_2O. Solids present in the waste, depositing as a layer on the bottom, act as an anaerobic phase. Here, organic matter is oxidized by anaerobic bacteria to CH_4, CO_2 and NH_3. The optional area exists near the anaerobic phase oxidation treatment of waste is cheap and operation and maintenance are simple [26].

The process can be used for all types of waste and can have any degree of purification. The process can bear loads of organic and hydraulic shock. Heavy metal ions present in the wastewater are precipitated as hydroxides (due to the high pH of the reflux in the oxidation pool) depositing in the form of mud. However, oxidation ponds require more space. The effluent from the oxidation ponds may require disinfection or extra water treatment treatment separately maturing before the final discharge [34].

6.3.3.5 *Anaerobic Digestion Process*

Anaerobic process is being used for sludge digestion since very long. It is a residue containing water from 1^0sedimentation to 2^0 sedimentation treatment. The particles-containing sludge experiences digestion and low level of fermentation using anaerobic bacteria in a sludge tank where watery residue is controlled at 34 °C and pH 6–8 for 30 days. CO, CH, and NH_3 are enlightened as final product[36].Type of Pseudomanas, *Aerobactor, Flavo bacterium, Alcalagenes*such as change into multifaceted organic compounds to less molecular weight organic alcohols and acids.

Methanococcus and Methanosarcina methano bacterium, kinds of bacteria are answerable for the group of methane. *Desulfo vibrio* bacteria decreases sulfates thus hydrogen sufide is reduced. The anaerobic sludge dissolved method is faster at high temperature. The sludge from tank may consist about 90–93% of water. The sludge having no water in dehydrating beds included vacuum filters and filter presses. The dehydrating sludge, after the process of chlorination and may be directed for the final disposal. The different methods used for the final disposal such as incineration, dumping at selected area, removal in land fillings or utilizing as a low quality grade using fertilizer [8].

6.3.3.6 *Biogenic Enzymatic Wastewater Treatment*

All living organisms have specialized type of proteins known as enzymes. In many cells a lot of enzymes are present and synthesized constantly. For biochemical processes enzymes is used as catalyst. The rate of reaction increase 1000000 times as compare to reaction occurred without catalyst and carry out reaction in temperature range which is bearable to living cells [23]. The non-protein part of enzyme is called cofactor. Within the cofactor organic molecules like biotin, metallic activators include Cu, Mg, Co, K and Zn along with coenzyme. The rate of reaction is enhanced in the presence of cofactor [8].

Enzymes are used again and again in the reaction because no structural change is produced in enzyme during the whole reaction. Only the active site of enzymes is used as a stage for reaction to proceed. This active site makes a weak bond with substrate and result in the formation of product. But each enzyme is specific to each specific substrate which got react with enzyme and form product [24].

An enzyme substrate complex is made when specific substrate binds with active site of enzyme. When the enzyme substrate complex is made the bonds within the substrate becomes weak and result in dissociation of molecule into simpler substances as shown in Figure 6.2 or they combine with each other to form complex structure. During catabolism complex is broken down e.g. the breakdown of stored food in bacteria. While in anabolism complex structure is formed e.g. the transformation of sugar into bacterial cell [18].

High level of specificity is present in enzymes. Enzymes are very specific for their (1) substrate attachment and discard (2) the ions or compound which they form e.g. enzymes have specific property for degenerating the number of carbohydrates [27].

It is due to the shape of active site that enzyme can catalyze specific reaction. The shape of enzyme is related to the binding of thiol group and electrical charge is produced in enzyme because of presence of ionized hydrogen bond. If two of more substrate can bind with same enzyme it is due to its same functional group that can be carboxyl, hydroxyl or chemical of same nature. Just like proteolytic enzyme which breaks down peptide bind of proteins. The name of enzyme is written by using suffix –ase related to the name of substrate which attach with it. Within the wastewater treatment plant hydrolases play an important role for break down in biochemical oxygen demand. The enzyme causes to convert the complex molecules into simpler substances which can be easily digested by bacterial cell [28].

Infact all the enzymes are produced within the cell. Related to the activity which enzymes perform they are categorized into two groups exoenzymes and endoenzymes. Endoenzymes work within the cell while exoenzymes work in the cell after entering in the cell membrane. Specific time is required to the cell for production of exoenzyme. Due to change in structure or electric charge in active site enzyme activity is reduced. This process is called denaturation and occur when disulfide bond or hydrogen bond breaks [28]. The denaturation of enzyme can be temporary or permanent. It is caused due to change in pH, elevation intemperature, inhibitory waste including oxidizing agents like Cl_2, H_2O_2, $KMnO_4$or due to the presence of heavy metals [7].

6.4 Conclusion

Most common sources of wastewater are discharges of textile industry, pulp and paper industry, electroplating industry and miscellaneous included domestic wastes, organic pollutants, radioactive pollutants, thermal pollutants, nutrients, pathogen, inorganic pollutants, suspended solids and sediments. Organic pollutants are further categorized as follows; Oxygen-demanding wastes, *oil pollutant,* sewage and agricultural runoff, synthetic organic compounds and disease-causing wastes. Microbial diversity of wastewater consists of protoza and bacteria. Protozoa included amoebae. Flagellates, ciliates, rotifers and nematodes that add weight to floc particles, improve their settleability, consume dispersed cells, cleanse the waste stream, produce and release secretions that coat and remove fine solids like colloids, dispersed cells, particulate material from the bulk solution to the surface of floc particles and recycle nutrients (nitrogen and phosphorus) through their excretions. In addition to these benefits provided by the ciliated protozoa, the metazoa burrow into floc particles. The burrowing action promotes acceptable bacterial activity for the degradation of substrates in the core of the floc particle by permitting the penetration of dissolved oxygen, nitrate, substrates, and nutrients. Substrates are the carbon and energy sources used by bacteria for cellular growth and activity. With exceptions, substrates consist of carbonaceous, biochemical oxygen demand compounds and nitrogenous, biochemical oxygen demand compounds while eukaryotic perform their function in various groups like hydrolytic, acetogenic, coliforms, cyanobacterial, denitrifying, fermentative, floc-forming, nocardio forms, methane-forming, nitrifying, denitrifying, phasphorous solubilizing,sulfur oxidizing and reducing communities. The comprehensive understanding of adequate and active microbial population, microbial-waste material interection, mutual association of biocommunity and their biological role is essential for efficient degradation of organic waste and limited growth of photosynthetic organisms.

References

1. Dara, S.S., Mishra, D.D., *A text book of enviromental chemistry and pollution control.* S. Chand Company, India, 2015.
2. Johns, L.S., Kelly, H., Ruttenberg, R.,*The US textile and apparel industry: a revolution in progress.* Research Report, Office of Technology Assessment, Washington, DC, 1987.

3. Moss, B., Water pollution by agriculture. *Philos. Trans. R. Soc. Lond., B, Biol. Sci.,* 363, 1491, 2008.
4. Laws, E.A. *Aquatic pollution: an introductory text 4th edition*, John Wiley & Sons. 2017.
5. National Management Measures to Control Nonpoint Source Pollution from Agriculture *(Report), EPA-841-B-03-004, July 2003.*
6. Clescerl, L.S., Greenberg, A.E., Eaton, A.D., *Standard methods for the examination of water and wastewater* (20th ed.), American Public Health Association, Washington, DC, ISBN: 0–87553-235–7, 1998.
7. Lawrence, A.W., McCarty, P.L., The role of sulfide in preventing heavy metal toxicity in anaerobic treatment. *J. Water Poll. Control Fed.*, 37, 1965.
8. Britton, G., *Wastewater microbiology*. Wiley-Liss, New York. ISBN: 0–471-30985–0, 1994.
9. Abraham, J.V., Butler, R.D., Sigee, D.C., Ciliate populations and metals in an activated-sludge plant. *Water Res.*, 31, 1103–1111, 1997.
10. Al-Shahwani, S.M., Horan, N.J., The use of protozoa to indicate changes in the performance of activated sludge plants. *Water Res.*, 25, 633–638, 1991.
11. Antonietti, R., Madoni, P., Ghetti, P.F., Some notes on the biological self-purification process: Biological water-treatment plants as man-made ecosystems. *SocietàItaliana di Ecologia, Atti*, 1, 379–382, 1981.
12. Ettl, M., The ciliate community (Protozoa: Ciliophora) of a municipal activated sludge plant: Interactions between species and environmental factors. *Protozoological Monographs*, 1, 1–62, 2001.
13. Hul, M., The composition of microfauna of the biofilm of a rotating biological disc and rotating biological multicage under conditions of high loading with pollutants. *ActaHydrobiologica.*, 34, 129–138, 1992.
14. Trinci, A.P.J., Thurston, C.F., *Transition to the non-growing stage in eukaryaticmicrorganisms.* In Gray, TRG, Postgate, JR, editors, The survival of vegetative microbes, 26th Symposium of the Society of General Microbiology, Univ of Cambridge, Cambridge Univ Press, Cambridge, 1976.
15. Klimowicz, H., Microfauna of activated sludge. Part I: Assemblage of microfauna in laboratory models of activated sludge. *ActaHydrobiologica.*, 12, 357–376, 1970.
16. Zehnher, A.B.J., *Biology of anaerobic microorganisms.* John Wiley and Sons, New York 1988.
17. Doetsch, R.N., Cook, T.M., *Introduction to bacteria and their ecobiology.* XII, 371 S., zahlreiche Abb., 23 Tab. Lancaster, Baltimore, MTP Medical and Technical Publ. Co. Ltd., University Park Press, 1973.
18. Taylor, G.T., The methanogenicbacteria. *Prog. Ind. Micro*, 16, 1982.
19. White, D., *The physiology and biochemistry of prokaryotes.* Oxford University Press, New York, 2000.
20. Salvadò, H., Gracia, M.P., Amigò, J.M., Capability of ciliated protozoa as indicators of effluent quality in activated sludge plants. *Water Res.*, 29, 1041–1050, 1995.

21. Gottschalk, G., *Bacterial Metabolism.* Springer-Verlag, New York, 1979.
22. Slater, J.H., *Mixed cultures and microbial communities.* In: Bushell, ME, Slater, JH, editors. *Mixed culture fermentations*, Academic Press, London, pp. 1–24, 1981.
23. Anderson, J.W., *Sulphur in biology.* University Park Press, Baltimore, 1978.
24. Gerardi, M.H., *Wastewater bacteria*, John Wiley & Sons, Inc., Hoboken, New Jersey, 2006.
25. Gerardi, M., *The microbiology of anaerobic digesters.* Wiley-Interscience, New York, 2003.
26. Leschine, S.B., Cellulose degradation in anaerobic environements. *Annu. Rev. Microbiol.*, 49, 1995.
27. Mara, D., Horan, N., eds., *Handbook of water and wastewater microbiology*, Academic Press, New York, 2003.
28. Warren, R.A.J., Microbial hydrolysis of polysaccharides, *Annu. Rev. Microbiol.*, 50, 1996.
29. Mudrack K., Kunst, S., *Biology of sewage treatment and water pollution control.* Ellis Horwoood Limited, Chichester, England, 1986.
30. McCarty, P.L., Smith, D.P., Anaerobic wastewater treatment. *Env. Sci. Tech.*, 20, 1986.
31. Gerardi, M., *Nitrification and denitrification in the activated sludge process.* Wiley-Interscience, New York, 2002.
32. Guentzel M., Baron, N., *Escherichia, Klebsiella, Enterobacter, Serratia, Citrobacter, and Proteus.* In: *Baron's medical microbiology* (4th ed.). Univ of Texas Medical Branch. ISBN: 0–9631172-1–1, 1996.
33. Hvited-Jacobsen, T., Sewer Processes: *Microbial and chemical process engineering of sewer network*, CRC Press, Boca Raton, FL, 2002.
34. Speece, R.E., Anaerobic wastewater treatment. *Env. Sci. Tech.*, 9, 1983.
35. Wentzel, M.C., Ekama, G.A., Marais, G.R., Processes and modelling of nitrification-denitrification biological excess phosphorus removal systems—a review. *Water Sci. Technol.*, 25, 1992.
36. McCarty, P.L., McKinney, R.E., Volatile acid toxicity in anaerobic digestion. *J. Water Poll. Control Fed.*, 33, 1961.
37. Zeikus, J.G., The biology of methanogenic bacteria. *Bacteriol. Rev*, 41, 1977.
38. Petropoulos, P., Gilbride, K.A., Nitrification in activated sludge batch reactors is linked to protozoan grazing of the bacterial population. *Can. J. Microbiol.*, 51, 791–799, 2005.
39. Pogue, A.J., Gilbride, K.A., Impact of protozoan grazing on nitrification and the ammonia – and nitrite-oxidizing bacterial communities in activated sludge. *Can. J. Microbiol.*, 53, 559–571, 2007.
40. Davelaar, D., Davies, T.R., Wiechers, S.G., The significance of an anaerobic zone for biological removal of phosphate from wastewaters, *Water S.A.*, 2, 54–60, 1978.
41. Ekama, G.A., Siebritz, I.P., Marais, G.V.R., *Considerations in the process design of nutrient removal activated sludge processes.* Presented at the I.A.W.P.R. post-conf. seminar on Phosphate removal, Pretoria, April 1982.

42. Rehman, A., Shuja, R., Shakoori, A.R., Tolerance of two ciliates, Stylonychiamytilus and Paramecium caudatum, isolated from industrial effluents to an organophosphate, ndosulfan and their potential use in bioremediation of insecticide-contaminated wastewater. *Pakistan J. Zool.*, 40, 255–261, 2008.
43. McCarty, P.L., McKinney, R.E., Volatile acid toxicity in anaerobic digestion. *J. Water Poll. Control Fed.*, 33, 1961.
44. Marcus Van Sperling, *Activated sludge and aerobic biofilm reactors.* IWA Publications, ISBN: 1-84339-165-1, 2007.
45. Liao, B.Q., Allen, D.G., Droppo, I.G., Surface properties of sludge and their role in bioflocculation and settleability. *Water Res.*, 35, 2001.

7

Role of Plant Species in Bioremediation of Heavy Metals from Polluted Areas and Wastewaters

Mayerly Alexandra Oyuela Leguizamo*

Faculty of Environmental and Natural Resources, Campus El Vivero, Natura Building, Universidad Distrital Francisco José de Caldas, Bogota-Colombia

Abstract

Natural resource pollution caused by the mobility and solubility of toxic elements significantly damages the environment. In the decontamination of heavy metals, phytoremediation is used as one common *in situ* method. This technology usually involves the use of species that may exhibit an invasive behavior, thereby affecting the environmental and ecological dynamics of the ecosystem into which they are introduced. This chapter focuses on the identification and research of keywords related to heavy metal decontamination and ecology. In addition, the author gathered and analyzed relevant information that allows the comprehension of the phytoremediation process. The author suggests a list of native or endemic species regarding their behavior toward heavy metal contamination. It is a contribution to ecology of possible phytoremediation bio-tools worldwide. Suggested plants were selected from a survey of native and endemic plants that belong to predominant families in heavy metal phytoremediation processes. Recent investigations focus on the identification of new species able to decontaminate substratum polluted with heavy metals, in the development of phytoremediation process. Finally, the use of native and endemic biodiversity contributes to the research of the ecology and environment of nature's remnants in the ecosystems.

Keywords: Phytoremediation, native plant species, endemic plant species, heavy metals, phytotechnologies

**Corresponding author*: mayerly.oyuela@gmail.com

Shahid-ul-Islam (ed.) Advanced Materials for Wastewater Treatment, (223–262)

7.1 Introduction

The ecosystems in the earth suffer the consequences of the vast level of organic and inorganic pollutants in the environment, which spread substrates, mainly via dry/wet depositions; most of them include the intervention of humans. Within the important ecosystemic services, is the filtering function that depends on the behavior of pollutants in the soil and the biological, chemical, hydrological and geological processes [1]. The final place of the contaminants depends on the specific adsorption, degradation and leaching processes to which they have been subjected.

Among other factors, heavy metals (HM) are spread by anthropogenic interferences such as mining, industry, agriculture and construction activities [2, 3]. Despite the essential functions of many metals for living beings, others can be toxic in high concentrations [4], when they are found in large quantities of ecosystems around the globe. Heavy metals exceed toxicity levels in nature, that means, the geochemical background limits. Therefore, the main problem lies in the toxicity of heavy metals and in the bioaccumulative behavior that harms humans, plants and animals' health [5, 6], among others, with deleterious effects on the environment like pollution, degradation of resources in quantity and quality.

As a result of the mentioned problems, HM must be immobilized or removed. One of the techniques used in this decontamination process involves phytoremediation, also called phytocorrection and phytocleaning, and are promising technological approaches in the decontamination process [2, 7–15]. This chapter presents detailed information about heavy metal phytoremediation and the advantages and disadvantages of its use.

Most of the ecological restoration, bioremediation and revegetation processes increase biological invasion due to the introduction of exotic species or it is caused indirectly by the creation of an artificial or altered environment in which exotic or invasive species can thrive [16]. From this fact, to implement phytoremediation projects, it is necessary to determine the ecological functions of native or endemic species into the ecosystems [17].

This chapter focuses in the study of native and endemic plant species, especially in terms of studies of plant behavior with the HM in growth substrate. Species were taken from the study developed by Oyuela Leguizamo *et al.* [18], among others. One the developed topics is the presence of HM worldwide, the implications of allochthonous and autochthonous vegetation in an ecosystem, and delves into the phytoremediation of HM. Finally, after a detailed summary of native and endemic plants species and the corresponding research, authors recommend a contribution to ecology of possible phytoremediation bio-tools. This, in order to contribute to the research of native and endemic species, their contribution within the services that the ecosystems provide, and the prospect of its use in bioremediation processes, as an initiative to contribute to the care and preservation of the existing remnants in many ecosystems degraded by substances in high concentration levels such as heavy metals, specially in ecosystems located in urban areas.

7.2 Heavy Metals (HM) Worldwide

In a biologically language, "heavy" designates a series of metals, in some cases metalloids, that even in low concentrations can be toxic for plants and animals. From a chemical perspective, heavy metals are transition metals with an atomic mass greater than 0.002 Kg, and a specific weight greater than 5 N/m3 [15]. These elements can exceed the levels of the soil's naturally occurring chemical elements, which is known as natural or geochemical background or geochemical baselines; this geochemical background is the average local content of chemical elements in rock soil, natural waters, surface atmosphere, plants, water, soil and sediments. Different degrees of metal accumulation have been reported, from slightly higher than the geochemical background, to extreme cases in which the metal exceeds 2% of the plant's dry material [20]. The origin of heavy metal contamination consists of activities that contribute to soil, air and water contamination. Common examples include chemical industries, metal processing industries [21], industrial activities as dumping and inadequate waste disposal, mining for waste processing [22, 23], traffic jams, construction materials and agriculture (inadequate agricultural practices), among others [24, 25].

Metalliferous soils have high concentration of metals elements, of natural origin, as well as sites polluted with HM due to anthropogenic activities. These soils are geographically distributed in New Caledonia [26], Saudi Arabia, Sudan, Papua New Guinea's [27], The Democratic Republic of Congo (50% of the world's cobalt reserves), Chile (30% of the world's copper reserves), Ukraine and South Africa (world's manganese reserves) [28]. Some others are located in Switzerland, Eastern France, Northern Italy [29]. Serpentine, ultramafic, and calamine soils are sources characterized by high concentrations of HM of natural origin [30]. Some authors refer areas with these characteristics in Southern Europe, Asia Minor, Portugal, West of Iraq, Turkey, Iran, northwestern North America, Zimbabwean, Cuba, New Caledonia, Brazil [31–33]. Also in Australia, Canada, Russia, Congo and Zambia, [28], Puerto Rico, Brazil, Costa Rica, California, the Philippines, Indonesia [34, 35], Albania [36]. Ultramafic rocks, places like

Cuba and New Caledonia, where serpentine-endemic hyperaccumulators can be concentrated in one locality [37]. Phytomining, the use of plants in mining, was the result of the discovery of HM, accompanied by considering their use in commercial processes, as well as the increase of HM in plant growth substrate [9, 13, 14, 38].

Soils in serpentinized or ultramafic areas are rich in Ni, and calamine (mineral with high levels of Zn and Cd), whether naturally occurring or the product of anthropic - or mining-related contamination. Ultramafic rocks describe, among others, piroxenite, dunite and igneous (such as peridotites) or metamorphics (such as serpentinites) [39], that are less than 45% silica (SiO2) and have high concentrations of Mg and Fe. Often, these soils present, in addition, high levels of Cr, Co and concentrations of Ni, raised Mg/Ca quotient and low levels of P, K and Ca [32, 40]. Consequently, soils that evolve on these types of ultramafic rocks are slightly acidic, have low nutrient contents (K, N, P), exhibit more erosion, contain high concentrations of HM (Fe, Co, Cr, Hg, Ni) and have a low Ca/Mg ratio. These conditions diminish the vegetation's productivity and represent a serious disadvantage for conditions for life of all organisms present [30, 41, 42].

The literature presents different studies about the presence of heavy metals in the environment, such as the work done by Schaller *et al.* [43] where he refers to many articles about the presence of metals/metalloids, such as Ni, Zn, U, Pb, Cu, and As, in sediments in wetlands around the world. He also showes the highest chemical elements content found for the sediment of different types of wetlands. Schaller *et al.* [43] compiles information on research done in sediments in Malaysia, Egypt, Czech Republic, Taiwan, Pakistan, China, Canada, Germany, Spain, UK, Zambia, USA, India, Paraguay, Poland, Brazil, Vietnam, Italy, Australia and Belgium. Some examples of these sediments are paddy fields, wetland near copper smelter, organic rich pond sediment, wetland effected by mine waters, urban lakes, lake wetlands receiving waste waters, stream sediments near Pb/Zn mine, anthropogenic lake sediments, saltmarsh sediments, river sediments, lake and wetland sediments, water spinach cultivation, urban riparian sediments, natural wetlands impacted by uranium mining and processing, lagoon and river sediment, Chesapeake Bay sediments, among others.

In order to decide the necessity/urgency of rehabilitation worldwide, actual and potential metal environmental risks were evaluated. Mench *et al.* [44] describes *in situ* metal immobilization and phytostabilization of contaminated soils, as follows: in many parts of Europe and North America, natural vegetation completely disappears due to HM contamination. In France, 2000 of the most polluted sites were rehabilitated. In northeast Belgium, the surface soil of more than 280 km^2 contains the highest background values levels of HM. In the United Kingdom, soils were contaminated by the industrial legacy from base metal mining, metal refining and smelting. In India, there was a contamination, product of industrial effluent and sewage sludge on agricultural land [21]; also, in this country, there are more than 3000 tanning industries [45]. Rail junction areas in Poland are contaminated with numerous compounds and chemical substances [46]. Sultan Marsh, one of the most important wetlands in Turkey, Middle East and Europe, is mainly polluted with Pb, Cd and Cr [47]. Distribution and phase association of Cd, Cr, Cu, Fe, Mn, Ni, Pb, Zn in Lake Nasser (Aswan, Egypt), were investigated [5]. In Spain, the potential use of spontaneous vegetation (*Festuca rubra* and *Juncus sp.*) was evaluated for phytoremediation and/or phytostabilization in the former serpentinite quarry of Penas Albas (Moeche, Galicia, NW Spain). The activity on this quarry left behind a large amount of waste material scattered over the surrounding area [48]. In the natural wetlands in Bogota, Colombia, the metals present include Al, Cu, Zn, Cn, (found in all wetlands), Cd, Cr, Ni and Pb. The presence of these HM can be attributed to flawed connections in the industrial sector and a flawed urban sewage system [49].

Many cases of contaminated substrates have been studied, such as heavy metal contamination in large areas due to mining and metallurgical operations, an environmental problem present in some industrial regions in Bulgaria. The Kremikovtsi region, investigated by Doichinova and Velizarova [50] located near the capital city of Sofia, is one of the most contaminated by heavy metals, in the soil as in the spoil banks formed by long-term deposition of steel production residues and foundries.

The research developed by Geissen *et al.* (2010), mentions the surface and groundwater contamination in a banana production region in tropical Mexico. In this study, the accumulation of the metabolite ethylenethiourea – ETU, and the heavy metals Mn and Zn, were measured.

7.3 Allochthonous and Autochthonous Plants

Allochthonous plants are non-native to a phytogeographic region; in contrast, autochthonous plants are native to a territory [52, 53].

Within autochthonous plants, we can find endemic species that are exclusive and unique to a specific geographic area; on the other hand, native species can be found in a wide geographic range in the world, and are species with no human intervention, just a natural process [54], and are species that contribute positively to the biodiversity of ecosystems and represent evolution and habitat in an area across millennia. Furthermore, these species are constitutive elements that help regulate the ecosystem, maintain equilibrium, and are capable of adjusting to the biogeographic conditions of their growth habitat [55].

The terminology for the allochthonous plants has been the object of various revisions [56–58], in which allochthonous plants have also been called non-native plants or exotic plants [53]. The allochtonous plants even includes plants that are:

1. Cultivated (intentionally introduced by humans to satisfy certain demands; they have specific care requirements) [52].
2. Non-cultivated (arrived with cultivated species without knowledge of their reproduction or if they constitute a population) [52].
3. Sub-spontaneous plants (also known as escaped cultivated plants that are found near cultivated areas) [52].
4. Some other revisions incorporate naturalized plants (allochthonous plants that establish themselves in the environment, without human assistance; these plants do not, over time, become native members of the local plant community) [55].
5. Established plants (those exotic species that do not exhibit invasive behavior, but can reproduce and produce a viable population) [59].
6. Other revisions include adventitious (which are established spontaneously in natural or artificial habitats, although they are not able to effectively prosper and are at risk of disappearing due to changes in the factors that favored their introduction) [52]
7. Invasive plants [57]. The work of the Swiss botanist Albert Thellung (1881–1928), a pioneer in invasion research, contributed to the understanding of this field [60]. Invasive species are introduced by humans in areas out of their natural distribution [53]; in some cases, they are used for biological control or in research [55, 61]. Biological control only represents 6.8% of cases of exotic species introduced to continental waters [62].

One of the effects that allochthonous species produce is the alteration of species diversity patterns in multiple spatial scales. In a study by Fukami *et al.*, (2013), it's clear that non-native species reduce the diversity of native species through the interruption of two processes of native community assembly: species immigration and community divergence. The way in which species affect each other in a local community depends on their immigration history. The results of this study make clear the importance of community assembly and succession, in order to understand the large-scale effects of the non-native species. In the same study, he presents evidence that allochthonous species can cause a simultaneous reduction in the diversity of native species.

In general, within the literature about the relationship of non-native species on the ecosystems where they are found, there's a tendency for a quantitative approach in the description of ecological impacts. Most of the research on species traits seeks to identify the characteristics that determine the invasion capacity of the species, instead of the impacts produced on the ecosystem [64].

In the analysis done by Pyšek *et al.*, [64], they present a global overview of frequencies of significant and non-significant ecological impacts, and their directions on outcomes related to the responses of resident populations, species, communities and ecosystems.

Exotic species with an invasive behavior cause disturbance in plant and animal species, consequently affecting the environment, which results in the displacement or extinction of the zone's characteristic species and biological communities, and may lead to native biota loss, mentioned by Yang *et al.* [65]. As noted by Tilman and Lehman [66], this reduction in biodiversity in various ecosystems all over the world poses one of the great ecological challenges to humankind. The introduction of exotic species may result in a native species becoming invasive [67]. Likewise, in accordance with Rodríguez [68], invasive species limit an ecosystem's natural capacity for resilience and recovery with the disruption of native mutualist edaphic networks. Also, they alter the structure and function of the ecosystem into which they are introduced [69].

Within the studies done with native and non-native species, it is important to look at the cultural relevance of non-native species in the urban landscape with the climate change that has made that many plant species be used in contemporary urban landscaping [70]. This not only allows for an interpretation of the bioremediation processes from a technical perspective, but also from the community's point of view that, directly or indirectly is involved, since at the end, it's the community, because of its perception of the landscape, economical or cultural, that can contribute to the success or failure of the remediation of degraded areas with the use of phytoremediation.

The use of native species in rehabilitation or restoration processes of ecosystems, in enormous scales in present projects, as a result of the vast loss of diversity, desertification, and environmental problems of our times, plus the accelerated climatic variations due to a destructive economic system, create opportunities and risks. Opportunities to significantly increase the use of native species, and the risk of failure associated with the use of inappropriate reproductive material, which can provide a short-term vegetable cover, probably cannot establish a self-sustaining ecosystem [71].

One of the most important aspects within phytoremediation studies is the identification and characterization of the ecosystem in which the decontamination process is being implemented. This way, it is possible to identify the native plant species in the ecosystem. In the study carried out by González *et al.*, [19], they identified and classified plant species according to their local use in the Duero Natural Park in Spain. The ethnofloristic catalogue used comprises 217 species belonging to 65 botanical families. The predominant families were: *Fabaceae*, *Asteraceae*, *Lamiaceae*, *Rosaceae*, *Apiaceae*, *Poaceae*, *Liliaceae*, and *Scrophulariace*. Of the 217 species, a total of 188 (87%) are native to the study area, whereas 29 (13%) are allochtonous plants.

In the work done by Oyuela Leguizamo *et al.* [18], they present the relative frequency of native herbaceous species and other status species (naturalized, introduced, invasive, cultivated and adventitious) that were studied in terms of their relation to HM. For the study, eight representative countries were selected. In the results, most species studied in terms of HM, these were catalogued as naturalized (*Marrubium vulgare*, *Chenopodium album*, *Chenopodium polyspermum*, *Hypochaeris radicata*, *Malva sylvestris*, *Portulaca oleracea*, *Silybum marianum*, *Taraxacum officinale*, *Trifolium pratense*, *Trifolium pratense*, *Cyperus difformis*), cultivated (*Armeria marítima*, *Pelargonium sp.*, *Allium schoenoprasum*), adventitious (*Sonchus oleraceus*, *Bidens pilosa*), naturalized and adventitious (*Sorghum bicolor*, *Brachiaria decumbens*), exotic with documented invasive behavior (*Cotula coronopifolia*) according to the Catalog of Invasive Exotic Species, and of invasive nature according to the Global Invasive Species Database (*Cynodon dactylon*, *Bidens pilosa*, *Sonchus oleraceus*, *Hypericum perforatum*, *Hypochaeris radicata*, *Taraxacum officinale*, *Trifolium repens*).

Within the research of rhizofiltration, phytodegradation, and phytoextraction, the ability to remove heavy metals from contaminated water was studied, using some aquatic plants, e.g. *Eichhornia crassipes* [72], *Lemna minor* [73, 74] and *Azolla pinnata* [45], *Myriophylhum aquaticum*, *Ludwigina palustris*, and *Mentha aquatic* [75]. These aquatic plant species are one of the most studied in phytoremediation processes and have been classified as having the greater heavy metal phytoremediation potential. Similarly, many aquatic species have been used in the bioremoval of

HM: *Azolla filliculoides*, *Azolla pinnata*, *Typha orientalis*, *Brassica juncea*, *Polygonum hydropiperoides* and *Salvinia molesta*, among others [76].

7.4 Phytoremediation of Heavy Metals (HM)

7.4.1 Phytoremediation

Remediation of a contaminated substrate and heavy metal decontamination relies on chemical, physical and biological techniques. These techniques fall into one or both of these two categories: *ex situ* and *in situ*. These techniques involve the extraction of the contaminant or the isolation and/or immobilization of the contaminated substrate. In *ex situ* methods, the contaminated substrate is removed, treated and returned. Common *ex situ* techniques include excavation, detoxification and/or destruction of contaminants by physical or chemical processes. In addition, these methods may subject the contaminants to stabilization, solidification, immobilization, incineration or destruction [77, 78].

Phytoremediation is a form of bioremediation that detoxifies by means of biological processes. This has been utilized since the eighteenth century [79]. It specifically consists of using plants to remove contaminants from the environment and render them harmless; it is also known as phytocleaning or phytocorrection [2, 7, 9, 10, 12, 80]. Phytoremediation entails the use of plants to sequester, bioaccumulate and transform hazardous substances in the environment in order to control contamination [81].

One example of this technology is the study of aquatic macrophytes to absorb HM [82, 83]. In phytoremediation, in situ or ex situ plants reduce the concentration of various compounds via their biochemical processes [84]. Furthermore, this technique employs associated micro-biotics, soil improvement and agronomic techniques to remove or retain contaminants or neutralize their effects [2]. Phytoremediation is currently used for the decontamination of soils that display increased levels of HM.

In sum, phytoremediation represents a promising solution to the decontamination of a number of contaminants across a variety of sites.

7.4.2 Phytoremediation Approaches and Technologies

Phytoremediation or phytocorrection encompasses a group of techniques and mechanisms premised on the use of vegetation species and their related microorganisms to remove, uptake, immobilize or transform soil or water contaminants [2, 10, 76, 79, 85]. Contaminants may be immobilized, stabilized or degraded in the rhizosphere, sequestered or degraded within the plant or volatilized [2, 86].

Within the field of phytoremediation of soils contaminated by HM, there are numerous sub-fields: phytostabilization or phytoimmobilization, phytovolatilization, phytodegradation, phytoextraction (principally soil, sediments and sludge's applications), rhizofiltration, rhizodegradation, hydraulic barriers, hydraulic control, vegetative caps, constructed wetlands (principally water applications), and phytorestauration [14, 79, 85, 87]. Vassilev *et al.* [13] summarize information concerning development, achievements, and research needs of metal phytoremediation concept, technological, and economical aspects, including description of phytoremediation subfields, and focuses on phytoextraction and phytovolatilization.

The following paragraphs describe some of the phytoremediation technologies, with few of the advantages and limitations of some phytoremediation technologies:

- **Phytostabilization**: (soil, sediment and sludge's media) is used to reduce or eliminated the mobility of toxic elements from the contaminated soil to the environment, thus, they are stabilized in the substrate or roots. These subfield transform soil metals to less toxic forms, without removing the metal from soil [9, 14, 87, 88].
  Advantages: It is not necessary soil removal, ecosystems restoration is enhanced by the revegetation, does not required the disposal of hazardous materials, reduces bioavailability for entry in food chains, prevents erosion and improves visual amenity in a derelict site [14, 87, 88].
  Limitations: It will be irrelevant if its focal point is the prevention of erosion and the improvement of visual amenity of a contaminated site. The land value is reduced due to the presence of toxic metals in the plants on sites to be used as a resource for biofuels or timber [88]. Another limitation is that vegetation and soil require long-term permanence in order to prevent discharge of the contaminants and leaching. Plants require extensive fertilization or the use of amendments. Phytostabilization might be considered as an interim measure. Soil and plant characteristics must be monitored to prevent the increase of metal solubility and leaching [87].
- **Phytovolatilization**: (groundwater, soil, sediment and sludge's media) is the extraction from media and transpiration of a contaminant or a modified form, by plants [14, 87].

Terry and Bañuelos [89] described investigations on phytovolatilization of heavy metals like As, Hg, and Se.
Advantages: pollutants could be transformed into less toxic forms; the contaminants forms released into the atmosphere might be subject to a natural degradation process [14, 87].
Limitations: There is evidence of low levels in plant tissues of the contaminant forms, which might have been released into the atmosphere or could accumulate in vegetation and be passed on in latter products as fruits [14, 87].

- **Phytodegradation**: (soil, sediments, sludge;s media) refers to the destruction and breakdown of the contaminant [87].
  Advantages: potentially could occur in soils where biodegradation cannot. This phytoremediation technology has the advantage that enzymes produced by a plant can occur in an environment free of microorganisms. Also, plants are able to grow in sterile soils, and in those with toxic levels for microorganisms [87].
  Limitations: contaminants destruction could be difficult to confirm; thus the presence or identity of metabolites may form degradation or intermediate products [87].
- **Rhizofiltration**: (groundwater, surface water media), also known as phytotransformation, is the use of plants' roots to remove contaminants from flowing water [9, 87].
  Advantages: can be used in terrestrial or aquatic plants. Terrestrial plants, through a platform, can accumulate more contaminants than aquatic species. Rhizofiltration is applicable to *in situ* and *ex situ* systems, the latter can be placed anywhere, it is not necessary to be at the original location of the contamination [87].
  Limitations: needs continuous adjustment of pH influent; interactions of species in the influent have to be understood; an engineered system is required to control influent concentration and flow rate; are required periodic harvesting and plant disposal [87].
- **Hydraulic control**: (groundwater, surface water media), is the use of plant uptake and consumption, to remove, contain or control the migration of contaminants [87].
  Advantages: avoids migration of leachate toward groundwater or receiving waters [14]. It is not necessary to install

an engineered system. The roots are in contact with a much greater volume of soil than a pumping well [87].

- **Phytoextraction**: (soil, sediments and sludge's media), also known as phytoaccumulation, uses the plants' ability to absorb and remove contaminants from the soil and "uptake" them into their leaves and stalks. Next, the plant parts storing contaminants are removed and destroyed or recycled; in either scenario, the metal is extracted from the soil [2, 13]. The first step in applying this method is the selection of plant species best suited for: a) the removal of the metals to be addressed and b) site characteristics. Once the plant is fully grown, it is cut and incinerated before finally being transported, in the form of ashes, to a disposal site [84].
  Advantages: This technology may be used to improve livestock nutrition [90], that is how plant biomass can be a resource of the extracted contaminant [14]. Species that hyperaccumulate metals have a massive biotechnological potential [91–93]. Plants capable of accumulating HM are grown on contaminated sites, and the aerial biomass rich in metals is collected when mature. As a result, part of the soil contaminant is removed. The success of phytoextraction as an environmental "purifier" hinges on factors such as availability of the metal for absorption, and the plant's capacity to absorb and uptake metals in its aerial parts. For cost-related purposes, the harvested biomass is generally incinerated or composted, and, less frequently, recycled for reuse [85] promising approaches in terms of commercialization. Patents for commercial phytoextraction were issued in Japan in 1980 [2].
  Limitations: success using hyperaccumulators has been hindered, in many cases, by the limited biomass of these species. Species with this limitation include that most heavily studied, *T. caerulescens*, which is probably feasible in moderately contaminated soils [12]. However, there are exceptions: hyperaccumulators with large biomasses include the Ni hyperaccumulator *Berkheya coddii* [40]. Therefore, there doesn't seem to be any intrinsic reason for low biomass in hyperaccumulators, i.e. there is no reason for all hyperaccumulator plants and/or plants capable of tolerating high metal concentrations to be competitively inferior or slow growth [78]. In any case, more biomass production by

> hyperaccumulator plants could significantly minimize the time needed for phytoextraction [12]. Robinson *et al.* (2015) explain the relation between soil conditions and the bioaccumulation coefficient, in which a small minority of plants with roots forage contaminant hotspots, would remove greater rate of metals. It is also mentioned that HM uptake is in function of the soil concentration, therefore, uptake and the metal concentration in soil decrease at the same time. That would represent a significantly efficiency into the field and laboratory experiments.

To develop phytoextraction mechanisms, there are numerous plants endowed with the natural ability to tolerate some toxic compounds, for they are perfectly adapted to their habitat's specific environmental conditions. Hyperaccumulators concentrate more than 1000 mg/kg of As, Co, Cr, Cu, Ni, Pb, Sb and Se; more than 10,000 mg/kg of Mn or Zn; more than 100 mg/kg of Cd in their aerial biomass [76, 94, 95], until exceeding concentrations of 2% of their dry material [20]. The behavior presented by hyperaccumulator species, confirms results of studies on serpentine substrates in Central America, Greater Antilles and South America, where most hyperaccumulators are focused on nickel [96]. In Baker *et al.* (2000), they indicate global distribution of 315 nickel hyperaccumulation, geographic areas and number of hyperaccumulators in Cuba, Dominican Republic, New Caledonia, Australia, South Europe/Asia Minor, SouthWest of Asia, U.S. (Pacific N.W. and California), Japan (Hokkaido), Brazil (Goias), Canada (Newfoundland) Zimbabwe and South Africa (Transvaal). Plants that can grow and develop in soils with high concentrations of HM are part of a specialized flora [97]. Those plants colonize soils characterized as lightly acidic.

Bioaccumulation refers to the mechanism for absorbing chemical compounds performed by an organism within the biotic (food resource) or abiotic (environment) medium. A quantitative form of expressing bioconcentration is via the bioconcentration factor (BCF), which is the ratio of the compound's concentration in the environment or in the growth substrate to that found in the living element. In the case of heavy metal phytoremediation, this factor is the ratio of the metal in the plant biomass, or tissue (mg/Kg dry weight) to the initial metal concentration in the soil or solution (Kg/mL) where the plant grows this ratio is a determining factor when it comes to identifying a plant's metal removal efficiency. Vegetation's species are considered to exhibit removal potential when they have a ratio greater than 1 [98].

Hyperaccumulation, per the definition provided by Jaffré *et al.* [94], describes plants that grow on serpentine soils, species able to concentrate up to 1000 mg/kg of Ni. The high concentration of HM in the vegetative organs of most plants reaches toxicity for Ni around 10e15 mg/kg. Consequently, hyperaccumulators can withstand up to 100 times higher metal concentrations than typical plants without reported accumulation processes [99]. Ni has been shown to reach its highest concentration in a plant between 100 and 1000 mg/kg of dry weight (accumulators) or more than 1000 mg/kg (hyperaccumulators) [15].

Heavy metal hyperaccumulators species occur in metal-rich soils, temperate and tropical zones (South America, North America, Europe and New Caledonia) [31]. These plants are chosen based on their high tolerance. Serpentine tolerance can be explained as part of three established physiological and evolutionary mechanisms: tolerance to a low Ca/Mg ratio, avoidance of Mg toxicity or high Mg requirement. The vast majority of known hyperaccumulator species belong to vegetation communities, characteristic of soils in serpentinized areas. Not all accumulative species concentrate metal equally. Small to large concentrations are tolerated in different plants. The former are considered "accumulators," while the latter are considered "hyperaccumulators" [94].

Research into hyperaccumulators identified more than 500 hyperaccumulator species [100], as shown on Table 7.1, that belong to a total of 45 families. The families Asteraceae, Brassicaceae, Caryophyllaceae, Cyperaceae, Cunouniaceae, Fabaceae, Flacourtiaceae, Lamiaceae, Poaceae, Violaceae and Euphobiaceae accounted for the majority of hyperaccumulator species and, subsequently, were the predominant families in metal accumulation processes [31, 101]. From hyperaccumulators, 79% are plant species found in Central and South America, mainly distributed throughout Cuba, 18% in Puerto Rico and Hispaniola, with values frequently between 1000 and 10000 mg/Kg. Families that stand out include: *Euphobiaceae*, *Asteraceae*, *Buxaceae*, *Acanthaceae*, *Ochnaceae*, *Clusiaceae*, *Tiliaceae*, *Turmeraceae*, *Boraginaceae*, *Scrophulariaceae* and *Violaceae* [102].

Within the advantages of phytoremediation, is the fact that it's the best alternative for the removal, decontamination and primary isolation of toxic substances in soils, for it does not destroy soil structure or fertility [76]. This method also boasts positive cost benefit and engineering-economy relations as an environmental [8, 105]. Horne [86] specified the historical background of wetlands and traditional remediation techniques, similarities and differences between conventional bioremediation,

Table 7.1 Number of metal hyperaccumulator plants.

Description	References
145 hyperaccumulators of Ni, distributed in 22 families	[31]
400 angiosperm hyperaccumulator species	[31, 37, 101]
415 species from 45 botanical families	[99]
317 species hyperaccumulate Ni, (28) Co, (37) Cu, (14) Pb, (1) Cd, (11) Zn and (9) Mn	[37]
320 species hyperaccumulate Ni, (30) Co, (34) Cu, (20) Se, (14) Pb, (1) Cd, (11) Zn and (10) Mn	[76]
450 species of angiosperms with accumulative behavior of: As, Cd, Co, Cu, Mn, Ni, Pb, Sb, Se, Tl, Zn	[15]
320 species hyperaccumulate Ni, (34) Cu, (34) Co, (20) Se, (18) Zn, (14) Pb, (9) Mn and (4) Cd	[103]
More than 450 species of known hyperaccumulators hyperaccumulate Ni, generally from ultramafic or serpentine soils	[104]
500 species that hyperaccumulate at least one heavy metal. (450) Ni, (32) Cu, (30) Co, (20) Se, (14) Pb, (12) Zn, (12) Mn, (5) As, (2) Cd.	[100]

phytoremediation, and wetlands phytoremediation Baker *et al.* [37] reported successful cases of metal removal by phytoremediation wetlands.

In general, phytoremediation also comes with its own obstacles, in this way within the limitation of phytoremediation we can find: this process is limited by the depth reached by the roots. It is restricted to surface area soil available for root uptake, surface and subterranean water. Moreover, phytoremediation is a slow method and one affected by the contaminant's bioavailability, which often means it ought to be employed in conjunction with other approaches so as to optimize performance [106]. As an example of this combination, there is the potential use of biochar and phytoremediation technologies on heavy metal polluted soils, with the goal that biochar can reduce the bioavailability and leachability of heavy metals in the soil (Paz *et al.* 2014). In spite of phytoremediation is an environmental friendly technology, it has its own impact, i.e., the decrease of resources, plus the risk for human health, as a result of exposure conditions, and viability of businesses [107].

7.5 Methodology

The authors identified and selected the following group of words. "Wetlands, Heavy Metals, phytoremediation, phytoremediation technologies, allochthonous and autochthonous plant species, native and exotic species, environmental, biodiversity and ecology." These words were searched individually and in groups in different databases all over the world. The search returned about 2.900 documents; of these, 142 were chosen based on relevance to the present meta-review's objectives. From the work done by Oyuela Leguizamo *et al.* [18], among others, the following information was searched: status, presence of invader character and global distribution. From this list, native, endemic, and non-invader plants were selected. Most of these plants also belong to predominant families in HM accumulation processes. Information about HM studies of all of these plants was researched at a family, gender or specie level. The result was 54 species, shown in Section 7.6. Based on the information found, the relevance of each of the topics presented in this chapter was discussed.

7.6 Analysis of Research on Heavy Metals (HM) and Native and Endemic Plant Species

Having as a reference the research done by Oyuela Leguizamo *et al.* [18] and the work by [19], the relationship of the native and endemic vegetal species studied was updated and analyzed, in terms of heavy metals, with non-invasive behavior. Of the wide range of extant native or endemic herbaceous species, the reseach carried out by [18] contains species that are found worldwide, and are commonly found in wetlands.

Table 7.2 includes species status (native and/or endemic) and the "Global Distribution," based on information from standard regional flora and online searches of herbarium records from the Missouri Botanical Garden, Calflora Database, Global Invasive Species Database, Catalogue of Plants and Lichens of Colombia, Catalog of Bogota's Invasive Wetland Plants, Navarra's Floristic Catalog, Catalogue of the Vascular Plants of Venezuela, Illustrated Catalog of Cundinamarca's Plants, Catalogue of the Vascular Plants of the Department of Antioquia, and Catalogue of the Southern Cone's Vascular Plants, "Comisión Nacional para el Conocimiento y Uso de la Biodiversidad (Conabio)."

Of the species described in [18], 5 are endemic and 36 are native (not one presents invasive behavior) and [19] a total of 188 (87%) are native to the study area. The species come from the following 18 families:

Table 7.2 Endemic and native vegetation species studied in terms of heavy metals.

Species	Species status		Endangered species	Global distribution	Information on publications about metals	Countries of cited publications	Refs.
	Native	Endemic					
Family: Amaranthaceae							
Amaranthus hybridus	x			The American continent	Pb	Mexico	[108]
Family: Apiaceae							
Hydrocotyle umbellata	x			Colombia, present throughout the American continent	Apiaceae Family Pb, Zn, Cu, Cd, As	India	[109]
Family: Asteraceae							
Baccharis latifolia	x			Colombia, Northern Andes	As, Hyperaccumulator of Pb	Brazil, USA, Mexico, Spain	[110–113]
Bidens triplinervia Kunth	x				Hyperaccumulator of Cu, Fe, Accumulator of Mn, Zn and in greater proportion of Pb	Spain, Peru	[110, 114]
Senecio carbonelli		x	x	Colombia-Bogota Endemic Plant	Senecio Genus Ni, Zn, Cu, Mn, Al	Cuba, Canada, Africa, Poland	[97, 99]
Gnaphalium chartaceum	x	x		Mexico	Cu, Mn, Zn, Pb	Mexico	[108]

(*Continued*)

Table 7.2 Cont.

Species	Species status		Endangered species	Global distribution	Information on publications about metals	Countries of cited publications	Refs.
	Native	Endemic					
Senecio salignus	x			From southern Arizona to El Salvador and Honduras	Zn	Mexico	[108]
Flaveria angustifolia		x		Mexico	As, Cu, Pb, Zn	Mexico	[108]
Simsia amplexicaulis	x			Mexico and Guatemala	Cu	Mexico	[108]
Brickelia veronicifolia	x			Texas, USA and Mexico	Pb, Zn	Mexico	[108]
Family: Brassicaceae							
Lepidium bipinnatifidum	x			Venezuela, Colombia, Ecuador, Peru, Bolivia, Brazil and Argentina	Accumulator of Pb	Spain	[110]
Family: Boraginaceae							
Wigandia urens	x			Mexico, Colombia and Venezuela	Zn	Mexico	[108]

Family: Cyperaceae							
Carex buchananii	x			Colombia and New Zealand	Cd, Zn, Ni	France, Spain, Ireland	[115–117]
Carex lanuginosa	x			Colombia, and reported in the American continent	Cd, Zn, Ni, Al, Co, Cr, Ni, Pb, Cu		
Carex lurida	x		x				
Cyperus bipartitus	x			Colombia, North, Central and South America	Mn, Zn, Cd, Ni, Pb, Cu, Cr	Malaysia, Egypt, Tanzania, USA	[118–121]
Cyperus rufus	x			Colombia, Argentina, Chile			
Eleocharis dombeyana	x			Colombia, Costa Rica, Central and South America	As	Mexico, Brazil, Argentina	[122]
Eleocharis montana	x			Colombia, the Antilles, South Eastern United States to Chile and Argentina			
Eleocharis palustris	x			Colombia, America, Europe, Asia and Africa			

(*Continued*)

Table 7.2 Cont.

Species	Species status		Endangered species	Global distribution	Information on publications about metals	Countries of cited publications	Refs.
	Native	Endemic					
Fuirena incompleta	x		x	Mexico to Argentina, Colombia and Uruguay	Apiaceae Family Pb, Zn, Cu, Cd, As	India	[123]
Kyllinga brevifolia	x			Colombia, Tropical America. Paraguay, India, Malaysia, the Philippines and China	U, Th, Sr, Ba, Ni and Pb	China	[124]
Family: Equisetaceae							
Equisetum bogotense	x			Colombia, Costa Rica and South America	As	Mexico, Brazil, Argentina	[122]
Scirpus americanus	x			South America	Pb, Cd, Cr, Mn	Mexico	[108]
Family: Euphorbiaceae							
Jatropha dioica	x			Texas in the United States as well as Mexico	Zn	Mexico	[108]

Family: Fabaceae							
Prosopis laevigata	x			Mexico, Bolivia, Peru, Argentina	As		[108]
Family: Hypericaceae							
Hypericum humboldtiana		x	x	Colombia-Bogota Endemic Plant	*Hypericum perforatum* was tolerant to Cr, presents effects to Ni	Italy, Canada	[125, 126]
Hypericum humboldtii		x	x	Colombia Endemic Plant			
Family: Juncaceae							
Juncus densiflorus	x			Colombia, Venezuela, Brazil, Argentina, Uruguay and Paraguay	Juncaceae Family Cu, Pb, Cd, studies on plant tolerance	Austria, Australia, Scotland	[127–129]
Juncus effusus	x			Colombia, common to large number of countries	Cd, Zn, Ni, Al, Co, Cr, Ni, Pb, Cu	France, Poland, China	
Juncus microcephalus	x		x	Colombia, Central Mexico to Bolivia and South West Brazil	Juncaceae Family Cu, genus juncus Pb, Cd, studies on plant tolerance	Austria, Australia, Scotland	

(*Continued*)

Table 7.2 Cont.

Species	Species status		Endangered species	Global distribution	Information on publications about metals	Countries of cited publications	Refs.
	Native	Endemic					
Juncus ramboi colombianus		x	x	Colombian Endemic Plant	Juncaceae Family with Cu, Pb, Cd, studies on plant tolerance		
Juncus ramboii Barros subsp. Colombianus Balslev	x			Colombia, Southeastern Brazil	Juncaceae Family with Cu, Pb, Cd, studies on plant tolerance		[115, 130, 131]
Family: Lamiaceae							
Hyptis capitata	x			Colombia, neotropics; introduced in tropical Asia and Pacific	Accumulator of Cd and Cu	Australia	[132]
Family: Marsilaceae							
Marsilea mollis	x			North America, Colombia to Argentina	Cu	Canada	[75]

Family: Oenotheraceae							
Ludwigia peploides	x			Colombia, South America, Greater Antilles and East Asia	Cu, Pb, Cr, Zn, Cd, Ni and Hg	India, Tanzania	[75, 121, 123]
Ludwigia peruviana	x			Colombia, South America; introduced in South Asia and Australia			
Family: Plantaginaceae							
Gratiola bogotensis	x		x	Colombia	Plantaginaceae Family Pb, Zn, Cu, Cd, As	Spain	[133]
Plantago orbignyana	x			Southern Colombia to Bolivia	Accumulator of Pb y Zn		[110]
Family: Plumbaginaceae							
Arabidopsis thaliana	x	x		Native to Europe, Asia, Africa, present in all five continents, poor in South America, Asia and Canada	Hg	Germany	[134]

(*Continued*)

Table 7.2 Cont.

Species	Species status		Endangered species	Global distribution	Information on publications about metals	Countries of cited publications	Refs.
	Native	Endemic					
Family: Poaceae							
Bromus catharticus	x			Colombia, Guatemala to South America	Cu, Zn	USA	[135]
Leersia hexandra	x			Colombia, the Antilles, Canada to Argentina and Uruguay	Hyperaccumulator of Cr		[136]
Bouteloua curtipendula	x			Mexico and South America	Ni	Mexico	[108]
Family: Polygonaceae							
Polygonum hydropiperoides	x			Colombia, the Antilles, Canada to South America	Cu, Pb	USA	[120]
Polygonum segetum	x			Mexico to Colombia, and the Antilles	Genus Polygonum accumulator of Cd	Japan	[137]
Polygonum aviculare	x			Europe	Zn	Mexico	[108]

Family: Potamogetonaceae							
Potamogeton paramoanus	x		x	Colombia, Venezuela to Bolivia	Potamogeton Cd, Pb, Cr, Ni, Zn y Cu	Turkey	[138]
Potamogeton pectinatus	x			Colombia, common to a large number of countries	Accumulator of Cd, Pb, Cr, Ni, Zn y Cu		
Family: Ranunculaceae							
Ranunculus flagelliformis	x		x	Colombia, Venezuela to Argentina	Genus Ranunculus Cr	Turkey	[47]
Family: Scrophulariaceae/ Calceolariaceae							
Calceolaria bogotensis	x		x	Colombian Native Endemic Plant	Cu	USA	[139]
Family: Solanaceae							
Cestrum buxifolium	x			Colombia, Venezuela to Peru	Solanaceae Family Cr, Hg, Pb, Zn, Cu, Cd, As	Poland, Spain	[105, 133]

(*Continued*)

Table 7.2 Cont.

Species	Species status		Endangered species	Global distribution	Information on publications about metals	Countries of cited publications	Refs.
	Native	Endemic					
Nicotiana tabacum	x			Colombia and South America. Native cultivated plant	Accumulator of Cd and Cu. Tolerance to Hg	Australia, Poland	[105, 132]
Physalis peruviana	x			Colombia, American continent. Native cultivated plant	Cr, Mn, Fe, Co, Ni, Cu, Zn, Cd, Hg, Pb	Kenya	[140]
Solanum corymbosum	x			Mexico and Peru	Cu	Mexico	[108]

Apiaceae, Asteraceae, Brassicaceae, Cyperaceae, Equisetaceae, Hypericaceae, Juncaceae, Lamiaceae, Marsilaceae, Oenotheraceae, Plantaginaceae, Plumbaginaceae, Poaceae, Polygonaceae, Potamogetonaceae, Ranunculaceae, Scrophulariaceae/Calceolariaceae and Solanaceae.

7.7 Results

According to the revision done, the global distribution of each species listed above includes the following countries: Colombia, New Zealand, Chile, Costa Rica, the Antilles, Paraguay, India, Malaysia, the Philippines, China, Uruguay, Venezuela, Brazil, Australia, Guatemala, Canada, Peru; in other words, they are found across Asia, Europe, North, Central and South America. Most of those countries are exposed to metalliferous, serpentine (ultramafic) or calamine soils, and in some of them, there are ecosystems with HM contamination problems. See Section 7.6 for more information.

All of these 54-species listed on Table 7.2 reported non-invasive behavior and have been studied in relation to HM, in countries such as Austria, Australia, Scotland, Brazil, USA, Mexico, Canada, China, Cuba, Poland, France, Spain, Ireland, India, Tanzania, Malaysia, Italy, Argentina, Turkey, Egypt, Switzerland, Portugal, Germany, Israel, Peru, Slovenia, Belgium, Lithuania, United States (specifically Alaska) and on the African Continent. Additionally, the majority of the studies compiled only record heavy metal concentrations; most do not analyze the plant processes involved in the phytoremediation.

As can be seen on Table 7.2, there are four predominant families in HM accumulation processes, as seen in Section 7.4.2, *Asteraceae, Cyperaceae, Scrophulariaceae,* and *Poaceae.*

7.8 Conclusion

As shown in the analysis of the consulted bibliography, the negative impacts generated in the resident ecosystems of the invasive plant species constitute a platform for analysis and debate about the implications of the use of an allochthonous species in a phytoremediation process, and the question arises, up to what point can all the possible phytoremediation biotools be a contribution to the decontamination process and to the ecology.

An integral evaluation is needed on the relationship between the characteristics of the species used in phytoremediation processes, as well as the study of the relationship between environmental settings of invasion on the characteristics of impacts and invasive species traits, in order to stablish a valid argument to justify the use of native and endemic species in the recovery of contaminated substrates with heavy metals.

It is not only necessary to talk about the decontamination of degraded areas, but also about the integration of concepts, techniques and technologies, such as the rehabilitation and restauration of ecosystems.

Phytoremediation studies can be based on the research on floristic analysis, as the one done by González *et al.*, [19]. Additionally, plants could be used as accumulation indicators of substratum quality and can be indicators of the ecosystem's grade of impact. One example is the work carried out by Zechmeister *et al.* [141] that mentioned bryophytes as bioindicators and biomonitors in terrestrial and aquatic habitats. The later stablishes that the role of the species within the ecosystem and their viability as phytoremediation bio-tools as a contribution to ecology. From this, one can see the need to not only see the potential of phytoremediation of native species, but also the role that it plays within the biogeochemical cycles, given that phytoremediation studies provide information on environmental aspects such as life cycles, the composition of growth substrate, morphological changes, chemical and biological structure, the ecological diversity. In addition to physiological, photochemical, ecological and biogeochemical cycles and the behavior of the contaminants in each substrate. In this way, this kind of investigations brings the option to establish ecological risks through the food chain in ecosystems contaminated with HM.

Even though the focal point of this article are heavy metals, there is a need for a future look at the research that has been done with the native and endemic vegetal species presented here, in terms of other parameters. In doing so, there will be a knowledge base to research the species in remediation processes of contaminated substrates. Hence, the need for studies on plant tolerance of metals becomes even greater, as well as the research that has looked into the toxic effects of metals in plants.

According to the researched bibliography, the presence of HM worldwide is discussed. On account of this, the use of friendly technologies with the environment that minimize the presence of HM concomitant toxicity becomes imperative. Therefore, special attention should be paid to the activities that contribute to soil, air and water contamination, and to the source of heavy metal contamination. The development of these activities has negative consequences on the ecosystems, affecting both biotic and abiotic components. On top of that, the antagonist technologies that do not adapt themselves to the natural conditions of the place where they are implemented, affect, in turn, the biodiversity, the landscape, the ecosystem's structure and functionality. Heavy metal contamination has been addressed with plant species in bioremediation processes as an effort to mitigate these deleterious effects, but what is needed is an approach with the natural dynamics of the ecosystem in order to reduce the negative impacts associated to the process.

From the potential of the species to be used in a phytoremediation process, it is imperative to contribute to the preservation of native and endemic species, as well as the sustainable use of phytoremediation as an alternative to decontaminate the environment. Here, the photoremediation processes involve the participation of diverse areas of science and engineering to achieve the integrality in the solution of the ecological quandary, which depends in great part on the customs, habits, and practices of the population of the zone where the ecological recovery and restoration processes are implemented. Lo anterior, requires an interdisciplinary study of the hydrosphere-hydrology, atmosphere, biosphere, lithosphere, biodiversity (mechanisms, organisms and soil properties), biogeochemical cycling, human health, social sciences and soil threats.

In the case of HM, and based on the wide studies on soil substrate compared to others, and the soil dependence on the dynamics of all the elements of an ecosystem, it is concluded that the use of the knowledge already acquired can improve the provision of ecosystem services, the restoration and conservation of biodiversity, including sustainable practices in the high impact activities on the soil or growth substrate.

Proctor (1999) argues that the phytobiography of regions with high concentrations of HM require further investigation, based on the lack of current knowledge on phytoremediation and the specific experiments and experimental plants, therefore, there is an imperative need to focus on the mechanism behind "phyto" technologies and the advantages and limitations of the phytoremediation process. Thus, phytoremediation can be studied based on the existing knowledge, such as in one of the most researched, hyperaccumulation.

One of the main benefits of developing future studies on native or endemic plant species, could help avoid the use of exotic and invasive species that have transformed biogeographical barriers. The accidental or deliberate spread of the exotic and invasive species, within new habitats, constitutes an environmental problem.

The relevance of biodiversity and the ecology of the ecosystems is gradually being considered for the heavy metals decontamination processes. An overriding investigation into native and endemic biodiversity contributes to the identification of possible phytoremediation bio-tools, as well as in the research of the ecology and environment of nature's remnants in the ecosystems.

References

1. Keesstra, S., Geissen, V., Mosse, K., Piiranen, S., Scudiero, E., Leistra, M., van Schaik, L., Soil as a filter for groundwater quality. *Curr. Opin. Environ. Sustain.*, 4, 507–516, 2012.
2. Cunningham, S., Berti, W., Huang, J., Phytoremediation of contaminated soils. *Trends Biotechnol.*, 13, 393–397, 1995.
3. Switras, S., The Potential of phytoremediation techniques for selenium removal. *J. Environ. Qual.* 5, 1–7, 1999.
4. Henry, J., *Overview of the phytoremediation of lead and mercury*, EPA, Washington, D.C., 2000. [5] Moalla, S., Awadallah, R., Rashed, M., Soltan, M., Distribution and chemical fractionation of some heavy metals in bottom sediments of Lake Nasser, *Hydrobiologia*. 364, 31–40, 1998.
6. Paz, J., Lu, H., Fu, S., Méndez, A., Gascó, G., Use of phytoremediation and biochar to remediate heavy metal polluted soils: a review, *Solid Earth*, 5, 65–75, 2014.
7. I. Raskin, Plant genetic engineering may help with environmental cleanup. *Proc. Natl. Acad. Sci.*, 93, 3164–3166, 1996.
8. Raskin, I., Smith, R., Salt, D., Phytoremediation of metals: Using plants to remove pollutants from the environment. *Curr. Opin. Biotech.*, 8, 221–226, 1997.
9. Chaney, R., Malik, M., Li, Y., Brown, S., Brewer, E., Angle, J., Baker, A., Phytoremediation of soil metals. *Curr. Opin. Biotech.* 8, 279–284, 199.
10. Pilon, E., Zayed, A., DeSouza, M., Lin, Z., Terry, N., Remediation of Selenium-Polluted Soils and Waters by Phytovolatilization, In: Terry, N, Bañuelos, G, editors, *Phytoremediation Contam. Soils Waters*, CRC Press LLC, Boca Raton, FL, USA, pp. 72–94, 2000.
11. Clemens, S., Molecular mechanisms of plant metal tolerance and homeostasis. *Planta.*, 212, 475–486, 2001.
12. McGrath, S., Zhao, F., Phytoextraction of metals and metalloids from contaminated soils. *Curr. Opin. Biotech.*, 14, 277–282, 2003.
13. Vassilev, A., Schwitzguebel, J., Thewys, T., Van der Lelie, D., Vangronsveld, J., The use of plants for remediation of metal-contaminated soils. *Sci. World J.*, 4, 9–34, 2004 .
14] Prasad, M., Phytoremediation of metals in the environment for sustainable development. *Proc. Indian Natn. Sci. Acad.*, B70, 70, 71–98, 2004.
15. Rascio, N., Navari, F., Heavy metal hyperaccumulating plants: How and why do they do it? And what makes them so interesting? *Plant Sci.*, 180, 169–181, 2011.
16. Castro, P., Valladares, F., Alonso, A., La creciente amenaza de las invasiones biológicas. *Ecosistemas.*, 13, 2004.
17. Budelsky, R., Galatowitsch, S., Establishment of Carex stricta Lam. seedlings in experimental wetlands with implications for restoration. *Plant Ecol.*, 175, 91–105, 2004.

18. Oyuela Leguizamo, M.A., Fernández Gómez, W.D., Sarmiento, M.C.G., Native herbaceous plant species with potential use in phytoremediation of heavy metals, spotlight on wetlands – a review. *Chemosphere*, 168, 2016 .
19. González, J., García, M., Ramírez, R., Bernardos, S., Amich, F., Ethnobotanical resources management in the Arribes del Duero Natural Park (Central Western Iberian Peninsula): relationships between plant use and plant diversity, ecological analysis, and conservation, *Hum. Ecol.*, 41, 615–630, 2013.
20. Kidd, P., Becerra, C., Garcia, M., Monterroso, C., Aplicación de plantas hiperacumuladoras de níquel en la fitoextracción natural : el género Alyssum L. *Ecosistemas*. 16, 1–18, 2007.
21. Athar, R., Ahmad, M., Heavy metal toxicity: effect on plant growth and metal uptake by wheat, and on free living azotobacter. *Water Air Soil Poll.*, 138, 165–180, 2002.
22. Pfeifer, H., Derron, M., Rey, D., Schlegel, C., Atteia, O., Piazza, R., Dubois, J.-P., Mandia, Y., Natural trace element input to the soil-sediment-water-plant system: examples of background and contaminated situations in Switzerland, Eastern France and Northern Italy, In: Markert, B, Friese, K, editors, *Trace Elem. - Their Distrib. Eff. Environ.*, Elsevier, pp. 33–86, 2000
23. Tanhan, P., Kruatrachue, M., Pokethitiyook, P., Chaiyarat, R., Uptake and accumulation of cadmium, lead and zinc by Siam weed [Chromolaena odorata (L.) King & Robinson]. *Chemosphere*., 68, 323–329, 2007.
24. Barba, L., Edith, L., Conceptos básicos de la contaminación del agua y parámetros de medición, 2002, /citations?view_op=view_citation&continue=/scholar%3Fq%3Dbarba%26hl%3Des%26as_sdt%3D0,5%26scilib%3D1%26scioq%3Dalarcon&citilm=1&citation_for_view=_h03bGoAAAAJ:Zph67rFs4hoC&hl=es&oi=p (accessed July 27, 2015).
25. Diez, J., *Fitocorrección de suelos contaminados con metales pesados : evaluación de plantas tolerantes y optimización del proceso mediante prácticas agronómicas*, Universidad de Santiago de Compostela, 2008.
26. Wei, C., Wang, C., Yang, L., Characterizing spatial distribution and sources of heavy metals in the soils from mining-smelting activities in Shuikoushan, Hunan Province, China. *J. Environ. Sci.*, 21, 1230–1236, 2009.
27. Bertram, C., Krätschell, A., O'Brien, K., Brückmann, W., Proelss, A., Rehdanz, K., Metalliferous sediments in the Atlantis II Deep—Assessing the geological and economic resource potential and legal constraints. *Resour. Policy.*, 36, 315–329, 2011.
28. USGS, *Mineral Commodity Summaries 2010*, Washington D.C. USA, 2010.
29. Mataba, G., Verhaert, V., Blust, R., Bervoets, L., Distribution of trace elements in the aquatic ecosystem of the Thigithe river and the fish Labeo victorianus in Tanzania and possible risks for human consumption. *Sci. Total Environ.* 547, 48–59, 2016 DOI:10.1016/j.scitotenv.2015.12.123.
30. Maleri, R., Reinecke, S., Mesjasz, J., Reinecke, A., Growth and reproduction of earthworms in ultramafic soils. *Arch. Environ. Contam. Toxicol.*, 52, 363–70, 2007.

31. Baker, A., Brooks, R., Terrestrial higher plants which hyperaccumulate metallic elements – a review of their distribution, ecology and phytochemistry. *Biorecovery.*, 1, 81–126, 1989.
32. Ghaderian, S., Mohtadi, A., Rahiminejad, M., Baker, A., Nickel and other metal uptake and accumulation by species of Alyssum (Brassicaceae) from the ultramafics of Iran. *Environ. Pollut.*, 145, 293–298, 2007.
33. Pollard, A., Reeves, R., Baker, A., Facultative hyperaccumulation of heavy metals and metalloids. *Plant Sci.*, 217–218, 8–17, 2014.
34. Reeves, R., Baker, A., Romero, R., The ultramafic flora of the Santa Elena peninsula, Costa Rica: a biogeochemical reconnaissance. *J. Geochem. Explor.*, 93, 153–159, 2007.
35. Van der Ent, A., Baker, A., Van Balgooy, M., Tjoa, A., Ultramafic nickel laterites in Indonesia (Sulawesi, Halmahera): Mining, nickel hyperaccumulators and opportunities for phytomining. *J. Geochem. Explor.*, 128, 72–79, 2013.
36. Shallari, S., Schwartz, C., Hasko, A., Morel, J., Heavy metals in soils and plants of serpentine and industrial sites of Albania. *Sci. Total Environ.*, 209, 133–142, 1998.
37. Baker, A., McGrath, S., Reeves, R., Smith, J., Metal hyperaccumulator plants: a review of the ecology and physiology of a biological resource for phytoremediation of metal-polluted soils, In: Terry, N, Bañuelos, G, editors, *Phytoremediation Contam. Soils Waters*, CRC Press LLC, Boca Raton, FL, USA,pp. 95–117, 2000.
38. Sheoran, V., Sheoran, A., Poonia, P., Phytomining: a review. *Miner. Eng.*, 22, 1007–1019, 2009.
39. Sánchez, M., *Contaminacion por metales pesados en el botadero de basuras de Moravia en Medellin: transferencia a flora y fauna y evaluacion del potencial fitorremediador de especies nativas e introducidas*, Pontificia Universidad Javeriana, 2010.
40. Robinson, B., Chiarucci, A., Brooks, R., Petit, D., Kirkman, J., Gregg, P., De Dominicis, V., The nickel hyperaccumulator plant Alyssum bertolonii as a potential agent for phytoremediation and phytomining of nickel. *J. Geochem. Explor.*, 59, 75–86, 1997.
41. Brady, K., Kruckeberg, A., Bradshaw, J., Evolutionary ecology of plant adaptation to serpentine soils. *Annu. Rev. Ecol. Evol. Syst.*, 243–266, 2005.
42. Garnier, J., Quantin, C., Martins, E., Becquer, T., Solid speciation and availability of chromium in ultramafic soils from Niquel, *J. Geochem. Explor.*, 88, 206–209, 2006.
43. Schaller, J., Vymazal, J., Brackhage, C., Retention of resources (metals, metalloids and rare earth elements) by autochthonously/allochthonously dominated wetlands: a review. *Ecol. Eng.*, 53, 106–114, 2013.
44. Mench, M., Vangronsveld, J., Clijsters, H., Lepp, N., Edwards, R., In situ metal immobilization and phytostabilization of contaminated soils, In: Terry, N, Bañuelos, G, editors, *Phytoremediation Contam. Soils Waters*, CRC Press LLC, Boca Raton, FL, USA, pp. 328–363, 2000.

45. Arora, A., Saxena, S., Sharma, D., Tolerance and phytoaccumulation of Chromium by three Azolla species. *World J. Microb. Biot.*, 22, 97–100, 2005.
46. Malawska, M., Wiołkomirski, B., An analysis of soil and plant (Taraxacum Officinale) contamination with heavy metals and polycyclic aromatic hydrocarbons (PAHs) in the area of the railway junction Iława Główna, Poland. *Water Air Soil Poll.*, 127, 339–349, 2001.
47. Aksoy, A., Demirezen, D., Duman, F., Bioaccumulation, detection and analyses of heavy metal pollution in Sultan Marsh and its environment, *Water Air Soil Poll.*, 164, 241–255, 2005.
48. Lago, M., Arenas, D., Rodríguez, A., Andrade, M., Vega, F., Cobalt, chromium and nickel contents in soils and plants from a serpentinite quarry. *Solid Earth*, 6, 323–335, 2015.
49. Alcaldía Mayor de Bogotá D.C., Informe de la IX Fase del programa de seguimiento y monitoreo de efluentes y afluentes al recurso hídrico de Bogotá, Convenio Interadministrativo 020/2008, suscrito entre la Secretaria Distrital de Ambiente (SDA) y la Empresa de Acueducto y Alcantarillado de Bogotá ESP (EAAB-ESP), del 2010, Bogotá D.C., Colombia, 2010.
50. Doichinova, V., Velizarova, E., Reuse of paper industry wastes as additives in phytoremediation of heavy metals polluted substrates from the Spoil Banks of the Kremikovtsi Region, Bulgaria. *Procedia Environ. Sci.*, 18, 731–736, 2013.
51. Geissen, V., Ramos, F., Bastidas, P., Díaz, G., Bello, R., Huerta, E., Ruiz, E., Soil and water pollution in a banana production region in tropical Mexico. *Bull Env. Contam. Toxicol.*, 85, 407–413, 2010.
52. Vigo, J., Sobre algunas plantas aloctonas. *Collect. Bot.*, 10, 351–364, 1976.
53. Elorza, M., Dana, E., Sobrino, E., Aproximación al listado de plantas alóctonas invasoras reales y potenciales en España, *Lazaroa.*, 22, 121–131, 2001.
54. Calvachi, B., Una mirada regional. La biodiversidad bogotana. *Rev. La Tadeo.*, 89–98, 2002.
55. Ojasti, J., Estudio sobre el estado actual de las especies exóticas, Caracas-Venezuela, 2001. http://www.comunidadandina.org/bda/docs/CAN-BIO-0012.pdf.
56. Schroeder, F., Zur klassifizierung der Anthropochoren. *Veg. Acta Geobot.*, 16, 225–238, 1968.
57. Pyšek, P., On the terminology used in plant invasion studies, In: Pyšek, P, Prach, K, Rejmánek, M, Wade, M, editors, *Plant Lnvasions – Gen. Asp. Spec. Probl.*, Academic Publishing, Amsterdam. The Netherlands, pp. 71–81, 1995.
58. Kasperek, G., Eine Bibliographie zur Klassifikation von Anthropochoren. *Braunschweiger Geobot. Arb.*, 2008, http://rzbl04.biblio.etc.tu-bs.de:8080/docportal/servlets/MCRFileNodeServlet/DocPortal_derivate_00012680/Kasperek-Bibliographie_Anthropochoren.pdf (accessed April 17, 2016).
59. McNeely, J.A., Mooney, H.A., Neville, L.E., Schei, P., Waage, J.K., (eds.) Global strategy on invasive alien species, IUCN, Gland, Switzerland and Cambridge,

UK in collaboration with the Global Invasive Species Programme 2001 IUCN on behalf of the Global Invasive Species Programme, Switzerland, and Cambridge, UK, 2001.
60. Kowarik, I., Pyšek, P., The first steps towards unifying concepts in invasion ecology were made one hundred years ago: revisiting the work of the Swiss botanist Albert Thellung. *Divers. Distrib.*, 18, 1243–1252, 2012.
61. Villaronga, B., Una amenaza para la biodiversidad: especies exóticas invasoras. *Ambient. La Rev. Del Minist. Medio Ambient.*, 23, 58–66, 2003.
62. Welcomme, R., International introductions of inland aquatic species, 1988, http://www.fao.org/docrep/x5628e/x5628e00.htm.
63. Fukami, T., Bellingham, P., Peltzer, D., Walker, L., Fukami, T., Bellingham, P., Peltzer, D., Walker, L., Non-native plants disrupt dual promotion of native alpha and beta diversity. *Folia Geobot.*, 48, 319–333, 2013.
64. Pyšek, P., Jarošík, V., Hulme, P., Pergl, J., Hejda, M., Schaffner, U., Vilà, M., A global assessment of invasive plant impacts on resident species, communities and ecosystems: the interaction of impact measures, invading species' traits and environment. *Glob. Chang. Biol.*, 18, 1725–1737, 2012.
65. Yang, X., Feng, Y., He, Z., Stoffella, P., Molecular mechanisms of heavy metal hyperaccumulation and phytoremediation. *J. Trace. Elem. Med. Bio.*, 18, 339–353, 2005.
66. Tilman, D., Lehman, C., Human-caused environmental change: Impacts on plant diversity and evolution. *Pnas.*, 98, 5433–5440, 2001.
67. Montes, C., Rendón, M., Varela, L., Cappa, M.J., Manual de restauración de humedales mediterráneos., *Sevilla*, 2007, http://www.juntadeandalucia.es/medioambiente/site/portalweb/menuitem.7e1cf46ddf59bb227a9ebe205510e1ca/?vgnextoid=d80a3139e13dd110VgnVCM1000001325e50aRCRD&vgnextchannel=http://www.juntadeandalucia.es/medioambiente/site/portalweb/menuitem.7e1cf46ddf59bb227a.
68. Rodríguez, S., Organismos del suelo: la dimensión invisible de las invasiones por plantas no nativas. *Ecosistemas.*, 18, 2009.
69. Pimentel, D., McNair, S., Janecka, J., Wightman, J., Simmonds, C., O'Connell, C., Wong, E., Russel, L., Zern, J., Aquino, T., Tsomondo, T., Economic and environmental threats of alien plant, animal, and microbe invasions. *Agric. Ecosyst. Environ.*, 84, 1–20, 2001.
70. Hoyle, H., Hitchmough, J., Jorgensen, A., Attractive, climate-adapted and sustainable? Public perception of non-native planting in the designed urban landscape. *Landsc. Urban Plan.*, 164, 49–63, 2017.
71. Thomas, E., Jalonen, R., Loo, J., Boshier, D., Gallo, L., Cavers, S., Bordács, S., Smith, P., Bozzano, M., Genetic considerations in ecosystem restoration using native tree species. *For. Ecol. Manage.*, 333, 66–75, 2014.
72. Farago, M., Parsons, P., The effects of various platinum metal species on the water plant Eichhornia crassipes (MART.). *Chem. Speciat. Bioavailab.*, 6, 1–12, 1994.
73. Wang, Q., Cui, Y., Dong, Y., Phytoremediation of polluted waters potentials and prospects of wetland plants. *Acta. Biotechnol.*, 22, 199–208, 2002.

74. Vardanyan, L., Ingole, B., Studies on heavy metal accumulation in aquatic macrophytes from Sevan (Armenia) and Carambolim (India) lake systems. *Environ. Int.*, 32, 208–218, 2006.
75. Kamal, M., Ghaly, A., Mahmoud, N., CoteCôté, R., Phytoaccumulation of heavy metals by aquatic plants. *Environ. Int.*, 29, 1029–1039, 2004.
76. Ghosh, M., Singh, S., A review on phytoremediation of heavy metals and utilization of it's byproducts. *Appl. Ecol. Env. Res.*, 3, 1–18, 2005.
77. Sadowsky, M.J., Phytoremediation: past promises and future practises. Proceedings of the 8th International Symposium on Microbial Ecology, Halifax, Canada, 1999.
78. Macnair, M., Tilstone, G., Smith, S., The genetics of metal tolerance and accumulation in higher plants, In: Terry, N, Bafiuelos, G, editors, *Phytoremediation Contam. Soil Water*, CRC Press LLC, Boca Raton, FL, USA, pp. 241–256, 1999.
79. Barceló, J., Poschenrieder, C., *Phytoremediation : principles and perspectives*, Barcelona, 2003.
80. Sarma, H., Metal hyperaccumulation in plants: a review focusing on phytoremediation technology. *J. Environ. Sci. Technol.*, 4, 118–138, 2011.
81. Rodríguez, D., El papel de los microorganismos en la biodegradación de compuestos tóxicos. *Ecosistemas.*, 12.
82. Celis, H., Junod, M., Sandoval, E., Recientes aplicaciones de la depuracion de aguas residuales con plantas acuáticas. *Rev. Theor. Ciencia, Arte Y Humanidades.*, 14, 17–25, 2005.
83. Robinson, B., Kim, N., Marchetti, M., Moni, C., Schroeter, L., Van Den Dijssel, C., Milne, G., Clothier, B., Arsenic hyperaccumulation by aquatic macrophytes in the Taupo Volcanic Zone, New Zealand. *Environ. Exp. Bot.*, 58, 206–215, 2006.
84. Delgadillo, A., González, C., Prieto, F., Villagómez, J., Acevedo, O., Fitorremediación: una alternativa para eliminar la contaminación. *Trop. Subtrop. Agroecosystems.*, 14, 597–612, 2011.
85. Prasad, M., Phytoremediation of metal-polluted ecosystems: hype for commercialization. *Russ. J. Plant. Physiol.*, 50, 686–701, 2003.
86. Horne, A., Phytoremediation by constructed wetlands, In: Terry, N, Bañuelos, G, (eds), *Phytoremediation Contam. Soils Waters*, CRC Press LLC, Boca Raton, FL, USA, pp. 25–51, 2000.
87. Adams, N., Carroll, D., Madalinski, K., Rock, S., *Introduction to phytoremediation*, United States Environmental Protection Agengy, Office of Research and Development, Washington D.C. USA, 2000, http://nepis.epa.gov/Adobe/PDF/30003T7G.PDF (accessed August 24, 2015).
88. Cunningham, S., Berti, W., Phytoextraction and phytostabilization: technical, economic and regulatory considerations of the soil-lead issue, In: Terry, N, Buñuelos, G, editors, *Phytoremediation Contam. Soils Waters*, CRC Press LLC, Boca Raton, FL, USA, pp. 363–380, 2000.
89. Terry, N., Bañuelos, G., editors, *Phytoremediation of contaminated soils and waters*, CRC Press LLC, Boca Raton, FL, USA, 2000.

90. Robinson, B., Anderson, C., Dickinson, N., Phytoextraction: Where's the action? *J. Geochem. Explor.*, 151, 34–40, 2015.
91. Lone, M., He, Z., Stoffella, P., Yang, X., Phytoremediation of heavy metal polluted soils and water: progresses and perspectives. *J. Zhejiang Univ-SC B.*, 9, 210–20, 2008.
92. Mengoni, A., Schat, H., Vangronsveld, J., Plants as extreme environments? Ni-resistant bacteria and Ni-hyperaccumulators of serpentine flora. *Plant Soil.*, 331, 5–16, 2010.
93. Hassan, Z., Aarts, M., Opportunities and feasibilities for biotechnological improvement of Zn, Cd or Ni tolerance and accumulation in plants. *Environ. Exp. Bot.*, 72, 53–63, 2011.
94. Jaffré, T., Brooks, R., Lee, J., Reeves, R., Sebertia acuminata: a hyperaccumulator of nickel from New Caledonia. *Science*, 193, 579–580, 1976.
95. Wei, S., Teixeira da Silva, J., Zhou, Q., Agro-improving method of phytoextracting heavy metal contaminated soil. *J. Hazard. Mater.*, 150, 662–668, 2008.
96. Berazain, R., Plantas serpentinícolas acumuladoras e hiperacumuladoras de níquel en América Tropical. IV Conferencia Internacional sobre Ecología de Serpentina., Cuba, 2003.
97. Reeves, R., Baker, A., Borhidi, A., Berazain, R., Nickel Hyperaccumulation in the Serpentine Flora of Cuba. *Ann. Bot.*, 83, 29–38, 1999.
98. Sun, Y., Zhou, Q., Liu, W., An, J., Xu, Z., Wang, L., Joint effects of arsenic and cadmium on plant growth and metal bioaccumulation: a potential Cd-hyperaccumulator and As-excluder Bidens pilosa L. *J. Hazard. Mater.*, 165, 1023–1028, 2009.
99. Boyd, R., Davis, M., Balkwill, K., Elemental patterns in Ni hyperaccumulating and non-hyperaccumulating ultramafic soil populations of Senecio coronatus. *S. Afr. J. Bot.*, 74, 158–162, 2008.
100. Van der Ent, A., Baker, A., Reeves, R., Pollard, A., Schat, H., Hyperaccumulators of metal and metalloid trace elements: facts and fiction. *Plant Soil*, 362, 319–334, 2013.
101. Prasad, M., Freitas, H., Metal hyperaccumulation in plants: Biodiversity prospecting for phytoremediation technology. *Electron. J. Biotechnol.*, 6, 285–321, 2003.
102. Marrero, J., Amores, I., Coto, O., Fitorremediación, una tecnología que involucra a plantas y microorganismos en el saneamiento ambiental, La Habana, Cuba, 2012.
103. Bhargava, A., Carmona, F., Bhargava, M., Srivastava, S., Approaches for enhanced phytoextraction of heavy metals. *J. Environ. Manag.*, 105, 103–120, 2012.
104. Chiarucci, A., Baker, A., Advances in the ecology of serpentine soils. *Plant Soil*, 293, 1–2, 2007.
105. Bennicelli, R., Stępniewska, Z., Banach, A., Szajnocha, K., Ostrowski, J., The ability of Azolla caroliniana to remove heavy metals (Hg(II), Cr(III), Cr(VI)) from municipal waste water. *Chemosphere*, 55, 141–146, 2004.

106. Monferrán, M., Wunderlin, D., Biochemistry of metals/metalloids toward remediation process magdalena, In: Gupta, D, Corpas, F, Palma, J, editors, *Heavy met. stress plants*, Springer-Verlag Berlin Heidelberg, Berlin, Heidelberg, pp. 43–72, 2013.
107. Cundy, A.B., Bardos, R.P., Church, A., Puschenreiter, M., Friesl-Hanl, W., Müller, I., Neu, S., Mench, M., Witters, N., Vangronsveld, J., Developing principles of sustainability and stakeholder engagement for "gentle" remediation approaches: the European context. *J. Environ. Manag.*, 129, 283–291, 2013.
108. Covarrubias, S., Peña, J., Contaminación ambiental por metales pesados en méxico: problemática y estrategias de fitorremediación. *Rev. Int. Contam. Ambie.*, 33, 7–21, 2017.
109. Sharma, R., Agrawal, M., Biological effects of heavy metals: an overview. *J. Environ. Biol.*, 26, 301–313, 2005.
110. Durán, A., *Transferencia de metales de suelo a planta en áreas mineras: Ejemplos de los Andes peruanos y de la Cordillera Prelitoral Catalana*, Universidad de Barcelona, 2010.
111. Haque, N., Peralta-Videa, J., Jones, G., Gill, T., Gardea-Torresdey, J., Screening the phytoremediation potential of desert broom (Baccharis sarothroides Gray) growing on mine tailings in Arizona, USA, *Environ. Pollut.*, 153, 362–368, 2008.
112. Menezes, A., Da Silva, J., Rossato, R., Santos, M., Decker, N., Da Silva, F., Cruz, C., Dihl, R., Lehmann, M., Ferraz, A., Genotoxic and biochemical changes in Baccharis trimera induced by coal contamination. *Ecotoxicol. Environ. Saf.*, 114, 9–16, 2015.
113. Santos, J., Castro, A., Huezo, J., Torres, L., Arsenic and heavy metals in native plants at tailings impoundments in Queretaro, Mexico. *Phys. Chem. Earth*, 37–39, 10–17, 2012.
114. Bech, J., Duran, P., Roca, N., Poma, W., Sánchez, I., Roca-Pérez, L., Boluda, R., Barceló, J., Poschenrieder, C., Accumulation of Pb and Zn in Bidens triplinervia and Senecio sp. spontaneous species from mine spoils in Peru and their potential use in phytoremediation. *J. Geochem. Explor.*, 123, 109–113, 2012.
115. Ladislas, S., Gérente, C., Chazarenc, F., Brisson, J., Andrès, Y., Floating treatment wetlands for heavy metal removal in highway stormwater ponds. *Ecol. Eng.*, 80, 85–91, 2014.
116. Matthews, D., Moran, B., Otte, M., Screening the wetland plant species Alisma plantago-aquatica, Carex rostrata and Phalaris arundinacea for innate tolerance to zinc and comparison with Eriophorum angustifolium and Festuca rubra Merlin. *Environ. Pollut.*, 134, 343–351, 2005.
117. Walker, D., Clemente, R., Bernal, M., Contrasting effects of manure and compost on soil pH, heavy metal availability and growth of Chenopodium album L. in a soil contaminated by pyritic mine waste. *Chemosphere*, 57, 215–224, 2004.

118. Akinbile, C., Yusoff, M., Ahmad, A., Landfill leachate treatment using subsurface flow constructed wetland by Cyperus haspan. *Waste Manag.*, 32, 1387–1393, 2012.
119. Ewais, E., Effects of cadmium, nickel and lead on growth, chlorophyll content and proteins of weeds. *Biol. Plant.*, 39, 403–410, 1997.
120. Jin-Hong, Q., Zayed, A., Zhu, Y.-L., Yu, M., Terry, N., Phytoaccumulation of trace elements by wetland plants: III. Uptake and accumulation of ten trace elements by twelve plant species. *J. Environ. Qual.*, 28, 1448–1455, 1999.
121. Mganga, N., Manoko, M., Rulangaranga, Z., Classification of plants according to their heavy metal content around North Mara Gold Mine, Tanzania: Implication for phytoremediation. *Tanz. J. Sci.*, 37, 109–119, 2011.
122. Litter, M., Alarcón, M., Arenas, M., Armienta, M., Avilés, M., Cáceres, R., Cipriani, H., Cornejo, L., Dias, L., Cirelli, A., Farfán, E., Garrido, S., Lorenzo, L., Morgada, M., Olmos, M., Pérez, A., Small-scale and household methods to remove arsenic from water for drinking purposes in Latin America. *Sci. Total Environ.*, 429, 107–122, 2012.
123. Das, M., Maiti, S., Comparison between availability of heavy metals in dry and wetland tailing of an abandoned copper tailing pond. *Environ. Monit. Assess.*, 137, 343–350, 2008.
124. Li, G., Hu, N., Ding, D., Zheng, J., Liu, Y., Wang, Y., Nie, X., Screening of plant species for phytoremediation of uranium, thorium, barium, nickel, strontium and lead contaminated soils from a uranium mill tailings repository in South China. *Bull. Environ. Contam. Toxicol.*, 86, 646–652, 2011.
125. Murch, S., Haq, K., Rupasinghe, H., Saxena, P., Nickel contamination affects growth and secondary metabolite composition of St. John's wort (Hypericum perforatum L.). *Environ. Exp. Bot.*, 49, 251–257, 2003.
126. Tirillini, B., Ricci, A., Pintore, G., Chessa, M., Sighinolfi, S., Induction of hypericins in Hypericum perforatum in response to chromium. *Fitoterapia.*, 77, 164–170, 2006.
127. Johnston, W., Mineral uptake of plants of serpentine and lead-mine soils, BA Thesis. Univ. Stirling. (1974).
128. Johnston, W., Proctor, J., A comparative study of metal levels in plants from two contrasting lead-mine sites. *Plant Soil*, 46, 251–257, 1977, 1.
129. Wenzel, W., Jockwer, F., Accumulation of heavy metals in plants grown on mineralised soils of the Austrian Alps. *Environ. Pollut.*, 104, 145–155, 1999.
130. Samecka, A., Kempers, A., Concentrations of heavy metals and plant nutrients in water, sediments and aquatic macrophytes of anthropogenic lakes (former open cut brown coal mines) differing in stage of acidification. *Sci. Total Environ.*, 281, 87–98, 2001.
131. Yanqun, Z., Yuan, L., Schvartz, C., Langlade, L., Fan, L., Accumulation of Pb, Cd, Cu and Zn in plants and hyperaccumulator choice in Lanping lead-zinc mine area, China. *Environ. Int.*, 30, 567–576, 2004.

132. Nedelkoska, T., Doran, P., Characteristics of heavy metal uptake by plant specie with potential for phytoremediation and phytomining. *Miner. Eng.*, 13, 549–561, 2000.
133. Del Río, M., Font, R., Almela, C., Vélez, D., Montoro, R., De Haro Bailón, A., Heavy metals and arsenic uptake by wild vegetation in the Guadiamar river area after the toxic spill of the Aznalcóllar mine. *J. Biotechnol.*, 98, 125–137, 2002.
134. Battke, F., Ernst, D., Fleischmann, F., Halbach, S., Phytoreduction and volatilization of mercury by ascorbate in Arabidopsis thaliana, European beech and Norway spruce. *Appl. Geochem.*, 23, 494–502, 2008.
135. O'Dell, R., Silk, W., Green, P., Claassen, V., Compost amendment of Cu-Zn minespoil reduces toxic bioavailable heavy metal concentrations and promotes establishment and biomass production of Bromus carinatus (Hook and Arn.). *Environ. Pollut.*, 148, 115–124, 2007.
136. Zhang, X., Liu, J., Huang, H., Chen, J., Zhu, Y., Wang, D., Chromium accumulation by the hyperaccumulator plant Leersia hexandra Swartz. *Chemosphere*, 67, 1138–1143, 2007.
137. Shinmachi, F., Kumanda, Y., Noguchi, A., Hasegawa, I., Translocation and accumulation of cadmium in cadmium-tolerant Polygonum thunbergii. *Soil. Sci. Plant Nutr.*, 49, 355–361, 2003.
138. Demirezen, D., Aksoy, A., Accumulation of heavy metals in Typha angustifolia (L.) and Potamogeton pectinatus (L.) living in Sultan Marsh (Kayseri, Turkey). *Chemosphere*, 56, 685–696, 2004.
139. Silk, W., Bambic, D., O'Dell, R., Green, P., Seasonal and spatial patterns of metals at a restored copper mine site II. Copper in riparian soils and Bromus carinatus shoots, *Environ. Pollut.*, 144, 783–789, 2006.
140. Maobe, M., Gatebe, E., Gitu, L., Rotich, H., Profile of heavy metals in selected medicinal plants used for the treatment of diabetes, malaria and pneumonia in Kisii Region, Southwest Kenya, Glob. *J. Pharmacol.*, 6, 245–251, 2012.
141. Zechmeister, H., Grodzińska, K., Szarek, G., Bioindicators & biomonitors – principles, concepts and applications, In: Markert, BA, Breur, AM, Zechmeister, HG, editors, *Bioindic. Biomonitors*, Elsevier, pp. 329–375, 2003.
142. Proctor, J., Toxins, nutrient shortages and droughts: the serpentine challenge. *Trends Ecol. Evol.*, 14, 334–335, 1999.

8

Bioremediation: A Green, Sustainable and Eco-Friendly Technique for the Remediation of Pollutants

Munawar Iqbal[1]*, Arif Nazir[1], Mazhar Abbas[2], Qudsia Kanwal[1] and Dure Najaf Iqbal[1]

[1]*Department of Chemistry, The University of Lahore, Lahore, Pakistan*
[2]*Institute of Molecular Biology and Biotechnology, The University of Lahore, Lahore, Pakistan*

Abstract

Manganese peroxidase (MnP), Lignin peroxidase (LiP) and Lacasse (Lac) have diverse application in biofuel production, biopulping, biobleaching, paper and pulp, organic pollutants degradation, stabilizer, detergent, biosensors, animal feed, textile, beverage processing, cosmetics, antibiotics and steroids fields. Enzymatic treatment of pollutants is a green, sustainable, cost effective and an eco-friendly technique versus physicochemical remediation methods. The biodegradation of model compounds and industrial wastes using enzymes have been studied well and enzymes have also been successfully utilized for the bioremediation of environmental concern pollutants. This review article addresses the applications of enzymatic system as environmental biocatalysts for the remediation of pollutants like dye and other organic pollutants in industrial wastewater. The State of the art of bioremediation is discussed in detail and process variables effect on degradation and water quality parameters is also discussed.

Keywords: Bioremediation, enzymes, dyes, lignin proxidase, manganese peroxidase, laccase, degradation, water quality parameters

**Corresponding author*: bosalvee@yahoo.com

Shahid-ul-Islam (ed.) Advanced Materials for Wastewater Treatment, (263–312)

8.1 Introduction

In the recent past, ligninolytic (LiP, MnP, Lac) enzymes attracted the attention of scientific community due to their vast applications in bioremediation, delignification of biomass for biofuel production, biopulping, biobleaching, organic pollutants degradation, detergent manufacturing, animal feed, stabilization in food industry, biosensors, textile, beverage processing, cosmetics, paper and pulp, steroids and transformation of antibiotics etc. [1–3]. Despite appreciable biotechnological potential, the use of native microbial enzymes encounter a number of problems practically at industrial scale, that is, cost and purification, structural instability in media, activity inhibition and reusability [4]. Thermo-stable enzymes are being widely used since they can withstand under harsh conditions, that is, high temperatures and alkaline pH conditions [5] with the advancement in existing techniques, it is now possible to modify the properties of enzymes. Thus, enzymes with desired properties can be molded through mutations, genetic engineering, chemical modifications and immobilization [6].

Among all advanced techniques, the immobilization of enzyme is regarded as excellent and is also a versatile which uses different matrix as a support and have been successful used at commercial scale. It is an innovative green biotechnology that could provide additional benefits over free enzymes such as facile recovery, reusability, longer half-lives, protection from shear damage and reduction in protease activity [7–8]. Enzyme-based approaches with positive edge of high efficiency and multifarious use replaced the conventional chemical methods to utilize the enzymes at industrial scale and this technique has been entered in exciting new phase [6]. To date, many studied have been conducted on the production of the product, its purification and followed by characterization of enzymes but very scarce data is published on immobilization to improve the catalytic efficiency of enzymes.

In present review, a comprehensive and fully documented knowledge of immobilization of fungal ligninolytic enzymes with particular emphasis on environmental applications are discussed. Entrapment, adsorption, covalent bonding, encapsulation and self-immobilization techniques are discussed for the immobilization of LiP, MnP and Lac and their applications for dye decolorization and other xenobiotics compounds.

8.2 Immobilization

Immobilization means unite the enzyme with an insoluble matrix (carrier), which enables the enzyme to retain its proper geometry in a reactor under specific conditions [6]. Numerous recent studies have been focused on ligninolytic enzyme immobilization, particularly Lac [9], LiP [2] and MnP [10]. Due to this property, ligninolytic enzyme received much attention versus economically competent and highly efficient biotechnological procedures in numerous fields including biotransformation, environmental monitoring, diagnostics, food industries and pharmaceutical or biomedical applications [11]. The immobilized enzymes can overcome some of the afore-mentioned limitations by providing additional advantages over free enzymes such as facile recovery of enzymes from the reaction mixture, reusability, less degradation rate by free enzyme and longer half-lives and protection from shear damage [7–8]. In addition, enzymes in an immobilized form are more irrepressible to ecological factors including pH, temperature and exposure to chemically toxic substances [5].

8.3 Enzyme Immobilization Strategies

There are various approaches that can be employed for immobilization of enzymes; nevertheless the process of enzyme immobilization is quiet in beginning [12]. Theoretically, the two basic methods that can be distinguished for immobilization of enzymes, as the link for enzyme-support may take place through various processes including physical or chemical interactions. So, the existence of several types of methodologies may have role during these interactions. In physical methods, the entrapment of the enzyme in a tri-dimensional matrix or its encapsulation may occur whether in an organic or inorganic membrane structure [13]. On the other hand, chemical method can process through covalent binding, adsorption or self-immobilization [14]. The enzymes immobilization techniques and the properties of the reported strategies for enzymes immobilization are given (Table 8.1).

8.4 Adsorption

In adsorption technique, the enzyme is adsorbed on the outer surface of the supporting matrix. The adsorption mechanism results from weak attractive forces such as hydrophobic interactions/Van der Waal's forces/electrostatic forces. In general, adsorption is a simple, comparatively inexpensive and non-destructive method of immobilization for enzyme activity and may therefore reveal higher profitable potential than other techniques [27]. In this technique, the catalytic site of the adsorbed enzyme may be blocked

Table 8.1 Reported strategies of enzyme immobilization and effect of immobilization on enzymes.

Strategy	Effect/improvement	Refs.
Covalent binding- by multipoint attachment	Enhanced stability by increasing bonds	[15]
Immobilized Cross-linked enzyme crystals (CLECs)	Enhanced Stability	[16–19]
Sol-gel entrapment	Activity & selectivity improvement	[2]
Poly(lactic-co-glycolic acid) (PLGA) nanofibers	Stability, feasibility of continuous operation	[20]
Cross-linked enzyme aggregates (CLEAs)	Recovery, stability & improved recyclability	[17–19]
Adsorption entrapment	Thermostability & resistance against detergents	[21]
Adsorption on nanoporous gold (NPG) particles	Enhanced thermostability	[22]
Covalent binding on cellular foam	Enhanced thermostability and activity	[23]
Covalent entrapment in nanoparticles	pH & thermostability	[24]
Entrapment in nanofibrous polymers	Intrinsic specific surface area, inter-fiber porosity, & mechanical strength	[25]
Porous cross-linked polymers using starch as pore making agent	Enhanced activity &catalytic efficiency increased (Vmax)	[26]

by the inert support, matrix or bead thus greatly reducing the efficacy of the enzyme. The ionic strength of the medium, pH and hydrophobicity must be considered in adsorption technique during the immobilization process [28]. In this process, the enzyme in solution confronted to the solid support for a static period of time followed by washing with buffer to remove unbound enzyme. Although, this immobilization strategy causes little or no obvious enzyme inactivation, however, non-specific adsorption of other proteins or substances and desorption of the enzyme due to change in in temperature, weak bonding of the enzyme with solid support, pH and ionic strength might be problematic. Consequently, poor operational and storage stability can be caused by the biosensors based on adsorbed enzyme. However, the enzymes (adsorbed) are sheltered from proteolysis, aggregation and interaction with hydrophobic borders [7].

In the past, various researchers have employed various environment friendly adsorptive supporting materials, for example, husk fibers of coconut having huge hygroscopic capacity and high cat-ion exchanger property; Carboxymethyl cellulose (CMC) with irreversible binding capacity; kaolinite (hydrated aluminum silicate) with high enzyme retain-ability and micro porous materials having thiol functional group, preferably suited for redox reactions [29]. The presence of silanols on pore walls facilitates enzyme immobilization by hydrogen bonding in silanized molecular sieves and also have been effectively used as supports for enzyme adsorption. Numerous adaptations and chemical alterations in presently used supporting matrix would significantly help in better check for economic production process. Environment friendly, green supports of biological origin cut down the production costs and prevent cropping up of ethical issues. Most recently, due to their reliable robustness and competence, microporous silica nanoparticles (MSNs) were used as biocompatible support material for enzyme immobilization in energy applications [30].

Various studies related to adsorption have been linked with the immobilization of different enzymes. [31] reported silicate-based porous material to immobilize *T. versicolor* Lac through physical adsorption. Forde *et al.* [28] observed promising results when Lac was immobilized onto mesoporous silicate particles, Santa Barbara amorphous (SBA-15) and cyano-modified silica (CNS) with surface modification by bi-functional glutaraldehyde (GLU) and ethylene glycol-N-hydroxysuccinimide (NHS) and mono-functional citraconic anhydride methods. A novel and efficient immobilization of Lac from *P. sanguineus* has been investigated using magnetic Cu^{2+} nanoparticles as carrier and it was found that following immobilization of Lac on magnetic Copper nanoparticles, will definitely enhance the stability and catalytic capacity of the enzyme in the direction of many types of physical and chemical denaturants. The ion exchange resins such as agarose, dextran and chitosan with several functional groups have also been used for adsorption [32].

8.5 Entrapment

Entrapment is caging of enzymes within porous solid matrix, that is, Ca-alginate beads, chitosan, polyvinyl alcohol or gels comprising of propyletrimethoxysilane (PTMS) and tetramethoxysilane (TMOS) and in this way entrapped enzyme is prevented from direct contact with the surroundings [6, 33]. The salient features of this technique are easy to prepare, cheap and there is no mechanical alteration of the enzyme. Nevertheless, this approach is characterized by low enzyme loading and mass transfer limitations [27]. Nanostructured supports used in entrapment, for example, lectrospun nanofibers and pristine materials have revolutionized the immobilization process by this technique [34]. Prevention of friability, augmentation of entrapment efficiency, leaching and enzyme activity entrapped ligninolytic enzymes in chitosan has been described. Additionally, the support has been considered to be nontoxic, amenable to chemical modification, biocompatible and hydrophilic in nature and is highly affinitive to protein. Ligninolytic enzymes entrapped in Sol-gel matrices have also been used with highly thermal stability and tolerance toward organic solvent [2]. In another study, *C. unicolor* lignase enzyme showed better thermal behavior when entrapped in a hydrogel matrix of poly (N-isopropyl acrylamide) [35–36], whereas Lac immobilized retained 80% higher activity onto amine-terminated magnetic nanocomposites up to 5 cycles. Therefore, the enzyme performance depends on support used for immobilization. Generally, two types of supports are used for enzyme immobilization, that is, hydrophobic and hydrophilic. Hydrophobic materials are preferable because of their ability to bind large amount of enzyme with higher activity. The use of gels for enzyme entrapment is receiving extensive attention due to preservation of more vigorous and thermostability of enzymes [6].

8.6 Encapsulation

This technique is similar to entrapment, also protect enzyme from outside environment, but the process of mass transfer is a serious hurdle in encapsulation and entrapment techniques during process of immobilization [27]. In microencapsulation the enzyme is restricted in the centre of micron-sized spheres composed of semipermeable material. There are very few studies have been done to find out different methodologies and applications of encapsulated enzymes. In another stydu, observed a significant change in optimum pH (4–5) of Lac from *Trametes sp.*, and the authors demonstrated that the activity is influenced by the quantity of the immobilized Lac. Another microencapsulation method that has also been tested is layer-by-layer (LbL) technique [37–39]. The supports for this technique include gold electrodes and alumina pellets. Szamocki *et al.* [39] microencapsulated *T. trogii* Lac by LbL method by making use of gold electrodes that had been previously functionalized with dithiobis-N-succinimidyl propionate. Efficient encapsulation accomplished with different carriers prevents enzyme leaching and provide enhanced automated permanence [40].

8.7 Covalent Binding

It is the most promising method used for the immobilization of enzymes for at industrial scale for various applications and lots of studies have been performed to immobilize the enzyme onto support by covalent bonding. Hence, it has been the most extensively employed method during the last decade for ligninolytic enzyme immobilization [41]. It involves a chemical reaction through which enzyme is either covalently bonded with supports or using a bi-functional reagent which attach the enzyme at one side and immobilization matrix on the other side. As the chemical reaction confirms that the covalent binding does not mask the enzyme's catalytic site, the activity of the enzyme remains unaffected by covalent bonding. In this technique enzyme can be immobilized via multipoint covalent attachments with support to enhance activity, stability and reusability of enzymes [42]. The bonding occurs between nucleophilic groups on the enzymes and reactive chemical groups on the support surface. Mostly, enzymes are attached by covalent bonds using their lysine amino groups because these are frequently present on the surface of the protein and also have high reactivity [27]. Generally, mechanically stable, nontoxic, chemically inert, insoluble in water are used as support, which must be readily available inexpensive [43]. In this regard, covalent binding of enzymes to silica gel carriers results in highly stable and hyperactive biocatalysts, modified by the process of silanization with elimination of unreacted aldehyde groups and the binding to SBA-15 supports having cage-like pores lined by Si–F linkages [44]. Covalent coupling of enzymes with mesoporous silica and chitosan increase half-life and thermal stability of enzymes [45].

Since covalent bonding is a versatile and most suitable technique, therefore, various different carriers have been designed for support, that is, Kaolinite (silica-based supports) or mesoporous silica nanoparticles [46]. Other researchers, tried fibers and polymeric materials, that is, Eupergit and Sepabeads (epoxy-activated resins) have been frequently employed [42]. A comparison between this kind of support and other entrapment method like covalent binding of immobilization and observed that immobilized *T. villosa* Lac with Eupergit C also showed higher activity versus triggered carbon and entrapment on a Copper-alginate matrix. Variety of electrodes based on glass, carbon, silver, gold or graphite have been intended and used for Lac immobilization [47–48, 39].

Ceramic-chitosan support was used for A. bisporus Lac by cross-linking and resultantly, operational, thermal and storage stabilities enhanced [49]. In another study, polyvinyl alcohol cryogel particles (PVA) were employed for the enzyme immobilization. Stability and operational pH range were enhanced meaningfully. Therefore, it is concluded that cross-linking is the most widely employed method in enzyme covalently binding offered stability under different pH and temperature with excellent activity [50].

8.8 Self-Immobilization

The stabilization of enzymes can be achieved not only by using immobilization supports, but also by carrier-free immobilization, that is, cross-linked enzyme aggregates or crystals formation. The specific and volumetric activity can be reduced by use of solid supports for enzyme immobilization. Self-immobilization is conceivable with the use of cross-linking agents such as gluter-aldehyde, to muddle enzymes with each other without resorting to a support [27]. Cross-linked enzyme aggregates (CLEAs) and Cross-linked enzyme crystals (CLECs) have been reported to be physically stronger, more recoverable, stable and recyclable compared to their soluble counterparts. The CLECs could function in organic solvents as they show remarkable activities and mechanical stabilities with the only precaution of extensive purification. Currently, this research area is of great interest of researchers and might be the future drive of biotechnology of commercial importance [51]. Sheldon's laboratory developed a cheap method of increasing the proximity of enzyme for cross-linking, commercialized with the name of CLEA Technologies. The formation of CLEAs involves the combination of precipitation, purification and immobilization of the enzyme into a single procedure [27].

CLEAs are eco-friendly and environmental beneficial in terms of industrial applications that can be prepared easily from enzyme extracts. Moreover, the promises exist to immobilize two or more than to enzymes to provide CLEAs that are accomplished of catalyzing multiple biotransformation reactions, self-sufficiently or in arrangement as catalytic cascade progressions. The aggregation methods can be improved up to 100% activity yield for any enzyme [6]. The properties of a CLEA can be adjusted significantly through indirect modification of the cross-linking conditions. There are many cross-linkers, that is, glutaraldehyde-ethylene diamine polymers, glutaraldehyde or dextran aldehyde that may be designated for optimal activity of a particular enzyme [52].

Pchelintsev *et al.* [53] found that the activity and microstructure of penicillin acylase CLEA may be influenced by the variation in the duration of the precipitation step prior to cross-linking. Wilson *et al.* [54] reported that stability, productivity and conversion yield of enzymes may be reduced by excess cross-linking agents. Wilson *et al.* [55] also demonstrated that multimeric enzymes, for example tetrameric catalases, retain significant activity with negligible damage of protein under denaturing settings of surfactant (SDS) and thermal conditions and can be immobilized using the CLEA method. Later on, a new disparity on the subject of self-immobilization has been established, that is, spherical catalytic macro-particles or spherezymes (SZs). This advanced method comprises of macroparticles that are in turn composed of the enzyme cross-linked in an oil emulsion generate spherical particles. The lone instance of this method applied to Lac. Table 8.2 shows the benefits and shortcomings of basic immobilization methods.

8.9 Properties of Immobilized Enzymes

8.9.1 Immobilized LiP

Different solid supports have been used for immobilization of LiPs from different WRF. In most of the cases, the immobilized LiPs had higher catalytic activities, thermostabilities and reusability as compared to their free counterparts. Asgher *et al.*, [2] revealed that by the xerogel entrapment LiP showed results comparable with those attained by other methods, for example, the Sepabeads EC-EP3 and Dilbeads NK supports, covalent binding of the enzyme on siliceous cellular foams (McFs) or cross-linked enzyme aggregates immobilization and this technique was also declared as cheap versus others since it needs only coupling agent such as glutareldehyde. Silica xerogels of different hydrophobicity are new materials acting as solid support for the entrapment of LiP to make it more industrially suitable catalyst [56]. The hydrophobicity of the solgels prepared using PTMS and TMOS in different molar ratios increases with increase in propyl groups and solgels enhance the relative activity and thermostability of LiP from *P.* chrysosporium [57]. Xerogel matrix enzyme immobilization technique gives high immobilization efficiency up to 88.6% and improves the kinetic and catalytically characteristics of LiP. Therefore, immobilized LiP remains steady over wide pH and thermal range as compare to free enzyme

Table 8.2 Advantages and drawbacks of the four basic immobilization methods.

Technique	Binding nature	Advantages	Drawbacks
Adsorption	Weak bonding	Simple and easy	Ease desorption
		Relatively inexpensive	Non-specific adsorption of other proteins
		Nondestructive method	
		Limited loss of enzyme activity	
Covalent coupling	Covalent linkages between functional groups of the enzyme and immobilization support	No diffusion barrier	Supporting matrix not regeneratable
		Stable	Coupling with toxic product
		Short response time	
Entrapment	Physically trapping of the enzyme within support matrix	High enzyme activity involve no chemical reaction	Diffusion barrier
		Induces no structural modification of the enzyme	Mass transfer limitation
		Several types of enzymes can be immobilized within the same polymer	Low enzyme loading
			Enzyme leakage
			High concentrations of enzyme needed
Cross-linking	Bond between enzyme/cross-linker (e.g., glutaraldehyde)/inert molecule (e.g., BSA)	Simple	High enzyme activity loss

to fulfill the necessities of the modern enzyme biotechnology at industrial scale [2]. The working of free LiP in the existence of xerogel increases up to 2-fold with a considerable level of hyper activation, lessens with increasing hydrophobic charisma of the xerogels. Xerogel LiP interactions produce a stabilizing affect and lead to the enhanced conformational stability and can be linked with the hyper activation. This may be associated with the fact that LiP may undergo a beneficial conformational change in the presence of these hydrophobic sites though being entrapped within the silica (SiO_2) network. LiP restrained on CNBr-Sepharose 4B support improved the decolorization from Kraft effluent by factor of 2.9 in 48 h.

Asgher *et al.*, [2] entrapped LiP from *T. versicolor* IBL-04 with 88.6% yield onto xerogel matrix of PTMS and TMOS. The entrapped LiP showed stability up to 80°C, while free LiPs deactivated beyond 60°C. The immobilized LiP showed higher activity and persisted stability for wide pH range. Kinetic constants (K_m and V_{max}) were 70 and 56 μM and 588 and 417 U/mg for the free and restrained LiP, correspondingly. This novel extra thermostable LiP was stimulated with regard to activity to variable extent in the presence of Copper, Manganese and iron, whereas Cystein, EDTA and silver showed inhibitory effects on the activity of LiP.

8.9.2 Immobilized MnP

Diverse practices were described for increasing the characteristics of MnP. Most often these enzymes were physically entrapped in gel carriers or their covalent binding were employed for immobilization. The trap of MnP into a sol-gel matrix of TMOS and PTMS and mixed alginate-pectin gel enhanced its storage, pH stability and thermo-stability along with reusability and efficiency [10, 58].

Bódalo *et al.* [59] used allophone as supports for MnP immobilization from *A. discolor* an enzyme valuable in the decomposition of polycyclic aromatic hydrocarbons (PAHs). Nanoclay-immobilized MnP were more effective at high thermal range and has shown good range of stability over wider pH range. Modification of pH profile versus immobilized enzyme activity has shown that the enzymes immobilization on surface-charged materials may lead to the dislocation of the pH-activity profile to acidic or alkaline regions [60]. The shift of the ideal pH of the immobilized enzyme activity may results from the fact that the pH in the region of the active site of the immobilized enzyme is lower than the pH in the solution bulk or pH dependent modification of MnP structure. It also helps to increase storage time of immobilized MnP suggestively. 20% loss of immobilized enzyme activity was observed after 6 months of storage at 4 °C, whereas free MnP enzyme under the similar conditions lost about 87% of its primary activity [60]. Li and Wen [61] immobilized MnP from *P. chrysosporium* onto multi walled carbon nanotubes through amino-based nonspecific interactions and enzyme retained its original three-dimensional shape with 23% higher activity versus free MnP. In another study, Bilal *et al.* [62] co-immobilized MnP from *P. radiata* on porous silica beads. Immobilization resulted in improved stability of MnP against H_2O_2 or high pH along with storage stability. Brady and Jordan, [27] employed hetero-functionalized supports to improve the support-based immobilization strategies which increase the binding ability and stability of the enzyme due to multiplex attachment. More recently, different new and advance methods for enzyme self-immobilization (CLEC, CLEA, Spherezyme) and carrier materials (Dendrispheres), encapsulation (PEI Microspheres) have been developed. Enzyme immobilizations have many benefits including enhanced enzyme activity, enantio-selectivity and multi-enzyme reactions along with retention, better recovery and stabilization. These advances have potential to increase the characters of immobilization enzymes in industry, as well as an introductory set up for new biotechnological applications [63].

In order to perform better functions, enzymes catalytic properties have to be usually improved before their implementation in industrial sectors where repeated cycles of high yielding processes are required. Generally, native enzymes have to be engineered via immobilization in industrial reactors to be reused for long times and, in addition some other critical enzymatic characteristics like thermo-stability, activity, feedback reticence by reaction products, discrimination toward artificial substrates have to be upgraded. In most of the investigated cases, immobilized MnP has been readily improved in terms of degree of immobilization, higher activities, thermo-stabilities, pH stability and reusability's as compared to their free counterparts [6].

8.9.3 Immobilized Lac

The Lac immobilization have also been studied various researchers, Arica *et al.* [64] immobilized Lace from *T. versicolor* on carbon nano-materials including carboxylated multiwalled carbon nanotubes (MWNT-COOHs), multiwalled carbon nanotubes (MWNTs) and graphene oxides (GOs) by using the process of physical adsorption lacking coupling agents and resultantly, activity of the immobilized Lac enhanced degradation of reactive dyes. Rekuc *et al.* [23] investigated covalently cellulose-based carrier using glutaraldehyde for Lac immobilization of *C. unicolor* and fine fold higher activity was achieved and also tolerance to pH was increased because Lac work properly in wider pH range [65]. Qiu *et al.* [22] demonstrated that although immobilized Lac had higher *KM* than soluble form, yet immobilized Lac had excellent reusability, thermo-stability and equipped permanence due to high *V*max. Rekuc *et al.* [23] restrained Lac in cellular foams and found *KM* for captured Lac was lower and Kcat was higher than free Lac. Immobilization results in modification of some functional groups on the catalytic site as well as on surface of enzyme resulting in lower *KM*, higher catalytic efficiency and increasing thermo-stability [64]. Half-lives for immobilized Lac are significantly higher as compared to those for free Lac. Hubert *et al.* [66] immobilized *C. polyzona* Lac covalently bonded on the diatomaceous earth support Celite® R-633 activated with aminopropyltriethoxysilane, followed by the reaction of functionalized surface with glutaraldehyde (GLU) or glyoxal (GLY). Internal cross-linked and bovine serum albumin added immobilization was studied. The highest Lac activity and the greatest degree of activity recovery tested using 2, 2'-azino-bis-(3-ethylbenzthiazoline-6-sulfonic acid) (ABTS) as the substrate were achieved by the sequential immobilization procedure using GLU as the cross-linking agent. The high efficiency for this substrate was shown by Michaelis-Menten kinetic parameters with respect to ABTS. The solid catalysts containing internal cross-linking of the protein showed meaningfully advanced permanence for numerous denaturants. Bautista *et al.* [67] immobilized Lac by attachment in covalent mode on functionalized SBA-15 prepared by co-condensation method and by physical adsorption on several SBA-15 with different textural properties and activity was enhanced significantly. Rekuc *et al.* [23] also covalently immobilized extracellular Lac produced by *C. unicolor* via glutaraldehyde to cellulose-based carrier Granocel and immobilized Lac showed five-fold higher activity.

In another study, Lac was restrained on adapted silica using glutaraldehyde as cross-linking reagent, with good working permanence. Activity and stability of immobilized enzyme was improved related to the free enzyme. The kinetics Vmax for immobilized Lac in the presence of ionic liquid was also greater compared to free Lac at the similar conditions, or even immobilized Lac in buffer. Both Vm and Km increased with ionic liquid concentration, resulting in increased catalytic competence of the immobilized enzyme [68].

The immobilized Lac was very operative for elimination of textile reactive dyes from aqueous solution in a batch reactor. CLEAs for Lac with BSA showed more resistance against proteases, sodium azide, methanol, EDTA and acetone and they also showed higher thermo resistance than free Lac. However, cross-linking with albumin did not show any improvement in Lac stability. By gel entrapment the pH optima of Lac is mostly shifted from less acidic toward more acidic when compare with free enzymes. Possibly the method of immobilization and hydrogen bindings or ionic bindings is accountable for this transference [2].

8.10 Enzymes Sources

Many species of bacteria showing lipolytic activity have been isolated from different sources and characterized. These isolated bacterial strains have tremendous lipid degrading activity, however different bacterial strains have variable ability to degrade lipid-based pollutants (Table 8.3). For instance Matsumiya *et al.* [69] isolated microorganisms that have ability to degrade lipids from different environmental sources for the construction of waste water treatment system containing lipids. of which, *Burkholderia sp.* DW2-1 strain showed maximum rate of degradation of 1% (w/v) salad oils.

Many investigators evaluate the degrading ability of bacterial strain on different substrates, which revealed that different bacterial strains play significant role in lipid degradation; however the ability varied in different strains of bacteria. For example *Alcaligenes piechaudii* SRS and *Ralstonia picketti* SRS, both of these strains degrade crude oil and 7 diverse distillation products [70]. 5 strains viz., F2, UP2, K11, E13 and N3 isolated from grease wastes and are effective to degrade olive oil and sunflower oil, tallow and lard; however two strains such as *Arthrobacter* sp and *Enterobacter aerogenes* E13 has ability to degrade both saturated and unsaturated fatty acids and triglycerides having these fatty acids. Prasad and Manjunath, [71] carried out studies on biodegradation of high waxy oil wastewater by lipase-producing bacteria such as *Bacillus subtilis, B. licheniformis, B. amyloliquefaciens, Serratiamarsescens, Pseudomonas aeruginosa* and *Staphylococcus aureus* in wastewater released from palm oil mill, slaughter house, dairy, soap industry and domestic wastewater with both individual and mixed culture (consortia). After 12 d of BOD and lipid content was observed to be decreased in consortia.

8.11 Conditions for Lipid Degradation

The growth and activity of bacterial strains depends upon number of factors, which influence the lipid biodegradation process. The nutrients requirements viz., phosphorus, nitrogen, carbon, oxygen etc. and suitable growth conditions like redox potential, salinity, pH and temperature are

Table 8.3 Bacterial strains, isolated from different sources, indicating lipolytic activities for lipids.

Source of bacteria	Bacterial strain	Carbon sources	% Degradation	Ref.
Oil polluted soil	*Bacillus stearothermophilus* SB-1	Neem oil		[72]
Oil polluted soil	*Bacillus licheniformis* SB-3	Neem oil		[72]
Icelandic hot spring	*Bacillus thermoleovorans* IHI-91	Olive oil (triolein)	93%	[73]
oil mill effluent	*Geobacillus zalihae T1T*	Olive oil		[74]
Oil polluted soil	*Burkholderia* sp. Y1	1% Salad oil	83.10%	[75]
Oil polluted soil	*Alcaligenes piechaudii* SRS	Crude oil	80%	[70]
Oil polluted soil	*Ralstonia picketti* SRS	Crude oil	80%	[70]
Oil polluted soil	*Enterobacter aerogenes* E13	Sunflower oil		[76]
Soil samples	*Burkholderia cepacia* S31	Olive oil	226.1 u/ml	[77]
Oil polluted soil	*Staphylococus sps*	1% olive oil		[78]
oil polluted soil	*Burkholderia cepacia*	Peanut oil	10.5 U mL^{-1}	[79]
wastewater	*S. aureus*	wastewater	320 mg/L	[71]
Food wastewater	*Bacillus sp N-09*	Noodles soup	73.50%	[81]
Palm oil effluent	*P. aeruginosa*	Palm oil effluent	325 mg/L	[71]
Dairy effluent	*S. marsescens*	Dairy effluent	280 mg/L	[71]
Oil processing plant	*Pseudomnas aeruginosa* KM110	2 % Olive oil		[80]
Oil mill effluent	*Bacillus* sp	Oil mill effluent		[82]
Oil mill waste	*Bacillus tequilensis* NRRLB41771	Oil mill waste		[83]
Soil samples	*Pseudomonas fluorescens KE38*	1% olive oil		[84]

the significantly imperative factors, which determines the rate of biodegradation process [80]. The temperature and pH are the most imperative factors, which determine the biodegradation process. However, temperature has a significant role in governing the efficiency and nature of microbial dilapidation of hydrocarbons. Furthermore degradation of long chain hydrocarbons by mesophiles at temperatures between 25 and 28 °C has been comprehensively studied [83]. Bacillus cereus N-09 strain shows maximum degradation of lipids and detergents at temperature of 30 °C, pH of 6, and agitation speed of 130 rpm [81]. Extremely operative degradation rate was detected at 35 °C and pH of 4, and the degradation rate after 24 hours was 62.5 ± 10.5 % for mixed lipids having concentration of 3000 ppm. Table 8.4 shows the optimized conditions for some lipolytic bacterial strains.

Lipolytic activity of bacteria can be enhanced in the presence of exogenous surfactants. It was reported the acceleration of oil degradation by genus Rhodococcusin the presence of exogenous surfactants that has been produced by Pseudomonas sp. Various bacterial species capable of growing at high temperatures have also shown lipolytic activities such as a novel

Table 8.4 Optimized conditions for some lipolytic bacterial strains.

	Optimized conditions		
Bacterial strain	**Temperature**	**pH**	**Ref.**
Bacillus stearothermophilus SB-1	50 °C	3	[72]
Bacillus licheniformis SB-3	50 °C	3	[72]
Bacillus thermoleovorans IHI-91	65 °C	6	[73]
Geobacillus zalihae T1T	70 °C	–	[74]
Burkholderia cepacia. S31	70 °C	9	[77]
Staphylococussps	36 °C	7	[78]
Burkholderia cepacia	37 °C	8	[79]
Pseudomnasaeruginosa KM110	45 °C	10	[80]
Bacillus sp N-09	30 °C	6	[81]
Bacillus sp	35 °C	8	[82]
*Bacillus tequilensis*NRRL B-41771	34 °C	7	[83]
Pseudomonas fluorescens KE38	45 °C	8	[84]

oil degrading bacteria identified as Pseudomonas aeruginosa has been secluded from hot spring, inhabiting at temperature ranges between 15 °C and 55 °C [84].

At present, microorganisms such as Bacillus sp. strain 398, Bacillus sp. strain A30–1, Bacillus sp. THL027, Bacillus thermocatenulatus, Bacillus thermoleovorans ID-1, Bacillus spp., Bacillus sp. RSJ-1 have been conveyed as thermo-stable lipase producers. The optimum growth of thermophilic bacterial strains were found at temperature range of 65–70 °C, lipases secluded from such strains are decent applicants for wax alterations [81–84].

8.12 Environmental Applications of Ligninolytic Enzymes

8.12.1 Degradation and Decolorization of Industrial (Textile) Dyes

Filamentous fungi are ubiquitous in the environment. White rot fungi (WRF) are the microorganisms that most extensively studied for industrial dye-degradation. The survival of the fungi is linked with their ability to rapidly adapt their metabolism to varying sources. The WRF have the capability to reduce a wider range of several complex organic pollutants such as PAH, dye effluents, organic waste and steroid compounds [85]. To date, many fungal strains have been reported for effective degradation of diverse artificial (textile) colors and wastes [86]. Most previous work on biodegradation of the textile dyes have dedicated on fungoid cultures from WRF which have been employed to improve bioprocesses for the mineralization of azo dyes [57]. *P. chrysosporium* is the most widely studied WRF, but others have also received considerable attention, such as *A. ochraceus*, *T. versicolor*, *Pleurotus* and *Phlebia* species etc [85].

Without any claims to comprehensiveness, the existing literature regarding textile dye decolorization summarizes that wherever possible, the ligninolytic enzymes of WRF enzymes appears to be tangled in the dye degradation procedure.

Immobilized ligninolytic enzymes proved to be most authentic and promising to substitute the conservative processes being employed in various industries. Though, these methods are processed for decolorization of dyes at initial level but the major concern is their capacity to reduce the toxicity. It urges on the development of well-organized, cost effective and green (eco-friendly) technique for the detoxification and decolorization of dyes [87]. Consequently, the field of investigation is almost entirely open for new findings and it is quite sensible to suggest that many novelties will be seen on the horizon of the world in the near future.

8.12.2 Dye Decolorization with Free Ligninolytic Enzymes

Reported dyes decolorization efficiency with the free and immobilized enzymes is shown in Table 8.5. Several researchers have attempted the production of lignin enzymes and used it for degradation and mineralization of dyes. Chagas and Durrant, [88] investigated MnP as the chief ligninolytic enzyme tangled in dye decolorization from *P. chrysosporium*. Phugare *et al.* [89] examined the role of lignin degrading enzymes in decolorization of dyes in textile wastewater. Initially the wastewater was characterized for quality assurance parameters, that is, BOD, COD, color, TSS, pH, heavy metals and alkalinity. During the decolorization experiment, high concentration of LiP and MnP was observed with no detectable Lac activity and as a result of degradation water quality parameter improved significantly. Singh *et al.*, [90] studied four *Pleurotus* species of WRF for color removal of direct blue 14 (DB14). Among all four species of *Pleurotus*, *P. flabellatus* displayed the rapid dye decolorization of up to 90.39% in 6 h of incubation, whereas other 3 *Pleurotus* species acquired additional time for dye degradation. Investigations revealed that the occurrence of influential enzymatic machinery of WRF could effectively destroy the recalcitrant and toxic dyes. Yao *et al.* [91] determined the in-vitro decolorization capability of three physically different azo dyes by the extracellular ligninolytic enzymes. The results designated that only basic enzyme extract MnP, but not LiP and Lac, played a decisive part in the decolorization of azo dyes. The dye decolorization ability of orange IV and orange G along with crude MnP was markedly heightened to 76 and 57% respectively, after 20 min of reaction at pH 4 and temperature level of 35°C. However, very less (8%) decolorization of Congo red dye was noted. Medium containing higher concentration of MnP activities was more vibrant in the detoxification of the various industrial dyes than those with high Lac activity [92]. Gomes and coworkers, [93] investigated the potential of crude ligninolytic enzymes obtained from cultivation of different WRF for the decolorization of physically dissimilar textile dyes and obtained 70% of colour removal of RBBR and cybacron blue 3GA after incubation for 6 hours. The findings specified that enzymes from these basidiomycetes have competence to decolorize diverse chemical classes of dyes, since efficient color removal was found for RBBR, cybacron 3GA, orange II and crystal violet, which belong to azo dye, antraquinone, triazyne and tri methyl methane, respectively. Narkhede *et al.* [94] screened

Table 8.5 Reported dyes decolorization efficiency of free and immobilized enzymatic systems.

Dyes name	D (%)	Time	Enzyme free	Support	Source	Ref.
Direct Fast Orange S, Nippon Fast Red BB, Sumilight Supra Blue, Suminol Milling SG, Suminol Fast Red B, Suminol Milling Green G, Suminol Fast Blue PR, Sunchromine Yellow MD, Sunchromine Fast Blue MB, Sunchromine Black ET, Sumifix Briliant Orange, Sumifix Black B	84.9, 99.5, 98.7, 98, 98.8, 98.9, 99.1, 99.5, 99.4, 98, 99.6, 99.3	14 d	MnP		newly isolated WRF strain L-25	[96]
RBBR, Congo red, Trypan blue, Amido black, Methylene blue, Poly R478, Ethyl violet, Methyl violet, Methyl green, Brilliant cresyl blue	92, 98, 84, 61, 95, 98, 77, 98, 52	18 d	MnP, Lip, Lac		*L. edodes*	[95]
Orange II, Reactive Red 120, RBBR, Cybacron Blue 3GA, Azure B,	74, 27, 66, 66, 22	48 h	LiP, MnP, Lac		*L. strigellus*	[93]
Methylene Blue	32	24 h				–
Methyl orange, cotton blue, methyl violet, amido black, orange HE2R, reactive blue 25, direct blue 6, reactive yellow 81, red HE4B, green HE4B, reactive green 19 A	100, 60, 80, 60, 90, 100, 70, 80, 90, 100, 60	5 h	LiP		*K. rosea MTCC 1532*	[115]

(*Continued*)

Table 8.5 Cont.

Dyes name	D (%)	Time	Enzyme free	Support	Source	Ref.
Sitara Textile (SIT)	100	2 d	LiP, MnP, Lac		*P. ostreatus+ C. versicolor*	[2]
RBBR	98.53	5 d	Lac		*Basidiomycete sp.* L-168	[94]
Direct yellow (DY106)	87	2 min	FCP		*Cucurbita pepo*	[116]
Dispersered 19+disperse black 9	81.10%	2 h	CBP		Cauliflower (*Brassica oleracea*) bud	[10]
Direct blue 14 (DB14)	90.39%	6 h	lac, MnP		*P. flabellatus*	[90]
Congo Red, Orange G, Orange IV	8, 57, 76	20 min	MnP		Schizophyllum sp. F17	[91]
Solar brilliant red 80 direct dye	84.8%	7 d	LiP, MnP, Lac		*S. commune*	[2]
Acid Blue 80	85%	120 min	Lac		*L. Polychrous*	[97]
Sandal-fix Red C4BL, Sandal-fix Turq Blue GWF, Sandal-fix Golden Yellow CRL, Sandal-fix Black CKF, Sandal-fix Foron Blue E2BLN	57.34%, 63.98%, 65.66%, 65.09%, 74.24%	12 h	MnP		*G. lucidum*	[7]
Textile wastewater (10 units)	37–63%	–	–		–	–
			Immobilized			

Methyl violet, Acid orange, Vat magenta, Congo red, Acid green, Methylene blue, Acid red 114	95.17, 90.14, 95.9, 85.35, 66.83, 86.87, 83.33	120 h, 48 h, 48 h, 120 h, 120 h, 96 h, 120 h	LiP, MnP, Lac	calcium alginate beads	*P. chrysosporium*	[117]
Reactive Red 120	91	10 h	Lac	poly(GMA/EGDMA) beads	*Rhus vernificera*	[64]
Procion Red, Reactive Green 5, Reactive Brown 10, Reactive Green 19, Cibacron Blue F3GA, Alkali Blue 6B, Brilliant Blue 6	81.53, 59.81, 73.59, 66.33, 62.95, 59.71, 61.68	10 min	Lac	Cryogels	*T. versicolor*	[113]
CRT effluent, Sitara Textile effluent	100, 98.5	5 h	MnP	sol-gel	*P. ostreatus*	[118]
Crescent textile effluent, Magna textile effluent, Arzoo textile effluent, Chenab textile effluent, Drimarine blue K2RL, CRT effluent, SIT efflunt	99.2, 94.6, 89.6, 78.5, 100, 100, 97	4 h, 4 h, 4 h, 4 h, 4 h, 5 h, 5 h, 5 h	MnP	Xero-gel	*G. lucidum*	[119]
Direct yellow (DY106)	75	15 min		Sodium alginate	ICP	[116]
Disperse red 19	90.60%	2h	CBP	Calcium alginate gel	Cauliflower (*Brassica oleracea*) bud	[10]

(*Continued*)

Table 8.5 Cont.

Dyes name	D (%)	Time	Enzyme free	Support	Source	Ref.
Lefavix Blue 16, Reactive Remazol Violet 9, Reactive Remazol Navy 4	84, 51, 46	24 h	LiP, MnP, Lac	*T. versicolor* U97	double layer of alginate bead	[108]
Lefavix Blue 16, Reactive Remazol Violet 9, Reactive Remazol Navy 4	53, 48, 44	–	–	Pestalotiopsis sp. NG007	–	–
Acid Blue 25, Acid Orange 7	76, 64	65 min	Lac	*Paraconiothyrium variabile*	porous silica beads	[114]
Sandal-fix Red C4BL, Sandal-fix Turq Blue GWF, Sandal-fix Golden Yellow CRL, Sandal-fix Black CKF, Sandal-fix Foron Blue E2BLN	78.14%, 86.81%, 89.31%, 87.63%, 92.29%	12 h	MnP	PVA	*G. lucidum*	[7]
Textile wastewater (10 units)	61–80%	–	–	–	–	–

13 phenol tolerant isolates from 51 rotted wood samples ligninolytic enzyme production. Finally, the isolates were studied for Remazol brilliant blue R decolorization with promising outcome. The role of MnP in dye decolorization was examined by some in-vitro tests with MnP enzymes. For this purpose, the dyes were made to react with partly purified MnP. Decolorization was detected only in the existence of manganese ions, MnP and H_2O_2. On the other hand, degradation of dyes was not observed in the absence of any of these components in the reaction mixture. Although some dyes showed very less decolorization with MnP that could be enhanced by optimizing reaction parameters for each dye. Enzymatic decolorization can also be improved by using ionic surfactants, such as Tween 80, Tween 20 etc.

For example, the activity was raised to 84.6% by the use of sol-gel matrix encapsulated MnP from *G. lucidum* versus free MnP of its initial activity after 10 periodic oxidation cycles and decolorized Magna (94.6%), Crescent (99.2%), Chenab (78.5%) and Arzoo (89.6%) textile effluents. An increase reaction time was another very crucial factor in the process of decolorization process [58]. Jamal *et al.* [10] immobilized MnP onto calcium alginate pectin gel. The result was again in favor of increased storage stability in comparison with the free MnP. The complex (entrapped) retained 59.6% of the initial activity after 50 days of development. Furthermore, the restrained complex decolorized Disperse Red 19 and Disperse Red 19 with Disperse black 9 mixtures (91.2% and 82.1%, correspondingly).

Boer *et al.* [95] studied the decolorization and degradation of several synthetic textile dyes by WRF *Lentinula* (Lentinus) *edodes* enzymes to decolorize. After incubation period of 18 days the fungus entirely decolorized congo red, amido black, methyl green, trypan blue, methyl violet, remazol brilliant blue R, ethyl violet and Poly R478 at 200 ppm. High MnP activity with very low LiP and Lac activities were perceived in the cultures. The presence of manganic ions and H_2O_2 in the medium markedly increased the decolorization rate in vitro. These data suggest that MnP appears to have capability to decolorize artificial dyes.

Kariminiaae-Hamedaani *et al.* [96] decolorized 12 different diazo, azo and anthraquinone dyes using a lately secluded WRF, strain L-25. A high decolorization efficiency of 84.9–99.6% was achieved by fungus-producing highly efficient MnP enzyme as its major ligninolytic enzyme. At the beginning the fungal cells became colored by absorbing dye during the decolorization. However, this color vanished when MnP was concealed by the strain. At the end of cultivation, the activity of MnP in the cultures was over 1.0 U/mL. Temporarily, MnP produced by strain L-25 was used for the enzymatic decolorization of the dyes thus authorizing the aptitude of the enzyme for decolorization.

Of most recent, Ratanapongleka and Phetsom, [97] carried out a study to degrade different dye structures by crude ligninolytic enzyme. The enzyme showed good decolorization efficiency of 85% in 2 h toward Acid Blue 80. They also determined the reaction kinetics and Michaelis-Menten behavior of the enzyme on Acid Blue 80 and resulted that the initial rate of decolorization depended on the dye concentration. In a variation on theme, Rodriguez and coworkers, [98] tested 23 industrial dyes for decolorization by enzymatic machinery of WRF and found that only Lac activity was associated with decolorization. Furthermore, two Lac iso-enzymes from *Trametes hispida* were purified, and applied for decolorization study. Thus from these findings it can be concluded that crude Lac from *L. polychrous* have obvious potential for decolorization especially anthraquinonic dye.

8.12.3 Dye Removal by Immobilized Ligninolytic Enzymes

Dyes are coloring substances with varied biochemical constructions. More than 100,000 different dyes and dye-stuff pigments are available around the world in different industrial units and most of them are unaffected to water, light exposure or chemical substances. It is the responsibility of textile industry to market dyes and the industrial wastes are measured by governments due to ecological concerns. Consequently, ligninolytic enzymes are of particular interest and are preferred for the development of biological processes because of their potential to degrade various types of toxic dyes. Therefore, the immobilized lignonolytic enzymes could be subjugated in dye decolorization, for other reasons like easy operation, recycling of the enzyme and the cost effective nature [99].

Many stimulating examples of dye decolorization are described below using different immobilization supports by ligninolytic enzymes. Inspite of the various applications of ligninolytic enzymes, the most important one is its ability to decolorize and degrade industrial dyes. Kandelbauer *et al.* [100] specifics the decolorization of several dyes with *Trametes modesta* Lac restrained by covalent binding to Al_2O_3 pellets. The alumina was previously silanized with 3-aminopropyltriethoxysilane (APTES) and activated with glutaric dialdehyde. The decolorization was trailed online via spectroscopic sensors engrossed in the employed reactor. Several anthraquinonic, that is, Lanaset Blue 2R, Terasil Pink 2GLA and azo dyes were proficiently decolorized, indicating that this system of bioremediation is not limited to few structural groups. Preferentially, azo dyes containing having hydroxyl groups in the ortho or para position compared to the azo bonded dyes were oxidized. On the similar grounds, alumina was also employed for the Lac immobilization from *T. hirsuta* and applied for the decolorization of other dyes like methyl green and (RBBR). Another way to use Al_2O_3 pellets were silanization with APTES following by activation with GLU to immobilize Lac covalently, coated with PAH and poly Sodium 4-styrenesulfonate (PSS). Using such immobilized Lac, methyl green was decolorized to a greater level than RBBR. However, RBBR was also decolorized by using immobilized Lac that was covalently bonded in pre-silanized controlled porosity-carrier beads (CPC-silica) previously activated by GLU. It was the enzymatic activity inspite of the adsorption property of the dye onto the carrier that was responsible for the decolorization of RBBR. Additionally, the permanence of the Lac was also reasonably upgraded [101].

Another Lac immobilization support that has been tested for dye treatment was magnetic chitosan beads. In this case, *T. versicolor* Lac was covalently bonded through epichlorhydrin, to improve its thermal and operational stability. The Reactive Yellow 2 (80%) and Reactive Blue 4 (55%) were effectively decolorized by this Lac in batch reactors. Several anthraquinonic dyes, namely Disperse Blue 3, RBBR and the indigoid dye Acid Blue 74 were decomposed better than the azo dyes like Reactive Black 5 and Acid Red 27 using Lac immobilized on CPC-silica beads through GLU. However, the toxicity of the solution was increased during decolorization of the anthraquinonic dyes as compared to azo and indigoid dyes that were proven to be less toxic. Consequently, it must be taken into account while making use of these dyes for decolorization and detoxification [102].

The consideration of another immobilization direction from environmental prospects by using entrapment of Lac is on the way. Lu *et al.* [103] selected an alginate-chitosan complex membrane as a supporting matrix to immobilize Lac. The best settings for immobilization were 0.3% chitosan, 2% $CaCl_2$ and a 1:8 of enzyme: alginate, causing in a charging efficiency and immobilized return of 88.12% and 46.93%, correspondingly. The lower activity and substrate affinity was presented by immobilized Lac but along with the better stability to thermal and acidic denaturation. In fact, the increase in the stability gave us an advantage in Alizarin Red decolorization either in the presence or absence of a Lac mediator. Under batch conditions, entrapped alginate with *Polyborus rubidus* Lac despoiled about 80% of a dye in 5 days. Likewise, a profitable Lac from Denilite II S using alginate-gelatin matrix with polyethylene glycol (PEG) was also immobilized to effectively decolorize Reactive Red B-3BF. The Lac stability was improved by the addition of PEG [31]. Phetsom *et al.* [33] immobilized *Lentinus polychrous* Lac in Copper, Zinc and Calcium alginate. The immobilized Lac was tested for high thermal conditions compared to free enzyme. Copper and Zinc alginate bio-particles has shown relatively good approach toward stability and activity for RBBR, indigo carmine, methyl red and bromophenol blue than free Lac. Zhao *et al.* [104] employed magnetic properties of Fe_3O_4 particles covered by alginate, modified with acyl chloride groups on the surface. The higher thermal and reusability properties were shown by immobilized Lac. After treatment of 12 hours, 39% and 22% of Congo red dye was oxidized by immobilized and free Lac, correspondingly. In the same way, poly acrylamide-N-isopropylacrylamide (P AAm-NIPA) hydrogel with semi-interpenetrating networks (semi-IPNs) of alginate was employed to immobilize *T. versicolor* Lac, that gave improvements in storage stability and in the decolorization of Acid Orange 52. Later on, Makas *et al.* [105] used semi-IPNs prepared from carrageenan with either poly acrylamide-acrylic acid (P AAm-AA/κ-car) or poly acrylamideitaconic acid (P AAm-IA/κ-car). The immobilized enzyme in the above-mentioned hydrogel systems retained more than 80% of its original activity when stored for 42 days, while the activity reduces to 50% after ten uses in a batch system. Methyl orange resulted in 35% decolorization after 6 hours of treatment for both systems as it was selected as a target dye, which rose up to 70% if ABTS was employed as reaction mediator.

The non-porous polymer poly glycidyl methacrylate/ethyleneglycol dimethacrylate (poly GMA/EGDMA) with the insertion arm 1, 6-diaminohexane (DAH) was activated by GLU and covalently bound to the *R. vernicifera* Lac. The efficiency of the resulted immobilized Lac was reduced for the sake of gain in stability. Additionally, Reactive Red 120 was competently degraded [64]. An operative decolorization of indigo carmine was also performed by *C. unicolor* Lac immobilized in meso-structured siliceous cellular foams (MCFs). The *T. versicolor* and *Pycnoporus cinnabarinus* Lac were immobilized in magnetic macroporous cellulose beads by making use of aldehydic groups as a support. This system of Lac immobilization was proved better and has shown positive results in anthraquinone and azo dye decolorization.

It should be kept in mind that only few examples of Lac adsorption are cited in the literature. For instance, *T. versicolor* Lac was adsorbed on magnetic beads modified with poly (4- vinylpyridine) or chelated with Copper ions to decolorize Reactive Green 19, Reactive Brown 10 and Reactive Red 2. The immobilized Lac on the glass beads Lac derived from *Pseudomonas sp.* SU-EBT displayed 97% and 90% Congo red (100 mg/L) decolorization activity after 12 and 60 hours, correspondingly, at pH 8 and 40 °C. It has also shown lasting storage stability at 40 °C for 192 hours. The immobilized Lac showed high thermal stability arises from the presence of covalent bonded nature that is associated with the attachment of enzyme to the glass beads, which basically enhances biocatalyst conformational rigidity compared to non-immobilized Lac [106].

Immobilization of the *T. hirsuta* Lac on alumina primarily increases the thermal stabilities and its tolerance against some enzyme inhibitors like halides, dyeing additives and copper chelators. At 50 mM NaCl, the Lac lost 50% of its activity while the 50% inhibitory concentration (IC50) of the immobilized enzyme was 85 mM. The triarylmethane, indigoid, azo and anthraquinonic dyes were degraded by the Lac. The decolorized wastes with immobilized Lac could be employed for staining tenacity.

For the first time, Cristovao *et al.* [107] used green coconut fiber (cheaply available) successfully to immobilize marketable Lac by adsorption in the improved settings. A high decolorization percentage of all tested reactive textile dyes RB5, RB114, Remazol Black B, Reactive Yellow 15 (RY15), Levafix Brilliant Blue E-BRA, Reactive Yellow 176 (RY176), Remazol Yellow GR, Reactive Red 239 (RR239), Remazol Yellow 3RS, Remazol Brilliant Red 3BS and Reactive Red 180 (RR180) for decolorization by the immobilized enzyme were obtained. The technique of immobilization permitted enzyme reuse with high operational stabilities and enhanced its stability that can be further employed as a potential lead in wastewater treatment.

Lu *et al.* [77] employed alginate-chitosan microcapsules for Lac immobilization. Though, the substrate affinity and activity displayed by immobilized Lac was lower than the free enzyme, it upgraded its stability to other parameters counting on thermal and acidic grounds. The immobilized Lac catalyzation efficiently decolorized Alizarin Red. A capability of immobilized enzymes obtained from *Trametes versicolor* U97 or *Pestalotiopsis* sp. NG007 showed high decolorization efficiency for three textile dyes Lefavix Blue 16 (LB16), Reactive Remazol Navy 4 (RRN4) and Reactive Remazol Violet 9 (RRV9) in a vertical bioreactor system. Immobilization was conducted using a double layer of alginate bead (1.5% w/w) and crude enzymes. The decolorization rate was accelerated when a mediator 2, 2, 6, 6-tetramethylpiperidin-1-yl oxyl (TEMPO) was used in immobilized enzyme of U97 or a Tween 80 was used in NG007. Reaction with glutaraldehyde (0.6%) for 4 h maintained the longevity and reusability of the beads. The study suggests that double layer immobilized enzyme in a bioreactor system has a terrific potential strategy for treating the textile dye effluents [108].

Shamsuria *et al.* [109] reported the degradation of dye malachite green oxalate using sol-gel Lac at the 3rd hour were 45% and 31% respectively. The optimum pH and temperature for sol-gel Lac against their degradation capacity respectively was at pH 5 and 40°C. Pezzella *et al.* [110] successfully immobilized a crude Lac from *Pleurotus ostreatus* on an inexpensive permeable silica material and tested for RBBR decolorization in a fluidized bed reprocess container. A maximum conversion degree of 56.1%

was accomplished at reactor space-time of 4.2 hours in the continuous laboratory equipment. Stability and catalytic parameters of the immobilized Lac were also measured in association with the soluble counterparts. It revealed an increase in stability despite of reduction in the catalytic presentations.

The partially purified Lac from *Fusarium* sp. immobilized in sodium alginate beads efficiently decolorized nine different textile dyes with maximum in Blue M2R (BM2R), Black-B (BB) and Orange M2R (OM2R) followed by minimum decolorization activity in Yellow MR(YMR), Red BSID (RBSID), Manenta MP (MMP), Blue MR (BMR), Orange 3R (O3R) and Brown GR (BGR) decolorization. Sadighi and Faramarzi, [111]accomplished decolorization of Congo red by making use of immobilized Lac via chitosan nanoparticles on the surface of glass beads. Glass surface functionalization was attained by using NaOH, glutaraldehyde and 3-aminopropyltriethoxysilane (APES). Fourier Transform Infrared Spectroscopy (FT-IR) was applied to investigate the presence of different functional groups attached on the glass beads. The immobilization technique caused an increased thermal stability, reusability, reliability and longer lifetime of the enzyme. Thermal stability of the nano-scaled immobilized enzyme augmented up to near boiling temperature. Additionally, 98% of initial decolorization activity of the nano-particulated immobilized enzyme was reserved after 25 consecutive sequences.

Thakur *et al.* [112] entrapped extracellular Lac enzyme produced from *Cercospora* sp. SPF-6, in calcium alginate beads with maximum immobilization efficiency of 72.42% and enzyme activity of 0.617 U/bead. The maximum 84.31% decolorization of methyl orange dye was observed at 30 C at pH 2.5 in 30 min. Immobilized enzyme system preserved 50% of its competence even after the span of seven consecutive feedback cycles.

The Lac from *Myceliophthora thermophila* was covalently immobilized on Sepabeads EC-EP3 and Dilbeads NK triggered with epoxy groups. The enzyme immobilized on EC-EP3 exhibited distinguished activity (203 U/g) along with highly improved features, that is, pH stability, temperature stability and better storage time but resistance to organic solvents was remain unaffected. In addition, good operational stability has been shown by the biocatalyst, maintaining 84% of its initial activity even after the period of 17 cycles of oxidation of ABTS. The Lac immobilized in EC-EP3 retained 41% activity in the decolorization of Methyl Green in a fixed-bed reactor after span of five cycles [42]. The immobilized Lac was applied to the decolorization of Reactive Black 5, Methyl Orange, Acid Blue 25, Methyl Green, Remazol Brilliant Blue B and Acid Green 27 in the presence or absence of the redox mediator.

Poly (methyl methacrylate-co-glycidyl methacrylate) (poly (MMA-co-GMA)) cryogels were used for covalent attachment of Lac from *Trametes versicolor*. The thermal-stability of the immobilized Lac > free Lac. Immobilized Lac could be used for 10 time's successive reuse with no significant decrease in its activity. Also, these Lac immobilized cryogels were successfully used for the decolorization of Procion Red, Reactive Brown 10, Reactive Green 5, Reactive Green 19, Alkali Blue 6B, Cibacron Blue F3GA and Brilliant Blue 6 [113].

Peralta-Zamora *et al.* [99] investigated by making use of immobilized Lac, the decolorization of Remazol Brilliant Blue R, Reactive Orange 122, Remazol Black B and Reactive Red 251 dyes within 30 min and decolorization capacities were found to be 35–45%, 10%, 10–30%, and 5–55%, respectively. Jamal *et al.* [10] evaluated the effectiveness of Ca-alginate entrapped peroxidase for dye color removal from aqueous solution. The resulted entrapped enzyme removed dyes color up to 90.6% and 81.1% from dispersed 19 and dye mixture (dispersered 19 + disperse black 9) respectively in repeated two reactor batch system. Entrapped enzyme was successfully recyclable for ten cycles in dye color removal. The findings suggested the usefulness of immobilized enzyme for dye color removal from industrial effluent.

Qiu *et al.* [22] exploited nano-porous gold (NPG) as a support material for LiP immobilization. The ideal temperature of the immobilized LiP was recorded as 40 °C, which is 10 °C higher than that of free LiP. After incubation span of 2 hours at 45 °C, free LiP was completely deactivated while 55% of the initial activity of the immobilized LiP was still reserved. Furthermore, a huge and long lasting LiP activity was attained through in-situ release of H_2O_2 by a co-immobilized enzyme glucose oxidase. The present co-immobilization system was established to be very operative for the decolorization of rhodamine B, fuchsine and pyrogallol red.

Mirzadeh *et al.* [14] immobilized Lac (purified from *Paraconiothyrium variabile*) on permeable silica globules and appraized for the decolorization of Acid Blue 25 and Acid Orange 7. Immobilized Lac represented higher temperature and pH stability 76% and 64% for Acid Orange 7 and Acid Blue 25 was decolorized, correspondingly after 65 minutes period of incubation. The activity of Acid Blue 25 and Acid Orange 7 was reduced by 26% and 31% respectively after the span of 9 cycles of decolorization.

8.12.4 Degradation of Lipids

Lipases are widely distributed in prokaryotes and eukaryotes in nature, but only microbial lipases have commercial importance [80]. Bacterial lipases are greatly used in food and dairy industry for milk fat hydrolysis, flavor enhancement, cheese ripening, and lipolysis of cream and butter fat. Moreover used in textile industry to enhance fabric absorbency, while in detergent industry as additive or supplement in washing powder [120], for various trans esterification reactions [121] and for synthesis of biodegradable compounds or polymers. Simultaneously, the enzymes are being employed in paper and pulp industry, as a catalyst for manufacturing various products used in cosmetic industry, in biodiesel synthesis [122], in pharmaceutical industry and in degreasing of leather.

Lipid-based pollution is a major issue in recent years not only in Pakistan but also in all over the world. Furthermore Oil spills are causing major hazards to the environment. Petroleum fuel spills from tank failure, pipeline ruptures, different production storage and transportation accidents are speculated as the most frequent organic pollutant of soil and aquatic environment and have been grouped as hazardous wastes due to their cytotoxic, carcinogenic and mutagenic effects on human. Micro-organisms are considered marvelous agents, who play a significant character in the degradation of lipids pollutants, found in soil and water etc. [123]. The application of lipase-producing microorganisms for degradation of waxy materials into wastewater treatment system (Table 8.6) is found satisfactory. A number of microorganisms like molds, bacteria and yeasts have been found to be capable of total degradation of oil wastewater [124]. These microorganisms can be applied for the treatment of industrial as well as domestic wastewater containing lipids pollutants.

Biodiesel has been considered the most effective energy alternative. At present biodiesel has been commercially carried out using alkaline catalyst. Lipases have been found to be active in trans esterification process leading to production of biodiesel, which may be proved to be environmental friendly and economical. This involves use of immobilized lipases through various techniques [125].

Boundaries of the industrial usage of these enzymes have chiefly been due to their high production costs. The production of these enzymes at high levels and in a virtually purified form can be abridged by molecular technologies.

8.12.5 Degradation of Miscellaneous Compounds

The applications of enzymes in various sectors have been shown in Table 8.7. Lipases have found versatile applications in pharmaceutical industry. Specifically, preparation of homochiral compounds being used against HIV and synthesis of alkaloids, anti-tumor agents, vitamins and

Table 8.6 Industrial wastes and microbial degradation.

Industry	Chemical (waste)	Microbial strains	Degree of degradation	Ref.
Leather industry	Lipidic concentrate of sheep fleshing	*Saccharomyces cerevisiae*	2749 U/mg	[126]
Slaughterhouse waste	Tallow	*Pseudomonas gessardii*	1,473 U/mg Lipase	[127]
Food-processing		*Rhizopus oryzae*	–	[128]
Petrobras Research Center using castor bean seeds	Toxic and alkaline waste	*Penicillium simplicissimum*	155.0 U/g after 96 h	[129]
Leather, detergent and chemical industries		*P. aeruginosa* KM110	–	[80]
Fish processing		*Enterococcus faecium NCIM5363 (EF-63)*	22 U/mL Lipase activity	[130]
Olive oil processing industry	Olive oil cake	*Candida utilis*	25 U/g (lipase activity)	[131]
Oil mill wastewater	Fatty acids and triglycerides	*Yarrowia lipolytica*	–	[132]
Food industry	Grease	*Yarrowia lipolytica*	–	[132]

Table 8.7 Application of enzymes in different industrial sectors.

Food Industry		
Lac	Phenolic emission from the food and beverage, Ascorbic acid determination Sugar beet pectin gelation	[135]
LiP	Source of natural aromatics Production of vanillin	[136]
MnP	Production of natural aromatic flavors	[136, 137]
Pulp and paper industry		
Lac	Depolymerization of lignin, Delignify wood pulps, Bleaching of kraft pulps	[138, 139]
LiP	Decoloriment of kraft pulp Mill effluents	[138]
MnP	Kraft pulp bleaching	[137]
Textile industry		
Lac	Textile dye degradation and bleaching	[42, 93, 137]
LiP		
MnP		
Bioremediation		
Lac	Biodegradation of xenobiotics, Polycyclic aromatic hydrocarbons(PAHs)degradation	[137, 139]
LiP	Degradation of azo, heterocyclic, reactive and polymeric dyes, Mineralization of environmental contaminants, Xenobiotic and pesticides degradation	[139, 138]
MnP	PAH's degradation, Synthetic dyes, Bleach from paper-producing plants, DDT, PCB, TNT	[139]
Organic synthesis, Medical, Pharmaceutical, Cosmetics and Nanotechnology Applications		
Lac	Polymers production, Coupling of phenols and steroids, Medical agents, Carbon-nitrogen bonds construction, Complex natural products synthesis, Personal higienic products, Biosensors and bioreporters	[41, 42, 139]

(*Continued*)

Table 8.7 Cont.

Organic synthesis, Medical, Pharmaceutical, Cosmetics and Nanotechnology Applications		
LiP	Functional compounds synthesis Cosmetics and dermatological for skin Bioelectro-catalytic activity at atomic resolution	[140]
MnP	Acrylamide polymerization Polymer styrene degradation Direct electron transfer (DET)	[20]

antibiotics are fundamental applications of lipases [133]. Furthermore, bacterial lipase treatment has been found to intensify the designing of cotton fabrics. Bacterial lipases have been analyzed for the efficiency as a scouring agent for raw cotton fabrics in order to eliminate the natural hydrophobic substances found in the fiber [134].

8.12.6 Xenobiotics and Industrial Effluents

It is normally decided in current scenario that xenobiotic substances are posing huge hurdles in wastewater treatment because these novel substances are nonbiodegradable and habitually resilient to the process of degradation by chemical and biological methods. Owing to high competence and cost-effective degradation of the environmental pollutant, ligninolytic enzymes (LiP, MnP and Lac) well-thought-out to be an alternative bioremediation treatment and are often used for the management of polluted environments [140]. The immobilization of these enzymes produce greater working stability and resilience of the enzyme and may leads to the possibility, in some cases, of its use in a constant process thus permitting biocatalysts to be used at larger scale [67]. The degradation of aromatic compounds by immobilized ligninolytic enzymes are discussed in below mentioned text to illustrate this point. For example, immobilized LiP is quite well-known for their ability to mineralize a diversity of intractable aromatic compounds and to oxidize a number of PAH and other phenolic compounds. Previously, Cornwell *et al.* [140] testified an immobilization of LiP on porous ceramic supports did not unfavorably affect LiP's stability and showed a incredible prospective for degradation of ecologically tenacious aromatics compounds. Color removal from kraft waste by "CNBr-sepharose 4B" immobilized LiP was investigated. The immobilized LiP upgraded the decolorization by a factor of 2.9. LiP immobilized on activated silica produced 65% chemical oxygen demand reduction (COD), 20% mineralization and 12% decolorization. The immobilized LiP exhibited 70% decolorization, 55% total phenol removal and a total organic carbon (TOC) reduction of 15% in 3 hours of treatment on Amberlite IRA-400 resin [99]. Grabski *et al.* [141] immobilized MnP isolated from *Lentinula edodes* on azlactone-functional copolymer covalently, derivatized with ethylenediamine and 2-ethoxy-1-ethoxycarbonyl-1, 2-hihydroquinoline as a coupling reagent. This enzyme was engaged in a 2-stage immobilized MnP bioreactor for catalytic production of chelate Mn-III and oxidation of chlorophenol subseqently.

Of most recent, Asgher and Iqbal. [118] immobilized purified MnP (2.1-fold) secluded using sol-gel matrix of trimethoxysilane and proplytetramethoxysilane from solid state culture of *Pleurotus ostreatus* IBL-02. They achieved 100% decolorization within a short span of time textile industry effluents, for example, Nishat textile (NIT), Sitara textile (SIT) and Crescent textile (CRT). Lower K_M, higher V_{max}, hyper-activation, improved acidic and thermal stability up to 70 °C, were increased catalytic structures of the presently sol-gel immobilized MnP.

8.12.7 Degradation of Aromatic Compounds

Liu *et al.* [142] reported the use of carbon-based mesoporous magnetic composites (CMMC) in phenol for immobilization of Lac and for the removal of *p*-chloro-phenol. Magnetic bio-separation technology was combined with the immobilized enzyme made the separation easier and also helped in easy recovery of the catalyst and made the process cost-effective. Significantly broader pH and temperature profile was shown by Lac immobilized on CMMC than the free enzyme. The similar kind of behavior was noticed for Lac immobilized on polyacrylonitrile (PAN), one of the most widely processed support having good mechanical attributes, high tensile strength, solvent and abrasion resistance [143]. The kinetic parameters of reactions catalyzed by Lac immobilized on CMMC designated less similarity for the substrate compared to free enzyme. However, the immobilized enzyme during 12 hours incubation period utilized 78% and 84% of phenol and *p*-chlorophenol, correspondingly [143]. A similar kind of statements regarding high efficiency of degradation was perceived by [144]. Lac immobilized on magnetic Copper chelated silica support detached penta-chlorophenol with 82.9% competence, whereas 87% efficiency of 2,4,6-trichlorophenol removal in 4 hours conjugation of the enzyme onto the surface of polyacrylonitrile electrospun fibrous membrane was noticed [144].

The dilapidation of particular hydroxylated aromatic compounds was directed by *R. vernicifera* Lac immobilized in fiber membranes. With the exception of 2-hydroxy-decahydronaphthalene, the effective degradation of 3, 4-dimethylphenols, 2-hydroxy-1, 2, 3, 4- tetrahydronaphthalene, 4-ethylphenol and 4-hydroxybiphenyl was achieved, extending from 50 to 100% with 2 days of handling. Mohajershojaei *et al.* [145] studied phenol degradation with immobilized Lac from *R. vernicifera* in a polypropylene membrane that was adapted with chromic acid and consequently activated by ethylenediamine and GLU. The distinguished outcomes of this work were the improvement in the pH and thermal stability of the insoluble enzyme and narrower pH-activity outline of the soluble Lac compared with the immobilized Lac. Lac from *T. versicolor* was immobilized in kaolinite that was functionalized by APTES and GLU and was then tested for its ability to degrade PAH. In the existence of the mediator ABTS, 80% of anthracene and benzo[a]pyrene (BaP) was corroded. Later, the same authors used kaolinite and SBA-15 for the immobilization of *T. versicolor* Lac, attaining high thermal stability and pH along with the effective oxidation of BaP, signifying again the potential of Lac for PAH remediation. Similarly, SBA-15 support was later used for the immobilization of *T. versicolor* Lac by adsorption and covalent binding methods. The covalently attached Lac to aminopropyl and aminobutyl functionalized SBA-15 degraded naphthalene with 35% and 39% in 5 hours correspondingly [99].

The degradation of phenolic compounds was established by immobilized Lac. For the degradation of 2, 4-dichlorophenol, *Coriolus versicolor* Lac was immobilized by adsorption on activated carbon followed by entrapment into calcium alginate gels. Better thermal and pH stability were presented by immobilized Lac compared to free enzyme and achieved a dechlorination efficiency of 99.5% during 8 recurrent batch reactions. Bisphenol A (BPA), an endocrine disruptor regularly found in effluents from industry was also degraded by *T. versicolor* Lac that had been covalently bonded to nylon membranes attached with glycidyl methacrylate and phenylenediamine as a insertion. In another study, Lac was encapsulated in the sol-gel, and its thermal and storage stability was improved. The kinetic parameters indicated that the tested Lac had higher affinity for Trichlorophenol than Dichlorophenol, and the maximum concentration of chlorophenol that could be tolerated by the immobilized Lac was comparatively high [146].

The oxidation of phenol, p-chlorophenol and aniline was verified with a Lac from *T. versicolor* that had been immobilized by ionic adsorption using poly (hydroxyethylmethacrylate)-glycidylmethacrylate aminated poly (HEMA-g-GMA) as a support material, ensuing the effective phenol degradation with recovery of 71% of the activity. Ordered mesoporous materials (OMMs) were employed to immobilize Lac for the oxidation of caffeic acid, protocatechuic acid, sinapic acid and ferulic acid from OMW [46]. The use of the cross-linked C. polyzona Lac was also reported in the literature, for the degradation of nonyl phenol p353NP, BPA and triclosan [17]. CLEAs of *T. versicolor, T. villosa* and *Agaricus bisporus* formed with PEG and GLU have also been lately considered mainly in the aerobic oxidation of linear aliphatic alcohols [13]. Chitosan was also chosen as a support material for the immobilization of *C. versicolor* Lac through covalent bonding using GLU. A degradation efficiency of 65% was attained for 2, 4-dichlorophenol (2, 4-DCP) after 6 sets of operation. The activity of the immobilized Lac was found less than compared to the free Lac but the stability was undoubtedly improved [146].

The *Streptomyces psammoticus* Lac entrapped in calcium alginate beads reserved 42.5% Lac activity. On the other hand, copper alginate beads proved a superior support for Lac immobilization by retaining 61% of the activity. Reusability of the immobilized matrix was studied for up to 8 consecutive runs for duration of 6 hours for each. After the initial run, the immobilized system removed 72% of the color and 69.9% of total phenolics from the phenol model solution [147].

The immobilized *P. ostreatus* MnP fraction by making use of hydrophobic sol-gel entrapment comprising tetramethoxysilane (T) and propyltrimethoxysilane (P) reserved 82% and 75% of its distinctive activity at pH 4 and temperature of 70 °C, correspondingly. The immobilized MnP exhibited 100% decolorization 4 dissimilar textile industrial wastes within the shortest span of time. A lower *K*M, higher *V*max, hyper-activation, and improved acidic and thermal resistance up to 70 °C were the novel catalytic features of the sol-gel immobilized MnP, signifying that it may be a probable contender for biotechnological applications predominantly for textile bioremediation purposes [148].

Combined versatile cross-linked enzyme aggregate (combi-CLEA) was formed by oncomitantly cross-linked after aggregation of 3 oxidative enzymes Lac from *Trametes versicolor,* versatile peroxidase (VP) from *Bjerkandera adusta* and glucose oxidase (GOD) from *Aspergillus niger* c. The constructed biocatalyst exerted higher resistance to harsh conditions (high temperature and low pH) and activity at room temperature were improved after insolubilization. Then, these combi-CLEAs were used for the elimination of pharmaceutically active compounds (PhACs) present in a filtered effluent of a community wastewater handling plant. The combi-CLEA has shown the ability to transform efficiently the nonsteroidal anti-inflammatory drugs acetaminophen, naproxen, mefenamic acid, diclofenac and indometacin. Globally, combining these enzymes able to realize cascade conversions of different substrates allows the generation of biocatalysts with expanded substrate ranges as well as operating conditions [149].

Very recently, Rahmani *et al.* [150]) measured the immobilization of Lac (*Trametes versicolor*) on CPC silica beads and removal of two sulfonamides, sulfathiazole (STZ) and sulfamethoxazole (SMZ). At the raised temperature of 70 °C, immobilized Lac detached 71.7% of STZ and 53% of SMZ compared to free Lac. 63.3% and 82.6% of the initial activity of the immobilized Lac toward SMZ and STZ persisted, correspondingly after 10 sets of removal tryouts. Long-established higher affinity of immobilized Lac toward STZ was confirmed by the lower Km value for STZ (0.056 mM) equated to that of SMZ (0.096 mM).

The suitability of these biocatalysts for continuous treatment of different industrial effluents was evaluated from different features of immobilized ligninolytic enzymes like stability and reusability on different carriers. In a recent study, Iqbal and Asgher, [119] used TMOS and PTMS composed xerogels of different hydrophobicities to immobilize a purified novel MnP from an aboriginal WRF strain *Trametes versicolor* IBL-04 and achieved 92.2% immobilization efficiency. The resulting immobilized MnP retained a high activity of 82 and 75% at pH 4 and 80°C when incubated at varying pH and temperatures for 24 h. The stability and reusability of the immobilized MnP was demonstrated by oxidation of MnSO4 for up to 10 continuous cycles. Shelf life contor discovered that enzyme may keep their activity without losing a bit of it for time period of 60 days at 25 °C. The immobilized MnP was verified against dissimilar textile effluents to explore the industrial applicability of MnP produced by G. lucidum. The industrial effluents were decolorized to different extents after being incubated for 4 hours (with a maximum efficiency of 99.2%). The maximally decolorized waste was examined for formaldehyde and nitroamines and it was declared that the toxicity parameters were underneath the acceptable boundaries.

8.13 Conclusions

Lipases are ubiquitous enzymes found in both prokaryotes and eukaryotes; however microbial enzymes have significant importance due to their fundamental role in lipids biodegradation. Lipases transmit innovative reactions in both non-aqueous and aqueous media. Moreover these enzymes have distinctive properties like easy management and availability, high stability, broad substrate tolerance under different conditions of temperature and types of solvents, which make Lipases more popular. Therefore they have immense industrial and biotechnological applications where they reveal amazing versatility in their catalytic behavior. A diversity of genes have been isolated, recognized and characterized in microorganisms (bacteria, fungi, yeast), by which lipases are produced in them.

A variation in artificial dyestuffs unrestricted by the textile industry carriage a threat to ecological safety. The greatest immediate environmental concern with regard to water quality appears due to release of textile and dye-house effluent may cause abnormal coloration of the surface water and directly affects the aquatic vegetation and wildlife.

Many biological systems play a significant character in the biodegradation of textile dye effluents and reduce its toxicity. Lately, several reports emphasized that microbial systems not only help in degradation and decolorization but also detoxify dyes, thus giving an essential green procedure over chemical decomposition procedures.

The unique direction for application of this enzyme technology has attracted much curiosity with regard to decolorization, degradation and detoxification of azo dyes from wastewater. This attempt is an excellent alternative strategy to conventional physicochemical treatments. The production of free radicals by oxido-reductive enzymes is responsible for complex series of spontaneous cleavage reactions. It is probable that enzymatic treatment will be the most effective for the highest concentrations of target contaminants. This delivers a complete foundation for the development of cost-effective Lac-based processes that are equally good for use in industrial and ecofriendly biotechnology applications.

The degradation of textile dyes by Lac immobilized on surface of porous glass beads revealed that anthraquinone and indigoid dyes were degraded more quickly than the azo dyes. Though, Lac treatment was more effective at decolorizing the anthraquinone dyes while the less for azo and indigoid dyes. These results validate (to enzymatically decolorize a range of different dye classes and reduce dye toxicity in a single step) the potential and limitations of using immobilized Lac. Lac from *Trametes modesta* immobilized on aluminum oxide pellets established that this remediation system is not restricted to a certain structural group of dyes. The maximally decolorized effluent toxicity level was found to be below the permissible limits indicated by formaldehyde and nitroamines toxicity parameters.

References

1. Kaur, S., Varsha Nigam., Production and application of laccase enzyme in pulp and paper industry. *Int. J. Res. Appl., Nat. Social Sci..*, 2(4), 153–158, 2014.

2. Asgher, M., Kamal, S., Iqbal, H.M.N., Improvement of catalytic efficiency, thermo-stability and dye decolorization capability of *Pleurotus ostreatus* IBL-02 laccase by hydrophobic sol gel entrapment. *Chem. Cent. J.*, 2012, DOI:10.1186/1752–153X-6–110.
3. Nouren, S., Bhatti, H.N., Iqbal, M., Bibi, I., Kamal, S., Sadaf, S., Sultan, M., Kausar, A., Safa, Y. By-product identification and phytotoxicity of biodegraded direct yellow 4 dye. *Chemosphere,* 169, 474–484, 2017.
4. Bilal, M., Asgher, M., Iqbal, M., Hu, H., Zhang, X., Chitosan beads immobilized manganese peroxidase catalytic potential for detoxification and decolorization of textile effluent. *Int. J. Biol. Macromol.*, 89, 181–189, 2016.
5. Cho, S-J, Park, S.J., Lim, J.S., Rhee, Y.H., Shin, K.S., Oxidation of polycyclic aromatic hydrocarbons by laccase of Coriolus hirsutus. *Biotechnol. Lett.*, 24, 1337–1340, 2002.
6. Bilal, M., Iqbal, M., Hu, H., Zhang, X. Mutagenicity and cytotoxicity assessment of biodegraded textile effluent by Ca-alginate encapsulated manganese peroxidase. *Biochem. Eng. J.*, 109, 153–161, 2016.
7. Bilal, M., Asgher, M. Dye decolorization and detoxification potential of Ca-alginate beads immobilized manganese peroxidase. *BMC Biotechnol.*, 15, 1, 2015.
8. Wang, Y., Zhang, D., He, F.R., Chen, X.C. Immobilization of laccase by Cu2+ chelate affinity interaction on surface modified magnetic silica particles and its use for the removal of pentachlorophenol. *Chin. Chem. Lett.*, 23, 197–200, 2012.
9. Fernández-Fernández M, Sanromán M.Á., Moldes D. Recent developments and applications of immobilized laccase. *Biotechnol. Adv.*, 31, 1808–1825, 2013.
10. Jamal, F., Singh, S., Khatoon, S., Mehrotra, S. Application of immobilized pointed gourd (*Trichosanthes dioica*) peroxidase-concanavalin a complex on calcium alginate pectin gel in decolorization of synthetic dyes using batch processes and continuous two reactor system. *J. Bioproces. Biotechniq*. 2013. Xu, L.Q., Wen, X.H., Ding, H.J., Immobilization of lignin peroxidase on spherical mesoporous material. 31(10), 2493–2499, 2010.
11. Catana, R., Ferreira, B., Cabral, J., Fernandes, P., Immobilization of inulinase for sucrose hydrolysis. *Food Chemistry,* 91, 517–520, 2005.
12. Aucoin, M.G., Erhardt, F.A., Legge, R.L., Hyperactivation of Rhizomucor miehei lipase by hydrophobic xerogels. *Biotechnol. Bioeng.*, 85, 647–655, 2004.
13. Matijošyte, I., Arends, I.W.C.E., de Vries, S., Sheldon, R.A., Preparation and use of cross-linked enzyme aggregates (CLEAs) of laccases. *J. Mol. Catal. B.*, 62, 142–148, 2010.
14. Šulek, F., Fernández, D.P., Knez, Ž., Habulin, M., Sheldon, R.A. Immobilization of horseradish peroxidase as crosslinked enzyme aggregates (CLEAs). *Proc. Biochem.*, 46, 765–769, 2011.
15. Guisán, J.M., Alvaro, G., Fernandez-Lafuente, R., Rosell, C.M., Garcia, J.L., Tagliani, A. Stabilization of heterodimeric enzyme by multipoint covalent

immobilization: penicillin G acylase from Kluyvera citrophila. *Biotechnol. Bioeng.*,42, 455–464, 1993.

16. Cao, L., van Langen, L., Sheldon, R.A. Immobilised enzymes: carrier-bound or carrier-free? *Curr. Opin. Biotechnol.*, 14, 387–394, 2003.
17. Cabana, H., Jones, J.P., Agathos, S.N., Preparation and characterization of cross-linked laccase aggregates and their application to the elimination of endocrine disrupting chemicals. *J. Biotechnol.*,132, 23–31, 2007.
18. Sheldon, R.A., Characteristic features and biotechnological applications of cross-linked enzyme aggregates (CLEAs). *Appl. Microbiol. Biotechnol.*,, 92(3), 467–477, 2011.
19. Persichetti, R.A., Clair, N.L., Griffith, J.P., Navia, M.A., Margolin, A.L., Cross-linked enzyme crystals (CLECs) of thermolysin in the synthesis of peptides. *J. Amer. Chem. Soc.*,117, 2732–2737, 1995.
20. Lee, J.-W., Lee, S.-M., Hong, E.-J., Jeung, E.-B., Kang, H.-Y., Kim, M.-K., Choi, I.-G., Estrogenic reduction of styrene monomer degraded by Phanerochaete chrysosporium KFRI 20742. *J. Microb.*, 44, 2, 177–184, 2006.
21. Otero, M.G.S., Alfaro, G.V., Galindo, H.S.G., Ros, R.M.O., Immobilization in the presence of Triton X-100: modifications in activity and thermostability of Geobacillus thermoleovorans CCR11 lipase. *J. Ind Microbiol. Biotechnol.*, 35, 1687–1693, 2008.
22. Qiu, L., Huang, Z. The treatment of chlorophenols with laccase immobilized on sol–gelderived silica. *World J. Microbiol. Biotechnol.*, 26, 775–781, 2010.
23. Rekuc, A., Bryjak, J., Szymanska, K., Jarzebski, A.B., Laccase immobilization on mesostructured cellular foams affords preparations with ultra-high activity. *Process Biochem.*, 44,, 191–198, 2009.
24. Kalkan, N.A., Aksoy, S., Aksoy, E.A., Hasirci, N., Preparation of chitosan-coated magnetite nanoparticles and application for immobilization of laccase. *J. Appl. Polym. Sci.*, 123, 707–716, 2012.
25. Ignatova, M.G., Manolova, N.E., Toshkova, R.A., Rashkov, I.B., Gardeva, E.G., Yossifova, L.S., Alexandrov, M.T., Electrospun nanofibrous mats containing quaternized chitosan and polylactide with in vitro antitumor activity against HeLa cells. *Biomacromolecules*, 11(6), 1633–1645, 2010.
26. Wang, F., Guo, C., Yang, L.R., Liu, C.Z., Magnetic mesoporous silica nanoparticles fabrication and their laccase immobilization performance. *Bioresour. Technol.*, 101, 8931–8935, 2010.
27. Nisha, S., Karthick, S.A., Gobi, N., A review on methods, application and properties of immobilized enzyme. *Chem. Sci. Rev. Lett.*,1, 148–155, 2012.
28. Forde, J., Tully, E., Vakurov, A., Gibson, T.D., Millner, P., Ó'Fágáin, C. Chemical modification and immobilisation of laccase from Trametes hirsute and from Myceliophthora thermophile. *Enzym. Microb. Technol.*, 46, 430–437, 2010.
29. Mitchell, S., Ramirez, J.P., Mesoporous zeolites as enzyme carriers: synthesis, characterization, and application in biocatalysis. *Catal Today*, 168, 28–37, 2011.

30. Popat, A., Hartono, S.B., Stahr, F., Liu, J., Qiao, S.Z., Lu, G.Q., Mesoporous silica nanoparticles for bioadsorption, enzyme immobilisation and delivery carriers. *Nanoscale*, 3, 2801–2818, 2011
31. Wang, P., Fan, X., Cui, L., Wang, Q., Zhou, A., Decolorization of reactive dyes by laccase immobilized in alginate/gelatin blent with PEG. *J. Environ. Sci.*, 20, 1519–1522, 2008.
32. Brady, D., Jordaan, J., Advances in enzyme immobilisation. *Biotechnol. Lett.*, 31, 1639, 2009.
33. Phetsom, J., Khammuang, S., Suwannawong, P., Sarnthima, R., Copper–alginate encapsulation of crude laccase from Lentinus polychrous lev. and their effectiveness in synthetic dyes decolorizations. *J. Biol. Sci.*, 9, 573–583, 2009.
34. Wen, H., Nallathambi, V., Chakraborty, D., Barton, S.C., Carbon fiber microelectrodes modified with carbon nanotubes as a new support for immobilization of glucose oxidase. *Microchim. Acta,* 175, 283–289, 2011.
35. Klis, M., Maicka, E., Michota, A., Bukowska, J., Sek, S., Rogalski, J., *et al.* Electroreduction of laccase covalently bound to organothiol monolayers on gold electrodes. *Electrochim. Acta,*52, 5591–5598, 2007.
36. Galperin, A., Long, T.J., Ratner, B.D., Degradable, thermo-sensitive poly (N-isopropyl acrylamide)-based scaffolds with controlled porosity for tissue engineering applications. *Biomacromolecules*, 11, 2583–2592, 2010.
37. El-Naas, M.H., Al-Muhtaseb, S.A., Makhlouf, S., Biodegradation of phenol by Pseudomonas putida immobilized in polyvinyl alcohol (PVA) gel. *J. Hazard. Mat.*, 164, 720–725, 2009.
38. Crestini, C., Perazzini, R., Saladino, R., Oxidative functionalisation of lignin by layer-by-layer immobilised laccases and laccase microcapsules. *Appl. Catal. Gen.*, 372, 115–123, 2010.
39. Szamocki, R., Flexer, V., Levin, L., Forchiasin, F., Calvo, E.J., Oxygen cathode based on a layer-by-layer self-assembled laccase and osmium redox mediator. *Electrochim. Acta*, 54, 1970–1977, 2009.
40. Ryoo, H.-w., Kim, Y.-s., Lee, J.-h., Shin, W.-s., Myung, N.-s., Hong, H.-G., Immobilization of horseradish peroxidase to electrochemically deposited gold-nanoparticles on glassy carbon electrode for determination of H_2O_2. *Bull. Korean Chem. Soc.*, 27, 672–678, 2006.
41. Durán, N., Rosa, M.A., D'Annibale, A., Gianfreda, L., Applications of laccases and tyrosinases (phenoloxidases) immobilized on different supports: a review. *Enzyme Microb. Technol.*, 31, 907–931, 2002.
42. Kunamneni, A., Ghazi, I., Camarero, S., Ballesteros, A., Plou, F.J., Alcalde, M., Decolorization of synthetic dyes by laccase immobilized on epoxy-activated carriers. *Process Biochem.*, 43, 169–178, 2008.
43. Silva, C., Silva, C.J., Zille, A., Guebitz, G.M., Cavaco-Paulo, A., Laccase immobilization on enzymatically functionalized polyamide 6,6 fibres. *Enzyme Microb. Technol.*, 41, 867–875, 2007.
44. Datta, S., Christena, L.R., Rajaram, Y.R.S., Enzyme immobilization: an overview on techniques and support materials. *Biotech.*, 3, 1–9, 2013.

45. Ispas, C., Sokolov, I., Andreescu, S., Enzyme-functionalized mesoporous silica for bioanalytical applications. *Anal. Bioanal. Chem.*, 393, 543–554, 2009.
46. Salis, A., Pisano, M., Monduzzi, M., Solinas, V., Sanjust, E., Laccase from Pleurotus sajor-caju on functionalised SBA-15 mesoporous silica: immobilisation and use for the oxidation of phenolic compounds. *J. Mol. Catal. B.*, 58, 175–80, 2009.
47. Cornejo, A., Fraile, J.M., García, J.I., Gil, M.J., Luis, S.V., Martínez-Merino, V., *et al.*, A flexible and versatile strategy for the covalent immobilization of chiral catalysts based on pyridinebis (oxazoline) ligands. *J. Org. Chem.*, 70, 5536–5544, 2005.
48. Balland, V., Hureau, C., Cusano, A.M., Liu, Y., Tron, T., Limoges, B., Oriented immobilization of a fully active monolayer of histidine-tagged recombinant laccase on modified gold electrodes. *Chemistry*, 14, 7186–7192, 2008.
49. Shang, W., Liu, W., Wang, L., Characterization and study of Agaricus bisporus laccase immobilized on a ceramic-chitosan composite support. Xiandai Huagong. *Mod. Chem. Ind.*, 29, 45–47, 2009a.
50. Stanescu, M.D., Fogorasi, M., Shaskolskiy, B.L., Gavrilas, S., Lozinsky, V.I., New potential biocatalysts by laccase immobilization in PVA cryogel type carrier. *Appl. Biochem. Biotechnol.*, 160, 1947–1954, 2010.
51. Bayramoğlu, G., Yilmaz, M., Arica, M.Y., Reversible immobilization of laccase to poly (4-vinylpyridine) grafted and Cu (II) chelated magnetic beads: biodegradation of reactive dyes. *Biores. Technol.*, 101, 6615–6621, 2010.
52. Stanescu, M.D., Gavrilas, S., Ludwig, R., Haltrich, D., Lozinsky, V.I., Preparation of immobilized Trametes pubescens laccase on a cryogel-type polymeric carrier and application of the biocatalyst to apple juice phenolic compounds oxidation. *Europ. Food Res. Technol.*, 234, 655–662, 2012.
53. Pchelintsev, N., Youshko, M., Švedas, V., Quantitative characteristic of the catalytic properties and microstructure of cross-linked enzyme aggregates of penicillin acylase. *J. Mol. Catal. B: Enzym.*, 56, 202–207, 2009.
54. Wilson, L., Illanes, A., Soler, L., Henríquez, M.J., Effect of the degree of cross-linking on the properties of different CLEAs of penicillin acylase. *Proc. Biochem.*,44, 322–326, 2009.
55. Wilson, L., Illanes, A., Abián, O., Pessela, B.C., Fernández-Lafuente, R., Guisán, J.M., Co-aggregation of penicillin G acylase and polyionic polymers: an easy methodology to prepare enzyme biocatalysts stable in organic media. *Biomacromolecules*, 5, 852–827, 2004.
56. Clifford, J.S., Legge, R.L. Use of water to evaluate hydrophobicity of organically-modified xerogel enzyme supports. *Biotechnol. Bioeng.*, 92(2), 2005, 231–237.
57. Asgher, M., Asad, M.J., Bhatti, H.N., Legge, R.L., *World J. Microbiol. Biotechnol.*, 23, 525, 2007.
58. Iqbal, H.M.N., Asgher, M. Decolorization applicability of sol-gel matrix immobilized manganese peroxidase produced from an

indigenous white rot fungal strain *Ganoderma lucidum*. *BMC Biotechnol.*, 2013, DOI:10.1186/1472-6750-13-56.

59. Bódalo, A., Bastida, J., Maximo, M.F., Montiel, M.C., Gómez, M., Murica, M.D. A comparative study of free and immobilized soybean and horseradish peroxidases for 4-chlorophenol removal: protective effects of immobilization. *Bioprocess. Biosyst. Eng.*, 31, 587–593, 2008.
60. Acevedo, F., Pizzul, L., Castillo, M.D., Gonzales, M.E., Cea, M., Gianfreda, L., Diez, M.C. Degradation of polycyclic aromatic hydrocarbons by free and nanoclay-immobilized manganese peroxidase from *Anthracophyllum discolor*. *Chemosphere*, 80, 271–278, 2010.
61. Wen, X., Jia, Y., Li, J., Enzymatic degradation of tetracycline and oxytetracycline by crude manganese peroxidase prepared from Phanerochaete chrysosporium. *J. Hazard. Mat.*, 177, 924–928, 2010.
62. Bilal, M., Asgher, M., Dye decolorization and detoxification potential of Ca-alginate beads immobilized manganese peroxidase. *BMC Biotechnol.*, 15, 1, 2015.
63. Kirby, N., Marchant, R., McMullan, G., Decolorisation of synthetic textile dyes by Phlebia tremellosa. *FEMS Microbiol. Lett.*, 93–96, 1881.
64. Arica, M.Y., Altintas, B., Bayramoglu, G., Immobilization of laccase onto spacer-arm attached non-porous poly(GMA/EGDMA) beads: application for textile dye degradation. *Bioresour. Technol.*, 100, 665–669, 2009.
65. Asgher, M., Iqbal, H.M.N., Enhanced catalytic features of sol–gel immobilized MnP isolated from solid state culture of Pleurotus ostreatus IBL-02. *Chinese Chem. Lett.*, 24, 344–346, 2013.
66. Arsenault, A., Hubert Cabana, Peter Jones, J., Laccase-based CLEAs: chitosan as a novel cross-linking agent. *Enzym. Res.*, 1–10, 2011.
67. Fernando Bautista, L., Morales, G., Sanz, R., Immobilization strategies for laccase from Trametes versicolor on mesostructured silica materials and the application to the degradation of naphthalene. *Bioresour. Technol.*, 101, 8541–8548, 2010.
68. Tavares, A.P.M., Rodriguez, O., Fernandez-Fernandez, M., Dominguez, A., Moldes, D., Sanroman, M.A., Macedo, E.A., Immobilization of laccase on modified silica: Stabilization, thermal inactivation and kinetic behaviour in 1-ethyl-3-methylimidazolium ethylsulfate ionic liquid. *Biores. Technol.*, 131, 405–412, 2013.
69. Matsumiya, M., Enzymatic production of N-acetyl-D-glucosamine using crude enzyme from the liver of squids. *Food Sci. Technol. Res.*, 10, 296–299, 2007.
70. Plaza, M., Santoyo, S., Jaime, L., Reina, G.G.-B., Herrero, M., Señoráns, F.J., *et al.*, Screening for bioactive compounds from algae. *J. Pharmaceut. Biomed. Anal.*, 51, 450–455, 2010.
71. Prasad, M., Manjunath, K., Comparative study on biodegradation of lipid-rich wastewater using lipase producing bacterial species, *Indian Journal of Biotechnology*, 10, 121–124, 2011.

72. Bradoo, S., Saxena, R., Gupta, R., Two acidothermotolerant lipases from new variants of Bacillus spp. *World J. Microb. Biotechnol.*, 15, 87–91, 1999.
73. Markossian, S., Becker, P., Märkl, H., Antranikian, G., Isolation and characterization of lipid-degrading Bacillus thermoleovorans IHI-91 from an Icelandic hot spring. *Extremophiles*, 4, 365–371, 2000.
74. Rahman, M.A., Noh, H-., Shim, Y-., Direct electrochemistry of laccase immobilized on Au nanoparticles encapsulated-dendrimer bonded conducting polymer: application for a catechin sensor. *Anal. Chem.*, 80, 8020–8027, 2008.
75. Muraoka, W., Nakashima, T., Tabira, Y., Eguchi, H., Imagawa, K., Mastumura, Y., *et al.*, Characterization of Burkholderia sp. Y1 isolated from oil polluted soil. *J. Environ. Biotechnol.*, 8, 43–47, 2008.
76. Čipinytė, V., Grigiškis, S., Baškys, E., Selection of fat-degrading microorganisms for the treatment of lipid-contaminated environment. *Biologija.*, 55, 84–92, 2009.
77. Lu, L., Zhao, M., Wang, Y., Immobilization of laccase by alginate–chitosan microcapsules and its use in dye decolorization. *World J. Microbiol. Biotechnol.*, 23, 159–166, 2006.
78. Sirisha, E., Rajasekar, N., Narasu, M.L., Isolation and optimization of lipase producing bacteria from oil contaminated soils. *Adv. Biol. Res.*, 4, 249–252, 2006.
79. Ma, Q., Sun, X., Gong, S., Zhang, J., Screening and identification of a highly lipolytic bacterial strain from barbecue sites in Hainan and characterization of its lipase. *Ann. Microbiol.*, 60, 429–437, 2010.
80. Mobarak-Qamsari, E., Kasra-Kermanshahi, R., Moosavi-Nejad, Z., Isolation and identification of a novel, lipase-producing bacterium, Pseudomnas aeruginosa KM110. *Iran. J. Microbiol.*, 3, 92, 2011.
81. Hidayat, N., Optimization of pH, temperature and agitation rate on biodegradation of lipids and detergents in food wastewater by Bacillus sp N-09. *J. Agric. Food Technol.*, 1, 59–62, 2011.
82. Mukesh Kumar, D., Rejitha, R., Devika, S., Balakumaran, M., Immaculate, A., Kalaichelvan, P., Production, optimization and purification of lipase from Bacillus sp. MPTK 912 isolated from oil mill effluent. *Adv. Appl. Sci. Res.*, 3, 930–938, 2012.
83. Kalyani, D., Dhiman, S.S., Kim, H., Jeya, M., Kim, I.W., Lee, J.K., Characterization of a novel laccase from the isolated Coltricia perennis and its application to detoxification of biomass. *Process Biochem.*, http://dx.doi.org/10.1016/j.procbio.2012.01.013.
84. Karnchanatat, A., Petsom, A., Sangvanich, P., Piaphukiew, J., Whalley, A.J., Reynolds, C.D., *et al.*, Purification and biochemical characterization of an extracellular β-glucosidase from the wood-decaying fungus Daldinia eschscholzii (Ehrenb.: Fr.) Rehm. *FEMS Microbiol. Lett.*, 270, 162–170, 2007.
85. Humnabadkar, R.P., Saratale, G.D., Govindwar, S.P., 'Decolorization of Purple 2R by Aspergillus ochraceus (NCIM-1146).' *Asian J. Microbiol. Biotechnol. Environ. Sci.*, 10, 693, 2008.

86. McMullan, G., Meehan, C., Conneely, A., Kirby, N., Robinson, T., Nigam, P., Banat, I.M., Smyth, W.F., Microbial decolourisation and degradation of textile dyes. *Appl. Microbiol. Biotechnol.*, 56, 81, 2001.
87. Mahmood, R., Sharif, F., Ali, S., Hayyat, M.U., Enhancing the decolorizing and degradation ability of bacterial consortium isolated from textile effluent affected area and its application on seed germination. *Sci. World J.*, 2015.
88. Chagas, E.P., Durrant, L.R., Decolorization of azo dyes by Phanerochaete chrysosporium and Pleurotus sajorcaju. *Enzyme Microb. Technol.*, 29, 473–477, 2001.
89. Phugare, S.S., Kalyani, D.C., Surwase, S.N., Jadhav, J.P., Ecofriendly degradation, decolorization and detoxification of textile effluent by a developed bacterial consortium. *Ecotoxicol. Environ. Safety*, 74, 1288–1296, 2011.
90. Singh, N.R., Narinesingh, D., Singh, G., Immobilization of β-galactosidase onto sepharose and stabilization in room temperature ionic liquids. *J. Mol. Liq.*, 152, 19–27, 2010.
91. Yao, J., Jia, R., Zheng, L., Wang, B., Rapid decolorization of azo dyes by crude manganese peroxidase from Schizophyllum sp. F17 in solid-state fermentation. *Biotechnol. Bioprocess Eng.*, 18, 868–877, 2013.
92. Bilal, M., Asgher, M., Dye decolorization and detoxification potential of Ca-alginate beads immobilized manganese peroxidase. *BMC Biotechnol.*, 15, 1, 2015.
93. Gomes, E., Aguiar, A.P., Carvalho, C.C., Bonfá, M.R.B., Da Silva, R., Boscolo, M., Ligninases production by Basidiomycetes strains on lignocellulosic agricultural residues and their application in the decolorization of synthetic dyes. *Brazilian J. Microbiol.*, 40(1), 31–39, 2009.
94. Narkhede, M., Mahajan, R., Narkhede, K. Ligninolytic enzyme production and Remazol Brilliant Blue R (RBBR) decolorization by a newly isolated white rot fungus: Basidiomycota spp. L-168. *Int. J. Pharm. Bio. Sci.*, 4, 220–228, 2013.
95. Boer, C.G., Obici, L., de Souza, C.G.M., Peralta, R.M., Decolorization of synthetic dyes by solid state cultures of Lentinula (Lentinus) edodes producing manganese peroxidase as the main ligninolytic enzyme. *Biores. Technol.*, 2004, 94, 107–112.
96. Kariminiaae-Hamedaani, H.-R., Sakurai, A., Sakakibara, M., Decolorization of synthetic dyes by a new manganese peroxidase-producing white rot fungus. *Dyes Pigm.*, 72, 157–162, 2007.
97. Ratanapongleka, K., Phetsom, J., Decolorization of synthetic dyes by crude laccase from lentinus polychrous lev. *Int. J. Chem. Eng. Appl.*, 5, 26, 2014.
98. Rodríguez-Rodríguez, C.E., Castro-Gutiérrez, V., Chin-Pampillo, J.S., Ruiz-Hidalgo, K., On-farm biopurification systems: role of white rot fungi in depuration of pesticide-containing wastewaters. *FEMS Microbiol. Lett.*, 2013, 345, 1–12.
99. Peralta-Zamora, P., Pereira, C.M., Tiburtius, E.R.L., Moraes, S.G., Rosa, M.A., Minussi, R.C., *et al.*, Decolorization of reactive dyes by immobilized laccase. *Appl. Catal. B Environ.*,, 42, 131–144, 2003.

100. Kandelbauer, A., Maute, O., Kessler, R.W., Erlacher, A., Gübitz, G.M., Study of dye decolorization in an immobilized laccase enzyme-reactor using online spectroscopy. *Biotechnol. Bioeng.*, 87, 552–563, 2004.
101. Champagne, P.-P., Ramsay, J.A., Dye decolorization and detoxification by laccase immobilized on porous glass beads. *Bioresource Technol.*, 101, 2230–2235, 2010.
102. Osma, J.F., Toca-Herrera, J.L., Rodríguez-Couto, S., Transformation pathway of remazol brilliant blue R by immobilised laccase. *Biores. Technol.*, 101, 8509–8514, 2010.
103. Lu, L., Zhao, M., Wang, Y., Immobilization of laccase by alginate–chitosan microcapsules and its use in dye decolorization. *World J. Microbiol. Biotechnol.*, 23, 159–166, 2007.
104. Zhao, M., Wang, W., Li, X., Wei, X-., Lu, L., Yang, C., Preparation of surface modified alginate nanoparticles with acyl chloride groups and its use in immobilization of laccase. *Zhongguo Zaozhi Xuebao/Trans China Pulp Pap*, 23, 94–99, 2008.
105. Makas, Y.G., Kalkan, N.A., Aksoy, S., Altinok, H., Hasirci, N., Immobilization of laccase inκ-carrageenan based semi-interpenetrating polymer networks. *J. Biotechnol.*, 148, 216–220, 2010.
106. Kim, M.I., Ham, H.O., Oh, S-.D., Park, H.G., Changa, H.N., Choi, S.H., Immobilization of Mucor javanicus lipase on effectively functionalized silica nanoparticles. *J. Mol. Catal. B Enzym.*, 39, 62–68, 2006.
107. Cristovao, R.O, Tavares, A.P.M., Brigida, A.I., Loureiro, J.M., Boaventura, R.A.R., Macedo, E.A., Coelho, M.A.Z., Immobilization of commercial laccase onto green coconut fiber by adsorption and its application for reactive textile dyes degradation. *J. Mol. Catal. B: Enzym.*, 72, 6–12, 2011.
108. Yanto, D.H.Y., Tachibana, S., Itoh, K., Biodecolorization of textile dyes by immobilized enzymes in a vertical bioreactor system. 4th International Conference on Sustainable Future for Human Security, SustaiN 2013.
109. Nur Afiqah Ahmad Shamsuri, Nur Atikah Mohidem, Azmi Fadziyana Mansor, Hanapi Mat Biodegradation of Dye Using Free Laccase and Solgel Laccase. 59, 39–42, 2012.
110. Cinzia Pezzella, Maria Elena Russo, Antonio Marzocchella, Piero Salatino, Giovanni Sannia, Immobilization of a Pleurotus ostreatus laccase mixture on perlite and its application to dye decolourisation. *BioMed. Res. Int.*, 308613–308624, 2014, http://dx.doi.org/10.1155/2014/308613.
111. Sadighi, A., Faramarzi, M.A., Congo red decolorization by immobilized laccase through chitosan nanoparticles on the glass beads. *J. Taiwan Inst. Chem. Eng.*, 44, 156–162, 2013.
112. Thakur, V., Kumar, P., Verma, A., Chand, D., Decolorization of dye by alginate immobilized laccase from *Cercospora* SPF-6: using compact 5 stage plug flow reactor. *Int. J. Curr. Microbiol. App. Sci.*, 4(1), 183–200, 2015.
113. Uygun, M.,Preparation of laccase immobilized cryogels and usage for decolorization. *J. Chem.*, 1–7, 2013.

114. Mirzadeh, S-.S., Khezri, S-.M., Rezaei, S., Forootanfar, H., Mahvi, A.H., Faramarzi, M.A., Decolorization of two synthetic dyes using the purified laccase of Paraconiothyrium variabile immobilized on porous silica beads. *J. Environ. Health Sci. Eng.*, 12, 6, 2014.
115. Parshetti, G., Telke, A., Kalyani, D., Govindwar, S., Decolorization and detoxification of sulfonated azo dye methyl orange by Kocuria rosea MTCC 1532. *J. Hazard. Mat.*, 176, 503–509, 2010.
116. Boucherit, N., Abouseoud, M., Adour, L., Degradation of direct azo dye by Cucurbita pepo free and immobilized peroxidase. *J. Environ. Sci.*, 25, 1235–1244, 2013.
117. Radha, K., Regupathi, I., Arunagiri, A., Murugesan, T., Decolorization studies of synthetic dyes using Phanerochaete chrysosporium and their kinetics. *Process Biochem.*, 40, 3337–3345.
118. Asgher, M., Iqbal, H.M.N., Enhanced catalytic features of sol–gel immobilized MnP isolated from solid state culture of Pleurotus ostreatus IBL-02. *Chinese Chem. Lett.*, 24, 344–346.
119. Asgher, M., Bashir, F., Iqbal, H.M.N., A comprehensive ligninolytic pretreatment approach from lignocellulose green biotechnology to produce bioethanol. *Chem. Eng. Res. Design*, 92, 1571–1578, 2013.
120. Noureddini, H., Gao, X., Philkana, R., Immobilized Pseudomonas cepacia lipase for biodiesel fuel production from soybean oil. *Bioresource Technol.*, 96, 769–777, 2005.
121. Pardal, A.C., Encinar, J., González, J., Martínez, G., Transesterification of rapeseed oil with methanol in the presence of various co-solvents. 2010.
122. Hasan, F., Shah, A.A., Hameed, A., Industrial applications of microbial lipases. *Enzyme Microb. Technol.*, 39, 235–251, 2006.
123. Lacayo Romero, M., *Microbial degradation of toxaphene*, Lund University, Sweden, 2005.
124. Pérez, J., Munoz-Dorado, J., de la Rubia, T., Martinez, J., Biodegradation and biological treatments of cellulose, hemicellulose and lignin: an overview. *Int. Microb.*, 5, 53–63, 2002.
125. Mateo, C., Palomo, J.M., Fernandez-Lorente, G., Guisan, J.M., Fernandez-Lafuente, R., Improvement of enzyme activity, stability and selectivity via immobilization techniques. *Enzyme Microb. Technol.*, 40, 1451–1463, 2007.
126. Teles, F.R., Cabral, J.M., Santos, J.A., Enzymatic degreasing of a solid waste from the leather industry by lipases. *Biotechnol. Lett.*, 23, 1159–1163, 2001.
127. Ramani, K., Chockalingam, E., Sekaran, G., Production of a novel extracellular acidic lipase from Pseudomonas gessardii using slaughterhouse waste as a substrate. *J. Industr. Microbiol. Biotechnol.*, 37, 531–535, 2010.
128. Papagianni, M., Recent advances in solid-state fermentation applications for the food industry. *Curr. Biochem. Eng.*, 1, 2–8, 2014.
129. Godoy, M.G., Gutarra, M.L., Castro, A.M., Machado, O.L., Freire, D.M., Adding value to a toxic residue from the biodiesel industry: production of

two distinct pool of lipases from Penicillium simplicissimum in castor bean waste. *J. Industr. Microb. Biotechnol.*, 38, 945–953, 2011.
130. Ramakrishnan, V., Balakrishnan, B., Rai, A.K., Narayan, B., Halami, P.M., Concomitant production of lipase, protease and enterocin by Enterococcus faecium NCIM5363 and Enterococcus durans NCIM5427 isolated from fish processing waste. *Int. Aqu. Res.*, 4, 14, 2012.
131. Moftah, O.A.S., Grbavčić, S., Žuža, M., Luković, N., Bezbradica, D., Knežević-Jugović, Z., Adding value to the oil cake as a waste from oil processing industry: production of lipase and protease by Candida utilis in solid state fermentation. *Appl. Biochem. Biotechnol.*, 166, 348–364, 2012.
132. Fickers, P., Benetti, P-.H., Wache, Y., Marty, A., Mauersberger, S., Smit, M., *et al.*, Hydrophobic substrate utilisation by the yeast Yarrowia lipolytica, and its potential applications. *FEMS Yeast Res.*, 5, 527–543, 2005.
133. Minussi, R.C., Miranda, M.A., Silva, J.A., Ferreira, C.V., Aoyama, H., Marangoni, S., Rotilio, D., Pastore, G.M., Durán, N., Purification, characterization and application of laccase from Trametes versicolor for colour and phenolic removal of olive mill wastewater in the presence of 1-hidroxybenzotriazole. *African J. Biotechnol.*, 6(10), 1248–1254, 2007.
134. Barbosa, E.S., Perrone, D., Vendramini, A.L.A., Leite, S.G.F., Vanillin production by Phanerochaete chrysosporium grown on green coconut agro-industrial husk in solid state fermentation. *BioResources*, 3(4), 1042–1050, 2008.
135. López-Nicolás M, García-Carmona F., *Enzymatic and nonenzymatic degradation of polyphenols. Fruit and vegetables phytochemicals.* Wiley-Blackwell Publishing, Ames, Iowa, USA, 101–103, 2010.
136. Widsten, P., Kandelbauer, A., Laccase applications in the forest products industry: a review. *Enzyme Microb. Technol.*, 42(4), 293–307, 2008.
137. Rodriguez Couto, S., Toca Herrera, J.L., Industrial and biotechnological applications of laccases: a review. *Biotechnol. Adv.*, 24, 500–513, 2006.
138. Sigoillot, C., Camarero, S., Vidal, T., Record, E., Asther, M., Pérez-Boada, M., Martínez, M.J., Sigoillot, J.-C., Asther, M., Colom, J.F., Martínez, Á.T., Comparison of different fungal enzymes for bleaching high-quality paper pulps. *J. Biotechnol.*, 115(4), 333–343, 2005.
139. Maijala, P., Mettãlã, A., Kleen, M., Westin, C., Poppius-Levin, K., Herranen, K., Lehto, J.H., Reponen, P., Mãentausta, O., Hatakka, A., Treatment of softwood chips with enzymes may reduce refining energy consumption and increase surface charge of fibers. In: 10th International Congress on Biotechnology in the Pulp and Paper Industry. Madison Wisconsin, Book of Abstracts. p. 65, 2007.
140. Anastasi, A., Coppola, T., Prigione, V., Varese, G., Pyrene degradation and detoxification in soil by a consortium of basidiomycetes isolated from compost: Role of laccases and peroxidases. *J. Hazard. Mat.*, 165(1–3), 1229–1233, 2009.
141. Robles-Hernández, L., Gonzales-Franco, A.C., Crawford, D.L., Chun, W.W.C., Review of environmental organopollutants degradation by white-rot basidiomycete mushrooms. *Tecnociencia Chihuahua*, 2(1), 32–39, 2008.

142. Liu, Y., Zeng, Z., Zeng, G., Tang, L., Pang, Y., Li, Z., Liu, C., Lei, X., Wu, M., Ren, P., *et al.*, Immobilization of laccase on magnetic bimodal mesoporous carbon and the application in the removal of phenolic compounds. *Bioresour. Technol.*, 115, 21–26, 2012.
143. Krings, U., Berger, R., Biotechnological production of flavours and fragrances. *Appl. Microb. Biotechnol.*, 49, 1–8, 1998.
144. Grabski, A.C., Grimek, H.J., Burgess, R.R., Immobilization of manganese peroxidase from *Lentinula edodes* and its biocatalytic generation of Mn-III-chelate as a chemical oxidant of chlorophenols, *Biotechnol. Bioeng.*, 60, 204, 1998.
145. Mohajershojaei, K., Mahmoodi, N.M., Khosravi, A., Immobilization of laccase enzyme onto titania nanoparticle and decolorization of dyes from single and binary systems. *Biotechnol. Bioprocess Eng.: BBE*, 20, 109, 2015.
146. Georgieva, S., Godjevargova, T., Portaccio, M., Lepore, M., Mita, D.G., Advantages in using non-isothermal bioreactors in bioremediation of water polluted by phenol by means of immobilized laccase from Rhus vernicifera. *J. Mol. Catal. B.*, 55, 177–184, 2008.
147. Zhang, S., Gao, E., Xia, L., Dechlorination of dichlorophenol in waste water by immobilized laccase. *Huagong Xuebao/J. Chem. Ind. Eng. (China)*, 57, 359–362, 2006.
148. Niladevi, K.N., Prema, P., Immobilization of laccase from Streptomyces psammoticus and its application in phenol removal using packed bed reactor. *World J. Microbiol. Biotechnol.*, 24, 1215–1222, 2008.
149. Imad E. Touahar, Lounès Haroune, Sidy Ba, Jean-Phillipe Bellenger, Hubert Cabana. Characterization of combined cross-linked enzyme aggregates from laccase, versatile peroxidase and glucose oxidase, and their utilization for the elimination of pharmaceuticals. *Sci. Total Environ.*, 481, 90–99, 2014.
150. Rahmani, K., Faramarzi, M.A., Mahvi, A.H., Gholami, M., Esrafili, A., Forootanfar, H., Farzadkia, M., Elimination and detoxification of sulfathiazole and sulfamethoxazole assisted by laccase immobilized on porous silica beads. *Int. Biodet. Biodeg.*, 97, 107–114, 2015.

9

Role of Plant-Based Biochar in Pollutant Removal: An Overview

D.S. Malik[1], C.K. Jain[2], Anuj K. Yadav[1*] and Sushmita Banerjee[3]

[1]Department of Zoology and Environmental Sciences, Gurukula Kangri Vishwavidyalaya, Haridwar, India
[2]Environmental Hydrology Division, National Institute of Hydrology, Roorkee, India
[3]Department of Chemistry, Indian Institute of Technology- BHU, Varanasi, India

Abstract

Biochar is a porous material that is produced by pyrolysis, that is, thermal heating under oxygen limited conditions. The potential of biochar has been investigated in various fields like wastewater treatment, soil treatment/improvement, organic/inorganic pollutant (like metals, dyes, fluoride, phenol, dyes and pesticide etc.). The cost of preparation of biochar is high, so, the researchers have drawn their attention toward the preparation of low cost plant derived biochar. Biochar as adsorbent for pollutant removal have been proved in various studied due to high carbon content, low ash content, large surface area, and high porosity. The present review explores the biochar application in various pollutants removal from aqueous solutions and soil treatments.

Keywords: Biochar, pollutant, heavy metal, soil treatment, pyrolysis

9.1 Introduction

Biochar is carbonic material which is synthesized through pyrolysis of plant and agricultural derived biomass having oxygen functional and aromatic groups [1]. Recently, biochar has been emerged as multipurpose product for pollutant remediation from aqueous solution and for soil amendment. The wide applicability of biochar is depends on its structural and chemical properties. Biochar is generally considered as a precursor of activated

**Corresponding author*: anujyadav2007@rediffmail.com

Shahid-ul-Islam (ed.) Advanced Materials for Wastewater Treatment, (313–330)

carbon [2]. Slow pyrolysis of biochar leads to its low cost in nature. The chemical modification and impregnation increases biochar efficiency as adsorbent. The low cost of biochar production may be a good alternative source of costly activated carbon, graphene and carbon nanotubes [3]. The feedstock of biochar materials are generally derived from plant, agricultural, forest and industrial waste. In addition, the use of waste material as biochar conversion not only solve waste disposal problem but other hand improve plant management and environmental protection [4].

Most of the studies have revealed that biochar has excellent ability to remove pollutant like heavy metals, dyes, fertilizer, organic waste from aqueous solution [5, 6]. The biochar is produced in solid material by pyrolysis (slow and fast), gasification and hydrothermal carbonization.

Heavy metals are introduced in the environment due to unsuitable treatment of industrial wastewater. These are highly toxic, non-biodegradable and pose threat to living organisms even at lower concentrations [7, 8]. The metals such as cadmium, lead, zinc, nickel, copper, mercury and chromium are considered as hazardous pollutants. The metal oxidant species like Cr(VI), Cr(III), Cu(II),Cd(II), Hg(II), Pb(II) and Zn(II) are present in industrial wastewater. These metal pollutant species are released by industries like leather and tanning, energy production, pigment and battery manufacturing, mining, pharmaceutical, metallurgical and electroplating. Industrial wastewater bearing metal species should be treated before release into environment. Conventional technologies like chemical precipitation, electrolysis, electrolytic recovery, ion exchange resins, reverse osmosis, flotation, coagulation, electrochemical are not capable in removing metal ion when it present in lower concentration moreover, the techniques are highly cost prohibitive [9, 10].

Biosorption technique successfully remediate heavy metals, dyes, pesticide, phenol, and other toxic pollutants from soil, sediment, surface and groundwater and industrial wastewater. The biochar also act as a potential precursor as soil fertilizer and also aids in sequestering atmospheric CO_2 concentration [11]. According to Ahmed *et al.* [12], the average price range of biochar was \$ 0.09 Kg^{-1} (Philippines) to \$8.85 Kg^{-1} (UK). Biochar can be prepared by pyrolysis, that is, thermal heating. Due to cost effective nature of biochar preparation there is need to focus on low cost plant derived biochar. Cellulosic and lignin rich woody biomass (e.g., wood, bark, shells, peels, straw, bagasse) derived biochar have high aromatic composition, low ash content, large surface area and highly porous in nature [13] which makes the biochar as a potential adsorbent. The current chapter focuses and summarizes recent investigation and applications of plant-based biochar in various type of pollutant removal.

9.2 Preparation Methods of Biochar

9.2.1 Pyrolysis

Pyrolysis is the most popular method of biochar preparation. It is the process by which organic matter (solid or liquid) undergo thermal decomposition in the absence of oxygen or in presence of little oxygen. It is chemical process where heat treatment generally breakdown larger molecule into smaller ones. The conversion process of biochar either slow or fast is depends on parameter like temperature, heating rate and vapor residence time [14, 15].

9.2.2 Slow Pyrolysis

This process leads to char production where biomass is heated slowly. The maximum char yield is obtained upto 30–35% by low biomass heating rate (<50 °C/min) with residence time for biomass upto minutes to several hours. The moderate pyrolysis occurs between temperature ranges from 300 to 700 °C.

9.2.3 Fast Pyrolysis

Fast pyrolysis done at different operating conditions (short residence time, <2 s, high heating rate, above 200 K/min). It results into low char yield but high yield of bio-oil [16].

9.2.4 Gasification

Gasification is process to convert biomass into gaseous mixture (viz., syngas, CO, CO_2, H_2, CH_4 and hydrocarbons) by treatment of oxidising agent under high temperature (>700 °C) [16]. Gasification agent (air, steam and oxygen) or mixture of gases used primarily. The operating conditions of gasification are biomass feed rate, type of oxidising agent, and temperature. In this process, syngas and ash are the main by-products. The advantage of gasification is the conversion of biomass source into liquid fuels.

9.2.5 Hydrothermal Carbonization

Hydrothermal carbonization (HTC) involves the conversion of biomass into biochar at moderate temperature 180–370 °C with high pressure (1.5–22 MPa) in water [17]. The reaction time is generally 1 min to several hours, but most of the material conversion performed upto 20 min reaction time [18]. The hydrochar is final product in HTC, has higher energy density equivalent to coal (15–30 MJ/Kg). Hydrochar can be utilized in carbon sequestration, wastewater treatment and soil improvement. The biochar yield, volatile matter production decreases at higher temperature but char carbon content increases [19].

9.3 Physico-chemical Characterization of Plant-Based Biochar

The application of any biochar depends on its physico-chemical properties which are governed by pyrolysis conditions and feedstock. The plant materials are used as biosorbent for remediation of toxic pollutants (Table 9.1). The main physico-chemical parameter are pH, ash content, moisture content, bulk density, elemental analysis, BET (Brunauer, Emmett and Teller),

Table 9.1 Plant material used for biochar production.

Agricultural products	Forest waste	Industrial plant waste	Weeds	Other
Corn straw	Pine wood	Sugarcane bagasse	Sea weed	Aquatic plants
Peanut shell	Pine needle	Paper waste	Lantana spp.	Salt marsh
Coir pith	Cactus fiber	Hemp fiber	*Onopordom heteracanthom*	*Miscanthus sacchariflorus*
Rice husk	Douglas fir	Coffee grounds	*Leersia hexandra*	Black cherry
Cotton seed hull	Willow chips	Sawdust	*Ageratum* spp	
Soyabean stover	Eucalyptus	Pomace	*Artemisia vulgaris*	
Wheat straw				
Cassawa waste				
Zea mays				

FTIR (Fourier Transform Infrared Spectroscopy), SEM (Scanning Electron Microscopy), EDX (Energy Dispersive X-ray). The physico-chemical characterization of various biochar materials are given in Table 9.2.

9.3.1 pH

pH plays important role in the uptake of pollutants from aqueous media. The anions and cations, dye, and metal uptake are depends on pH characteristics of biochar. Anions removal occurs rapidly with decrease in pH and cation uptake increases with increasing pH. At lower pH, H^+ ions compete with metals, which decreases metal uptake. At higher pH, the soluble hydroxyl formation takes place which decreases metal uptake [20]. The competitions of H_3O^+ ions are also depend on solution concentrations and type of metal ions. In aqueous media, exchange of anions and cations, and availability of active ites on adsorption is depends on solution pH. pH of biochar found alkaline (8.7–12) and hydrochar possess acidic (4.4–6.0) nature [21].

9.3.2 Ash Content

Ash content in biochar constitutes inorganic matter as residue (metals, dyes) and impurities after pyrolysis. Percentage of ash content in biochar is highly dependent on type of feedstock and pyrolysis temperature. Santos *et al.* [22] observed that biochar of pine needle has lower ash content (1.23–1.80%). Mary *et al.* [23] found ash percentage in biomass ranged from 3.50–18.86%.

9.3.3 Moisture Content

Biochar adsorption capacities are higly interrelated on moisture content. The higher level of moisture in biochar increases cost for transportation per unit of biochar. Santos *et al.* [22] prepared biochar by slow pyrolysis, they evaluated lower moisture content (0.9–1.8%). The moisture content in argan shells biochar was 0.97% [24].

9.3.4 Bulk Density

Bulk density is the mass of biochar occupied divided by its total volume. Bulk of biochar decreases with temperature, due to porous formation on surface [25]. Biochar employed to decrease the bulk density and increase moisture content in agricultural soil [26]. According to Rafiq *et al.* [27] temperature increases in pyropysis generally increase in carbon content

Table 9.2 Physicochemical characteristics of plant-derived biochar.

Biochar type	Pyrolysis temperature °C	Moisture content	C %	H%	N%	S%	O%	Ash%	BET surface area	Reference
Cotton seed hull	200		51.9	6.0	0.60	–	40.50	3.1		[31]
Oak bark	450		71.25	2.63	0.46	–	12.99	11.1		[32]
Soyabean straw	400		44.10	–	2.38	–	–	–		[33]
Bamboo (BBn)	–		65.17	0.05	0.96	0.88	–	9.1		[34]
Bagasse (BBg)	–		59.19	0.46	0.19	0.81	–	17.2		[34]
Rubber wood sawdust	700		43.27	6.83	0.39	–	–	–		[35]
Dew melon peel	450		83.2	3.12	0.2	–	13.47	14.23		[36]
Pine cone	500		67.88	3.89	0.55	<0.01	22.07	2.13		[38]
Orange peel	–	–	41.8	6.4	1.0	–	47.9	2.9	–	[38]
Residual wood	–	–	45.8	6.1	0.3	–	44.0	3.8	–	[38]
Corn stover	500	1.7	29.5	1.6	0.6	0.02	7.9	59.3	8.5	[39]
Corn stover	600	1.8	31.4	2.0	0.6	0.03	9.2	55.7	3.3	[39]
Corn stover	732	1.0	21.8	0.1	0.4	0.02	3.4	73.6	14.3	[39]
Switch grass	450	2.6	37.5	2.2	0.5	0.16	8.9	49.6	15.6	[39]

Switch grass	500	2.4	40.7	1.9	0.5	0.13	1.0	54.5	16.8	[39]
Switch grass	550	2.9	42.2	1.9	0.5	0.17	4.9	49.0	26.2	[39]
Red oak	500	2.2	62.0	2.7	0.6	0.02	9.8	23.4	3.8	[39]
Mixed hardwood	~400	3.6	79.2	2.4	0.5	0.01	11.0	6.7	8.1	[39]
Wood waste	~800	4.0	76.6	1.3	0.5	0.01	4.2	16.7	5.8	[39]
Rapeseed	–	–	52.25	8.06	3.91	–	35.78	5.73	–	[41]
Rapeseed	400	–	57.95	3.43	5.43	–	33.13	19.28 ± 0.48	–	[41]
Rapeseed	500	–	61.98	1.92	4.32	–	31.78	23.35 ± 0.55	–	[41]
Rapeseed	550	–	67.29	1.75	4.75	–	26.21	24.94 ± 0.64	–	[41]
Rapeseed	700	–	70.41	1.24	3.24	–	25.11	26.8 ± 0.72	–	[41]
Raw PC biochar	216	3.12	67.88	3.89	0.55	<0.01	22.07	2.13	6.60	[37]
Zn-loaded PC biochar	256	3.08	71.21	3.03	0.51	<0.01	20.43	2.14	11.54	[37]
Pea pod	–	7.0 ± 1.40	39.32	4.75	2.40	0.23	53.30	3.50 ± 1.40	–	[23]
Cauliflower	–	18.86 ± 1.30	31.80	3.20	4.01	1.59	59.40	18.86 ± 1.30	–	[23]
Orange peel	–	5.50 ± 0.70	40.43	4.83	1.56	0.27	52.90	5.50 ± 0.70	–	[23]

(45.5 to 64.5 %), vice versa, oxygen content decreases. Higher pyrolysis temperature leads to loss of oxygen content but increase in nitrogen content [28].

9.3.5 Elemental Analysis

Elemental analysis represents the elemental profile of biochar. C, H, N, S (O) are generally analyzed by elemental analyser.

9.3.6 BET (Brunauer, Emmett, and Teller)

The surface area of biochar was determined by using the BET method. Ahmad *et al.* [29] found that pyrolysis process increased the BET surface area of biochar upto 99.8%. The maximum surface area for MBC and BC were 198.359 and 164.731 m^2/g, respectively. High temperature treated biochar exhibit higher BET surface area, but not more than activated carbon [30].

9.3.7 SEM and EDX

SEM (Scanning Electron Microscopy) is effective tool to study the surface morphology of biochar while, EDX (Energy Dispersive X-Ray) characterize elemental composition of material. SEM-EDX studies help in the understanding of organic and inorganic chemistry of biochar. SEM studies give the detail of porous structure formation on biochar by the reduction of volatile and metal content in char particle.

9.3.8 FTIR

FTIR (Fourier Transform Infrared Spectroscopy) helps in the identification of the functional groups present on the surface of the biochar. Generally, FTIR spectra recorded at wavelength between 400 and 4000 cm^{-1}.

9.4 Biochar for Heavy Metal Removal

Karim *et al.* [4] used banana derived biochar for Cr(VI) removal in aqueous solutions. The banana biochar yield was obtained 66 and 40 % at 300 °C and 500 °C, respectively. They observed that maximum monolayer adsorption was 91.18 and 54.06 % (at pH-2, 50 ml/L initial concentration and 2 g/L dose) for BC-300 and BC-500, respectively. Gharbani- Lee *et al.* [43] compared heavy metal adsorption by peat and peat moss derived biochar. It was found that peat moss biochar showed highest uptake for metal like Pb(81.3 mg/g), Cd(39.8 mg/g) and Cu(18.2 mg/g), respectively. Peat moss biochar were prepared by different range of pyrolysis (Temperature, 400–1000 °C) for 30–90 min. The biochar prepared at 800 °C was most efficient for Cu and Pb removal. Komkiene and Baltrenaite [44] found that biochar produced by scots pine and silver birch efficiently reduces concentration of heavy metal contaminated water. The copper(II) removal was observed to be 128.7 μg/g on silver birch and for lead(II) on silver birch, and scot pines were varied from 1.29 to 3.77 μg/g and 2.37 to 4.49 μg/g, respectively. The zinc removal on scot pine was found to be 107.0 μg/g. Usman *et al.* [45] have taken date palm biochar for Cd(II) removal from aqueous solutions. They prepared date palm biochar at two temperature viz. 300 °C(BC-300) and 700 °C(BC-700) as sorbents. The highest Cd uptake capacity was found by BC-700 (43.58 mg/g) than BC-300 (26.96 mg/g) at pH 2. Shen *et al.* [46] performed removal of nickel metal ions from aqueous solution by four standard biochar. The biochar prepared at 500 and 700 °C from wheat straw pellets (WSP 500, WSP 700) and rice husk (RH 500, RH 700). The maximum uptake capacities were found in the order of WSP700>WSP500>RH700>RH550. *Astragalus membranaceus* redidue was utilized as magnetized biochar for uptake of chromium in batch experiments. The maximum Cr (VI) removal was found to be 23.85 ± 0.23 mg/g at pH 2 [46]. *Canna indicia, Pennisetum purpureum* schum derived biochar were investigated for the sorption of Cd(II) ions. The wetland derived plant biomasses were heated upto 500 °C for biochar preparation. Similar metal removal efficiencies were noticed by *C. Indica* (125.8 mg/g) and *P. Purpurreum* Schum (119.3 mg/g) [47]. The applications of plant-based biochar are given in Figure 9.1.

9.5 Biochar for Dye Removal

Mi *et al.* [48] utilized cetyltrimethylammonium bromide (CTAB) modified corn stalk biochar for adsorption of orange (II) in aqueous solutions. They found that the removal efficiency of anionic dye orange (II) uptake increased from 46.9 % of the raw corn stalk biochar to 99.7 % of the CTAB modified cornstalk biochar. The maximum monolayer adsorption capacity was found to be 26.9 mg/g by Langmuir model at 298 K. Leng *et al.* [49] prepared biochar from rice husk by using different solvents. The biochar produce with water (WBC) or water/ethanol (WEBC) and ethanol as solvent (EBC) was found rich in functional groups. EBC adsorption capacities were found higher (32.5–67.6 mg/g) than WBC and WEBC. It was observed that electrostatic attraction was major force for dye uptake.

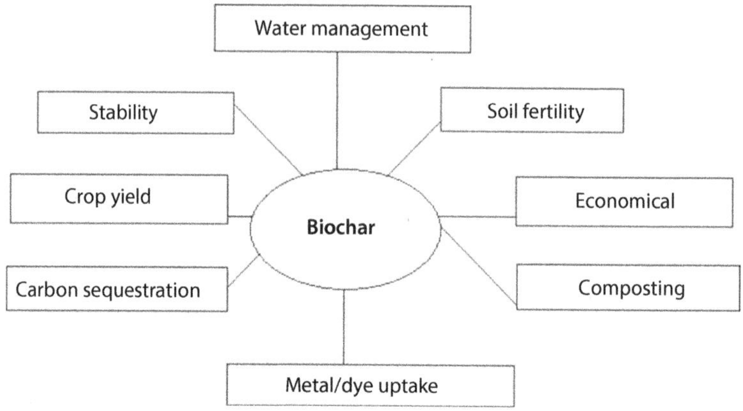

Figure 9.1 Schematic overview of plant-based biochar applications.

The peanut shell KOH activated biochar was explored in the removal of methylene blue. The impregnation ratio of biochar was 1.5:1, at 800 °C of activation temperature and 90 min of the holding time in activation. The monolayer adsorption capacity of peanut shell was 208 mg/g [50]. A stable cornstalk biochar modified with cetyltrimethylammonium bromide (CTAB) for removal of orange (II) as a target organic pollutant [48]. Sewu *et al.* [51] prepared biochars from rice straw, wood chip and Korean cabbage for cationic dye removal. Congo red (CR) and crystal violet (CV) were used as sorbate. They observed that Korean cabbage biochar has 4.8 times more adsorption capacity than activated carbon. Langmuir adsorption capacity of Korean cabbage, rice straw and wood chips were 1304, 620.3 and 195.6 mg/g, respectively. Biochar from sugarcane bagasse was tested for adsorptive removal of 2,4,6-trichlorophenol. The biochar exhibits 361.77 m^2/g methylene blue surface area and 50.47 % fixed carbon. Further, SEM images confirm the cylindrical, porous structure formation on surface of biochar. The maximum uptake capacity, q_m, was found to be 253.38 mg/g [52].

9.6 Biochar for Fluoride Removal

Fluoride is present in groundwater naturally and it is essential element for human health, play key role in development of bones and dental enamel [53]. Excessive fluoride consumption causes adverse effects like bone and dental fluorisis, lesions of the endocrine glands, thyroid and liver toxicity [54]. Raw and activated biochar prepared from *Colcocasia esculenta* stem was assessed for fluoride uptake by Mukherjee and Halder [55].

The fluoride uptakes were shown by raw biomass and biochar were 33.00 and 72.82 %, respectively at laboratory conditions: pH 4.25, temperature 63.75 °C, adsorbent dosage of 20 g/L and contact time of 180.75 minutes. Mohan *et al.* [56] studied the effect of slow pyrolysis for the removal of fluoride from groundwater by using magnetic and non-magnetic corn stover biochar. Maximum fluoride removal occurs at pH 2. The raw corn biomass biochar showed the adsorption capacity 6.42, 5.17 and 5.00 mg/g at 25, 35, 45 °C, respectively, while magnetic biochar adsorption capacities were 4.11, 3.45 and 3.41 at 25 °C temperature.

9.7 Biochar for Persistent Organic Pollutant Removal

Persistent organic pollutants (POPs) are based chlorinated toxic organic compounds which are harmful to the environment due to persistant in nature. These are generally low water solubility, hydrophobic, lipophilic chemicals. POPs substances are also called "Dirty dozen". It includes 12 POPs substances (aldrin, endrin, chlordane, DDT, HCB, PCBs, dieldrin, heptachlore, mixer, toxaphene, polychlorinated dibenzo-p-dioxins, polychlorinated dibenzofurans). The application of POPs in current world is in pesticide and pharmaceutical industries.

Uncontrolled or excessive use of pesticide can pose threat to the ecosystem. The pesticide sorption technique is a efficient technique that applied in the treatment of soil, sediment, surface and groundwater [57]. Various biochar material hav been tested for pesticide sorption by many authors [58, 59].

9.8 Biochar for Other Pollutant Removal

Liu *et al.* [60] applied biochar for removal of tetracycline from aqueous solution. In the study, they found that the alkali treated biochar showed maximum removal performance (58.8 mg/g) than other biochar. Hafshejani *et al.* [61] prepared biochar from sugarcane bagasse and tested for nitrate removal from aqueous solution. The biochar prepared at 300 °C temperature having higher stable organic matter. The maximum nitrate uptake was 28.21 mg/g at pH 4.64, 60 min contact time and with an adsorbent dose of 2 g/L. In the adsorption study, they also found that carbonate ions have shown maximum and chloride ions shown minimum influence on nitrate removal. Essandoh *et al.* [62] used pine wood fast pyrolysis biochar for removal of salicylic acid and ibuprofen from aqueous solutions. The biochar was prepared by incorporated Mg-Fe layered double hydroxide

(MgFe-LDH) particle into wheat straw biochar in aquoues phase. An X-ray diffraction pattern confirms the LDH particle deposition on straw biochar surface. The found the maximum nitrate removal capacity of 24.8 mg/g at doses of 5–10 g/L. Vithanage *et al.* [63] investigated the potential of biochar produced from tea waste and rice husks for removal of carbofuron in batch studies. The rice husk and tea waste into biochar at 700 °C. The experimental study was performed at 25, 35 and 45 °C, 5–100 mg/L, dose 1 g/L at pH 5.0. The removal capacities of rice husk biochar (RHBC) and tea waste biochar (TWBC) were 25.2 and 10.2 mg/g by pseudo-second order kinetic model, respectively.

9.9 Biochar for Soil Treatment/Improvement

Biochar is excellent tool for soil amendment, it can enhance soil structure, reduce methane emission from soil and nitrous oxide, decreases farm chemicals leaching into aquifers [64], For soil enhancement, compost prepared biochar have advantages over wood derived biochar [64]. Biochar also helps in the improvement of soil biological and microbial community. Ultimately, the enhance soil properties increases the nutrition, water retention capacity and crop productivity [65].

Bamminger *et al.* [66] studied the biochar application for the carbon sequestration and mitigation of greenhouse gas (GHGs) emissions. They performed incubation experiment with an arable soil with and without N-rich addition. After 37 days, they found that biochar significant uptake CO_2 (~42%), as well as NH_4^+N and NO_3^- concentrations. Sun *et al.* [65] conducted experiments to study the effects of biochar on *Saeda salsa* growth and properties of saline soil in Yellow river delta. They used three types of biochar viz. wheat straw biochar(WS), corn stalk biochar (CS), and peanut shell biochar (PS) with application rates (5, 10, 20 g/Kg). They observed that *S. salsa* growth rate increased from 11.7 to 115% under WS application at a application range of 5–10 g/Kg. The biochar effects on *S. salsa* growth varied in the order of PS > WS > CS.

9.10 Conclusion

Biochar has been emerged as precursor of activated carbon. The biochar can be widely applied in removal of various types of pollutants such as heavy metals, dyes, POPs, and also for soil amendments. The feasibility and applicability of biochar also depends on pyrolysis temperature, retention time, type of feedstock and pollutant concentration. Physico-chemical characterizations revealed that biochar possess good characteristics and properties close to that of activated carbon. In soil improvement and CO_2 sequestration, biochar exhibit good potential. The chemical modification like treatment with acid, alkali and impregnations can improve its efficiency as adsorbents. Further, pilot scale researches are required to enhance its wide applications and to reduce its negative impacts on environment.

Acknowledgments

One of author(s) (Anuj Kumar Yadav) acknowledges the financial support of University Grant Commission, New Delhi, India for UGC-BSR Fellowship.

References

1. Tan, X., Liu, Y., Zeng, G., Wang, X., Hu, X., Gu, Y., Yang, Z. Application of biochar for the removal of pollutants from aqueous solutions. *Chemosphere*, 125, 70–85, 2015.
2. Azargohar, R., Dolai, A.K., Biochar as a precursor of activated carbon. *Appl. Biochem. Biotechnol.*, 129–132, 762–773, 2006.
3. Ahmed, M.B., Zhou, J.L., Ngo, N.H., Guo, W., Chen, M., Progress in the preparation and application of modified biochar for improved contaminant removal from water and wastewater. *Bioresour. Technol.* PMID 27241534, DOI: 10.1016/j.biortech.2016.05.057, 2016.
4. Dong, X., Ma, L.Q., Zhu, Y., Li, Y., Gu, B., Mechanistic investigation of mercury sorption by Brazilian pepper biochar of different pyrolytic temperatures based on X-ray photoectron spectroscopy and flow calorimetry. *Environ. Sci. Technol.*, 47, 12156–12164, 2013.
5. Xue, Y., Gao, B., Yao, Y., Inyang, M., Zhang, M., Zimmerman, A., Rao, K.S., Hydrogen peroxide modification enhances the ability of biochar (hydrochar) produced from hydrothermal carbonization of peanut hull to remove aqueous heavy metals: batch and column tests. *Chem. Eng. J.*, 200, 673–680, 2012.
6. Zhang, M., Gao, B., Yao, Y., Xue, Y., Inyang, M., Synthesis of porous MgO biochar nanocomposities for removal of phosphate and nitrate from aqueous solutions. *Chem. Eng. J.*, 210, 26–32, 2012.
7. Kalavathy, H., Karthik, B., Miranda, L.R., Removal and recovery of Ni and Zn from aqueous solution using activated carbon from *Havea brasiliensis*: Batch and column studies. *Colloid Surf. B: Biointerf.*, 78, 291–302, 2010.
8. Mohan, S., Sreelakshmi, G., Fixed bed column study for heavy metal removal using phosphate treated rice husk. *J. Hazard Mater.*, 153, 75–82, 2008.

9. Soares, E.V., Soares, H.M.V.M., Cleanup of industrial effluents containing heavy metals: a new opportunity of volarising the biomass produced by brewing industry. *Appl. Microbiol. Biotechnol.*, 97(15), 6667–6675, 2013.
10. Malik, D.S., Jain, C.K., Yadav, A.K., Removal of heavy metals from emerging cellulosic low-cost adsorbents: a review. *Appl. Water. Sci.*, DOI: 10.1007/s13201-016-0401-8, 2016.
11. Kwapinsk, W., Byrne, C.M.P., Eryachko, E., Wolfram, P., Adly, C., Leahy, J.J., Novotny, E.H., Hayes, M.H.B., Biochar from biomass and waste. *Waste Biomass Volar.*, 1, 177–189, 2010.
12. Ahmed, M.B., Zhou, J.L., Ngo, N.H., Guo, W., Adsorptive removal of antibiotics from water and wastewater: progress and challenges. *Sci. Total Environ.* 532, 112–116, 2015.
13. Yavari, S., Malakahmad, A., Sopri, N.B., Biochar efficiency in pesticide sorption as a function of production variables-a review. *Environ. Sci. Pollut. Res.*, 22, 13824–13841, 2015.
14. Mohan, D., Pittman, Jr. C.U., Steele, P.H., Pyrolysis of wood/biomass for bio-oil: a critical review. *Energy Fuels,* 20(3), 848–889, 2006.
15. Czernik,S., Bridgewater, A.V., Overview of applications of biomass. *Fast Pyrolysis Oil,* 18(2), 590–598, 2004.
16. Qian, K., Kumar, A., Zhang, H., Bellmer, D., Huhnke, R., Recents advances in utilization of biochar. *Renew. Sust. Energy Rev.*, 42, 1055–1064, 2015.
17. Kim, D., Lee, K., Bae, D., Park, K.Y., Characterization of biochar from hydrothermal characterization of exhausted coffee residue. *J. Mater. Cycles. Waste. Manag.*, DOI: 10.1007/s10163-016-0572-2, 2016.
18. Lynam, J.G., Reza, M.T., Yan, W., Vasquez, V.R., Coronella, C.J., Hydrothermal carbonization of various lignocellulosic biomass. *Biomass Conv. Bioref.*, 5, 173–181, 2015.
19. Kang, S., Li, X., Fan, J., Chang, J., Characterization of hydrochars produced by hydrothermal carbonization of lignin, cellulose, D-xylose and wood meal. *Ind. Eng. Chem. Res.*, 51, 9023–9031, 2012.
20. Tashauoei, H.R., Hashemis, Ardani, R., Yavari, Z., Asadi-Ghalhari, M., Adsorption of lead from aqueous solution by modified beech sawdust. *J. S. of Environ. Health. Res.*, 1(1), 11–16, 2016.
21. Bargmann, I., Rilling, M.C., Buss, W., Kruse, A., Kuecke, M., Hydrochar and biochar effects on germination of spring barley. *J. Agron. Crop Sci.*, 199, 360–373, 2013.
22. Santos, L.B., Striebeck, M.V., Crespi, M.S., Ribeiro, C.A., Julio, M.D., Charcaterization of biochar pine pellet. *J. Therm. Anal. Calorim.*, 122, 21–32, 2015.
23. Mary, G.S., Sugumaran, P., Niveditha, S., Ramalakshmi, B., Ravichandran P, Seshadri S, Production characterization and evaluation of biochar from pod (*Pisum sativam*) leaf (*Brassica oleracea*), peel (*Citrus sinensis*) waste. *Int. J. Recycl. Org. Waste Agricult.* 5, 43–53, 2016.

24. Bouqbis, L., Daoud, S., Koyro, H.W., Kammann, C.I., Ainlhout, L.F.Z., Harrouni, M.C., Biochar from argan shells: production and characteriation. *Int. J. Recycl. Org. Waste Agricult.*, 5, 43–53, 2016.
25. Belcher, C.M., eds., Fire phenomenon and the earth system: An interdisciplinary guide to fire science. Wiley-Blackwell, 2013.
26. Al-Wabel, M., Usman, A.R.A., El-Naggar, Ah., Aly, A.A., Ibrahim, H.H., Elmaghraby, S., Al-Omran, A., *Conocarpus* biochar as a soil amendmend for heavy metal availability and uptake by maize plants. *Saudi. J. Biol. Sci.*, 22(4), 503–511, 2015.
27. Rafiq, M.K., Bachmann, R.T., Rafiq, M.T., Shag, Z., Josheph, S., Lang, R., Influence of pyrolysis temperature on physic-chemical properties of corn stover (*Zea mays* L.), Biochar and feasibility for carbon capture and Energy balance. *Plos one,* 11(6), 1–17, 2016.
28. Yakout, S.M., Physico-chemical characterization of biochar produced from rice straw at different pyrolysis temperature for soil amendmend and removal of organics. *Proc. Natl. Acad, Sci.India Sec(a). Phys Sci.*, DOI: 10.1007/s40010-017-0343-z, 2017.
29. Ahmad, M., Ahmad, Usman A.R.F., Al-Faraj, As., OK, A.A.Y.S., Alwabel, M.I., Date palm waste derived biochar composites with silica and zeolites, sysntheis, characterization and implication for carbon stability and recalcitrant potential. *Environ Geochem Health*, DOI: 10.1007/s10653-017-9947-0, 2017.
30. Jiang, S., Nguyen, T.A.H., Huang, L., Characterization of hard and softwood, biochar pyrolyzed at high temperature. *Environ. Geochem. Health.*, 39, 403–415, 2017.
31. Uchimiua, M., Bannon, D.I., Wartelle, L.H., Lima, I.M., Klasson, K.T., Lead retention by broiler litter biochar in small arms range soil: impact of pyrolysis temperature. *J. Agric, Food Chem.*, 60, 5035–5044, 2012.
32. Mohan, D., Rajput, S., Singh, V.K., Steele, P.H., Pittman, Jr. C.U., Modelling and evaluation of chromium remediation from water using low cost biochar, a green adsorbent. *J. Hazard. Mat.*, 188, 319–333, 2011.
33. Tong, S.J., Li, L.Y.,Yuan, J.H., Xu, R.K., Adsorption of Cu(II) by biochars generated from three crop straws. *Chem. Eng. J.*, 172, 828–834, 2011.
34. Ramola, S., Mishra, T., Rana, G., Srivastava, R.K., Characterization and pollutant removal efficiency of biochar derived from bagasse, bamboo and tyre. *Environ. Monit. Assess.*, 186, 9023–9039, 2014.
35. Shaaban, A., Se, S.-M., Mitan, N.M.M., Dimin, M.F., Characterization of biochar derived from rubber wood sawdust through slow pyrolysis on surface porosities and functional groups. *Procedia Eng.*, 68, 365–371, 2015.
36. Ahmadi, M., Kouhgardi, E., Ramavandi, B., Physico-chemical study of dew melon peel biochar for chromium attenuation from silumated and actual wastewaters. *Korean J. Chem. Eng.*, 33(9), 2589–2601, 2016.
37. Vinn, N.V., Zafar, M., Behera, S.K., Park, H.-S., Arsenic(III) removal from aqueous solution by raw and zinc loaded pine cone biochar: equilibrium,

kinetics, and thermodynamic studies. *Int. J. Environ. Sci. Technol.*, 12, 1283-1294, 2015.

38. OH, T.K., Chois, B.S., Shinogi, Y., Chikushi, J., Characterization of biochar derived from three types of biomas. *J. Fac. Agric. Kyushu. Univ.* 57(1), 61–66, 2012.
39. Brewer, C.E., Unger, R., Schmidt-Rohr, K., Brown, R.C., Criteria to select biochar for field studies based on biochar. *Bioenergy Res.*, 4, 312–323, 2011.
40. Sensöz, S., Angin, D., Yorgun, S., Influence of positive size on the pyrolysis of rapeseed (*Brassica nopus* L.): fuel-properties of bio-oil. *Biomass Bioenerg.*, 19(4), 271–279, 2000.
41. Angin, D., Sensoz, S., Effects of pyrolysis temperature on chemical and surface properties of rapeseed (*Brassica napus* L.). *Int. J. Phytoremediation,* 16, 7–8, 684–693, 2014.
42. Karim, A.A., Kumar, M., Mohapatra, S., Panda, C.R., Singh, A., Banana peduncle biochar: Characteristics and adsorption of hexavalent chromium from aqueous solution. *Int. Res. J. Pure. Appl. Chem.*, 7(1), 1–10, 2015.
43. Gharbani-Khosrowshahi, S., Behnajady, M.A., Chromium(VI) adsorption from aqueous solution by prepared biochar from *Onpordom heterocanthum*. *Int. J. Environ. Sci. Technol.*, 13, 1803–1814, 2016.
44. Komkiene, J., Baltrenaite, E., Biochar as adsorbent for removal of heavy metal ions cadmium(II), copper(II), lead(II), zinc(II). from aqueous phase. *Int. J. Environ. Sci. Technol.*, 13, 471–482, 2016.
45. Usman, A., Sallan, A., Vithanage, M., Ahmad, M., Al-Farraj, A., OK, Y.S., Abduljabbar, A., Al-Wabel, M.,Sorption process of date palm biochar for aqueous Cd(II) removal: efficiency and mechanisms. *Water Air Soil Pollut.* 227, 449, 2016.
46. Shen, Z., Zhang, Y., McMillan, O., Jin, F., Al-Tabbaa, A., Characterization and mechanism of nickel adsorption on biochar produced from wheat straw pellets and rice husk. *Environ Sci. Pollut. Res.* DOI. 10.1007/s11356-017-8847-8, 2017.
47. Cui, X., Hao, H., Zhang, C., He, Z., Yang, X., Capacity and mechanism of ammonium and cadmium sorption on different wetland plant derived biochar. *Sci. Total Environ.*, 539, 566–575, 2016.
48. Mi, X., Li, G., Zhu, W., Liu, L., Enhanced adsorption of orange(II) using cationic surfactant modified biochar pyrolyzed from cornstalk. *J. Chem.*, 1–7, 2016.
49. Leng, L., Yuan, X., Zeng, G., Shao, J., Chen, X., Wu, Z., Wang, H., Peng, X., Surface characterization of rice husks biochar produced by liquefaction and application for cationic dye (Malachite green) adsorption. *Fuel*, 155(1), 77–85, 2015.
50. Han, X., Chu, L., Liu, S., Chen, T., Ding, C., Yan, J., Cui, L., Quan, G., Removal of methylene blue from aqueous solution using porous biochar obtained by KOH activated of peanut shell biochar. *Bioresources*, 10(2), 2836–2849, 2015.

51. Sewu, D.D., Boakye, P., Woo, S.H., Highly efficient adsorption of cationic dye by biochar produced with Korean cabbage waste. *Bioresour. Technol.*, 224, 206–213, 2017.
52. Mubarik, S., Saeed, A., Athar. M.M., Iqbal, M., Characterization and mechanism of the adsorptive removal of 2,4,6-trichlorophenol by biochar prepared from sugarcane baggase. *J. Ind. Eng. Chem.*, 33, 115–121, 2016.
53. Rao, C.R.N., Karthikeya, J., Removal of fluoride from water by adsorption onto lanthanum oxide. *Water, Air Soil Pollut.*, 231, 1101, 2011.
54. Tor, A., Dangagolu, N., Arslan, G., Cengeloglu, Y., Removal of fluoride from water bu using granular mud: Batch and column studies. *J. Hazard. Mater.*, 164, 271–278, 2009.
55. Mukherjee, S., Halder, G., Assessment of fluoride uptake performance of raw biomass and activated biochar of *Colocasia esculnta* stem: optimization through response surface methodology. *Environ. Prog. Sust. Energy*, 35, 1305–1316, 2016.
56. Mohan, D., Kumar, S., Srivastava A., Fluoride removal from ground water using magnetic and nonmagnetic corn stover biochars. *Ecol. Eng.*, 798–808, 2014, 2014.
57. Yavari S., Malakahmad A., Sapari NB., Biochar efficiency in pesticides sorption as a functionof production variables—a review. *Environ. Sci. Pollut. Res.*, 22, 13824–13841, 2015.
58. Clay, S.A., Malo, D.D., The influence of biochar production on herbicide sorption chara teristics. In: Hasaneen, MN, editor, *Herbicides properties, synthesis, and control of weeds*. In Tech, New York, pp 59–74, 2012.
59. Sun, K., Gao, B., Ro, K.S., Novak, J.M., Wang, Z., Herbert, S., Xing, B., Assessment of herbicide sorption by biochars as a function of molecular structure. *Bioresour Technol.*, 102, 9897–9903, 2012.
60. Liu, P., Liu, W.-J., Jiang, H., Chen, J-.J., Li, W.-W., Yu, H.-Q., Modification of biochar derived from fast pyrolysis of biomass and its application in removal of tetracycline from aqueous solution. *Bioresour Technol.*, 121, 235–240, 2012.
61. Hafshejani, L.D., Hooshmand, A., Naseri, A.A., Mohammadi, A.S., Abbasi, F., Bhatnagar, A., Removal of nitrate from aqueous solution by modified sugarcane bagasse biochar. *Ecol. Eng.*, 95, 101–111, 2016.
62. Essandoh, M., Kunwar, B., Pittman Jr, C., Mohan, D., Mlsna, T., Sorptive removal of salicylic acid and ibuprofen from aqueous solutions using pine wood fast pyrolysis biochar, *Chem. Eng. J.*, 265, 219–227, 2015.
63. Vithanage, M., Mayankaduwa, S.S., Herath, I., OK, Y.S., Mohan, D., Kinetic, thermodynamics and mechanistic studies of carbofuran removal using biochar from tea waste and rice husks. *Chemosphere*, 150, 781–789, 2016.
64. Shareef, T.M.E., Zhao, B., Review paper: the fundamentals of biochar as a soil ammendmend tool and management in agriculture scope: an overview for farmer and gardners. *J. Agric. Chem. Environ.*, 6, 38–36, 2017.

65. Ding, Y., Liu, Y., Liu, S., Li, Z., Tan, X., Huang, X., Zeng, G., Zhou, L., Zheng, B., Biochar to improve soil fertility: a review. *Agron. Sustain. Dev.*, 36, 36, 2016.
66. Bamminger, C., Zaiser, N., Zinsser, P., Lammers, M., Kammann, C., Marhan, S., Effects of biochar, earthworms, and litter addition on soil microbial activity and abundance in a temperate agricultural soil. *Biol. Fertil. Soil*, 50, 1189–1200. John Wiley & Sons, Oxford, 2014.
67. Sun, J., He, F., Shao, H., Zhang, Z., Xu, G., Effects of biochar application on *Saeda salsa* growth and saline soil properties. *Environ. Earth Sci.*, 75, 630, 2016.

10

A Review on Ferrate(VI) and Photocatalysis as Oxidation Processes for the Removal of Organic Pollutants in Water and Wastewater

Kyriakos Manoli[§], Malini Ghosh[§], George Nakhla and Ajay K. Ray*

Department of Chemical and Biochemical Engineering, University of Western Ontario, London, Canada

Abstract

Nowadays, there is an increasing concern related to the pollution of surface waters (e.g., rivers and lakes) with organic molecules such as pharmaceuticals and personal care products. Among the various methods that are being investigated, oxidation processes are receiving a great attention because of their ability to degrade organics in water. This chapter deals with two promising oxidation technologies: ferrate(VI) and photocatalysis. A review on the research progress related to the oxidation of organics by ferrate(VI) (FeO_4^{2-}, Fe(VI)) is presented, including synthesis and characterization of Fe(VI), stoichiometry of the reactions (Fe(VI) to organic molar ratios), kinetic studies (second-order reaction rate constants) and pH dependency. Application and performance of Fe(VI) in wastewater treatment is also reviewed. The second part of the chapter emphasizes on the relevant recent developments in photocatalytic degradation of water pollutants. The latest advances not only include the different innovative methods used for photocatalytic reactions but also the different materials and their modifications used as photocatalysts. The combination of Fe(VI) and photocatalysis to degrade organic contaminants is briefly reviewed and discussed. Future prospects and research directions are discussed for both oxidation processes.

Keywords: Oxidation, ferrate(VI), kinetics, stoichiometry, photocatalysis, semiconductors, dye sensitization, organic pollutants

**Corresponding author*: aray@eng.uwo.ca;
[§]*K.M. and M.G. contributed equally to this work.*

Shahid-ul-Islam (ed.) Advanced Materials for Wastewater Treatment, (331–390)

10.1 Introduction

The increasing disparity between the demand and availability of clean water is becoming a pressing issue across the globe. Population growth and industrialization have profoundly intensified water demand through time. There are around 4 billion people worldwide who are suffering from little or no clean water supply while millions of people die every year from severe water borne diseases [1]. Thus, there is an urgent need to develop sustainable wastewater treatment technologies to not only meet the current water needs but also to maintain clean water resources for the next generations.

Nowadays, there is an increasing discharge of "new" organic compounds, also known as emerging pollutants, in the water and wastewater streams, such as pharmaceuticals and personal care products (PPCPs) and endocrine disrupting compounds (EDCs). These PPCPs and EDCs have been detected in influents and effluents of wastewater treatment plants (WWTP) of many countries worldwide, ending up in surface waters (i.e., rivers, lakes and oceans), resulting in negative environmental and socio-economic impacts [2–6]. Thus, there is an increasing interest in developing technologies to be used as alternative or addition to conventional biological wastewater treatment processes to enhance the removal of emerging organic contaminants.

Among the various treatment technologies that are being studied, oxidation processes are receiving a great attention because of their ability to degrade organics in water. The last decades, chemical oxidants (e.g., ferrate(VI), chlorine, permanganate, etc.) and advanced oxidation processes (e.g., photocatalysis, UV/H_2O_2, UV/O_3, UV/H_2O_2/O_3, H_2O_2/Fe^{3+}, H_2O_2/Fe^{2+} (UV/H_2O_2/Fe^{3+}, UV/H_2O_2/Fe^{2+}, etc.) have been investigated for the degradation of organics in water [7–19]. This chapter deals with ferrate(VI) (FeO_4^{2-}, Fe(VI)) and photocatalysis as environmentally-friendly oxidation processes for the removal of emerging organic pollutants from water.

Several chemical oxidants have been studied for water and wastewater treatment applications because of their oxidation and disinfection properties. The presence of emerging organic pollutants in the secondary effluent of WWTP, increases the discussions on upgrading the WWTP with additional treatment steps such as absorption and oxidation treatment technologies [7], to reduce the discharge of emerging pollutants to the surface waters. Selective chemical oxidants such as chlorine, chlorine dioxide and ozone, have been studied for disinfection of drinking water and oxidation of micropollutants [7]. Chlorine is the most commonly used disinfectant worldwide [9]. Chlorine dioxide was suggested to reduce the formation of chlorinated by-products during the disinfection of wastewater by chlorine [7]. However, the accumulation of chlorite and chlorate as oxidation by-products of chlorine dioxide, resulted in regulations to minimize the dose of the oxidant that is applied for wastewater treatment [11]. Ozone, which also reacts with the same moieties (electron-rich) as other selective oxidants, has been also used to oxidize contaminants in water [15]. Both chlorine and ozone react with bromide to form carcinogenic bromate ion and brominated by products [11]. Fe(VI), an iron-based chemical, acts as oxidant, disinfectant and coagulant (Fe(VI) is reduced to Fe(III) initiating coagulation), without the formation of brominated and chlorinated by-products.

As already mentioned, many studies have been carried out on advanced oxidation processes (AOPs) for degradation of emerging organic pollutants. Advanced oxidation processes can be largely defined as the oxidation process where a highly reactive, non-selective radical is being used as the oxidants/reacting species [18]. For the water purification processes the radical which has congregated the greatest attention for AOPs is the hydroxyl (OH·) radical [18]. In most of the advanced oxidation processes the hydroxyl radical is being produced in-situ and has a strong non-selective oxidizing ability with standard potential of 2.80 V against standard hydrogen electrode [13]. Such AOPs involve the utilization of individual or combination of UV light, hydrogen peroxide, ozone, sunlight, Fe^{3+}, Fe^{2+}, etc. [13, 18]. Another group of compounds which are being significantly reviewed for their competency of producing hydroxyl radical in the presence of light energy, is the semiconducting materials. The efficacy of the semiconducting materials for advanced oxidation processes have been intricately conferred in the later sections of this chapter.

The individual use of chlorine and UV light is well known in water treatment processes for their disinfection properties. However, when used together, they are considered as an effectual means of AOP for treating organic hazardous materials [10]. It has been reported that in UV/chlorine system both hydroxyl and chlorine radicals are being formed as the reacting species [10]. Another example of AOP is Fenton or Fe^{II}/H_2O_2 [13]. When H_2O_2 reacts in the outer surface of Fe^{II}, it forms hydroxyl radical [12]. Whereas, the reaction between Fe^{III} and H_2O_2 produces hydroperoxyl radical (HO_2) along with the hydroxyl radical [12]. A more recent approach to advance oxidation processes is the use of metals or metalloids along with UV light. However, the efficiency of such processes to produce hydroxyl radical largely depend on the selected metals/metalloids and their ability to absorb light energy [18]. Several metal nanoparticles are being studied nowadays for their use in AOPs, for example, copper nanoparticles, selenium nanoparticles, etc. [18].

Figure 10.1 shows the road map of the chapter. In the first part of this chapter, the methods used to synthesize Fe(VI) salts as well as analytical techniques used to characterize solid and liquid Fe(VI) are reviewed. Also, the methodology usually used for the kinetic investigation of oxidation of organics by Fe(VI) is discussed. Moreover, a comprehensive literature review on the kinetics (second-order reaction rate constants) of oxidation of PPCPs and EDCs by Fe(VI) and stoichiometry of the Fe(VI)-organic reactions (Fe(VI) to organic molar ratio) as a function of pH is presented. Results of studies related to the application and performance of Fe(VI) in wastewater treatment are also presented.

The second part of this chapter encompasses another water purification technology: Photocatalysis. The detailed mechanism and the efficiency of photocatalysis have been discussed in this section. Besides, the principle behind a photocatalytic reactor and the comparisons between different reactors have also been explained. Finally, the different materials used as photocatalysts have been elucidated. Several new modifications of photocatalysts have also been included along with their several advantages and disadvantages.

The combination of Fe(VI) and photocatalysis to degrade organic contaminants is also briefly reviewed and discussed. Future prospective,

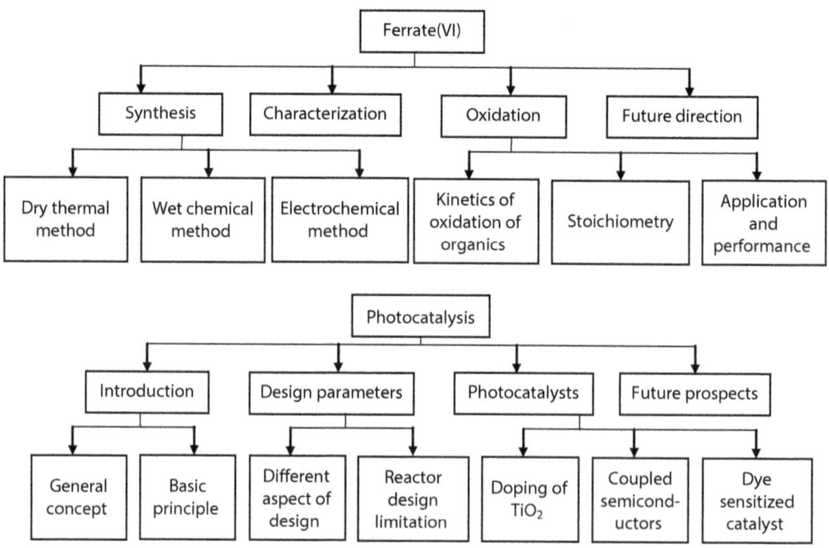

Figure 10.1 Road map of the chapter.

research directions and required improvements are discussed for both oxidation processes, that is, Fe(VI) and photocatalysis.

10.2 Ferrate(VI)

10.2.1 Introduction

Iron is one of the most abundant metals on earth, offering valence states of from 0 to VI, thus playing an important role in a wide range of areas from physiology to (bio)chemical and industrial processes [14]. In the past few decades, there has been an increasing interest in the tetraoxy high-valent iron(VI), known as ferrate(VI) (Fe(VI), FeO_4^{2-}), as a potential green molecule for organic synthesis, iron batteries, disinfection of viruses and bacteria, and water and wastewater treatment [14, 20–25]. Fe(VI) is a powerful oxidant, leaving Fe(III) after its application initiating a coagulation process. As a result, Fe(VI) is a multipurpose water treatment chemical [23], acting as an oxidant, disinfectant and coagulant with a single dose and mixing unit process. The redox potentials of Fe(VI) compared to common oxidants used in water and wastewater treatment is given in Table 10.1. Fe(VI) has the highest redox potential of +2.20 V under acidic conditions (Table 10.1).

Fe(VI) is unstable in aqueous solutions. In other words, Fe(VI) reacts with water and molecular oxygen is formed as a product of this reaction (Eqn. 10.1) [23]. It was recently shown that hydrogen peroxide (H_2O_2) is also formed from the self-decay of Fe(VI) in phosphate buffered solution [30]. The rate of Fe(VI) self-decomposition depends on the pH, temperature, initial Fe(VI) concentration, water constituents and the physical properties of the Fe(III) oxides/hydroxides formed during the Fe(VI) degradation [14]. The self-decomposition of Fe(VI) should be considered for kinetic investigation of Fe(VI) oxidation reactions [14], to measure true kinetic rate constants.

$$2Fe^{VI}O_4^{2-} + 5H_2O \rightarrow 2Fe^{III} + 1.5O_2 + 10OH^- \qquad (10.1)$$

Several reactions may occur during the oxidation of an organic compound (X) by Fe(VI) [23, 31]: i) 1-e^- and/or 2-e^- transfer to form Fe(V) and/or Fe(IV), ii) reaction of Fe(V) and/or Fe(IV) with X, iii) Fe(VI), Fe(V) and Fe(IV) self-decompositions, and iv) reaction of Fe(II)/Fe(III)/Fe(IV)/Fe(V) with oxygen species, that is, H_2O_2 formed from self-decompositions. It is known that Fe(V) is 2–4 orders of magnitude more reactive than Fe(VI) probably because of its partial free radical character [27].

Table 10.1 Redox potentials for oxidants/disinfectants used in water and wastewater treatment [26–29].

Oxidant/disinfectant	Reaction	E°, V
Chlorine	$Cl_2(g) + 2e^- \Leftrightarrow 2Cl^-$	1.36
Hypochlorite	$HClO + H^+ + 2e^- \Leftrightarrow Cl^- + H_2O$	1.48
	$ClO^- + H_2O + 2e^- \Leftrightarrow Cl^- + 2OH^-$	0.84
Chlorine dioxide	$ClO_2(aq) + e^- \Leftrightarrow ClO_2^-$	0.95
Perchlorate	$ClO_4^- + 8H^+ + 8e^- \Leftrightarrow Cl^- + 4H_2O$	1.39
Ozone	$O_3 + 2H^+ + 2e^- \Leftrightarrow O_2 + H_2O$	2.08
	$O_3 + H_2O + 2e^- \Leftrightarrow O_2 + 2OH^-$	1.24
Hydrogen peroxide	$H_2O_2 + 2H^+ + 2e^- \Leftrightarrow 2H_2O$	1.78
	$H_2O_2 + 2e^- \Leftrightarrow 2OH^-$	0.88
Dissolved oxygen	$O_2 + 4H^+ + 4e^- \Leftrightarrow 2H_2O$	1.23
	$O_2 + 2H_2O + 4e^- \Leftrightarrow 4OH^-$	0.40
Permanganate	$MnO_4^- + 4H^+ + 3e^- \Leftrightarrow MnO_2 + 2H_2O$	1.68
	$MnO_4^- + 8H^+ + 5e^- \Leftrightarrow Mn^{2+} + 4H_2O$	1.51
Ferrate(VI)	$FeO_4^{2-} + 8H^+ + 3e^- \Leftrightarrow Fe^{3+} + 4H_2O$	**2.20**
	$FeO_4^{2-} + 4H_2O + 3e^- \Leftrightarrow Fe(OH)_3 + 5OH^-$	0.72

10.2.2 Synthesis

Different approaches were investigated to produce Fe(VI) salts (e.g., $NaFe^{VI}O_4$ and $K_2Fe^{VI}O_4$). The main methods used for the synthesis of sodium and potassium Fe(VI) are the electrochemical, wet chemical and dry thermal methods that are discussed below.

10.2.2.1 Electrochemical Synthesis

In this method, highly concentrated KOH or NaOH is used as the electrolyte [32]. Cast iron is used as the anode (iron source) and it is dissolved and then it is oxidized to form K_2FeO_4 [26]. The reactions that take place are presented below (Eqns. 10.2–10.5) [26, 29, 32].

Table 10.2 Efficiency and operational conditions of electrochemical production of ferrate(VI).

Efficiency (%)	Operational conditions	Ref.
15	Raw iron current density = 10 A/m² [NaOH] = 16.5 M	[35]
27	Steel current density = 10 A/m² [NaOH] = 16.5 M	[35]
50	Cast iron current density = 10 A/m² [NaOH] = 16.5 M	[35]
35	Steel (0.1%C) current density = 36 A/m² [NaOH] = 16 M	[36]
38	Steel (0.3%C) current density = 46.19 mA/cm² [NaOH] = 12 M	[37]

Anode reaction:

$$Fe + 8OH^- \rightarrow FeO_4^{2-} + 4H_2O + 6e^- \quad (10.2)$$

Cathode reaction:

$$2H_2O \rightarrow H_2 + 2OH^- - 2e^- \quad (10.3)$$

Overall reactions:

$$Fe + 2OH^- + 2H_2O \rightarrow FeO_4^{2-} + 3H_2 \quad (10.4)$$

$$FeO_4^{2-} + 2K^+ \rightarrow K_2FeO_4 \quad (10.5)$$

The main advantage of the electrochemical method is the use of electrons as clean reactants to produce Fe(VI) [27]. However, the yield of the process is relatively low (<50%) and there are challenges related to the separation of the solid product [27]. Moreover, the addition of the produced Fe(VI) solution to water usually results in high pH (>11) [29]. A yield higher than 50% was obtained using mixed NaOH-KOH solutions on production of solid potassium Fe(VI) [33]. In Table 10.2, the efficiency and the operational conditions of typical electrochemical production of Fe(VI) reported in different studies, are presented. The in situ electrochemical generation of Fe(VI) has been also investigated [32, 34].

10.2.2.2 *Wet Chemical Method*

In the presence of sodium hydroxide, Fe(III) salt (e.g., $FeCl_3$) reacts with sodium hypochlorite producing sodium Fe(VI) (Na_2FeO_4) [38]. The reactions that take place are presented below (Eqns. 10.6 and 10.7) [27]:

$$2FeCl_3 + 3NaOCl + 10NaOH \rightarrow 2Na_2FeO_4 + 9NaCl + 5H_2O \quad (10.6)$$

$$Na_2FeO_4 + 2KOH \rightarrow K_2FeO_4 + 2NaOH \quad (10.7)$$

The yield (in terms of potassium Fe(VI)-K_2FeO_4) of this procedure is 10%–15% and after numerous separation steps, a solid potassium Fe(VI) (>90% purity) can be obtained. Because of the high solubility of Na_2FeO_4 in NaOH solutions, it is difficult to obtain solid sodium Fe(VI) (Na_2FeO_4) [29]. A yield of potassium Fe(VI) up to 75% was achieved by replacing the sodium hydroxide with the potassium hydroxide [39]. In that case the formation of Na_2FeO_4 was avoided. Ozone instead of hypochlorite can be also used to prepare Fe(VI) [40].

10.2.2.3 *Dry Thermal Method*

Potassium Fe(VI) can be produced by calcination of ferric oxide-potassium peroxide mixture at 350–370 °C or by oxidizing the iron oxide using sodium peroxide at 370 °C, under dry oxygen conditions [26, 27, 29]. Moreover, galvanizing wastes were used to produce sodium Fe(VI), by mixing them with ferric oxide at 800 °C. Then, the mixture was cooled down, stirred with solid sodium peroxide, and heated gradually for few minutes [26, 29]. The yield of the dry thermal method is usually less than 50% [27]. The dry synthesis is considered as an old and expensive method due to the required high temperatures [29]. Solid sodium Fe(VI) was produced using caustic soda (instead of potassium hydroxide) and sodium hypochlorite (instead of calcium hypochlorite), which are cheaper [41].

10.2.3 Characterization

Different analytical techniques were investigated to characterize solid and liquid Fe(VI). Fourier transform infrared (FTIR) spectroscopy, X-ray diffraction, X-ray absorption near edge structure (XANES) and Mössbauer spectroscopy are used to characterize solids salts of Fe(VI) [23, 31]. The reactions of Fe(VI) with chromium(III) and arsenic(III) to form chromate(VI) and arsenate respectively, can be used as volumetric methods to quantify

$Na_2Fe^{VI}O_4$ and $K_2Fe^{VI}O_4$ in solutions [40]. Furthermore, electrochemical techniques including cyclic voltammetry and potentiometry are also available methods for Fe(VI) quantification in solutions [42, 43]. In addition, a fluorescence method by means of Fe(VI) reaction with scopoletin (fluorescence agent) is also used to determine Fe(VI) in acidic solutions [44].

Among the various analytical techniques, three colorimetric methods (Iodide, ABTS and Direct) are mainly used for the quantification of Fe(VI) in solutions related to water treatment. Fe(VI) solutions have a characteristic purple color with a maximum absorbance at 510 nm and a molar absorptivity (ε) of 1150 M^{-1} cm^{-1} [40]. Thus, the direct colorimetric method is widely used for quantification of Fe(VI) in alkaline solutions where Fe(VI) is more stable, to investigate the kinetics of the reaction of Fe(VI) with organic molecules. A typical spectra of potassium Fe(VI) in Milli-Q water is shown in Figure 10.2. The interference with water constituents and the dependency on the pH and Fe(VI) stability are the main limitations of the direct colorimetric method [40].

The ABTS colorimetric method is based on the reaction of Fe(VI) with the 2,2′-azino-bis(3-ethylbenzothiazoline-6-sulfonate) (ABTS). A stable radical cation ($ABTS^{\bullet+}$) is formed showing peak absorbance at 415 nm [45]. The reaction of ABTS with Fe(VI) is very fast making the method suitable for determining Fe(VI) also in natural waters at a concentration range of 0.03–35 μM [45]. The Iodide colorimetric method is based on the reaction of Fe(VI) with sodium iodide (NaI), to produce I_3 – that shows peak

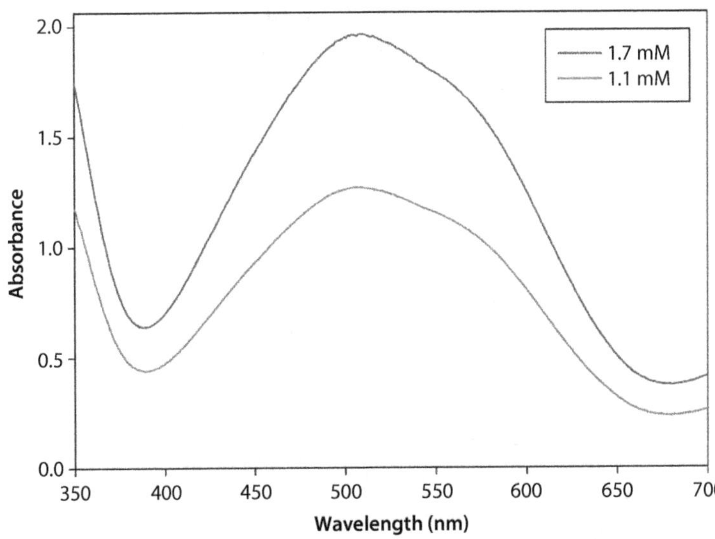

Figure 10.2 Spectra of ferrate(VI) in Milli-Q water.

Table 10.3 Spectrophotometric methods for ferrate(VI) determination in water [29, 40, 45].

Method	Reaction	Wavelength, λ (nm)	Molar absorptivity, ε (M^{-1} cm^{-1})	pH
Direct	–	510	1150	Basic
ABTS	Fe(VI) + 2ABTS → Fe(III) + $ABTS^{\bullet +}$ + $ABTS_{oxidized}$	415	34000	Acetic to neutral
Iodide	Fe(VI) + $3I^-$ → Fe(III) + I_3^-	351	29700	5.5–9.3

absorbance at 351 nm [40]. The minimum detection limit of this method is 0.25 μM Fe(VI) and the method is also suitable for determining Fe(VI) in tap water [40]. A comparison of the three commonly used colorimetric methods is given in Table 10.3.

10.2.4 Oxidation

10.2.4.1 Kinetics of the Oxidation of Organics by Ferrate(VI)

Kinetic studies on the oxidation of a broad range of inorganic and organic compounds have been carried out to understand different reactions and the Fe(VI)-related chemistry [31, 46–49]. Nowadays, there is an increasing interest in oxidation of emerging organic pollutants such as pharmaceuticals and personal care products (PPCPs) and endocrine distributing compounds (EDCs) by Fe(VI), because of the increasing concerns related to their negative effects on the aquatic environment [14, 29, 50].

Most of the kinetic studies of oxidation of organic compounds (X) by Fe(VI) have been carried out using a stopped-flow spectrometry technique. The reactions of Fe(VI) with X follow overall second order kinetics, that is, first order with respect to each reactant, Fe(VI) and X, as it has been shown in many studies with different organic molecules (Eqn. 10.8) [31]:

$$-\frac{d\left[\mathrm{Fe(VI)}\right]}{dt} = k_{app} \times \left[\mathrm{Fe(VI)}\right] \times \left[\mathrm{X}\right] \tag{10.8}$$

where $\frac{d\left[Fe(VI)\right]}{dt}$ is the rate of the loss of Fe(VI), [Fe(VI)] and [X] are the concentrations of Fe(VI) and X respectively, and k_{app} is the apparent reaction rate constant. The reactivity of Fe(VI) with X is mostly investigated under pseudo first-order conditions, with X in excess ([Fe(VI)] << [X]), following the Fe(VI) concentration with time [23]. The rate constants are mostly calculated as a function of pH.

Apparent second-order reaction rate constants (k_{app}) of the reaction of Fe(VI) with selected PPCPs and EDCs at room temperature (23 ± 2 °C) are presented in Table 10.4. Kinetic studies were mostly carried out at neutral to slightly basic pH range of 7.0–9.0 which is relevant to water treatment processes and regulatory criteria. In addition, the instability of Fe(VI) in acidic conditions prevents kinetic investigations at low pH levels.

Generally, the rate constants of Fe(VI) oxidation reactions increase with decreasing pH (Table 10.4). This pH dependency of k_{app} can be explained by the acid dissociation constants (K_a) expressed as pK_a (Eqns. 10.9–10.11) [31, 68]:

$$H_3FeO_4^+ \leftrightarrow H^+ + H_2FeO_4 \quad pK_{a1} = 1.6 \tag{10.9}$$

$$H_2FeO_4 \leftrightarrow H^+ + HFeO_4^- \quad pK_{a2} = 3.5 \tag{10.10}$$

$$HFeO_4^- \leftrightarrow H^+ + FeO_4^{2-} \quad pK_{a3} = 7.3 \tag{10.11}$$

Fe(VI) has triprotonated ($H_3Fe^{VI}O_4^+$), diprotonated ($H_2Fe^{VI}O_4$), monoprotonated ($HFe^{VI}O_4^-$) and deprotonated ($Fe^{VI}O_4^{2-}$) species depending on pH (Eqns. 10.9–10.11). The speciation of Fe(VI) at different pH values is given in Figure 10.3. Species (acid-base) of ionizable compounds are also used to explain the trend of increasing reaction rates by decreasing the pH [50].

The second-order reaction rate constants show that Fe(VI) (selective oxidant) reacts preferably with electron-rich organic moieties such as phenolic, organosulfur and polycyclic aromatic compounds [7]. The k_{app} of the oxidation of PPCPs and EDCs range from 0.1×10^0 (ibuprofen) to 7.9×10^3 (tetrabromobisphenol A) $M^{-1}\ s^{-1}$ at a pH relevant to wastewater treatment process and room temperature (Table 10.4).

10.2.4.2 Stoichiometry

Information regarding the kinetics of oxidation of pollutants is useful to understand these reactions and compare the ability of different oxidants to oxidize pollutants in water. However, a higher k of an oxidant than a different one, does not necessarily mean higher oxidative transformation of

Table 10.4 Apparent second-order rate constants (k_{app}) of the oxidation of PPCPs and EDCs by Fe(VI) at room temperature.

PPCP/EDC	Structure	pH	k_{app} (M^{-1} s^{-1})	Ref.
Triclosan	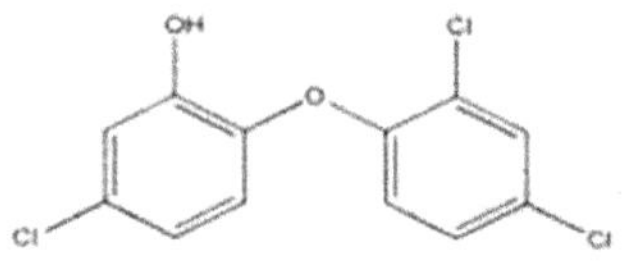	7.0	1.1×10^3	[50]
			7.5×10^2	[51]
Bisphenol A		7.0	6.4×10^2	[50, 52]
		8.0	4.1×10^2	[50]
Sulfamethoxazole	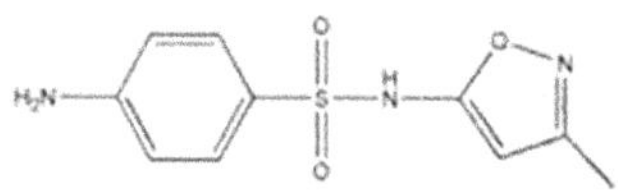	7.0	1.8×10^3	[50]
			1.3×10^3	[53]
		8.0	7.7×10^1	[50]
Atenolol	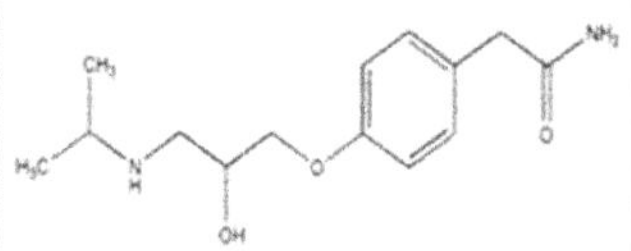	8.0	0.7×10^1	[7]
Diclofenac	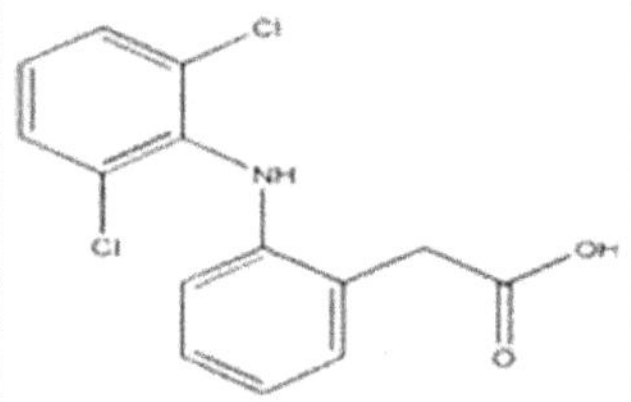	7.0	1.3×10^2	[50]
			1.2×10^1	[54]
		8.0	3.2×10^1	[50]
		11.0	2.5×10^0	[54]
Ibuprofen	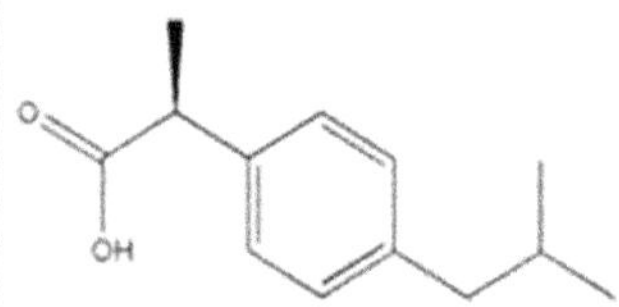	8.0	0.1×10^0	[55]
			$<0.1 \times 10^0$	[7]
			1.2×10^{-1}	[56]
		9.0	1.5×10^{-2}	[56]
Ciprofloxacin	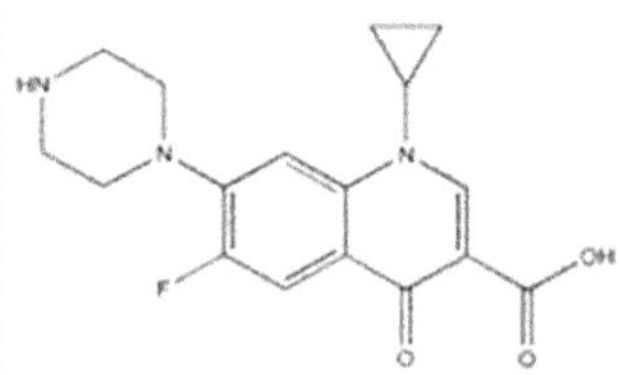	7.0	4.7×10^2	[50]
		8.0	1.7×10^2	[50]
			1.1×10^2	[56]
		9.0	6.4×10^1	[56]

(Continued)

Table 10.4 Cont.

PPCP/EDC	Structure	pH	k_{app} (M^{-1} s^{-1})	Ref.
Enrofloxacin	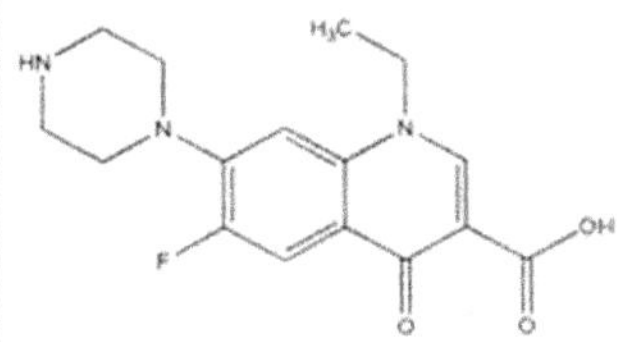	7.0	4.6×10^1	[50]
		8.0	2.4×10^1	
Carbamazepine	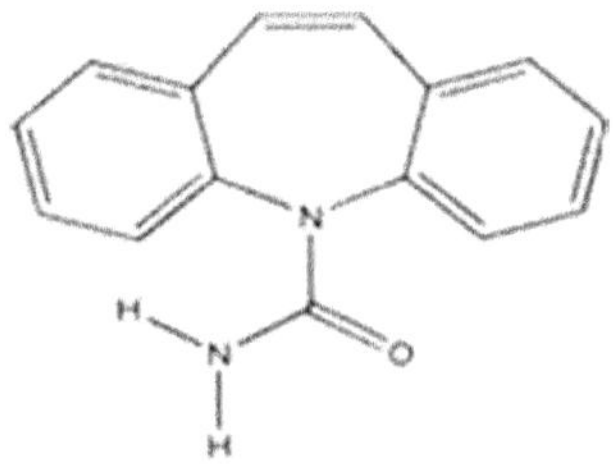	7.0	6.7×10^1	[50]
		8.0	1.6×10^1	
Octylphenol	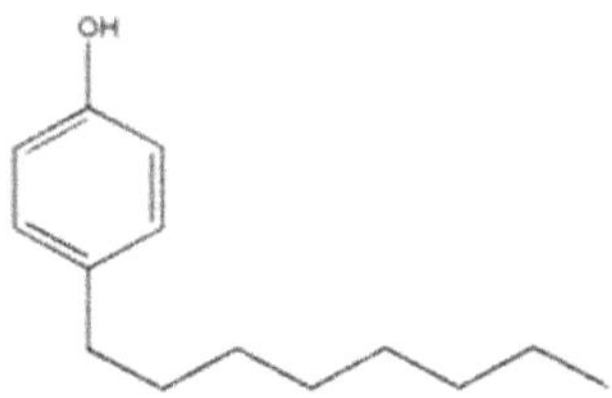	7.0	1.2×10^3	[57]
		8.0	0.3×10^3	
Nonylphenol	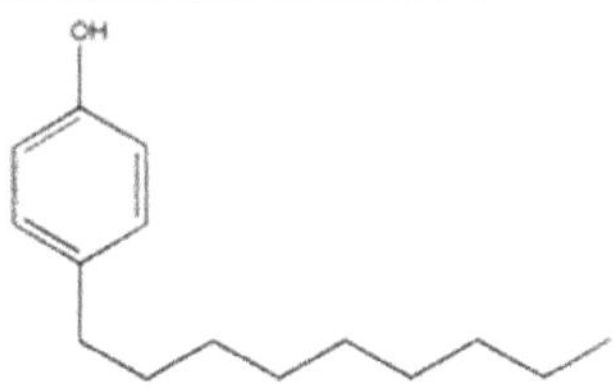	7.0	1.1×10^3	[58]
		8.0	2.7×10^2	
		9.0	1.0×10^2	
Tetrabromobisphenol A	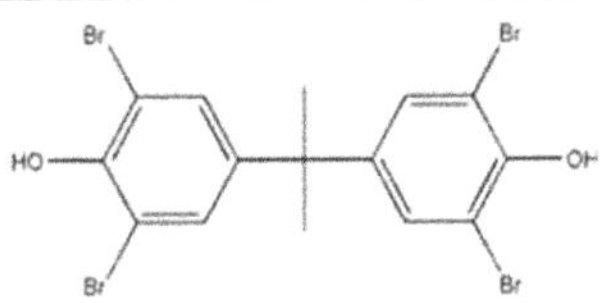	7.0	7.9×10^3	[59]
		10.0	3.3×10^1	
17α-Ethinylestradiol	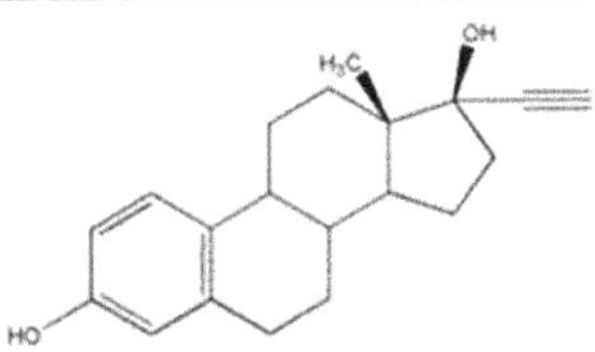	7.0	7.3×10^2	[50, 52]
			8.1×10^2	[14, 60]
		8.0	4.5×10^2	[50]

(*Continued*)

Table 10.4 Cont.

PPCP/EDC	Structure	pH	k_{app} ($M^{-1}\ s^{-1}$)	Ref.
Estrone	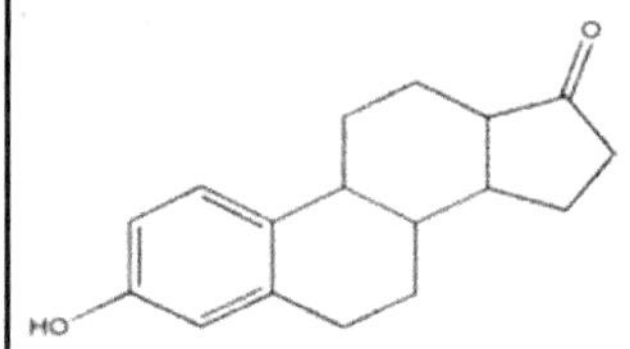	7.0	1.0×10^3	[14, 60]
17β-estradiol	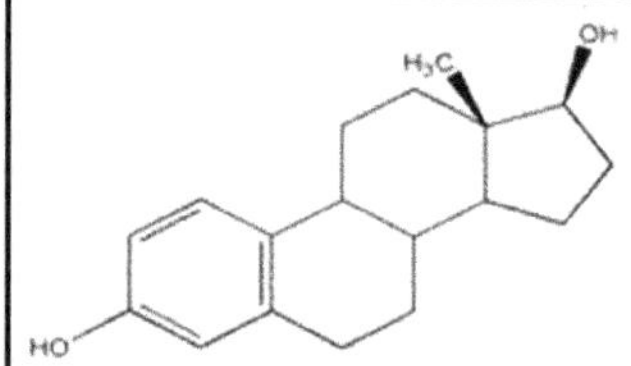	7.0	7.6×10^2	[50]
			1.1×10^3	[14, 60]
		8.0	4.6×10^2	[50]
Estriol	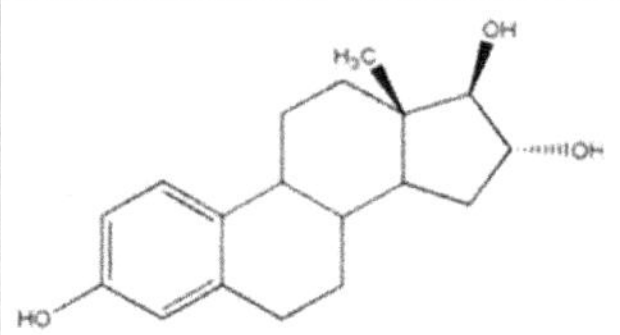	7.0	1.2×10^3	[14, 60]
Sulfisoxazole	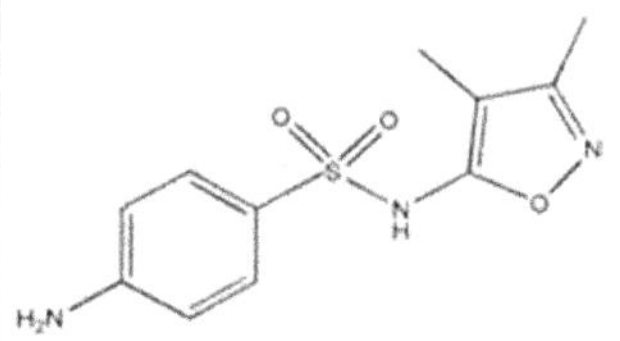	7.0	1.5×10^3	[14, 53]
Sulfamethazine	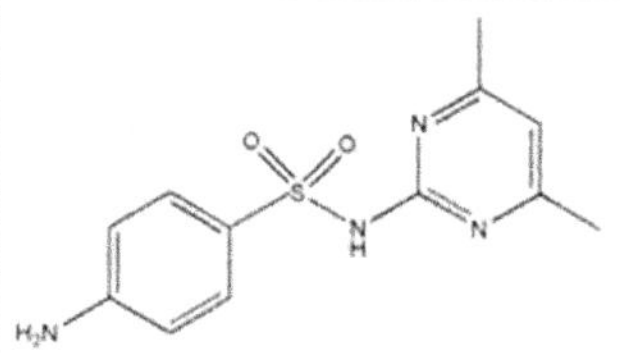	7.0	1.0×10^3	[14, 53]
Sulfamethizole	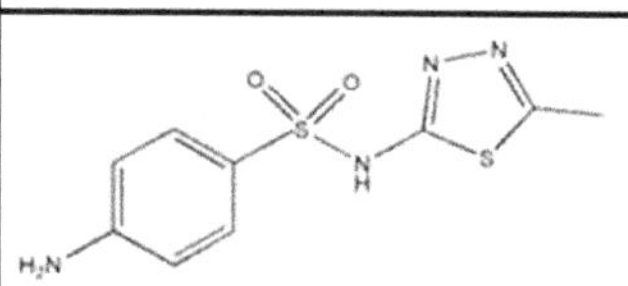	7.0	4.1×10^2	[14, 53]
Sulfadimethoxine	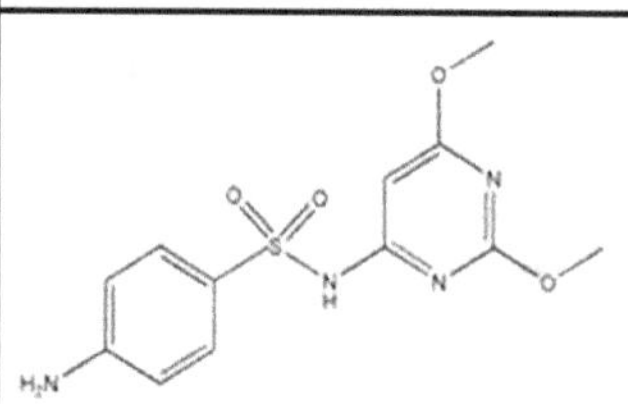	7.0	0.8×10^2	[46, 53]

(Continued)

Table 10.4 Cont.

PPCP/EDC	Structure	pH	k_{app} (M^{-1} s^{-1})	Ref.
Tetracycline	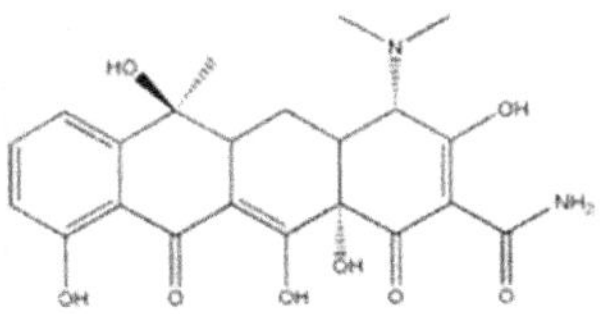	7.0	3.0×10^2	[14]
Amoxicillin	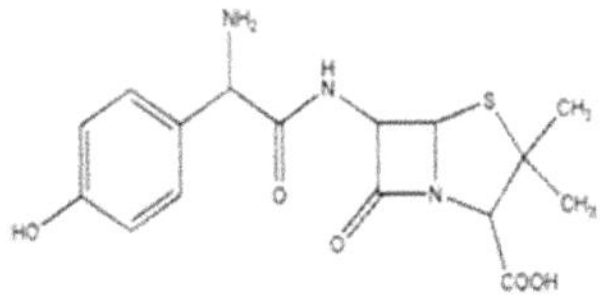	7.0	2.8×10^3	[61]
Ampicillin	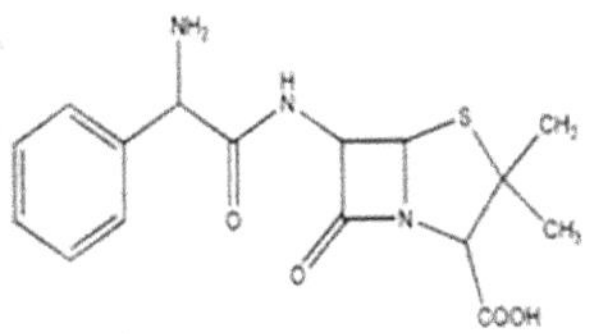	7.0	1.1×10^3	[61]
			4.2×10^2	[62]
		8.5	5.0×10^1	[62]
Cloxacillin	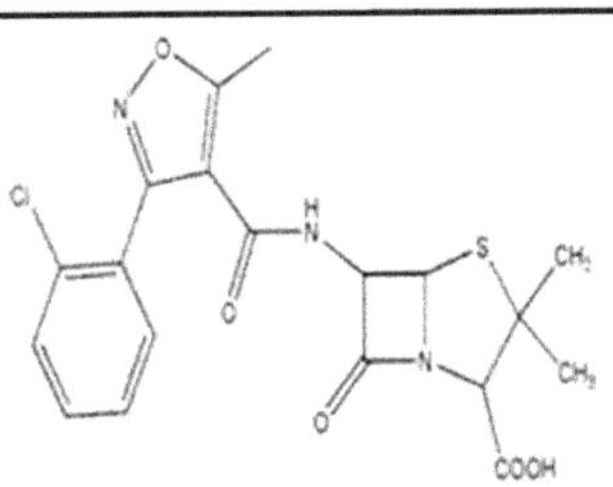	7.0	1.2×10^2	[62]
		8.5	2.4×10^1	
Penicillin G	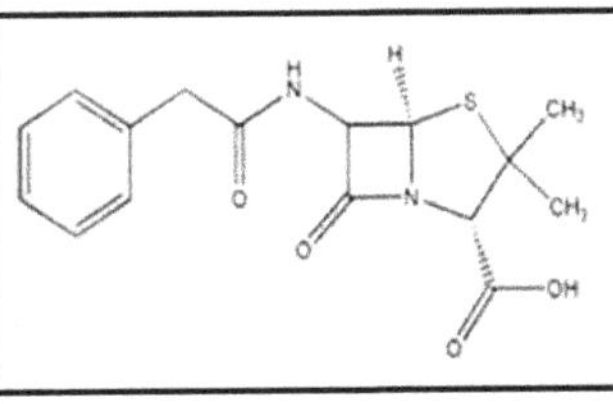	7.0	1.1×10^2	[62]
		8.5	1.8×10^1	
Cephalexin	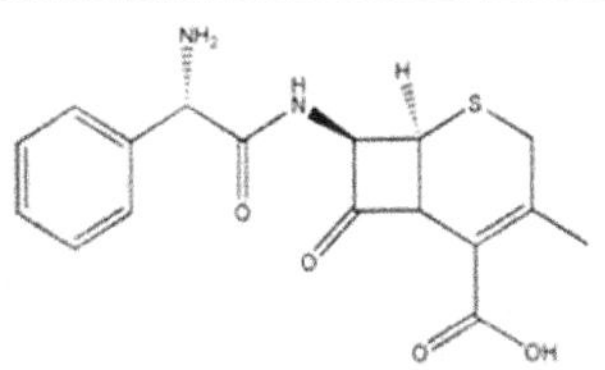	7.0	6.9×10^2	[62]
		8.5	7.4×10^1	
Amoxicillin	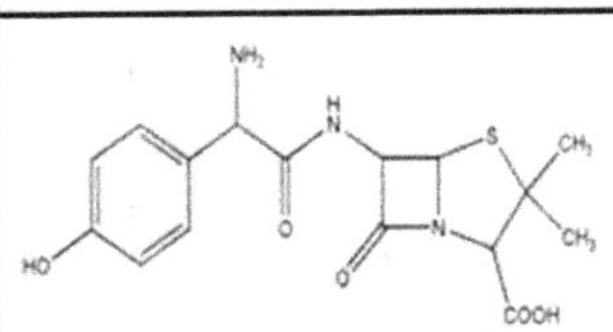	7.0	7.7×10^2	[62]
		8.5	1.6×10^2	

(*Continued*)

Table 10.4 Cont.

PPCP/EDC	Structure	pH	k_{app} ($M^{-1}\ s^{-1}$)	Ref.
2-amino-2-phenyl-acetamide	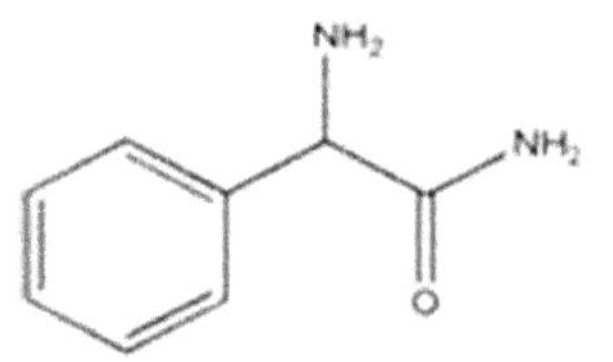	7.0	2.9×10^{2}	[62]
		8.5	3.4×10^{1}	
3-methylcrotonic acid	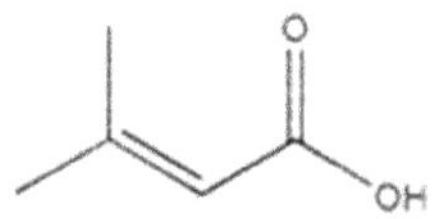	7.0	2.3×10^{0}	[62]
		8.5	0.5×10^{0}	
3,5-dimethylisoxazole	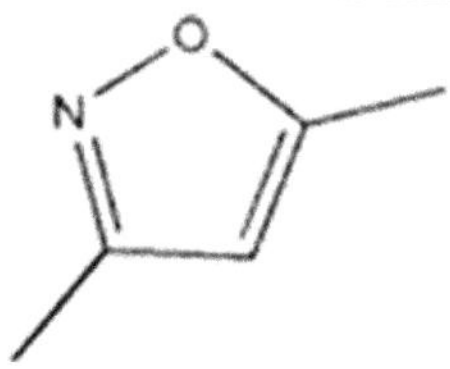	8.0	1.7×10^{-1}	[62]
Trimethoprim	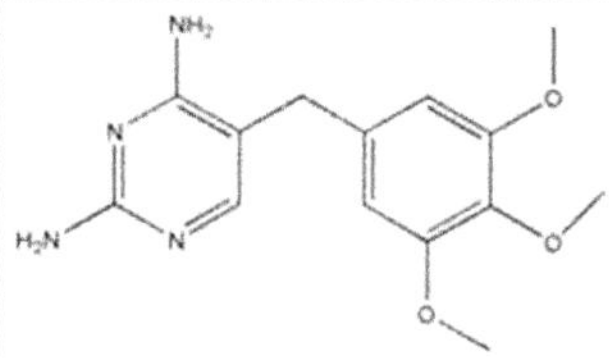	7.0	4.0×10^{1}	[14, 63]
Propranolol	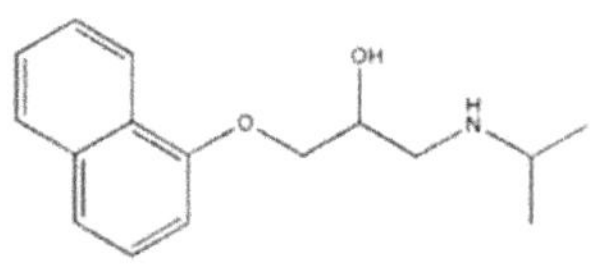	8.0	2.0×10^{1}	[64]
Tramadol	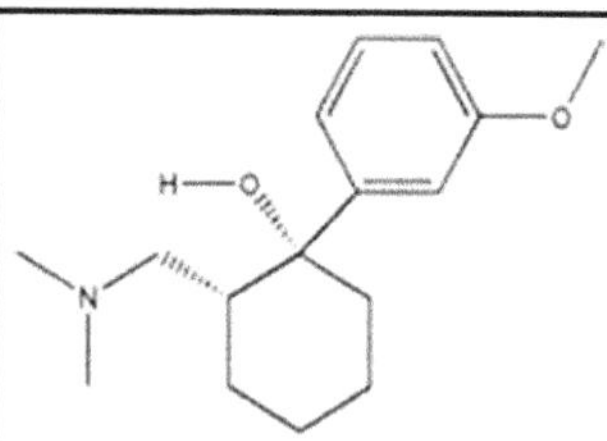	7.0	1.4×10^{1}	[65]
		8.0	7.4×10^{0}	
Diatrizoic acid	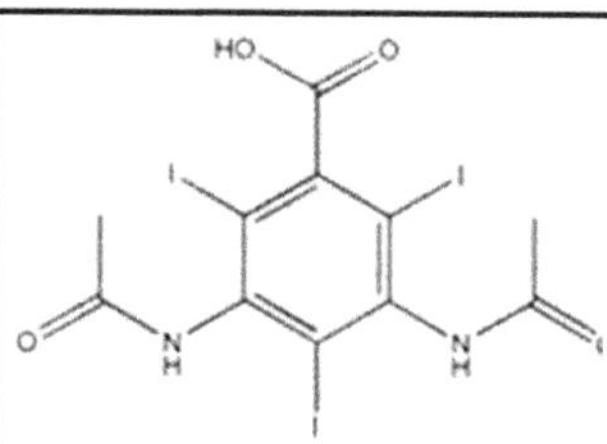	7.0	5.5×10^{0}	[66]

(Continued)

Table 10.4 Cont.

PPCP/EDC	Structure	pH	k_{app} ($M^{-1}\ s^{-1}$)	Ref.
Benzophenone-3	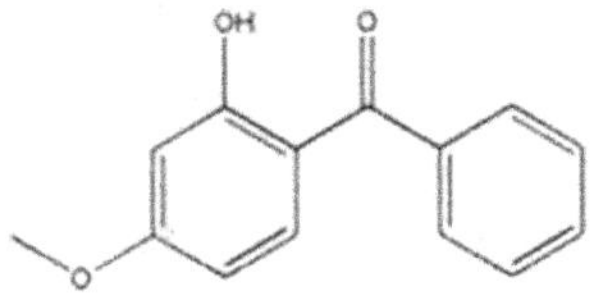	8.0	8.2×10^1	[67]
4-Methylphenol		7.0	6.9×10^2	[50]
		8.0	3.3×10^2	
Buten-3-ol	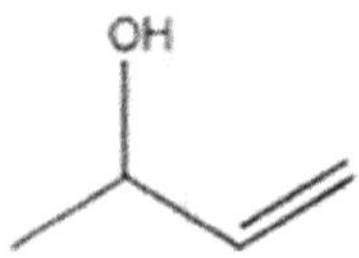	7.0	1.2×10^1	[50]
		8.0	3.0×10^0	
Phenol	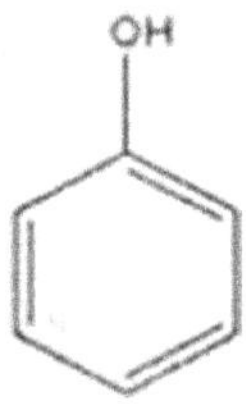	7.0	7.7×10^1	[52]

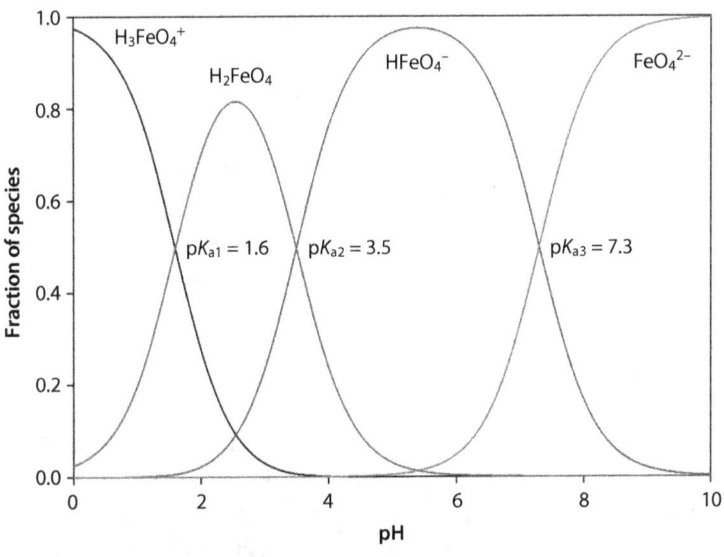

Figure 10.3 Speciation of ferrate(VI) [31].

the organic molecule, because of the competition for oxidants with target organic compound and wastewater matrix (i.e., ions and organic matter) [7]. For example, the hydroxyl radical, a powerful non-selective oxidant, generally reacts much faster with organics than Fe(VI) ($k_{OH\bullet} > k_{Fe(VI)}$). However, the aforementioned competition is more obvious for hydroxyl radicals and this competition remains unchanged during the oxidation process because hydroxyl radicals react non-selectively with other water constituents. That was shown in the oxidation of electron-rich moiety compounds by selective oxidants (including Fe(VI)) and hydroxyl radicals where the oxidative transformation of these organic molecules was higher in the case of selective oxidants than hydroxyl radicals [7]. Thus, the experimental investigation of the stoichiometry of the Fe(VI)-organic reactions is critical to optimize the required oxidant dose and to evaluate the ability of Fe(VI) to remove organics in water. Table 10.5 shows the required Fe(VI) to organic (X) molar ratios for complete removal of an organic in water, at a relevant to water treatment pH range and room temperature.

Although the effect of the pH on the rate constants of Fe(VI) reaction with contaminants is clear (the higher the pH the lower the rate constant due to the fact that Fe(VI) is a stronger oxidant upon protonation; Figure 10.3 and Table 10.4), the effect of the pH on the oxidative transformation of contaminants by Fe(VI), after complete reaction, is mostly negligible at the pH range of 7.0–9.0. However, three explanations have been reported in the literature so far to explain any observed effect of the pH on the removal of pollutants by Fe(VI): (i) when the removal is higher at the higher pH (e.g., pH 9), this is due to the fact that Fe(VI) is more stable at pH 9, and the Fe(VI)'s self-decomposition is diminished resulting in higher Fe(VI) exposure, (ii) when the removal is high at low pH (e.g., 7.0), this is due to the fact that Fe(VI) is a stronger oxidant upon protonation and this is consistent with the effect of the pH on the rate constants, and (iii) when the effect of the pH on the removal of the contaminant is not significant, this is probably due to the high reactivity of Fe(VI) with the contaminant [56], that makes the effect of the pH, usually at the pH range 7.0–9.0, to be negligible. The effect of the pH on the removal of contaminants by Fe(VI) can be seen in recently reported studies that show different or negligible effect of the pH on the oxidative transformation of organics by Fe(VI). Yang et al. studied the oxidation of tetrabromobisphenol A and bisphenol A by Fe(VI) [59]. In both cases, around 10% higher removal had been obtained at pH 8 compared to pH 7, and that was explained with the higher stability of Fe(VI) at pH 8 resulting in higher Fe(VI) exposure [59]. Casbeer et al. investigated the oxidation of tryptophan by Fe(VI) at pH 7 and 9 and around 10% higher removal was observed at pH 9 compared

Table 10.5 Stoichiometry of oxidation of organic molecules by Fe(VI) at room temperature.

Compound (X)	Structure	pH	[Fe(VI)]/[X] (mol/mol)	Ref.
Flumequine	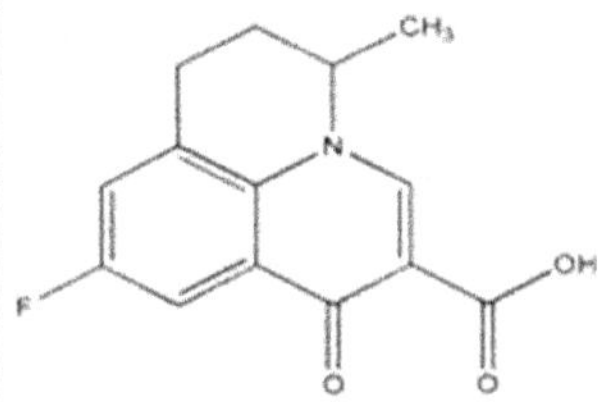	7.0	>50	[69]
Enrofloxacin	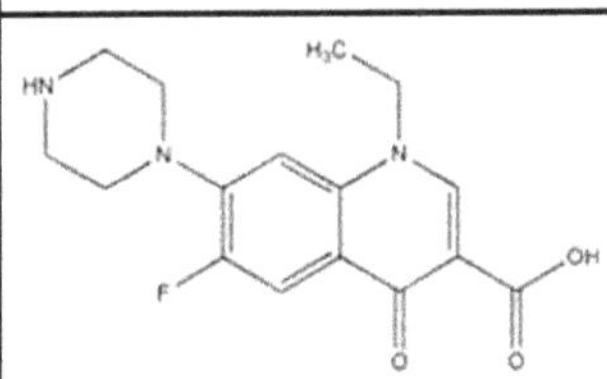	7.0	15	[69]
		7.0	17	[70]
Norfloxacin	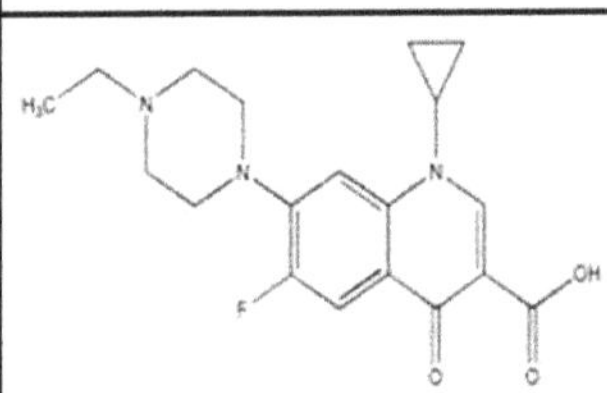	7.0	10	[69]
Ofloxacin	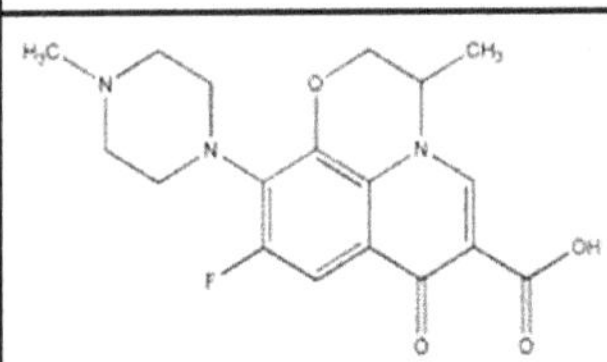	7.0	15	[69]
Marbofloxacin	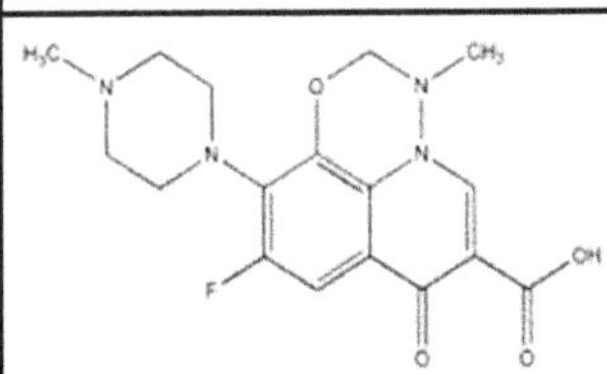	7.0	20	[69]
Ciprofloxacin	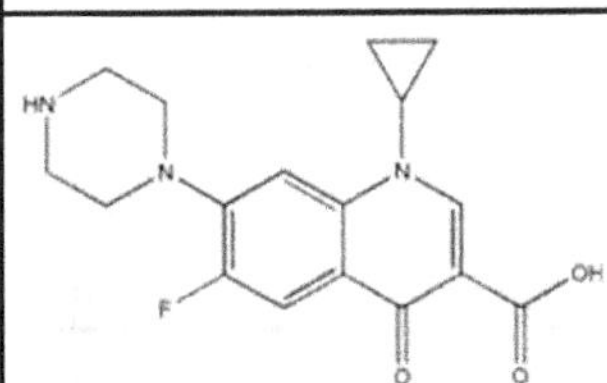	7.0	13	[70]

(Continued)

Table 10.5 Cont.

Compound (X)	**Structure**	**pH**	**[Fe(VI)]/[X] (mol/mol)**	**Ref.**
Acetaminophen	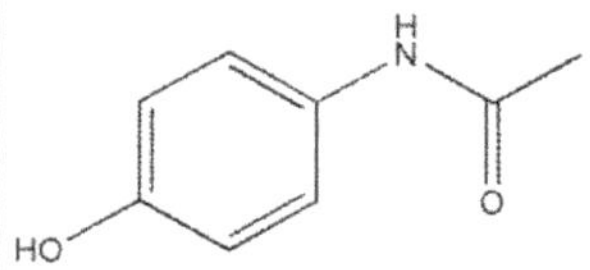	6.0–9.0	25	[71]
Sulfamethoxazole	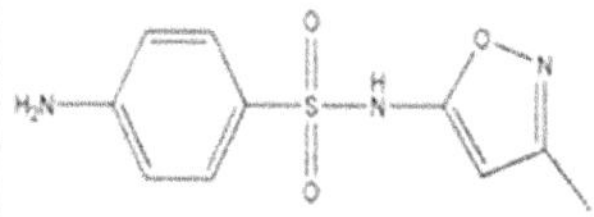	7.0–9.0	4	[72]
		9.0	4	[53]
Bisphenol A	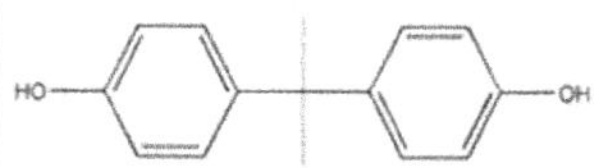	7.0	8	[73]
		7.0	15	[59]
		8.0	10	
		9.4	5	[74]
		9.2	4	[60]
Tetrabromobisphenol A	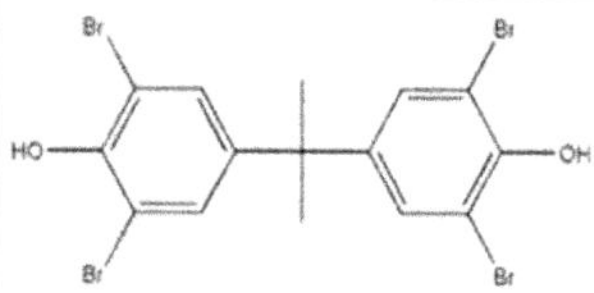	7.0	6.3	[59]
		8.0	3.8	
Tryptophan	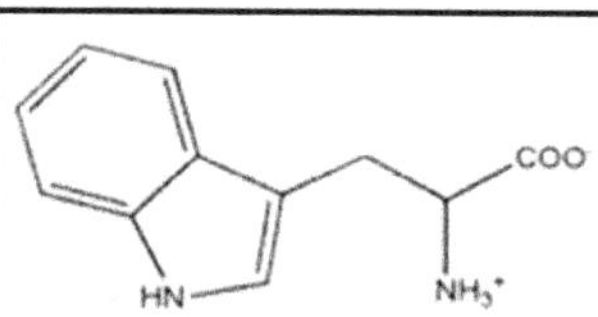	7.0 and 9.0	4	[75]
Benzophenone-3	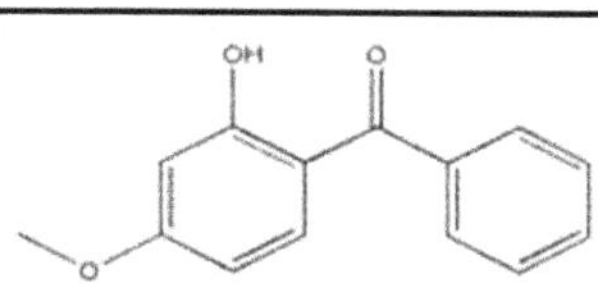	8.0	25	[67]
Amoxicillin	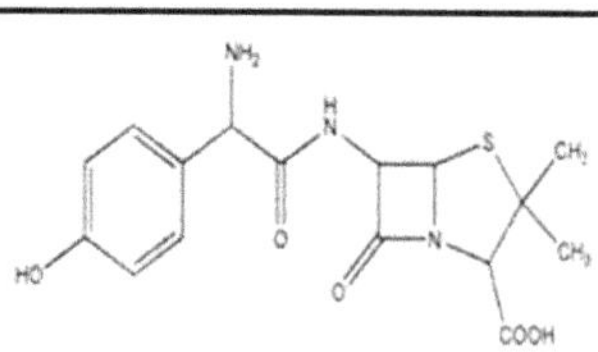	7.0	4.5	[76]

(*Continued*)

Table 10.5 Cont.

Compound (X)	Structure	pH	[Fe(VI)]/[X] (mol/mol)	Ref.
Ampicillin	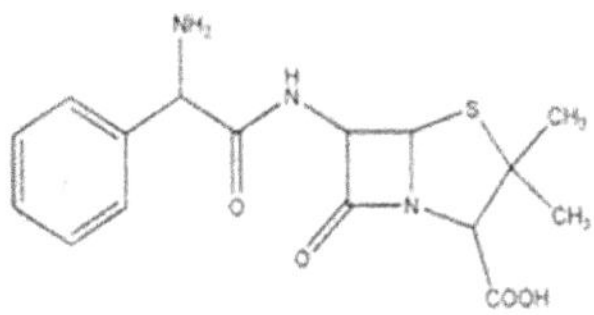	7.0	3.5	[76]
Propranolol	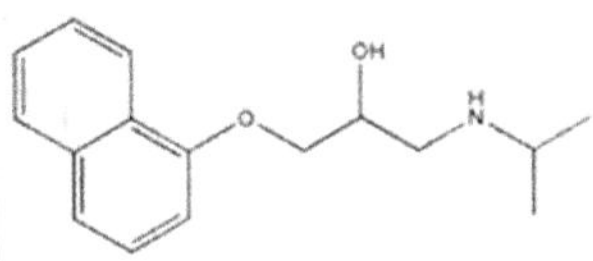	9.0	6	[64]
Trimethoprim	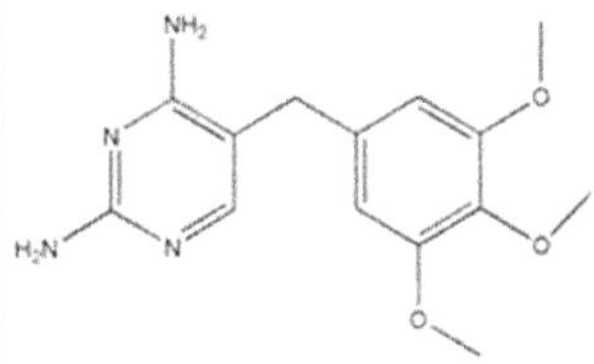	9.0	5	[63]
1H-benzotriazole	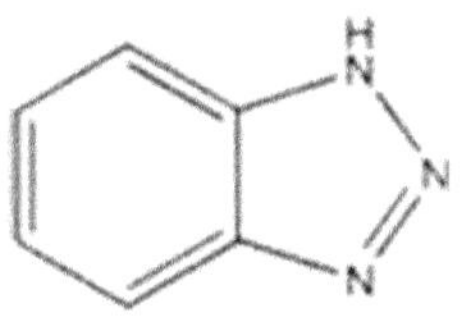	8.0	30	[77]
5-methyl-1H-benzotriazole	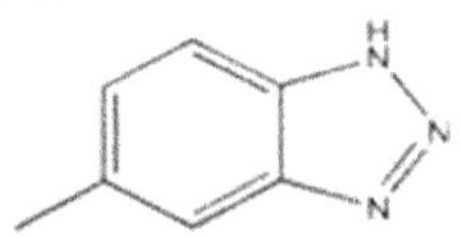	8.0	30	[77]
5,6-dimethyl-1H-benzotriazole hydrate	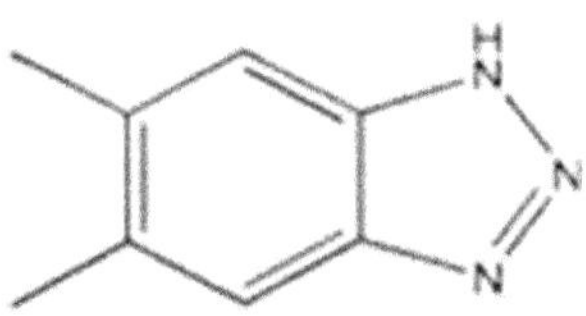	8.0	30	[77]
5-chloro-1H-benzotriazole	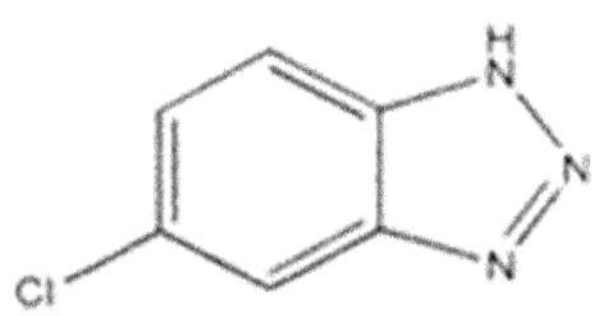	8.0	30	[77]

(*Continued*)

Table 10.5 Cont.

Compound (X)	**Structure**	**pH**	**[Fe(VI)]/ [X] (mol/ mol)**	**Ref.**
1-hydroxybenzotriazole	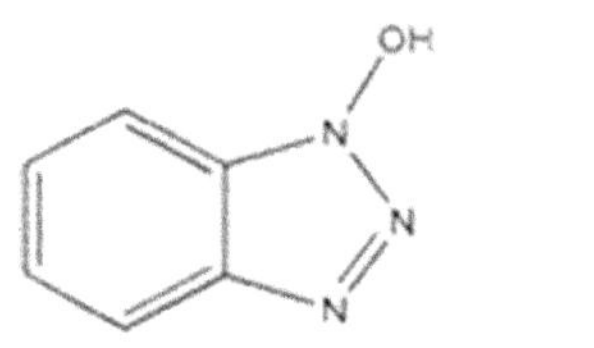	8.0	30	[77]
Triclosan	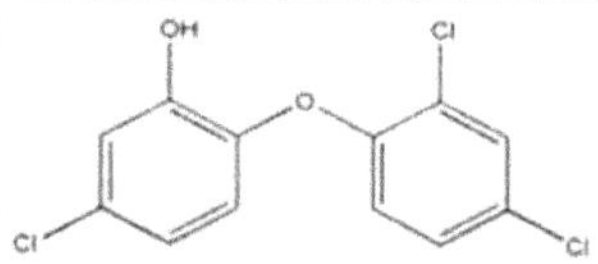	7.0	10	[51]
Cetylpyridinium chloride	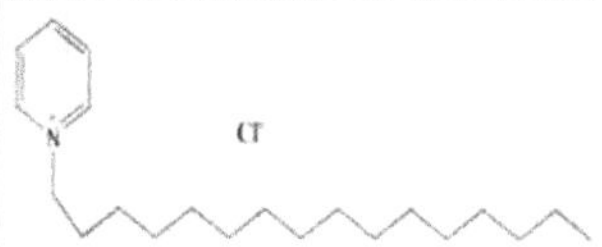	9.2	1	[78]
Glycine	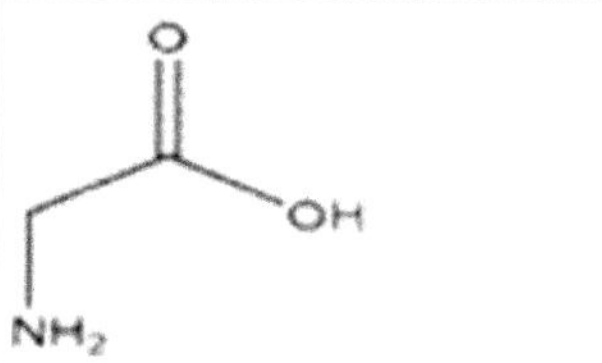	9.0	1	[79]
Methyl mercaptan	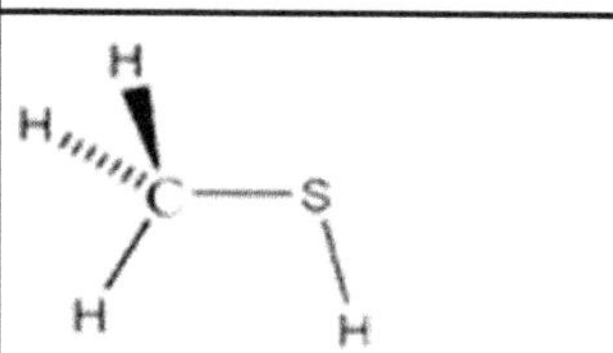	9.0	4.6	[80]
Thiosemicarbazide	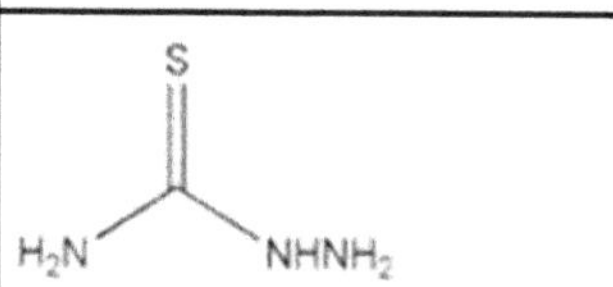	9.0	2	[80]
Thiourea	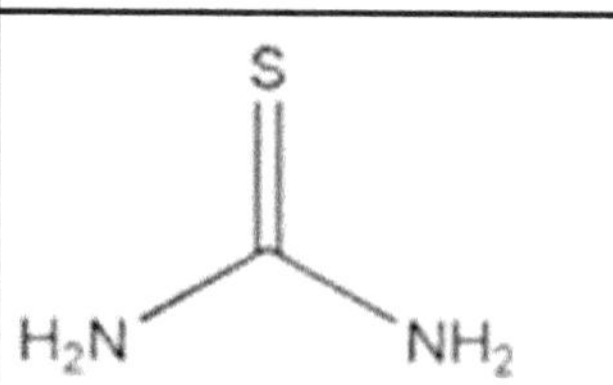	9.0	2.7	[81]

(Continued)

Table 10.5 Cont.

Compound (X)	Structure	pH	[Fe(VI)]/[X] (mol/mol)	Ref.
Thioacetamide	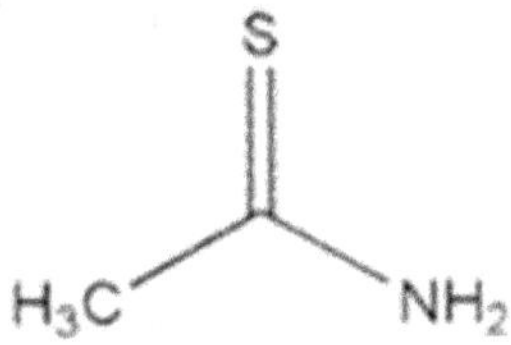	9.0	2.7	[27, 82]
4-Chlorophenol	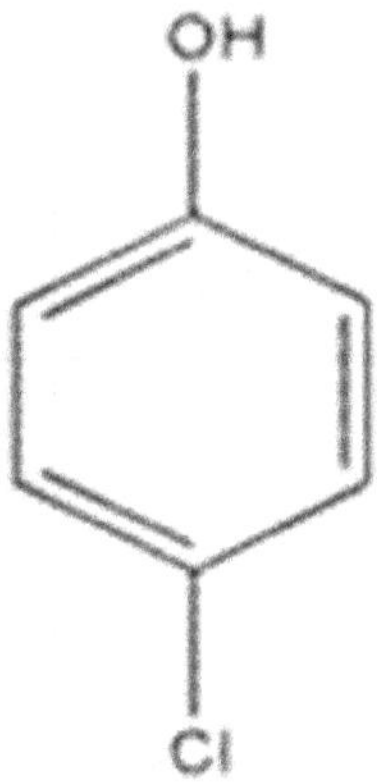	9.2	5	[83]
N-methylhydroxylamine	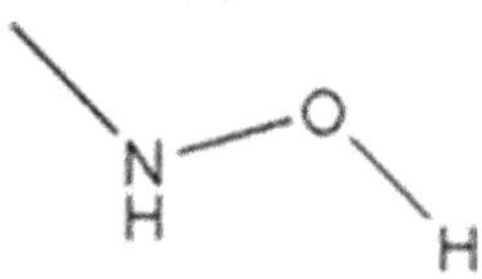	8.0–11.0	0.5	[84]
N-phenylhydroxylamine	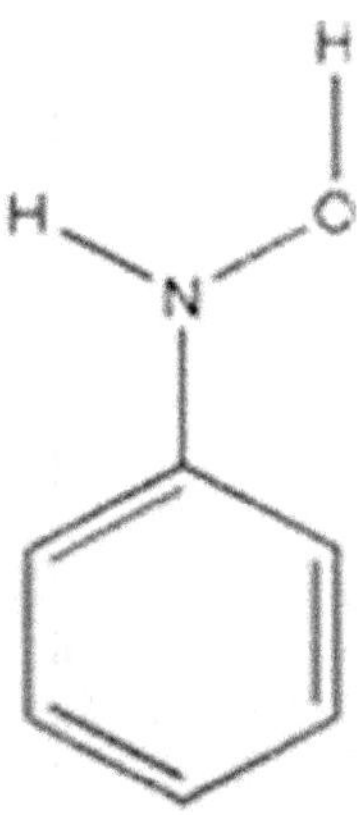	8.0–11.0	0.5	[84]

(Continued)

Table 10.5 Cont.

Compound (X)	Structure	pH	[Fe(VI)]/[X] (mol/mol)	Ref.
3-Mercapto-l-propane sulfonic acid	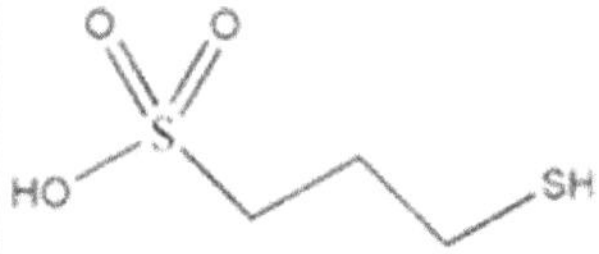	8.4–10.2	1	[85]
2-Mercaptonicotinic acid	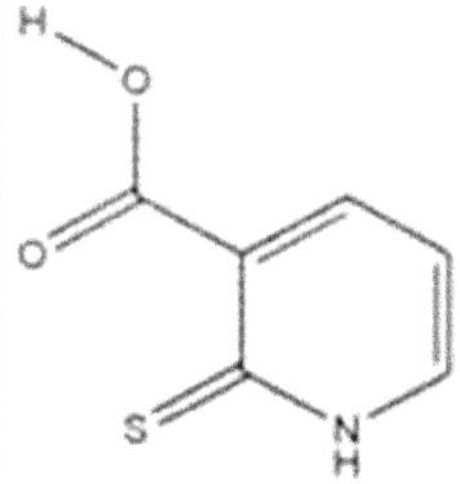	8.4–10.2	0.5	[85]
Benzenesulfinate	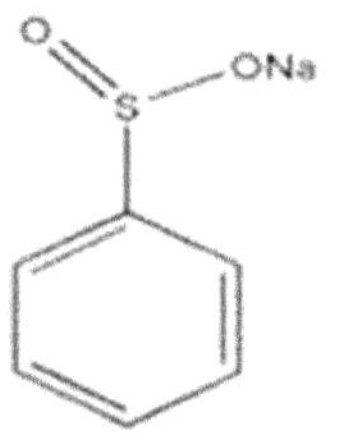	9.0	0.7	[86]
Methionine	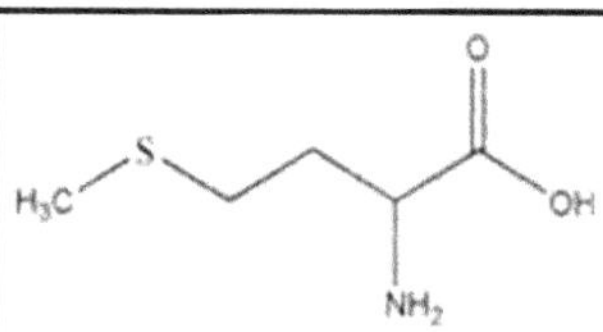	9.0	0.7	[86]
Dimethyl sulfoxide	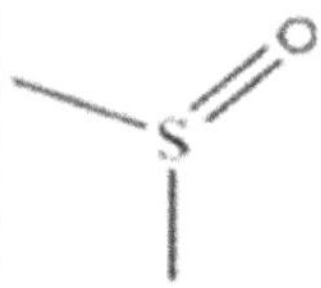	9.0	0.7	[86]
Methyl cysteine	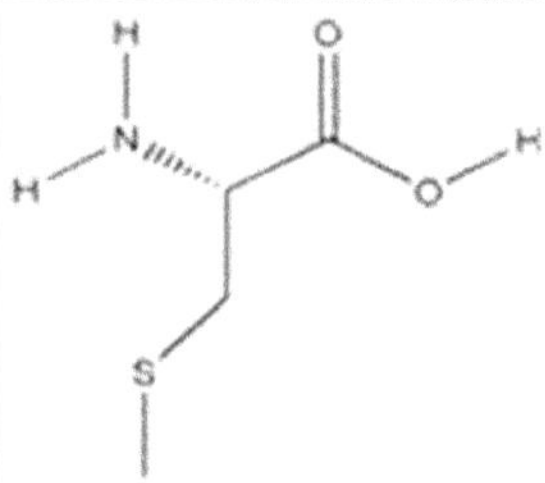	9.2–10.4	0.7	[87]

(*Continued*)

Table 10.5 Cont.

Compound (X)	Structure	pH	[Fe(VI)]/[X] (mol/mol)	Ref.
Cystine	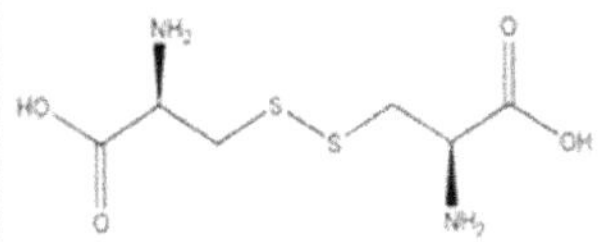	8.4–9.9	1.3	[87]
Cysteine	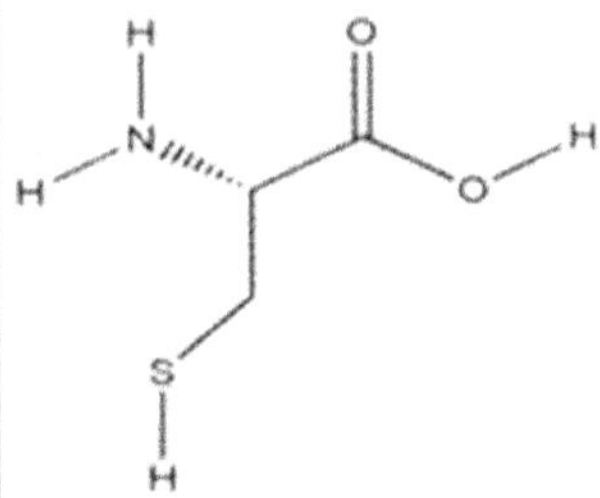	8.9–10.2	1	[87]
Aniline	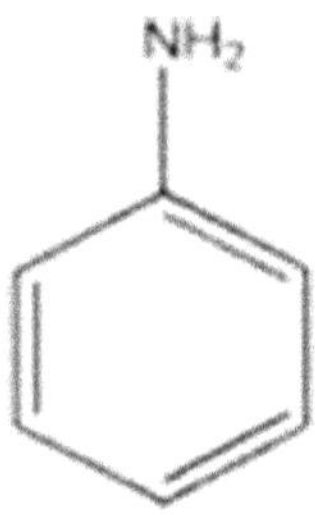	9.0–10.0	0.5	[88]
3-Mercaptopropionic acid	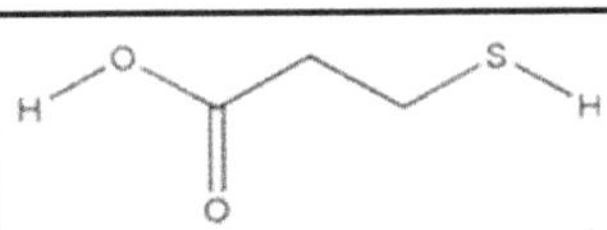	9.0–11.0	1.5	[89]
2-Mercaptoethanesulfonic acid	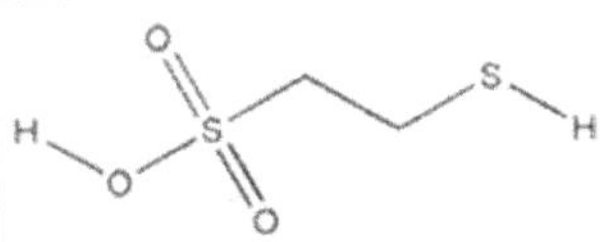	8.7–10.0	2	[89]
2-Mercaptobenzoic acid	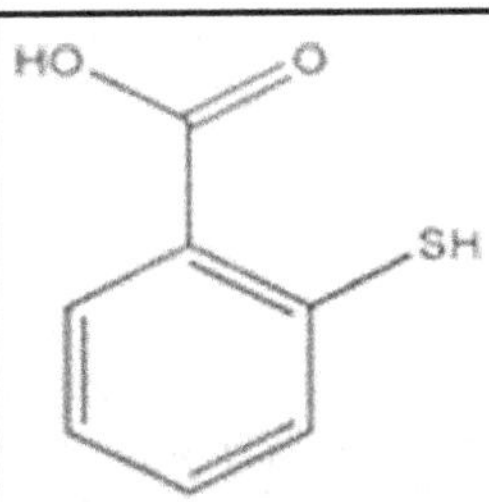	9.6–10.4	0.7	[89]

(Continued)

Table 10.5 Cont.

Compound (X)	Structure	pH	[Fe(VI)]/ [X] (mol/ mol)	Ref.
1,4-Thioxane	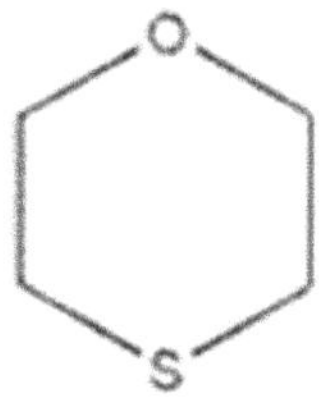	9.0–9.8	1	[90]
Amitriptyline	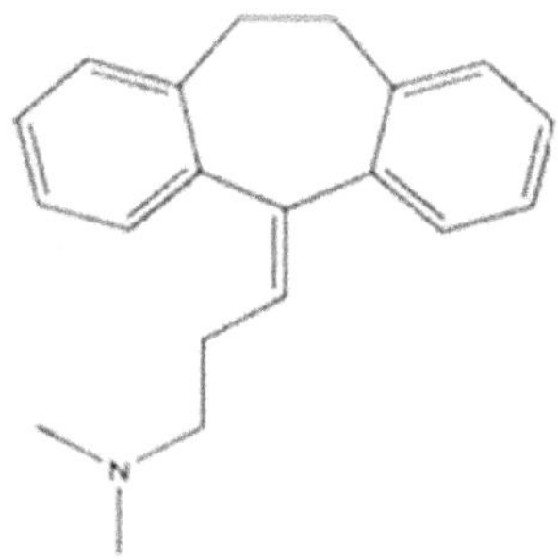	9.0	20	[91]
Nortriptyline	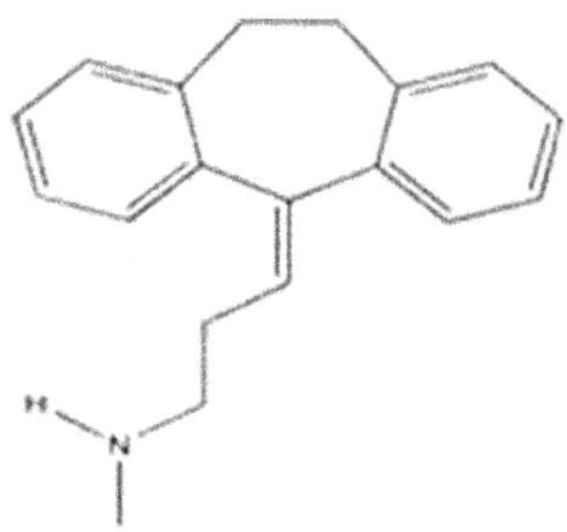	9.0	15	[91]

to pH 7 [75]. The oxidation of bisphenol A by Fe(VI) has been also investigated by Han et al. at various pHs and around 10% higher removal was reported at pH 7 compared to pH 9 [92], which is in contrary to the previous reported study related to bisphenol A. The authors stated that this is due to the fact that Fe(VI) is a stronger oxidant upon protonation [92]. The aforementioned differences in oxidative transformation of organics at the pH range of 7.0–9.0 are very small ($<10\%$) and within experimental and analytical errors. On the other hand, in all the cases mentioned above, the determined rate constants were higher at lower pH [59, 75, 92], as was expected. Zhou and Jiang have recently investigated the degradation of ciprofloxacin by Fe(VI) at the pH range of 6–9, and reported that the effect of the pH on the removal of ciprofloxacin is not significant due to the high reactivity of Fe(VI) with ciprofloxacin [56].

10.2.4.3 Application and Performance of Ferrate(VI) in Wastewater Treatment

With one dose of Fe(VI), oxidation, disinfection, and coagulation take place, which makes Fe(VI) very attractive for wastewater treatment. The results of selected studies that show the Fe(VI)'s ability to treat PPCPs and EDCs in wastewater are summarized in Table 10.6.

Fe(VI) can play a multipurpose role in the treatment of wastewater. In Tables 10.7–10.10, the performance of Fe(VI) in removal of TCOD (total chemical oxygen demand), TSS (total suspended solids), TBOD (total

Table 10.6 Removal of PPCPs and EDCs spiked in real wastewater by Fe(VI).

Contaminant (X)	Experimental conditions	% Removal	Ref.
17α-ethinylestradiol, β-estradiol and bisphenol-A in Lake Zurich water (DOC = 1.6 mg/L)	Dose: 0.5 mg Fe(VI)/L contact time = 30 min pH = 8 $[X]_0 = 0.15\ \mu M$	99	[52]
17α-ethinylestradiol, β-estradiol and bisphenol-A in Kloten wastewater (DOC = 5.3 mg/L)	Dose: 2 mg Fe(VI)/L contact time = 30 min pH = 8 $[X]_0 = 0.15\ \mu M$	99	[52]
naproxen, paracetamol, diclofenac, carbamazepine and triclosan in wastewater (DOC = 5 mg/L)	Dose: 10 mg Fe(VI)/L contact time = 1–5 min pH = 7 $[X]_0 = 100\ \mu g/L$	~100	[93]
sulfamethoxazole, diclofenac, and carbamazepine in wastewater (DOC = 5.1 mg/L)	Dose: 5 mg Fe(VI)/L contact time = 3–5 h pH = 7 $[X]_0 = 0.2–1\ \mu M$	>85	[50]

Table 10.7 Hailsham North Wastewater Treatment Plant of Southern Water Ltd of UK – Pilot scale – performance at 0.03 mg of online and electrochemically produced Fe(VI)/L [36].

0.03 mg Fe(VI)/L	**TSS**	**TCOD**	**TBOD**	**TP**
Untreated raw sewage (mg/L)	730	1125	388	18.5
Treated raw sewage (mg/L) (calculated)	153.3	562.5	271.6	8.1
Average %Removal	79	50	30	56

Contact time = 30 min

Table 10.8 Performance of commercial and electrochemically produced (using NaOH and KOH) Fe(VI) – samples taken from Wastewater Treatment Plant Degremont, in Culiacan city, in Mexico [37].

	Fe(VI) commercial (supplied by Sigma–Aldrich)	**Fe(VI) (obtained from NaOH)**	**Fe(VI) (obtained from KOH)**
Conditions	–	Steel (0.3%C) density = 46.19 mA/cm^2 [NaOH] = 12M	Steel (0.3%C) density = 46.19 mA/cm^2 [KOH] = 12M
Dose (mg/L)	40	40	16
%COD removal	17	50	51
pH of treated wastewater	7.4	10.8	12.6

Untreated wastewater: TCOD = 454 mg/L, Contact time = 15 min

Table 10.9 Performance (in removal of DOC) of Fe(VI) in comparison with ferric sulfate [24].

	Ferrate(VI)	**Ferric sulfate**
Dose (mg/L)	20	20
%DOC removal	70	45

Untreated fluvic acid model water: DOC=12.4 mg/L, Contact time = 20 min

Table 10.10 Performance of Fe(VI) in comparison with aluminum and ferric sulfate [94].

	Aluminum sulfate	Ferric sulfate	Ferrate(VI)
pH	6.75–7.48	6.75–7.48	7
Optimum dose (mg/L)	8	22	22
% TSS removal	91	95	94
% TCOD removal	7	18	32

biochemical oxygen demand), TP (total phosphorous) as well as DOC (dissolved organic carbon) is presented.

10.2.5 Future Directions

Nowadays, the activation of oxidants to enhance the oxidative transformation of organics is receiving a great attention. For example, researchers put efforts into investigating the activation of peroxymonosulfate, manganese, hydrogen peroxide and persulfate [95–98]. In the case of Fe(VI), limited work has been done so far related to its activation. Interestingly, the activation of Fe(VI) by acids (e.g., HCl and HNO_3) to enhance significantly the transformation of organics at lower Fe(VI) dose and shorter contact times, was recently demonstrated [99]. However, more investigation is needed to understand the mechanism of the acid-activation of Fe(VI) and test it on different compounds with different moieties.

Compared to the kinetic studies of oxidation of PPCPs and EDCs by Fe(VI) that are available in the literature, the effect of water constituents such as ions and organic matter on the oxidative transformation of these compounds was not excessively studied. Moreover, the toxicity of the Fe(VI)-treated water compared to the untreated water has to be evaluated by doing toxicity assessments, especially in the cases that no mineralization of the contaminant was observed. The identification of possible intermediates will help to understand the mechanisms of these reactions and the Fe(VI)-related chemistry in water.

Fe(VI) is a powerful oxidant that can be used to oxidize organics in water. The synthesis, instability and storage of Fe(VI) are still typical issues. Any improvement on these aspects will be beneficial for the further development and application of the Fe(VI) water treatment technology.

10.3 Photocatalysis

10.3.1 Introduction

Diverse methods of advanced oxidation processes are being studied for the degradation of various organic pollutants. Among them, heterogeneous photocatalysis is congregating an immense prominence due to its high competence for addressing enormous number of organic pollutants. The following sections will be concentrating on the basic principles behind photocatalysis, various approaches of the method and different types of photocatalysts used for the process.

10.3.1.1 General Concept of Photocatalysis

In heterogeneous photocatalysis a semiconducting material is being used as a catalyst. There are several semiconducting materials that can be used as catalyst for photocatalysis, such as TiO_2, ZnO, Fe_2O_3, CdS, GaP and ZnS. These compounds have shown their effectiveness in degrading several persistent organic pollutants, water pathogens and disinfection by-products [100]. In the photocatalytic processes, reactive hydroxyl radical and superoxide ions are formed which react with pollutants degrading them to carbon dioxide and water. Furthermore, few advantageous qualities of photocatalysis have increased its potential as a water treatment process such as the fact that the process can be carried out at ambient temperature and pressure and it is a low-cost process [100].

Among the different semiconductors used for photocatalysis, as mentioned earlier, TiO_2 is the most widely used one due to its high stability, non-toxicity, corrosion resistance and insolubility in water, excellent optical transparency (both visible and infrared light) and low cost [101]. TiO_2 is excited in the wavelength range from 330 nm to 390 nm, and remains stable even after few cycles of the photocatalytic process [100]. On the other hand, CdS and GaP both degrade after a single photocatalytic reaction [100]. However, there are few limitations related to the use of TiO_2 which hinders the process from being commercially used. First, the separation of the catalyst after its use is a major concern for TiO_2. Moreover, due to its very small particle size and high surface area-to-volume ratio, it has a high tendency to agglomerate, thereby reducing the active sites and its optical efficiency [100]. Another major disadvantage of TiO_2 is its wide band gap. All these issues related to catalysts and how they can be overcome are discussed in the following sections of this chapter.

10.3.1.2 Basic Principle of Photocatalysis

Photocatalytic reactions can be defined as a photo induced reaction that is being accelerated by the addition of a catalyst [102]. As already mentioned, in heterogeneous photocatalysis a solid semiconducting material is being used as a catalyst. When light energy is incident on these catalysts, they absorb the light energy. When the absorbed light energy (hυ) is higher than their bandgap (E_{bg}), an electron (e^-) from the conduction band of the catalyst gets transmitted to its valence band producing a hole (h^+). Hence a pair of e^- and h^+ is being produced [103]. This is called the excited state of the catalyst. The formed e^- acts as a strong reducing agent (+0.5 to −1.5V) and the valence band h^+ acts as a strong oxidizing agent (+10 to +3.5V) [101]. This h^+ reacts with either water or hydroxyl anion (OH^-) to form hydroxyl radical ($OH^{\bullet}$). Hydroxyl radical (non-selective oxidant) reacts with the target pollutant as well as with the possible intermediates formed, resulting in mineralization of the organic carbon (formation of CO_2, H_2O and mineral acids) [101]. Generally, the hydroxyl radical reactions occur at the surface of the catalyst [104, 105].

As it was stated earlier, TiO_2 is the most widely used semiconductor for photocatalytic processes. The bandgap of TiO_2 (anatase) is 3.2 eV and that of TiO_2 (rutile) is 3.0 eV [100]. When the absorbed light energy (hυ) on TiO_2 becomes higher than its band gap, it goes to the excited state. The reaction mechanism with TiO_2 is shown in Eqns. 10.12 and 10.13 [106]:

$$TiO_2 \xrightarrow{hv,\ \lambda<380\ nm} TiO_2(e^- + h^+) \tag{10.12}$$

$$h^+ + OH^-_{ad} \longrightarrow OH^{\cdot}_{ad} \tag{10.13}$$

The overall photocatalytic degradation process is summarized in Eqn. 10.14 [106]:

$$\text{Organic Pollutant} + O_2 \xrightarrow{\text{Semiconductor + UV-A Light}} CO_2 + H_2O + \text{mineral acid} \tag{10.14}$$

The first step of any photocatalytic process is the adsorption of pollutant molecule on the surface of catalyst where the actual reaction takes place. In order to differentiate between removal of the organic due to adsorption and photocatalytic oxidation, the molecule is first allowed to get adsorbed on the catalyst and once adsorption equilibrium is reached, the reaction solution is exposed to light. This type of adsorption usually follows one of the two types of adsorption isotherm: Langmuir or Freundlich adsorption isotherm. In Langmuir adsorption model the following assumptions are taken into consideration [107]:

a. All adsorption sites are identical
b. Each adsorption site can adsorb only one adsorbate molecule
c. The sites are not energetically independent on the adsorbed quantity.

This model is presented in Eqn. 10.15 [107]:

$$Q = Q_{max} \frac{Lc}{1 + Lc} \tag{10.15}$$

where, Q is the solid concentration of the adsorbate, Q_{max} is the maximum adsorption capacity of the adsorbent, L is the affinity of the adsorbate of the adsorbent and Q_{max}. L corresponds to a constant K_d, called distribution coefficient [107]. Equation 15 can be linearized to the following form (Eqn. 10.16) [107]:

$$\frac{Q}{c} = Q_{max} L * LQ \tag{10.16}$$

In most of the cases adsorption on solid surface follows Freundlich adsorption isotherm [21]. This isotherm is represented by Eqn. 10.17 [107]:

$$Q = F * C^n \tag{10.17}$$

Eqn. 10.17 can also be linearized to Eqn. 10.18:

$$\log Q = logF + nlogC \tag{10.18}$$

where, Q is the adsorbed quantity of the adsorbate, C is the remained quantity of the adsorbate, F and n are two constants [107].

In few recent studies, it has been shown that the efficiency of photocatalytic process is not only confined to the degradation of emerging organic pollutants, but it is also equally competent in microbial disinfection [108]. The fact that microorganisms can also be deactivated by photocatalysis was first established in 1985 [109]. The study was done on E. Coli bacteria. Till then, several studies have been carried out on photocatalytic disinfection processes. From these studies, though not very clear, but we have also come to know different mechanisms involved in these photocatalytic disinfection processes. Although microbial cells consist of different lipids, proteins, nucleic acids (DNA and RNA) and polysaccharides, which can be attacked by hydroxyl radical, in most of the cases these compounds are surrounded by a rigid cell wall. Hence it can be assumed that the first step of the disinfection process is the disintegration of the cell wall [108]. From a study on the components of microbial cells, it has come to our knowledge that around 96% of a cell's dry mass consists of organic macromolecule [110]. Around 3% of the dry weight includes all types of monomers, such as amino acids, sugars and nucleotides and rest 1% includes inorganic ions [110]. In different studies, it has been shown that the photocatalytically generated (OH·) can successfully attack and completely damage the primary structures of amino acids [111, 112] and DNA [113]. Moreover, few target sites have also been identified where the photocatalytic attack can take place. These target sites have been broadly classified as intracellular and extracellular target sites [108]. The major extracellular target sites are peptidoglycan, lipids and polysaccharides. The intracellular target sites mainly include enzymes, coenzymes and nucleic acid.

10.3.2 Design Parameters of Photocatalysis

The catalysts used in photocatalytic processes are mostly chemically inert, non-hazardous, inexpensive and can be reused for quite a few times. Moreover, the low energy requirement of the process makes it economically viable. The average energy requirement of a photocatalytic reaction is as low as 1–5 W/m^2 [114]. Sometimes, with few modifications of the catalyst, the activation energy can also be sourced from natural sunlight, which makes the process more cost effective and environmentally-friendly. Table 10.11 shows a comparison between estimated cost of different water treatment processes [115].

Despite of the many potentialities of photocatalysis, there are few factors for which the process has not yet been commercially utilized in industries.

In a broader view, two facts can be stated which hinders the use of this technology commercially: i) the low photocatalytic efficiency with respect to the photoreactor configuration and (ii) the lack of instances and guidelines of the process from being scaled up commercially [103]. Moreover, TiO_2 in a photocatalytic process can be used not only as a slurry but also as immobilized. Both design techniques have their own pros and cons. When TiO_2 is being used as a slurry then number of active sites are far higher than the immobilized catalyst. Hence the activation rate is higher. But filtration of the catalyst at the end of the process is a

Table 10.11 Estimated cost of different water treatment processes [115].

Capacity	Cost in USD		
M^3/hr	Activated carbon	UV/O_3	Photocatalysis
5.0	7.8	13.0	9.9
18.0	4.3	6.3	4.4
36.0	3.2	4.9	3.2
145.0	2.2	3.8	2.3
385.0	2.0	3.1	2.0

challenge due to the fine particle size of the catalyst. Hence the rate of recyclability decreases and the cost of a membrane increases. In case of immobilized TiO_2, no filtration is required after the application. However, due to the immobilization of the catalyst, the number of active sites decreases, thereby reducing the catalyst activity. There are several studies which addresses these reactor design and catalyst parameters to enhance the efficiency of the process [116–119]. The following sections will expound some of those design parameters for photocatalysis.

10.3.2.1 Different Aspects of Design Parameters

In conventional chemical reactor design the parameters which are mostly contemplated are mass transfer, mixing rate, flow pattern, reaction kinetics, catalyst fixing, temperature regulator, reactant – catalyst contact, etc. [114]. However, in photocatalytic reactor design the most important design factor is the catalyst illumination factor because the higher the illumination the more the catalyst activation, hence the water treatment efficiency of the process is higher [114]. According to a study the volume of a photocatalytic reactor can be expressed by Eqn. 10.19 [114]:

$$V_r = \frac{QXC_{in}}{\kappa \Re} \tag{10.19}$$

where, Q is the volumetric flowrate, V_r is the reactor volume, X is the fractional desired conversion, κ is the illuminated catalyst surface area in contact with reaction liquid and $\Re$ is the average mass destruction rate.

According to Eqn. 10.19, one can say that V_r will be small when κ and $\mathfrak{R}$ are substantially high [114].

In recent years, different types of reactors have been patented for photocatalysis. These reactors can be generally catagorized into four categories [116]: (i) the slurry type reactor where the catalyst is being suspended in a form of slurry [106, 120], (ii) the immersion type where the light source is being immersed into the reactor [117], (iii) the external type where the light source is placed outside the reactor [121], and (iv) the distributive type where the reactor includes a reflector and light conductors [118]. Most of the reactors established till date are a type of slurry reactor [116]. Few immobilized reactors are also being used. Most of these reactors are of the type of external reactor [116]. Generally in these types of external reactors the catalyst is being immobilized on the reactor walls, internal pipe lines, on semipermeable membranes, glass wool matrix or on the ceramic membranes [119]. Considering the shapes, the reactors can be either helical or shallow cross flow basins or spiral or optical fibre [116]. However, till date all these reactors are designed for small scale and for large scale applications only multiple number of small units can be used [116]. The κ values of these different types of reactors are shown in Table 10.12 [116].

10.3.2.2 Reactor Design Limitations Along with Proposed Solution

There are few challenges that need to be addressed before designing a photocatalytic reactor. As already mentioned, the volume of the reactor (V_r) can be minimized if the average mass destruction rate ($\mathfrak{R}$) is quite high. In other words we can say that if the destruction rate is high we will need less amount of catalyst to be activated and hence the required reactor volume will be less [114]. A high value of $\mathfrak{R}$ can be achieved by modifying the catalyst [114]. The catalyst modification for higher efficiency of photocatalysis is discussed in the following sections.

Another major challenge of designing a photoreactor is the requirement of a large amount of catalyst to be activated inside the reactor [114]. Even if the effective surface area of the catalyst is very high, then also the catalyst can be present in the reactor in the form of a thin film. This eventually restricts the efficiency of the reactor. Moreover, due to this limitation the retention time of the reactor also eventually increases [114]. Hence the efficacy of a reactor increases by addressing these two issues.

One of the new concepts of photocatalytic reactor is the hollow tube reactor or multiple tube reactor (MTR) [114]. In this type of reactor, hollow tubes have been used as light conductors and this reactor satisfies most

Table 10.12 Comparison of κ value in m^2/m^3 for different types of reactors [116].

Photocatalytic reactor	k, m^2/m^3	Parameters	k, m^{-1}	Remarks	Ref.
Slurry Reactor	$\left[\frac{6C_c}{\rho c}\right]\frac{1}{d_p}$	$d_p = 0.3$ µm $C_c = 0.5$ Kg/m³	2631 (the actual value is much lower than this)	Scale-up not possible	[106]
External type – annular reactor	$\frac{4d_0}{d_0^2 - d_i^2}$	$d_0 = 0.2$ m $d_i = 0.1$ m	27	Scale-up not possible	[121]
Immersion type – with classical lamps	$\left[\frac{4\in}{1-\in}\right]\frac{1}{d_0}$	$d_0 = 0.09$ m $\varepsilon = 0.75$	133	Scale-up possible but large V_r	[114]
Immersion type with novel lamps	$\left[\frac{4\in}{1-\in}\right]\frac{1}{d_0}$	$d_0 = 0.0045$ m $\varepsilon = 0.75$	2667	Scale-up possible with small V_r	[114]
Distributive types with hollow tubes	$\left[\frac{4\in}{1-\in}\right]\frac{1}{d_0}$	$d_0 = 0.006$ m $\varepsilon = 0.75$	2000	Scale-up possible with small V_r	[118]

Table 10.13 Comparison of reactor specifications of CAR, TLR, and MTR[114].

Specifications	CAR	TLR	MTR
Volume of reactor, m^3	3.48×10^{-3}	5.36×10^{-4}	1.23×10^{-3}
Catalyst surface are, m^2	0.18	0.15	0.51
Parameter, k, m^2/m^3	69	618	1087
Volumetric flow rate, m^3/S	8.42×10^{-5}	1.67×10^{-5}	3.00×10^{-5}
Electrical energy input, W	400	126	40
Scale-up possibilities	No	Yes	Yes
Ref.	[121]	[114]	[114]

of the conditions for scaling it up. In this reactor a high surface area of the catalyst have been achieved at a relatively low reactor volume. Infact, 70–100 folds of surface area has been accomplished per m^3 of the reactor volume as compared to the Classical Annular Reactor (CAR) [114]. In this innovative reactor design, the light enters through one side of the hollow tube and being reflected through out the length of the tube. During these reflections the light is being exposed to the immobilized catalyst coated on the outer surface of the tube. Since the cylindrcal surface area is quite huge (200 times than the light entrance area), a large amount of catalyst can be illuminated by this method [114].

There are other two conventional reactors, called Clasical Annular Reactor (CAR) and tube light reactor (TLR), which are also being used for photocatalyst. In CAR a single lamp is placed in a tube or annulus through which the reactant solution passes through and the outer layer of the inner annulus are being coated with the catalyst [114]. A CAR has 0.099 m outer daimeter, 0.065 m of inner diameter and 0.77 m of length. The TLR is an immersion type reactor which consists of 21 lamps. They have a very narrow diameter of 0.0045 m and catalyst is coated on its surface. The lamp is immersed inside the reactor liquid [114]. A comparison study of these three reactors is given in Table 10.13 [114].

10.3.3 Photocatalysts

Photocatalysts play the most crucial role in any photocatalytic reaction. This is the compound on which the overall process efficiency and capability is dependent on. The light energy requirement for the process and the type of light required (UV or visible) is also determined by the photocatalyst used in the process. From the inception of photocatalysis, semiconducting materials have been used as catalysts. Semiconducting materials can be difined as the compound whose electrical conductivty is in between that of a conductor and an insulator. The conduction band of any semiconducting material is either half filled or empty, so that when it absorbs light energy, an electron from their valence band gets excited and are promoted to the conduction band. Hence the conduction band produces one electron (e^-) and the valence band produces one hole (h^+). This is the basic principle of the excitation of semiconductors. The imparative factor here is the light energy. Every semiconductor has its characteristic band gap, that is, the energy difference between the valence band and conduction band. The absorbed light energy should be higher than the band gap of the semiconductor to get excited. Different semiconducting materials along with their different band gaps are presented in Table 10.14 [122].

So far, TiO_2 is the most widely used photocatalyst [122]. Another widely used catalyst is ZnO. However, regardless of these advantages of TiO_2 there are certain limitations of this compound which restricts its use in the commercial scale. In Table 10.14, we can see that the band gap of TiO_2 is quite high (3.2 eV) and hence the wavelength of light energy required to excite it is less than 387 nm. Consequently UV light is needed to excite TiO_2 molecule. The similar kind of constraints also exist for ZnO. Now, we can only make a process more environmentally friendly, by using sunlight as its energy source. However, sunlight contains only around 5% UV light and the rest of it is mostly visible light. Hence for solar photocatalysis, where the photocatalysis

Table 10.14 Common semiconductors used in photocatalysis [122].

Semiconductor	Band gap (eV)	Wavelength (nm)	Light absorption	Valence b and (V vs. NHE)	Conduction b and (V vs. NHE)
TiO_2	3.2	387	UV	+3.1	−0.1
SnO_2	3.8	318		+4.1	+0.3
ZnO	3.2	387		+3.0	−0.2
ZnS	3.7	335		+1.4	−2.3
WO_3	2.8	443	Visible	+3.0	+0.4
CdS	2.5	496		+2.1	−0.4
CdSe	2.5	729		+1.6	−0.1

is done solely by sunlight, only a very small portion of the sunlight can be utilized for catalyst excitation. That not only decreases the effciency of the process but also makes it energetically ineffcicient. Hence, further modification of TiO_2 is required so that it can absorb the visible region of solar spectrum. Several modification techniques on semiconductors have already been studied and their few details have been outlined in the following section.

10.3.3.1 Doping of TiO_2

One of the most expansively studied method to overcome the above mentioned inadequecy of TiO_2 is doping of the semiconductor with some other ions, that is, imparting defect on TiO_2 structures. It can be doped either by anions or cations to sensitize the catalyst toward visible light. The doping of TiO_2 not only helps to sinsitize it toward visible light but also helps prevention of electron (e^-) and hole (h^+) recombination. Different types of doping of TiO_2 are explained below.

10.3.3.1.1 Cation Doped TiO_2

One of the most extensively used method for doping TiO_2 is cation doping. Doping of TiO_2 with different types of cations like rare earth metals, noble metals, poor metals, and transition metals has been investigated in the past few years [123, 124]. The reasons behind the increased efficiency of photocatalysis of a cation doped catalyst are as follows[125–127]:

- It increases the light absorption range of the catalyst. Doped TiO_2 can be effciently activated by the visible range of solar spectrum, thereby making the process energetically efficient.
- The generated radicals by this process has a higher redox potential than the radicals generated from normal TiO_2.
- It prevents the recombination of conduction band electron and valence band hole, which eventually increases the quantum efficiency of the process.

Another phenomenon can also explain the increased photocatalytic activity of cation doped TiO_2 [128, 129]. If TiO_2 is being doped with a lower oxidation state cation (e.g., Fe^{3+}) (conduction p-type), to maintain the electrical neutrality, electrons from valence band are being removed. As a result, conduction band hole is produced eventually. This way, the number of e^- and h^+ produced increases, thereby increasing the efficacy of photocatalysis. In another way, if TiO_2 is being doped with higher oxidation state cation (e.g., Nb^{5+}) (condution n-type), to maintain the electrical neutrality, excess electron are placed in the empty conduction band of TiO_2. This eventually also increases effciency of photocatalysis.

One of the foremost example of cation doping is doping of rare earth metals on TiO_2. The rare earth metals essentially includes Scandium (Sc), Yttrium (Y) and 15 lanthanoids [122]. According to the previous studies, the presence of incomplete 4f and empty 5d orbitals in these metals actually increases the rate of photocatalysis [123]. According to another study [130], Cerium (Ce) can be successfully used as a rare earth metal for doping on TiO_2. There are few definite advantages of using Ce for this purpose. It helps to prevent the size increase of TiO_2 after doping. Hence, the crystallite size of TiO_2 decreases, thereby increasing the specific surface area of the catalyst and eventually increasing its photocatalytic activity [130]. Moreover, the formed redox pair (Ce^{3+}/Ce^{4+}) basically acts as an electron scavenger. Hence it can trap the electron generated from TiO_2, preventing the recombination of electron and h^+. Consequently that increases the efficacy of the photocatalysis process. Among rare earth metals, Gadolinium (Gd) has also been identified as a very efficient doping cation for TiO_2 [131]. When Gd^{3+}/TiO_2 pair is being prepared by sol-gel method, it provide a catalyst with lowest band gap and highest surface area [131].

Several studies have been carried out till date on doping of TiO_2 with several noble metals and almost all of these studies have shown promising results [122]. Noble metals like Osmium (Os), Palladium (Pd), Ruthenium (Ru), Rhodium (Rh), Silver (Ag), Iridium (Ir), Platinum (Pt) and Gold (Au) have shown their excellence in photocatalytic activity when doped on TiO_2 [122]. Besides, all these metals have excellent resistance to corrosion [123]. The activity of silver doped TiO_2 has been studied with different model compounds like Rhodamine 6G Dye [132], Methylene Blue [133], Methylene Orange [134]. In all these studies it has been shown that the activity of TiO_2 increases due to presence of silver and hence the rate of degradation of these model compounds also increases with Ag doped TiO_2 [132–134]. The increased activity of Ag doped TiO_2 can be attributed to the increased specific surface area of the doped catalyst [123]. Also, the Ag on TiO_2 surface acts as an electron trap which ultimately hinders the recombination of e^- and h^+ and increases the photocatalytic activity [132, 133].

Doping of TiO_2 has also extensively being studied with transition metals. Based on IUPAC nomenclature, a transition metal can de defined as an element which consists of an incomplete d subshell [122]. Transition metals like Iron (Fe), Cobalt (Co), Nickel (Ni), Manganese (Mn), Chromium (Cr), Vanadium (V), Copper (Cu), Zinc (Zn), and Zirconium (Zr) have been studied for their catalytic activity when doped with other photocatalysts [122]. It can be concluded that transition metals, when doped on

TiO_2, are effective in transferring the light absorption from UV to visible region. However, the photocatalytic activity of transition metal doped catalyst depends on few factors like the type of transition metal, the concentration of metal and their microstructural characteristics [135].

10.3.3.1.2 Anion Doped TiO_2

Another promising method for catalyst doping is the anion doping. Anion doping primarily increases the photocatalytic activity of TiO_2 by inhibiting the recombination of formed e^- and h^+. Anions like Carbon (C) [136, 137], Nitrogen (N) [138, 139], Sulfur (S) [140, 141], and Iodine (I) [142] have been extensively studied for their doping efficiencies on TiO_2. From these studies it is evident that minimising the recombination of e^- and h^+ is more promising in anion doped TiO_2 than in cation doped [143, 144]. It has also been shown that the morphology and eventually the photocatalytic activity of TiO_2 is greatly enhanced by anion dopings [145]. The band gap of anion doped TiO_2 has been found to be much lesser than that of normal TiO_2 [146]. As a result, these doped catalysts can be effectively activated by the use of visible spectrum of solar light.

Among these anions, carbon doping of TiO_2 has gathered much attention due to few reasons. When the carbon molecule is being introduced into the TiO_2 structure, it leads to a new state (C 2p) close to the valence band edge of TiO_2 (O 2p). Accordingly, the conduction band of TiO_2 shifts and results in a lower band gap [147–149]. Also, the specific surface area of the carbon doped catalyst becomes higher than the normal catalyst. Hence the photocatalytic activity increases due to increased number of active sites. This increased activity has been studied on tricholoro acetic acid, where it has been found that the degradation of the model compound is much more efficient under solar light with carbon doped TiO_2 than normal TiO_2 [149]. This increased efficiency of carbon doped TiO_2 is mainly due to the rapid formation of superoxide (O^{2-}) anion and hydroxyl radical ($OH\cdot$) under visible light [147–149]. Nitrogen is also gathering an immense importance for doping on TiO_2 because of its more environment friendliness than other doping anions [148].

10.3.3.2 Coupled Semiconductors

In recent years, another approach is being immensely studied which is called the coupled semiconductors [122], where a semconductor is being coupled with another one to decrease their band gap effectively. In this approach, a high band gap semiconductor like TiO_2 is being coupled with a low band gap semiconductor [150]. When a low band gap semiconductor having a more negative conduction band is being coupled with TiO_2, the formed electron can be injected from the low band gap semionductor to TiO_2, thereby activating the TiO_2 molecule [150]. However, the effectiveness of the process depends on the potential differences of conduction and valence bands of the two semiconductors [151]. It can also be said that this electron transfer will be more successful if the conduction band of TiO_2 is more anodic than the other semiconductor [152]. In other words, the valence band of the second semiconductor must be more cathodic than TiO_2 [152].

The coupled semiconductors which are primarily used for water treatment are M_xO_y/TiO_2 and M_xS_y/TiO_2 [152]. An example of M_xS_y/TiO_2 system is CdS/TiO_2. In this system, CdS has a lower band gap of 2.4 eV and it is one of the most extensively studied coupled semiconductor [153]. Both the conduction band and the valence band are higher than that of TiO_2 [150]. This system has been used to study the degradation of phenazopyridine where 75% degradation of the model compound has been achieved after 60 min of light irradiation [154]. In this study, a solar lamp (400–800 nm, 0.0212 W/cm^2) has been used as the source of light energy [154]. Another example of M_xS_y/TiO_2 system is $CdSe/TiO_2$. CdSe has a band gap as low as 1.74 eV and has been identified as another promising element for semiconductor coupling with TiO_2 [155]. The efficiency of the system has been studied for degradation of 4-Chlorophenol. More than 30% degradation and more than 20% removal of total organic carbon has been achieved under solar light [155].

An example of M_xO_y/TiO_2 system is SnO_2/TiO_2 or ZnO/TiO_2. Both the systems have been used for the degradation study of methyl orange dye [150, 156]. It has been found that SnO_2/TiO_2 can degrade methyl orange to up to 95% under visible light [150]. Similarly for ZnO/TiO_2 the degradation has been found to be 98.1% under visible light [156]. Another example of M_xO_y/TiO_2 system is WO_3/TiO_2 which received a lot of attention in photo-electrochemistry in recent years [153]. The band gap of WO_3 is 2.8 eV [157], hence it can be excited by the visible light. Besides, the upper edge of the valence band and the lower edge of the conduction band of WO_3 are lower than that of TiO_2. Hence, after light activation of WO_3 the generated h^+ can be transferred from WO_3 to TiO_2 [158]. However, the preparation technique of WO_3/TiO_2 is quite critical and for that reason SnO_2/TiO_2 system is gathering even more attention [150]. These results clearly show the potentiality of using coupled semiconductors over normal semiconductors.

10.3.3.3 Dye-Sensitized Catalyst

Another new approach for catalyst modification is the dye-sensitization of semiconductors. This is an incredibly promising approach that is being extensively studied in these days [159]. In this process, a dye molecule is being incorporated into the TiO_2 molecule. The dye so selected that it can absorb visible light. A dye molecule get chemisorbed on the surface of TiO_2. When visible light is being incident on the dye/TiO_2 system, the chemisorbed dye molecule absorbs the visible light and gets excited. The excited state of dye can be of two types. In one type the excited dye throws an electron to the conduction band of TiO_2, which is called the anodic dye sensitization [160]. In the second type, after the excitation of dye, it generates a hole (h^+) and injects it to the valence band of TiO_2. This type of dye-sensitization is called cathodic dye sensitization [161]. In most of the cases, anodic dye has been used for photocatalysis. The anodic dye sensitization process is shown in Eqns. 10.20 and 10.21. When the visible light energy is incident on a dye, it absorbs the light energy (hυ) and gets excited (S^*) (Eqn. 10.20). This excited dye then injects an electron (e^-) to the conduction band of TiO_2 (Eqn. 10.21), thereby exciting the TiO_2 molecule. Hence, the catalyst is getting excited by using visible light [159].

$$S + h \rightarrow S^* \tag{10.20}$$

$$S^* \rightarrow S^+ + e_{CB}^- \tag{10.21}$$

The excited dye molecule ultimately takes an electron from an electron donor present in the system. The electron donor can be a molecule or a mediating redox couple in a regenerative cell [159]. For example, Triethanol amine or EDTA can be used as an electron donating agent.

The type of dye effective for dye-sensitized photocatalysis depends on several factors. Few of them are listed below [160]:

- surface anchoring group
- energy levels
- ground state redox potential

Phosphates and carboxylic anchoring groups have been found to have strong covalent bond with the semiconductors. A strong covalent bond finally intensifies the electronic coupling between the molecular orbital of dye and the semiconductor, which eventually increases the electron transfer rate between them [162].

There are many dyes that can be used as sensitizers for semiconductors. Porphyrins, coumarins, phalocyanines, carboxylate derivatives of anthracine are few examples of such sensitizers [159]. Other than these, transition metal dyes are also being extensively studied nowadays due to their strong dye-sensitization properties. The transition metal dyes have been proved to be the best for dye-sensitization process [163]. Transition metals such as Ru(II), Fe(II), and Os(II) have shown their excellect properties as dyes. These metals can form a d^6 complex and can undergo a strong charge tranfer absoption through out the entire visible range [163]. Primarily, Ru(II) polypyridines complexes are used for photocatalysis [159]. However, all these transitions metal dyes consist of heavy metals which is not environmentally friendly. Besides, the systhesis process of these dyes is very complex and expensive. Hence, a new approach to dye sensitization is the use of natural dyes and replacement of heavy metal dyes. The natural dyes can be extracted from fruits, flowers, leaves and vegetables [159]. Table 10.15 shows few examples of these natural dyes [159].

Table 10.16 shows the comparative efficiencies of different modified photocatalysts on phenol degradation [167]. From Table 10.16, it is evident that by the process of dey-sensitization we can succesfully run the photocatalysis process under visible solar light. The efficiency of the process is also quite high as compared to the other photocatalytic processes.

Table 10.15 Extracts used as natural dyes for dye-sensitization [159].

Extract source	Ref.
Rosella	[164]
Blue Pea	[164]
Jaboticaba's skin	[165]
Chaste tree fruit	[166]
Mulberry	[166]
Cabbage – palm fruit	[163]
Java palm	[163]
Pomegranate seeds	[163]

Table 10.16 Photocatalytic degradation of Phenol with different modified photocatalysts [167].

Photocatalysts	Initial concentration of phenol	Light source	Degradation rate	Ref.
Zn(0.5%)/ Fe(1%) – TiO_2	5×10^{-4} mol/L	Solar light	Complete degradation in 6 hr	[168]
Au(2%) – TiO_2	30 ppm	Solar light	Complete degradation in 3.5 hr	[169]
Fe(0.5%) – TiO_2	3.56×10^{-4} mol/L	UV light	66% degradation in 8 hr	[170]
Pr(0.072%) – TiO_2	50 ppm	UV light	94% degradation in 2 hr	[171]
La(1%) – TiO_2	0.2 mmol/L	UV Light	27% degradation in 2 hr	[172]
Eosin Y – TiO_2/ Pt(0.5%)	40 ppm	Visible solar light	100% degradation in 2 hr	[173]
ZnO – G/ TiO_2 – G(1%)	40 ppm	Solar Light	100% degradation in 1 hr	[167]

10.3.4 Challenges and Future Prospects of Photocatalysis

Heterogeneous photocatalysis can gather an immense importance in future water treatment processes due its enormous potential of addressing several water contaminants, microbes and emerging pollutants. However, this process could not be commercialized till date due to few limiting factors. There are a few uncertainites that need to be addressed related to photocatalysis. For example, the replacement of UV light with solar light will significantly decreases the cost of the process while it will be more enviromentally friendly. Although a lot of work has been done for the development and understanding of the photocatalytic processes, specific improvements are needed particularly in the following areas:

- The generation of new catalyst systems that can be activated by visible spectrum of solar light, thereby making the process more sustainable.
- Increasing the environment friendliness and sustainability of the photocatalyst.
- Designing an effective solar photo reactor that can be scaled up to the commercial uses.

10.4 Combination of Photocatalysis (UV/TiO_2) and Ferrate(VI)

A limiting factor of photocatalysis is the potential recombination of conduction band electron (e_{cb}^-) and the valence band hole (h_{vb}^+), on TiO_2 surface, that result in process inefficiencies [174–177]. Fe(VI) can play the role of electron scavenger, solving the problem of the recombination of conduction band electrons with the holes, that happens in photocatalysis, to result in higher efficiency of the oxidation process [175]. In addition to this, the Fe(V) produced by the reduction of Fe(VI), which is 2 to 4 orders of magnitude more reactive than Fe(VI), has a great potential for remediation of pollutants [178]. Limited work related to the removal of organic pollutants in water using the combination of Fe(VI) and heterogeneous photocatalysis has been carried out [176, 177]. In Table 10.17, examples of enhanced oxidation of pollutants by photocatalysis-Fe(VI) system are presented. These studies show evidence for the enhancement of the photocatalytic oxidation in the presence of Fe(VI).

The concentration of dimethyl phthalate (DMP) decreased very slowly by TiO_2/UV photocatalysis or Fe(VI) alone indicating a clear selectivity of

Table 10.17 Pollutants are enabled to Fe(VI)-enhanced photocatalytic oxidation.

Organic (X)	Experimental conditions	% Removal	Ref.
Bisphenol-A (BPA)	Slurry system, 1 g TiO_2/L, 5.6 mg Fe(VI)/L, pH = 9, $[X]_0$ = 0.1 mM, reaction time = 10 min	Fe(VI): 33 TiO_2/UV: 18 Fe(VI)/TiO_2/UV: 65	[174]
Dibutyl Phthalate (DBP)	Slurry system, 20 mg TiO_2/L, 4.5 mg Fe(VI)/L, I = 0.40 mW/cm^2, pH = 9, $[X]_0$ = 5–7 mg/L, reaction time = 2 h	Fe(VI): 20 TiO_2/UV: 59 Fe(VI)/TiO_2/UV: 85	[175]
Dimethyl phthalate (DMP)	Slurry system, 40 mg TiO_2/L, 9 mg Fe(VI)/L, I = 0.40 mW/cm^2, pH = 9, $[X]_0$ = 10.3 mg/L, reaction time = 2 h, N_2 flow	Fe(VI): 5 TiO_2/UV: 12 Fe(VI)/TiO_2/UV: 65	[177]
Microcystin-LR (MCLR)	Slurry system, 2 g TiO_2/L, 4.5 mg Fe(VI)/L, 20.2 mg Fe(III)/L, I=40 W/m^2, pH=6, $[X]_0$=1–2 mg/L, reaction time=30 min, air flow	Fe(VI): 54 TiO_2/UV: 63 $Fe(VI)_{10min}$ then TiO_2/UV_{20min}: 79 Fe(III)/TiO_2/UV: 85 Fe(VI)/TiO_2/UV: 100 $Fe(VI)_{18.5mg/L}$/TiO_2/UV: 83 Fe(VI)/TiO_2/UV_{10min}: 80	[176]

Fe(VI) oxidation to DMP [177]. On the other hand, the concentration of DMP decreased rapidly by TiO_2/UV in the presence of Fe(VI). This dramatic enhancement effect of Fe(VI) is probably due to the oxidation of DMP by the produced Fe(V) and Fe(IV) species [175]. Moreover, these results demonstrate a synergetic effect of Fe(VI) and photocatalysis. In the case of DBP, the results emphasize the important role of Fe(VI) in enhancing the oxidation of DBP. BPA was oxidized by several oxidants (hydroxyl radicals, Fe(VI), and Fe(V)) and the recombination of electron/hole pairs was inhibited [174]. The degradation of ammonia using the Fe(VI)/photocatalysis system has been also studied. Although the %removal of ammonia was low, the study showed that the photocatalytic oxidation of ammonia has been enhanced by Fe(VI) [178].

10.5 Conclusions

There is an urgent need to develop greener wastewater treatment technologies due to the increased discharge of organic emerging pollutants to the water streams. The last few decades, Fe(VI) and photocatalysis received a great attention because of their ability to degrade organics in water. Researchers put a lot of efforts on the kinetic investigation of Fe(VI)-organic reactions as well as on the stoichiometry of the reactions to determine required Fe(VI) to organic pollutant molar ratios, showing the ability of Fe(VI) to remove a wide range of organics from water, leaving a non-toxic Fe(III) as a by-product initiating coagulation, in a multi-action Fe(VI) water treatment process. Studies related to the effect of water constituents, identification of oxidized products and the evaluation of the toxicity of the Fe(VI)-treated water are needed for the further development of the Fe(VI) technology. Compared to other chemical oxidants, little work has been done on the activation of Fe(VI) to enhance the oxidation process. Importantly, the activation of Fe(VI) by simple acids to enhance the oxidative transformations of organics at lower Fe(VI) to organic molar ratios and shorter contact time was recently reported [99]. At the same time, an enormous number of studies have also been carried out on photocatalysis due to its potential efficiency on the water pollutant degradation. The process can be carried out without the use of any hazardous toxic chemicals and in most of the cases the final product is innocuous carbon dioxide and water. This is the foremost reason behind the enormous prominence of photocatalytic processes. Moreover, few recent studies have also shown that by different modifications of the photocatalysts, the process can be made visible-solar-light driven, which further enhances the importance of the process in future.

References

1. Malato, S., Fernández-Ibáñez, P., Maldonado, M.I. Blanco, J. Gernjak, W., Decontamination and disinfection of water by solar photocatalysis: Recent overview and trends. *Catal. Today.*, 147, 1–59, 2009.
2. Pal, A., Gin, K.Y.-H., Lin, A.Y.-C., Reinhard, M., Impacts of emerging organic contaminants on freshwater resources: review of recent occurrences, sources, fate and effects. *Sci. Total Environ.*, 408, 6062–6069, 2010.
3. Wilkinson, J.L., Hooda, P.S., Barker, J., Barton, S., Swinden, J., Ecotoxic pharmaceuticals, personal care products, and other emerging contaminants: A review of environmental, receptor-mediated, developmental, and epigenetic toxicity with discussion of proposed toxicity to humans. *Crit. Rev. Environ. Sci. Technol.*, 46, 336–381, 2016.
4. Tijani, J.O., Fatoba, O.O., Babajide, O.O., Petrik, L.F., Pharmaceuticals, endocrine disruptors, personal care products, nanomaterials and perfluorinated pollutants: a review. *Environ. Chem. Lett.*, 14, 27–49, 2016.
5. Kim, M., Guerra, P., Shah, A., Parsa, M., Alaee, M., Smyth, S.A., Removal of pharmaceuticals and personal care products in a membrane bioreactor wastewater treatment plant. *Water Sci. Technol.*, 69, 2221–2229, 2014.
6. Cizmas, L., Sharma, V.K., Gray, C.M., McDonald, T.J., Pharmaceuticals and personal care products in waters: occurrence, toxicity, and risk. *Environ. Chem. Lett.*, 13, 381–394, 2015.
7. Lee, Y., von Gunten, U., Oxidative transformation of micropollutants during municipal wastewater treatment: Comparison of kinetic aspects of selective (chlorine, chlorine dioxide, ferrateVI, and ozone) and non-selective oxidants (hydroxyl radical). *Water Res.*, 44, 555–566, 2010.
8. Klementova, S. Kahoun, D. Doubkova, L. Frejlachova, K. Dusakova, M. Zlamal, M., Catalytic photodegradation of pharmaceuticals – homogeneous and heterogeneous photocatalysis. *Photochem. Photobiol. Sci.*, 16, 67–71, 2017.
9. Li, W., Jain, T., Ishida, K., Liu, H., A mechanistic understanding of the degradation of trace organic contaminants by UV/hydrogen peroxide, UV/persulfate and UV/free chlorine for water reuse. *Environ. Sci. Water Res. Technol.*, 3, 128–138, 2017.
10. Li, M., Qiang, Z., Hou, P., Bolton, J.R., Qu, J., Li, P., Wang, C., VUV/UV/chlorine as an enhanced advanced oxidation process for organic pollutant removal from water: assessment with a novel mini-fluidic VUV/UV photoreaction system (MVPS). *Environ. Sci. Technol.*, 50, 5849–5856, 2016.
11. Guan, X., He, D., Ma, J., Chen, G., Application of permanganate in the oxidation of micropollutants: a mini review. *Front. Environ. Sci. Eng. China.*, 4, 405–413, 2010.
12. He, J., Yang, X., Men, B., Wang, D., Interfacial mechanisms of heterogeneous Fenton reactions catalyzed by iron-based materials: a review. *J. Environ. Sci.*, 39, 97–109, 2016.
13. Wang, N., Zheng, T., Zhang, G., Wang, P., A review on Fenton-like processes for organic wastewater treatment. *J. Environ. Chem. Eng.*, 4, 762–787, 2016.

14. Sharma, V.K., Chen, L., Zboril, R., Review on high valent Fe^{VI} (Ferrate): a Sustainable Green Oxidant in Organic Chemistry and Transformation of Pharmaceuticals. *ACS Sustain. Chem. Eng.*, 4, 18–34, 2016.
15. Ning, B., Graham, N., Zhang, Y., Nakonechny, M., Gamal El-Din, M., Degradation of Endocrine Disrupting Chemicals by Ozone/AOPs. *Ozone Sci. Eng.*, 29, 153–176, 2007.
16. Ray, A.K., Photocatalytic Reactor Configurations for Water Purification. *Advances in Chemical Engineering*, 36, 145–184, 2009.
17. Xie, P., Guo, Y., Chen, Y., Wang, Z., Shang, R., Wang, S., Ding, J., Wan, Y., Jiang, W., Ma, J., Application of a novel advanced oxidation process using sulfite and zero-valent iron in treatment of organic pollutants. *Chem. Eng. J.*, 314, 240–248, 2017.
18. Kalidhasan, S., Ben-Sasson, M., Dror, I., Carmieli, R., Schuster, E.M., Berkowitz, B., Oxidation of aqueous organic pollutants using a stable copper nanoparticle suspension. *Can. J. Chem. Eng.*, 95, 343–352, 2017.
19. Jin, X., Peldszus, S., Huck, P.M., Reaction kinetics of selected micropollutants in ozonation and advanced oxidation processes. *Water Res.*, 46, 6519–6530, 2012.
20. Farmand, M., Jiang, D., Wang, B., Ghosh, S., Ramaker, D.E., Licht, S., Super-iron nanoparticles with facile cathodic charge transfer. *Electrochem. Commun.*, 13, 909–912, 2011.
21. Hu, L., Page, M.A., Sigstam, T., Kohn, T., Mariñas, B.J., Strathmann, T.J., Inactivation of bacteriophage MS2 with potassium ferrate(VI). *Environ. Sci. Technol.*, 46, 12079–12087, 2012.
22. Schink, T., Waite, T.D., Inactivation of f2 virus with ferrate (VI). *Water Res.*, 14, 1705–1717, 1980.
23. Sharma, V.K., Zboril, R., Varma, R.S., Ferrates: greener oxidants with multimodal action in water treatment technologies. *Acc. Chem. Res.*, 48, 182–191, 2015.
24. Jiang, J.-Q., Wang, S., Panagoulopoulos, A., The exploration of potassium ferrate(VI) as a disinfectant/coagulant in water and wastewater treatment. *Chemosphere*, 63, 212–219, 2006.
25. Jiang, J.-Q., Durai, H.B.P., Winzenbacher, R., Petri, M., Seitz, W., Drinking water treatment by in situ generated ferrate(VI). *Desalin. Water Treat.*, 55, 731–739, 2015.
26. Jiang, J.-Q., Lloyd, B., Progress in the development and use of ferrate(VI) salt as an oxidant and coagulant for water and wastewater treatment. *Water Res.*, 36, 1397–1408, 2002.
27. Sharma, V.K., Kazama, F., Jiangyong, H., Ray, A.K., Ferrates (iron(VI) and iron(V)): environmentally friendly oxidants and disinfectants. *J. Water Health.*, 3, 45–58, 2005.
28. Sharma, V.K., Oxidative transformations of environmental pharmaceuticals by Cl_2, ClO_2, O_3, and Fe(VI): Kinetics assessment. *Chemosphere*, 73, 1379–1386, 2008.

29. Jiang, J.-Q., Advances in the development and application of ferrate(VI) for water and wastewater treatment. *J. Chem. Technol. Biotechnol.*, 89, 165–177, 2014.
30. Lee, Y., Kissner, R., von Gunten, U., Reaction of ferrate(VI) with ABTS and self-decay of ferrate(VI): kinetics and mechanisms. *Environ. Sci. Technol.*, 48, 5154–5162, 2014.
31. Sharma, V.K., Ferrate(VI) and ferrate(V) oxidation of organic compounds: Kinetics and mechanism. *Coord. Chem. Rev.*, 257, 495–510, 2013.
32. Alsheyab, M., Jiang, J.Q., Stanford, C., On-line production of ferrate with an electrochemical method and its potential application for wastewater treatment - a review. *J. Environ. Manage.*, 90, 1350–1356, 2009.
33. Lapicque, F., Valentin, G., Direct electrochemical preparation of solid potassium ferrate. *Electrochem. Commun.*, 4, 764–766, 2002.
34. Ghernaout, D., Naceur, M.W., Ferrate(VI): In situ generation and water treatment - a review. *Desalin. Water Treat.*, 30, 319–332, 2011.
35. Bouzek, K., Roušar, I., Influence of anode material on current yields during ferrate(VI) production by anodic iron dissolution Part I: Current efficiency during anodic dissolution of grey cast iron to ferrate(VI) in concentrated alkali hydroxide solutions. *J. Appl. Electrochem.*, 26, 919–923, 1996.
36. Jiang, J.-Q., Stanford, C., Alsheyab, M., The online generation and application of ferrate(VI) for sewage treatment—A pilot scale trial. *Sep. Purif. Technol.*, 68, 227–231, 2009.
37. Taylor, P., Perez-sicairos, S., Carrillo-mandujano, A.J., Lopez-lopez, J.R., Lin-Ho, S.W., Wastewater treatment via electrochemically generated ferrate and commercial ferrate. *Desalin. Water Treat.*, 52, 37–39, 2014.
38. Thompson, G.W., Ockerman, L.T., Schreyer, J.M., Preparation and purification of potassium ferrate. VI. *J. Am. Chem. Soc.*, 425, 67–69, 1951.
39. Williams, D.H., Riley, J.T., Preparation and alcohol oxidation studies of the ferrate(VI) ion, FeO42−. *Inorganica Chim. Acta.*, 8, 177–183, 1974.
40. Luo, Z., Strouse, M., Jiang, J.-Q., Sharma, V.K., Methodologies for the analytical determination of ferrate(VI): a review. *J. Environ. Sci. Health. A. Tox. Hazard. Subst. Environ. Eng.*, 46, 453–460, 2011.
41. Ninane, L., Kanari, N., Criado, C., Jeannot, C., Evrard, O., Neveux, N., New Processes for Alkali Ferrate Synthesis, in: *Ferrates,* 985(6), pp. 102–111, American Chemical Society, 2008.
42. Golovko, D.A., Sharma, V.K., Suprunovich, V.I., Pavlova, O.V., Golovko, I.D., Bouzek, K., Zboril, R., A simple potentiometric titration method to determine concentration of ferrate(VI) in strong alkaline solutions. *Anal. Lett.*, 44, 1333–1340, 2011.
43. Venkatadri, A.S., Wagner, W.F., Bauer, H.H., Ferrate(VI) analysis by cyclic voltammetry, *Anal. Chem.*, 43, 1115–1119, 1971.
44. Noorhasan, N.N., Sharma, V.K., Baum, J.C., A Fluorescence technique to determine low concentrations of Ferrate(VI), in: *Ferrates,* 985(9), pp. 145–155, American Chemical Society, 2008..

45. Lee, Y., Yoon, J., von Gunten, U., Spectrophotometric determination of ferrate (Fe(VI)) in water by ABTS. *Water Res.*, 39, 1946–1953, 2005.
46. Sharma, V.K., Oxidation of nitrogen-containing pollutants by novel ferrate(VI) technology: a review. *J. Environ. Sci. Health. A. Tox. Hazard. Subst. Environ. Eng.*, 45, 645–667, 2010.
47. Sharma, V.K., Luther, G.W., Millero, F.J., Mechanisms of oxidation of organosulfur compounds by ferrate(VI). *Chemosphere.*, 82, 1083–1089, 2011.
48. Sharma, V.K., Oxidation of inorganic compounds by ferrate(VI) and ferrate(V): one-electron and two-electron transfer steps. *Environ. Sci. Technol.*, 44, 5148–5152, 2010.
49. Sharma, V.K., Potassium ferrate(VI): an environmentally friendly oxidant. *Adv. Environ. Res.*, 6, 143–156, 2002.
50. Lee, Y., Zimmermann, S.G., Kieu, A.T., von Gunten, U., Ferrate (Fe(VI)) application for municipal wastewater treatment: a novel process for simultaneous micropollutant oxidation and phosphate removal. *Environ. Sci. Technol.*, 43, 3831–3838, 2009.
51. Yang, B., Ying, G.-G., Zhao, J.-L., Zhang, L.-J., Fang, Y.-X., Nghiem, L.D., Oxidation of triclosan by ferrate: Reaction kinetics, products identification and toxicity evaluation. *J. Hazard. Mater.*, 186, 227–235, 2011.
52. Lee, Y., Yoon, J., von Gunten, U., Kinetics of the Oxidation of phenols and phenolic endocrine disruptors during water treatment with ferrate (Fe(VI)). *Environ. Sci. Technol.*, 39, 8978–8984, 2005.
53. Sharma, V.K., Mishra, S.K., Nesnas, N., Oxidation of sulfonamide antimicrobials by ferrate(VI) [$Fe^{VI}O_4^{2-}$]. *Environ. Sci. Technol.*, 40, 7222–7227, 2006.
54. Wang, Y., Liu, H., Liu, G., Xie, Y., Gao, S., Oxidation of diclofenac by potassium ferrate (VI): Reaction kinetics and toxicity evaluation. *Sci. Total Environ.*, 506–507, 252–258, 2015.
55. Sharma, V.K., Mishra, S.K., Ferrate(VI) oxidation of ibuprofen: a kinetic study. *Environ. Chem. Lett.*, 3, 182–185, 2006.
56. Zhou, Z., Jiang, J.-Q., Reaction kinetics and oxidation products formation in the degradation of ciprofloxacin and ibuprofen by ferrate(VI). *Chemosphere*, 119, S95–S100, 2015.
57. Anquandah, G.A.K., Sharma, V.K., Oxidation of octylphenol by ferrate(VI). *J. Environ. Sci. Heal. Part A.*, 44, 62–66, 2009.
58. Sharma, V.K., Anquandah, G.A.K., Nesnas, N., Kinetics of the oxidation of endocrine disruptor nonylphenol by ferrate(VI). *Environ. Chem. Lett.*, 7, 115–119, 2009.
59. Yang, B., Ying, G.-G., Chen, Z.-F., Zhao, J.-L., Peng, F.-Q., Chen, X.-W., Ferrate(VI) oxidation of tetrabromobisphenol A in comparison with bisphenol A. *Water Res.*, 62, 211–219, 2014.
60. Li, C., Li, X.Z., Graham, N., Gao, N.Y., The aqueous degradation of bisphenol A and steroid estrogens by ferrate. *Water Res.*, 42, 109–120, 2008.

61. Sharma, V.K., Liu, F., Tolan, S., Sohn, M., Kim, H., Oturan, M.A., Oxidation of β-lactam antibiotics by ferrate(VI). *Chem. Eng. J.*, 221, 446–451, 2013.
62. Karlesa, A., De Vera, G.A.D. Dodd, M.C., Park, J., Espino, M.P.B., Lee, Y., Ferrate(VI) oxidation of β-lactam antibiotics: reaction kinetics, antibacterial activity changes, and transformation products. *Environ. Sci. Technol.*, 48, 10380–10389, 2014.
63. Anquandah, G.A.K., Sharma, V.K., Knight, D.A., Batchu, S.R., Gardinali, P.R., Oxidation of trimethoprim by ferrate(VI): Kinetics, products, and antibacterial activity. *Environ. Sci. Technol.*, 45, 10575–10581, 2011.
64. Anquandah, G.A.K., Sharma, V.K., Panditi, V.R., Gardinali, P.R., Kim, H., Oturan, M.A., Ferrate(VI) oxidation of propranolol: kinetics and products. *Chemosphere*, 91, 105–109, 2013.
65. Zimmermann, S.G., Schmukat, A., Schulz, M., Benner, J., von Gunten, U., Ternes, T.A., Kinetic and mechanistic investigations of the oxidation of tramadol by ferrate and ozone. *Environ. Sci. Technol.*, 46, 876–884, 2012.
66. Anquandah, G., Ray, M.B., Ray, A.K., Al-Abduly, A.J., Sharma, V.K., Oxidation of X-ray compound ditrizoic acid by ferrate(VI). *Environ. Technol.*, 32, 261–267, 2011.
67. Yang, B., Ying, G.-G., Oxidation of benzophenone-3 during water treatment with ferrate(VI). *Water Res.*, 47, 2458–2466, 2013.
68. Sharma, V.K., Burnett, C.R. Millero, F.J., Dissociation constants of the monoprotic ferrate(VI) ion in NaCl media. *Phys. Chem. Chem. Phys.*, 3, 2059–2062, 2001.
69. Feng, M., Wang, X., Chen, J., Qu, R., Sui, Y., Cizmas, L., Wang, Z., Sharma, V.K., Degradation of fluoroquinolone antibiotics by ferrate(VI): effects of water constituents and oxidized products. *Water Res.*, 103, 48–57, 2016.
70. Yang, B., Kookana, R.S., Williams, M., Ying, G.-G., Du, J., Doan, H., Kumar, A., Oxidation of ciprofloxacin and enrofloxacin by ferrate(VI): products identification, and toxicity evaluation. *J. Hazard. Mater.*, 320, 296–303, 2016.
71. Wang, H., Liu, Y., Jiang, J.Q., Reaction kinetics and oxidation product formation in the degradation of acetaminophen by ferrate (VI). *Chemosphere*, 155, 583–590, 2016.
72. Kim, C., Panditi, V.R., Gardinali, P.R., Varma, R.S., Kim, H., Sharma, V.K., Ferrate promoted oxidative cleavage of sulfonamides: kinetics and product formation under acidic conditions. *Chem. Eng. J.*, 279, 307–316, 2015.
73. Han, Q., Wang, H., Dong, W., Liu, T., Yin, Y., Fan, H., Degradation of bisphenol A by ferrate(VI) oxidation: Kinetics, products and toxicity assessment. *Chem. Eng. J.*, 262, 34–40, 2015.
74. Li, C., Li, X.Z., Graham, N., A study of the preparation and reactivity of potassium ferrate. *Chemosphere*, 61, 537–543, 2005.
75. Casbeer, E.M., Sharma, V.K., Zajickova, Z., Dionysiou, D.D., Kinetics and mechanism of oxidation of tryptophan by ferrate(VI). *Environ. Sci. Technol.*, 47, 4572–4580, 2013.

76. Sharma, V.K., Liu, F., Tolan, S., Sohn, M., Kim, H., Oturan, M.A., Oxidation of β-lactam antibiotics by ferrate(VI). *Chem. Eng. J.*, 221, 446–451, 2013.
77. Yang, B., Ying, G.-G., Zhang, L.-J., Zhou, L.-J., Liu, S., Fang, Y.-X., Kinetics modeling and reaction mechanism of ferrate(VI) oxidation of benzotriazoles. *Water Res.*, 45, 2261–2269, 2011.
78. Eng, Y.Y., Sharma, V.K., Ray, A.K., Ferrate(VI): green chemistry oxidant for degradation of cationic surfactant. *Chemosphere*, 63, 1785–1790, 2006.
79. Noorhasan, N., Patel, B., Sharma, V.K., Ferrate(VI) oxidation of glycine and glycylglycine: Kinetics and products. *Water Res.*, 44, 927–935, 2010.
80. He, C., Li, X., Sharma, V.K., Li, S., Elimination of sludge odor by oxidizing sulfur-containing compounds with ferrate(VI). *Environ. Sci. Technol.*, 43, 5890–5895, 2009.
81. Sharma, V.K., Rivera, W., Joshi, V.N., Millero, F.J., O'Connor, D., Ferrate(VI) oxidation of thiourea. *Environ. Sci. Technol.*, 33, 2645–2650, 1999.
82. Sharma, V.K., Rendon, R.A., Millero, F.J., Vazquez, F.G., Oxidation of thioacetamide by ferrate(VI). *Mar. Chem.*, 70, 235–242, 2000.
83. Graham, N., Jiang, C., Li, X.-Z., Jiang, J.-Q., Ma, J., The influence of pH on the degradation of phenol and chlorophenols by potassium ferrate. *Chemosphere*, 56, 949–956, 2004.
84. Johnson, M.D., Hornstein, B.J., The kinetics and mechanism of the ferrate(VI) oxidation of hydroxylamines. *Inorg. Chem.*, 42, 6923–6928, 2003.
85. Read, J.F., Graves, C.R., Jackson, E., The kinetics and mechanism of the oxidation of the thiols 3-mercapto-1-propane sulfonic acid and 2-mercaptonicotinic acid by potassium ferrate. *Inorganica Chim. Acta.*, 348, 41–49, 2003.
86. Johnson, M.D., Read, J.F., Kinetics and mechanism of the ferrate oxidation of thiosulfate and other sulfur-containing species. *Inorg. Chem.*, 35, 6795–6799, 1996.
87. Read, J.F., Bewick, S.A., Graves, C.R., MacPherson, J.M., Salah, J.C., Theriault, A., Wyand, A.E.H., The kinetics and mechanism of the oxidation of s-methyl-l-cysteine, l-cystine and l-cysteine by potassium ferrate. *Inorganica Chim. Acta.*, 303, 244–255, 2000.
88. Huang, H., Sommerfeld, D., Dunn, B.C., Lloyd, C.R., Eyring, E.M., Ferrate(VI) oxidation of aniline. *J. Chem. Soc. Dalt. Trans.*, 1301–1305, 2001.
89. Read, J.F., Adams, E.K., Gass, H.J., Shea, S.E., Theriault, A., The kinetics and mechanism of oxidation of 3-mercaptopropionic acid, 2-mercaptoethanesulfonic acid and 2-mercaptobenzoic acid by potassium ferrate. *Inorganica Chim. Acta.*, 281, 43–52, 1998.
90. Read, J.F., Boucher, K.D., Mehlman, S.A., Watson, K.J., The kinetics and mechanism of the oxidation of 1,4-thioxane by potassium ferrate. *Inorganica Chim. Acta.*, 267, 159–163, 1998.
91. Chen, Y., Xiao, M., Wang, Z., Jiang, W., Guo, Y., Liu, Z., Oxidation of amitriptyline and nortriptyline by ferrate(VI): efficiency and reaction pathways. *Desalin. Water Treat.*, 57, 12882–12890, 2016.

92. Han, Q., Wang, H., Dong, W., Liu, T., Yin, Y., Fan, H., Degradation of bisphenol A by ferrate(VI) oxidation: Kinetics, products and toxicity assessment. *Chem. Eng. J.*, 262, 34–40, 2015.
93. Yang, B., Ying, G.-G., Zhao, J.-L., Liu, S., Zhou, L.-J., Chen, F., Removal of selected endocrine disrupting chemicals (EDCs) and pharmaceuticals and personal care products (PPCPs) during ferrate(VI) treatment of secondary wastewater effluents. *Water Res.*, 46, 2194–204, 2012.
94. Tiwari, D., Yang, J., Lee, S., Applications of ferrate(VI) in the treatment of wastewaters. *Environ. Eng. Res.*, 10, 269–282, 2005.
95. Ghanbari, F., Moradi, M., Application of peroxymonosulfate and its activation methods for degradation of environmental organic pollutants: review. *Chem. Eng. J.*, 310, 41–62, 2017.
96. Zhou, Y., Jiang, J., Gao, Y., Ma, J., Pang, S.Y., Li, J., Lu, X.T., Yuan, L.P., Activation of peroxymonosulfate by benzoquinone: a novel nonradical oxidation process. *Environ. Sci. Technol.*, 49, 12941–12950, 2015.
97. Sun, B., Guan, X., Fang, J., Tratnyek, P.G., Activation of manganese oxidants with bisulfite for enhanced oxidation of organic contaminants: the involvement of Mn(III). *Environ. Sci. Technol.*, 49, 12414–12421, 2015.
98. Anipsitakis, G.P., Dionysiou, D.D., Radical generation by the interaction of transition metals with common oxidants. *Environ. Sci. Technol.*, 38, 3705–3712, 2004.
99. Manoli, K., Nakhla, G., Ray, A.K., Sharma, V.K., Enhanced oxidative transformation of organic contaminants by activation of ferrate(VI): Possible involvement of Fe^{V}/Fe^{IV} species. *Chem. Eng. J.*, 307, 513–517, 2017.
100. Chong, N. M., Jin, B., Chow, C. W. K., Saint, C., Recent developments in photocatalytic water treatment technology: a review. *Water Res.*, 44, 2997–3027, 2010.
101. Mukherjee, D., Barghi, S., Ray, A.K., Degradation of methyl orange by TiO_2/polymeric film photocatalyst. *Can. J. Chem. Eng.*, 9, 92, 2014.
102. Mills, A., Le Hunte, S., An overview of semiconductor photocatalysis. *J. Photochen. Photobiol. A: Chem.*, 108, 1–35, 1997.
103. Gerven, T. V., Mul, G., Moulijn, J., Stankiewicz, A., A review of intensification of photocatalytic processes. *Chem. Eng. Process.*, 46, 781–789, 2007.
104. Tang, J., Durant, J.R., Klug, D.R., Surface tuning for oxide-based nanomaterials as efficient photocatalysts. *J. Am. Chem.*, 42, 9509–9549, 2013.
105. Peak Pure Air, http://www.peakpureair.com/TiO2-photocatalytic, 2009.
106. Chen, D., Ray, A.K., Photodegradation kinetics of 4-nitrophenol in TiO_2 suspension. *Water Res.*, 32, 3223–3234, 1998.
107. Limousin, G., Gaudet, J. P., Charlet, L., Szenknect, S., Barthes, V., Krimissa, M., Sorption isotherm: a review on physical bases, modeling and measurement. *Appl. Geochem.*, 22, 249–275, 2007.
108. Dalrymple, O. K., Stefanakos, E., Trotz, M. A., Goswami, D.Y., A review of the mechanism and modelling of photocatalytic disinfection. *Appl. Catal. B. Environ.*, 98, 27–38, 2010.

109. Matsunaga, T., Tomoda, R., Nakajima, T., Wake, H., Photoelectrochemical strerilization of microbial cells by semiconductor powders. *FEMS Microbiol. Lett.*, 29, 211–214, 1985.
110. Magdigan, M.T., Martinko, J.M., *Brock biology of microorganisms* (11th ed.), Pearson Education, Inc, Upper Saddle River, NJ, 2006.
111. Hidaka, H., Horikoshi, S., Ajisaka, K., Zhao, J., Serpone, N., Fate of amino acids upon exposure to aqueous titania irradiated with UV-A and UV-B radiation photocatalyzed formation of NH3, NO3-, and CO2. *J. Photochem. Photobiol. A: Chem.*, 108, 197–205, 1997.
112. Tran, T.H., Nosaka, A.Y., Nosaka, Y., Adsorption and photocatalytic decomposition of amino acids in TiO_2 photocatalytic systems. *J. Phys. Chem. B*, 110, 25525–25531, 2006.
113. Yang, X., Wang, Y., Build. Photocatalytic effect on plasmid DNA damage under different UV irradiation time. *Build Environ.*, 43, 253–257, 2008.
114. Ray, A.K., Beenackers, A.A.C.M., Novel swirl-flow reactor for kinetic studies of semiconductor photocatalysis. *AIChE J.*, 43(10), 2571, 1997.
115. Mills, A., Davies, R.H., Worsley, D., Water purification by semiconductor photocatalysis. *Chem. Soc. Rev.*, 417, 1993.
116. Mukherjee, P.S., Ray, A.K., Major challenges in the design of a large-scale photocatalytic reactor for water treatment. *Chem. Eng. Technol.*, 22, 3, 1993.
117. Ray, A.K., A new photocatalytic reactor for destruction of toxic water pollutants by advanced oxidation process. *Catal. Today*, 44, 357–368, 1998.
118. Ray, A.K., Development of a new photocatalytic reactor for water purification. *Catal. Today*, 40 (1), 73, 1998.
119. Ray, A.K., Design, modelling and experimentation of a new large-scale photocatalytic reactor for water treatment. *Chem. Eng. Sci.*, 54, 3113–3125, 1999.
120. Toor, A.P., Verma, A., Jotshi, C.K., Bajpai, P.K., Singh, V., Photocatalytic degradation of direct yellow 12 dye using UV/TiO_2 in a shallow pond slurry reactor. *Dyes PIGM*, 68 (1), 53–60, 2006.
121. Assink, J.W., Koster, T.P.M., Slaager, J.M., Fotokatalytische oxydatie voor afvalwater behandeling. Internal Report Ref. No. 93 ± 137, TNO ± Milieuen Energie, Apeldoorn (The Netherlands) 1993.
122. Daghrir, R., Drogui, P., Robert, D., Modified TiO_2 for environmental photocatalytic applications: A review. *Ind. Eng. Chem. Res.*, 52, 3581–3599, 2013.
123. Teh, C.M., Mohamed, A.R., Role of titanium dioxide and iondoped titanium dioxide on phtotocatalytic degradation of organic pollutants (phenol compounds and dyes) in aqueous solutions: A review. *Alloys Compd.*, 509, 1648–1660, 2011.
124. Yalcin, Y., Kiliç, M., Cinar, Z., Fe^{+3}-doped TiO_2: a combined experimental and computational approach to the evaluation of visible light activity. *Appl. Catal. B. Environ.*, 99, 469–477, 2010.
125. Robert, D., Photosensitization of TiO_2 by MxOy and MxSy nanoparticles for heterogeneous photocatalysis applications. *Catal. Today*, 122, 20–26, 2007.
126. Zhang, H., Chen, G., Behnemann, D.W., Photo-electrocatalytic materials for environmental applications. *J. Mater. Chem.*, 19, 5089–5121, 2009.

127. Liu, S., Chen, X., A visible light response TiO_2 photocatalyst realized by cationic S-doping and its application for phenol degradation. *J. Hazard. Mater.*, 152, 49–55, 2008.
128. Bally, A.R., Korobeinikova, E.N., Schmid, P.E., Lévy, F., Bussy, F., Structural and electrical properties of Fe-doped TiO_2 thin films. *Phys. D: Appl. Phys.*, 31, 1149–1154, 1998.
129. Furubayashi, Y., Hitosugi, T., Yamamoto, Y., Inaba, K., Kinoda, G., Hirose, Y., Shimada, T., Hasegawa, T., A transparent metal: Nb doped anatase TiO_2. *Appl. Phys. Lett.*, 86, 252101, 2005.
130. Fan, C., Xue, P., Sun, Y., Preparation of nano-TiO_2 doped with cerium and its photocatalytic activity. *J. Rare Earth*, 24, 309–313, 2006.
131. El-Bahy, Z.M., Ismail, A.A., Mohamed, R.M., Enhancement of titania by doping rare earth for photodegradation of organic dye (Direct blue). *J. Hazard. Mater.*, 166, 138–143, 2009.
132. Seery, M.K., George, R., Floris, P., Pillai, S.C., Silver-doped titanium dioxide nanomaterials for enhanced visible light photocatalysis. *Photochem. Photobiol. A.*, 189, 258–263, 2007.
133. Tomas, S.A., Luna-Resendis, A., Cortés-Cuautli, L.C., Jacinto, D., Optical and morphological characterization of photocatalytic TiO_2 thin films doped with silver. *Thin Solid Films*, 518, 1337–1340, 2009.
134. Binitha, N.N., Yaakob, Z., Reshmi, M.R., Sugunan, S., Ambili, V.K., Zetty, A.A., Preparation and characterization of nano-silverdoped mesoporous titania photocatalysts for dye degradation. *Catal. Today*, 147, S76–S80, 2009.
135. Kment, S., Kmentova, H., Kluson, P., Krysa, J., Hubicka, Z., Cirkva, V., Gregora, I., Solcova, O., Jastrabik, L., Notes on the photoinduced characteristics of transition metal-doped and undoped titanium dioxide thin films. *Colloid Interf. Sci.*, 348, 198–205, 2010.
136. Park, J.H., Kim, S., Bard, A.J., Novel carbon-doped TiO_2 nanotube arrays with high aspect ratios for efficient solar water splitting. *Nano Lett.*, 6, 24–28, 2006.
137. Khan, S.U.M., Al-Shahry, M., Ingler, W.B., Jr., Efficient photochemical water splitting by a chemically modified n-TiO_2. *Science*, 297, 2243–2245, 2002.
138. Valentin, C.D., Finazzi, E., Pacchioni, G., Selloni, A., Livraghi, S., Paganini, M.C., Giamello, E., N-doped TiO_2: theory and experiment. *Chem. Phys.*, 339, 44–56, 2007.
139. Asahi, R., Morikawa, T., Ohwaki, T., Aoki, K., Taga, Y., Visible light photocatalysis in nitrogen-doped titanium oxides. *Science*, 293(5528), 269–271, 2001.
140. Ohno, T., Mitsui, T., Matsumura, M., Photocatalytic activity of S-doped TiO_2 photocatalyst under visible light. *Chem. Lett.*, 32, 364–365, 2003.
141. Umebayashi, T., Yamaki, T., Itoh, H., Asai, K., Analysis of electronic structures of 3d transition metal-doped TiO_2 based on band calculations. *J. Phys. Chem. Solids*, 63, 1909–1920, 2002.

142. Hong, X., Wang, Z., Cai, W., Lu, F., Zhang, J., Yang, Y., Ma, N., Liu, Y., Visible light activated nanoparticle photocatalyst of iodinedoped titanium dioxide. *Chem. Mater.*, 17, 1548–1552, 2005.
143. Wang, H., Lewis, J.P., Second-generation photocatalytic materials: anion-doped TiO_2. *Phys.: Condens. Matter.*, 18, 421–434, 2006.
144. Chen, D., Jiang, Z., Geng, J., Wang, Q., Yang, D., Carbon and nitrogen Co-doped TiO_2 with enhanced visible light photocatalytic activity. *Ind. Eng. Chem. Res.*, 46, 2741–2746, 2007.
145. Tan, Y.N., Wong, C.L., Mohamed, A.R., An overview on the photocatalytic activity of nano-doped TiO_2 in the degradation of organic pollutants. *ISRN Mater. Sci.*, ID261219, 1–18, 2011.
146. Ren, W., Ai, Z., Jia, F., Zhang, L., Fan, X., Zou, Z., Low temperature preparation and visible light photocatalytic activity of mesoporous carbon-doped crystalline TiO_2. *Appl. Catal. B. Environ.* 69, 138–144, 2007.
147. Leary, R., Westwood, A., Carbonaceous nanomaterials for the enhancement of TiO_2 photocatalysis. *Carbon*, 49, 741–772, 2011.
148. Rehman, S., Ullah, R., Butt, A.M., Gohar, N.D., Strategies of making TiO_2 and ZnO visible light active. *J. Hazard. Mater.*, 170, 560–569, 2009.
149. Shen, M., Wu, Z., Huang, H., Du, Y., Zou, Z., Yang, P., Carbon-doped anatase TiO_2 obtained from TiC for photocatalysis under visible light irradiation. *Mater. Lett.*, 60, 693–697, 2006.
150. Zhang, H., Chen, G., Behnemann, D.W., Photo-electrocatalytic materials for environmental applications. *Mater. Chem.*, 19, 5089–5121, 2009.
151. Bessekhouad, Y., Robert, D., Weber, J.V., Bi2S3/TiO_2 and CdS/TiO_2 heterojunctions as an available configuration for photocatalytic degradation of organic pollutant. *J. Photochem. Photobiol. A*, 163, 569–580, 2004.
152. Robert, D., Photosensitization of TiO_2 by MxOy and MxSy nanoparticles for heterogeneous photocatalysis applications. *Catal. Today*, 122, 20–26, 2007.
153. Huo, Y., Yang, X., Zhu, J., Li, H., Highly active and stable CdSTiO_2 visible photocatalyst prepared by in situ sulfurization under supercritical conditions. *Appl. Catal. B. Environ.*, 106, 69–75, 2011.
154. Zyoud, A.H., Zaatar, N., Saadeddin, I., Ali, C., Park, D., Campet, G., Hilal, H.S., CdS-sensitized TiO_2 in phenazopyridine photo-degradation: catalyst efficiency, stability, and feasibility assessment. *J. Hazard. Mater.*, 173, 318–325, 2010.
155. Ho, W., Yu, J.C., Sonochemical synthesis and visible light photocatalytic behavior of CdSe and CdSe/TiO_2 nanoparticles. *J. Mol. Catal. A: Chem.*, 247, 268–274, 2006.
156. Ge, M., Guo, C., Zhu, X., Ma, L., Han, Z., Hu, W., Wang, Y., Photocatalytic degradation of methyl orange using ZnO/TiO_2 composites. *Front. Environ. Sci. Eng.*, 3, 271–280, 2009.
157. Gupta, S.M., Tripathi, M., A review of TiO_2 nanoparticles. *Chin. Sci. Bull.*, 56, 1639–1657, 2011.

158. Miyauchi, M., Nakajima, A., Watanabe, T., Hashimoto, K., Photoinduced hydrophilic conversion of TiO_2/WO_3 layered thin films. *Chem. Mater.*, 14, 4714–4720, 2002.
159. Chowdhury, P., Gomaa, H., Ray, A.K., Dye-sensitized photocatalyst: a breakthrough in green energy and environmental detoxification, ACS, p. 1124, 2013.
160. Gratzel, M.J., Dye-sensitized solar cell. *Photochem. Photobiol., C*, 4, 145–153, 2003.
161. Tennakone, K., Kumarasinghe, A.R., Sirimanne, P.M., Dye sensitization of low-bandgap semiconductor electrodes: cuprous oxide photocathode sensitized with methyl violet. *Semicond. Sci. Technol.*, 8, 1557–1560, 1993.
162. Galoppini, E., Linkers of anchoring sensitizers to semiconductor nanoparticles. *Coordin. Chem. Rev.*, 248 (13), 1283–1297, 2004.
163. Polo, A.S., Itokazu, M.K., Murakami Iha, N.Y., Metal complex sensitizers in dye-sensitized solar cells. *Coord.Chem. Rev.*, 248, 1343–1361, 2004.
164. Wongcharee, K., Meeyoo, V., Chavadej, S., Dye sensitized solar cell using natural dyes extracted from rosella and blue pea flower. *Sol. Energy Mater. Sol. Cells*, 91, 566–571, 2007.
165. Polo, A.S., Murakami Iha, N.Y., Blue sensitizers for solar cells: Natural dyes from calafate and jaboticaba. *Sol. Energy Mater. Sol. Cells*, 90, 1936–1944, 2006.
166. Garcia, C.G., Polo, A.S., Murakami Iha, N.Y., Fruit extracts and ruthenium polypyridinic dyes for sensitization of TiO_2 in photoelectrochemical solar cells. *J. Photochem. Photobiol., A*, 160, 87–91, 2003.
167. Malekshoar, G., Pal, K., He, Q., Yu, A., Ray, A.K., Enhanced solar photocatalytic degradation of phenol with coupled graphene-based titanium dioxide and zinc oxide. *Ind. Eng. Chem. Res.*, 53, 18824–18832, 2014.
168. Yuan, Z., Jia, J.H., Zhang, L.D., Influence of co-doping of Zn(II) and Fe(III) on the photocatalytic activity of TiO_2 for phenol degradation. *Mater. Chem. Phys.*, 73, 323, 2002.
169. Sonawane, R.S., Dongare, M.K., Sol-gel synthesis of Au/TiO_2 thin films for photocatalytic degradation of phenol in sunlight. *J. Mol. Catal. A: Chem.*, 243, 68, 2006.
170. Naeem, K., Ouyang, F., Preparation of $Fe3^+$-doped TiO_2 nanoparticles and its photocatalytic activity. *Physica B: Condens. Matter*, 405, 221, 2009.
171. Chiou, C.H., Juang, R.S., Photocatalytic degradation of phenol in aqueous solutions by Pr-doped TiO_2 nanoparticles. *J. Hazard. Mater.*, 149, 1, 2007.
172. Liqiang, J., Xiaojun, S., Baifu, X., Baiqi, W., Weimin, C., Honggang, F., The preparation and characterization of La doped TiO_2 nanoparticles and their photocatalytic activity. *J. Solid State Chem.*, 177, 3375, 2004.
173. Chowdhury, P., Moreira, J., Gomaa, H., Ray, A.K., Visible-solar-light-driven photocatalytic degradation of phenol with dye-sensitized TiO_2: parametric and kinetic study. *Ind. Eng. Chem. Res.*, 51, 4523, 2012.

174. Li, C., Li, X.Z., Degradation of endocrine disrupting chemicals in aqueous solution by interaction of photocatalytic oxidation and ferrate (VI) oxidation. *Water Sci. Technol.*, 55, 217 LP–223 LP, 2007.
175. Sharma, V.K., Graham, N.J.D., Li, X.-Z., Yuan, B.-L., Ferrate(VI) enhanced photocatalytic oxidation of pollutants in aqueous TiO_2 suspensions. *Environ. Sci. Pollut. Res.*, 17, 453–461, 2010.
176. Yuan, B., Li, Y., Huang, X., Liu, H., Qu, J., Fe(VI)-assisted photocatalytic degradating of microcystin-LR using titanium dioxide. *J. Photochem. Photobiol. A Chem.*, 178, 106–111, 2006.
177. Yuan, B.L., Li, X.Z., Graham, N., Aqueous oxidation of dimethyl phthalate in a Fe(VI)-TiO_2-UV reaction system. *Water Res.*, 42, 1413–1420, 2008.
178. Sharma, V.K., Chenay, B.V.N., Heterogeneous photocatalytic reduction of Fe(VI) in UV-irradiated titania suspensions: effect of ammonia. *J. Appl. Electrochem.*, 35, 775–781, 2005.

11

Agro-Industrial Wastes Composites as Novel Adsorbents

Haq Nawaz Bhatti[1]*, Amina Kamal[1] and Munawar Iqbal[2]

[1]*Environmental and Material Chemistry Laboratory, Department of Chemistry, University of Agriculture, Faisalabad, Pakistan*
[2]*Department of Chemistry, The University of Lahore, Lahore, Pakistan*

Abstract

The composites-based on agricultural waste gained much attention due to their efficient adsorption efficiencies. Composites were seen as an adsorbent for different types of pollutant remediation and sugarcane bagasse composite was exemplified for the adsorption of Congo Red (CR) dye. Sugarcane bagasse composites with chitosan aniline, polyaniline, polypyrrole, chitosan pyrrole and starch were prepared and used for dye adsorption. The effect of process variables, that is, composites doses, pH, contact time, CR initial concentration and temperature on dye adsorption were studied. The pseudo-second-order kinetic model and Freundlich isotherm fitted well to the CR dye adsorption data and intraparticle diffusion was the dye adsorption rate limiting step. The maximum dye adsorption was achieved using 180 mg/L CR initial concentration at 40 °C temperature for 60 min contact time using 0.05 g adsorbent doses in the pH range of 2–6 (pH was variable for different composites). Thermodynamic study revealed that the CR dye adsorption process was endothermic, spontaneous and energetically stable onto sugarcane bagasse composites. Results revealed that sugarcane bagasse-based composites have potential for the adsorption of dyes, which could possibly be used for the adsorption of dyes from textile wastewater.

Keywords: Biocomposites, biomolecules, sugarcane bagasse, adsorption mechanism

**Corresponding author*: hnbhatti2005@yahoo.com

Shahid-ul-Islam (ed.) Advanced Materials for Wastewater Treatment, (391–432)

11.1 Introduction

Biocomposites have been emerged as alternative, efficient and environmental friendly adsorbents for the remediation of water pollution and dyes from textile industry [1–3]. Due to limited resources, the wastewater treatment is a challenge, especially in developing countries. Worldwide annual production of dyes is approximately 0.7 million tons and textile industry generates huge quantities of complex chemical substances as a part of unused materials including dyes. To produce 1 kg of textile about 200 L of water is consumed and an average sized textile mill having a production of about 8000 kg of fabric per day consumed about 1.6 million L water [4]. The chemicals present in the wastewater cause harm to both human health and the environment (sunlight passage through the water, photosynthesis, increase the biological oxygen demand and affect the aquatic life) [5, 6]. Respiratory diseases, dermatitis, asthma, changes in their immunoglobulin, colon and rectum cancers and bladder cancer in textile workers is reported [7]. Discharge of the colored effluent into water bodies also reported to be toxic to aquatic organisms [8]. Recently, number of studies have been performed to evaluate the toxicity of textile wastewater and textile wastewater and dyes revealed toxic nature (cytotoxic, genotoxic and mutagenic) to living organisms [9, 10]. Therefore, to ensure environmental safety, the treatment of textile wastewater is the urgent need of the era. Different physico-chemical methods have been used for the remediation of textile wastewater [11–33]. However, adsorption using biocomposites have attracted the attention of scientific community for the removal of metals, dyes and other pollutant from wastewater since biocomposites are efficient adsorbents. A composite is a combination of two or more components, in which one component acts as a reinforcing agent and other provides compatible matrix [34, 35], which are also low cost, efficient and environmental friendly and have been prepared and successfully used for the adsorption of dyes, metal and various other inorganic and inorganic pollutants [34, 35].

Recently, biocomposites have gained much attention for the adsorption of dyes from wastewater. The biocomposites application to remove dyes, metal ions etc in wastewater is shown in Tables 11.1–11.3. To date, different types of composites have prepared and used for remediation of pollutants [34–36], that is, magnetic chitosan/poly (vinyl alcohol) hydrogel beads [37], glutamic acid modified chitosan magnetic composite [38], ferrofluid modified peanut husks [39], magnetic γ-Fe_2O_3 cross-linked chitosan composite [40], starch functionalized magnetic MWCNT [41], magnetically modified spent coffee grounds [42], chitosan suppoted CNT-MNPs

Table 11.1 Adsorption capacities of various composites for the adsorption of dyes and present investigation (polyaniline, starch, polypyrrole, chitoson/aniline and chitoson/pyrrole composites).

Adsorbents	Dyes	q_e (mg/g)	Ref.
Polyaniline-coated nylon-6 nanofibers	methyl orange	370	[55]
hydroxyapatite/chitosan composite	Congo red	769.0	[56]
Graphene oxide (GO) and polyacrylamide (PAM)	Methylene blue (MB) and Rhodamine 6G	292.84 & 288	[57]
Starch functionalized magnetic MWCNT	Methyl orange	135.6	[41]
Fe_3O_4@graphene nanocomposite	Methylene blue, Congo red	45.27, 33.66	[58]
Cross-linked polyhedral oligomeric silsesquioxane	Methyl orange	237.5	[59]
Polymer modified magnetic nanoparticles	Alkali blue 6B, Crystal violet	22, 208.3	[44]
Ethylenediamine-modified magnetic chitosan nanoparticles	Acid orange 10, Acid orange 7	1017, 1214	[60]
Nano-Fe3O4/carboxyl-functionalized baker's yeast composites	Methylene blue	141.75	[61]
Chitosan suppoted CNT-MNPs	Acid red 18	809.9	[43]
Magnetic biocomposite (metal chlorides and aquatic macrophytes)	Metanil yellow dye	90.91	[62]
magnetically modified spent coffee grounds	Acridine orange, Amido black 10B, Bismark brown, Congo red, Crystal red, Malachite green, Safranin-O	73.4, 1.24, 69.2, 9.43, 68.1, 43, 59,	[42]

(*Continued*)

Table 11.1 Cont.

Adsorbents	Dyes	q_e (mg/g)	Ref.
Core-shell nano-ZnO/ pollen grain (n-ZnO/ PG) biocomposite	Malachite green	145.9	[63]
Ferrofluid modified peanut husks	Bismarck brown, Safranin-O, Crystal violet, Acridine orange	95.3, 86.1, 80.9, 71.4	[39]
Aminoguanidine modified MNPs	Acid green 25, Acid violet 43, Acid orange 20, Acid red 27, Methyl blue	95.2, 137, 120, 83.3, 165	[46]
Mango stone biocomposite	crystal violet	-	[64]
Magnetic zeolite/iron oxide nanocomposite	Reactive orange 16	1.1	[65]
Magnetic N-lauryl chitosan nanocomposite	Remazol red 198	267	[66]
Magnetic γ-Fe_2O_3crosslinked chitosan composite	Methyl orange	29.83	[40]
Magnetic ferrite nanoparticle alginate composite	Basic red 18, Basic blue 41, Basic blue 9	56, 25, 106	[45]
polyurethane/chitosan composite foams	Acid Violet 48	30	[47]
ionic liquid-coated Fe3O4@chitosan@ graphene oxide	methylene blue	262	[48]
amphoteric chitosan/ gelatin composite microspheres	Acid red 337	748.50	[48]
Glutamic acid modified chitosan magnetic composite	Methylene blue, Crystal violet, Light yellow 7GL	180, 375.4, 217.3	[38]

(*Continued*)

Table 11.1 Cont.

Adsorbents	Dyes	q_e (mg/g)	Ref.
Bioadsorbent-modified ostrich bone wastes	Methyl orange	90%	[67]
Chitosan/ montmorillonite	Congo red	53.42	[49]
Chitosan/polyurethane	Acid violet 48	30	[47]
Polyaniline (PANI)-Fe_2O_3 nano-composite	Acid violet 19	7.7 mg/g	[68]
Chitosan/activated clay	Methylene blue	330	[51]
Chitosan/activated clay	Reactive dye RR222	1912	–
Chitosan/bentonite	Malachite green	435	[50]
Chitosan/oil palm	Reactive blue 19	909.1	[52]
Chitosan coated magnetic iron oxide	Congo red	56.66	[69]
Magnetic chitosan/poly (vinyl alcohol) hydrogel beads	Congo red	470.1	[37]
Calcium alginate/ multi-walled carbon nanotubes	Methylene blue and methyl orange	79.7% and 80.2%	[70]
Surfactant Doped Polyaniline/MWCNTs Composite	Brilliant green	434.78	[53]
Eggshell with a polymer mixture of alginate and polyvinyl alcohol	C.I. Remazol Reactive red 198	46.9	[71]
Composites of polyaniline, starch, polypyrrole, chitosan aniline and chitosan pyrrole using peanut waste	Crystal violet	100.6 mg/g	[72]
Polypyrrole Composite	Acid orange10	243.9	[73]

(Continued)

Table 11.1 Cont.

Adsorbents	Dyes	q_e (mg/g)	Ref.
Polyaniline Nano Composite	Reactive Dye	143.59	[74]
Modified corn pith	Malachite green	488.3	[75]
Gum ghatti and acrylic acid-based graft co-polymers	Methylene blue and rhodamine B	909.09, and 819.67	[76]
Polyaniline lignocellulose composite	Congo red	1672.5	[77]
Chitosan–silica composite aerogels	Congo red	150	[78]
Chitosan/Rectorite Composite	Congo red	73.8	[79]
Sugarcane bagasse composite with starch	Congo red	101.50	Present study
Sugarcane bagasse composite with polypyrrole	Congo red	89.10	Present study
Sugarcane bagasse composite with polyaniline	Congo red	90.20	Present study
Sugarcane bagasse composite with chitosan pyrrole	Congo red	93.50	Present study
Sugarcane bagasse composite with chitosan aniline	Congo red	105.20	Present study

[43], polymer modified magnetic nanoparticles [44], magnetic ferrite nanoparticle alginate composite [45], aminoguanidine modified MNPs [46], polyurethane/chitosan composite foams [47], chitosan/polyurethane [47], amphoteric chitosan/gelatin composite microspheres [48], chitosan/montmorillonite [49], chitosan/bentonite [50], chitosan/activated clay & chitosan/activated clay [51], chitosan/oil palm [52] and surfactant doped polyaniline/MWCNTs composite [53] have been reported to be an efficient

Table 11.2 Adsorption capacities of various composites for the adsorption of metal ions and present investigation (polyaniline, starch, polypyrrole, chitoson/aniline and chitoson/pyrrole composites).

Adsorbents	Metal ions	q_e (mg/g)	Ref.
Lignosilicate (LS) composite	Ni(II)	38.74%	[80]
Calcium alginate beads and compsoites with 1) glu-taraldehyde, 2) *Aspergillus caespitosus* and 3) *A. caespitosus*	Pb(II)	398.4, 413.8, 651.6 and 670	[81]
Guar gum–nano zinc oxide	Cr(VI)	55.56	[82]
Biocomposite of mango	Cr(VI)	–	[83]
alginate and gelatin cross-linked with Ca^{2+}, Ce^{3+} and Zr^{4+} ions namely Ca@AlgGel, Ce@AlgGel and Zr@AlgGel composites	Cr(VI)	19.40, 24.50 and 25.40	[84]
macro marine algae (*Jania rubens*) and yeast (*Saccharomyces cerevisiae*) immobilized on silica gel	U(VI)	–	[85]
Nano-hydroxyapatite (n-HAp) with chitin and chitosan	Fe(III)	4238, 5800 and 6.753 mg/kg	[86]
Bentonite composite with *Eriobotrya japonica*	Cu(II)	–	[34]
Alumina/alginate (AlAlg) composite	Cr(VI)	17.45	[87]
Chitosan-montmorillonite (KSF-CTS): Montmorillonite (KSF-Na), CTS and KSF-CTS biocomposite	Cu(II)	86%, 85% and 84%	[88]
Chitosan-nanoclay composite	Cr(VI)	128.43 and 21.83 mg/g	[89]

(Continued)

Table 11.2 Cont.

Adsorbents	Metal ions	q_e (mg/g)	Ref.
Bionanocomposites of fungus-Fe_3O_4	Sr(II), Th(IV) and U(VI)	100.9, 223.9 and 280.8	[90]
Cellulosic biopolymers-based graft copolymers	Cu^{2+}, Zn^{2+}, Cd^{2+}, Pb^{2+}	1.209, 0.9623, 1.2609 and 1.295 mmol/g	[91]
Nano-sized hydroxyapatite (nHAp) into alginate polymer	Pb(II)	270.3 mg/g	[92]
Alumina/chitosan (AlCs) composite	Cr	8.62 mg/g	[93]
Hydrogel modified by triethylene tetra amine (TETA-NBC) and β-cyclodextrin	Hg(II)	407.9 mg/g	[94]
Sulfonated plant gum-mushroom biocomposite	Ag(I) and Zn(II)	137.8 and 287.9 mg/g	[95]
Cellulose-*graft*-polyacrylamide/hydroxyapatite composite hydrogels	Cu(II)	175 mg	[96]
Fungal dead biomass composite with bentonite	Ni(II) and Zn(II)	161 and 78.5 mg/g	[35]
Chitosan (CT) and halloysite nanotubes biopolymer	Cu(II)	10.55 mg/g	[97]
Groundnut husk modified with Guar Gum	Pb(II), Cu(II) and Ni(II)		

absorbents. In view of excellent dye adsorption properties of composites [54], the biocomposites received much attention as an adsorbent. So far, it was hypothesized that the composites of biopolymer with biomass may offer excellent adsorption properties and the principle objectives were to prepare composites of polyaniline, starch, pyrrole, chitosan aniline and chitosan pyrrole with sugarcane bagasse biomass and to check their

Table 11.3 Adsorption capacities of various composites for the adsorption of ions and organic compounds.

Adsorbents	Adsorbate	q_e (mg/g)	Ref.
Chitosan supported bentonite (CSBent) composites (Zr@ CSBent, Fe@CSBent and Ca@ CSBent)	Phosphate ion	40.86, 22.15 and 13.44	[98]
Chitosan/bentonite, chitosan/ titanium oxide, and chitosan/ alumina	Nitrate ion	35.68 and 43.62, and 45.38	[99]
γ-Al_2O_3 with MgO nanocrystals	Fluoride ion	5.6	[100]
Alginate (Alg) bioencapsulating nano-hydroxyapatite	Fluoride ion	3870 mg/kg	[101]
Chitosan and activated carbon nanocomposite	Phenol	409	[102]
silica gel/chitosan (SGCS) and cerium loaded silica gel/chito-san (Ce-SGCS) composites	Fluoride ion	4821 ad 3470 mg/kg,	[103]
Chitin/chitin-based biocomposite	Fluoride ion	85%, 84%	[104]
Lanthanum complex onto imino-diacetic acid and chitosan	Fluoride ion	17.50 mg/g	[105]
Vineyard pruning waste entrapped in calcium alginate beads	P, K, N–NH_4, SO_4, TN, TC and PO_4	-	[106]
Graphene oxide doped calcium alginate	Ciprofloxacin	18.45 to 39.06 mg/g	[107]

adsorption behaviors toward CR dye (Figure 11.1(a)) dye. Different process variables, that is, pH, dye initial concentration, contact time, composites doses and temperature were optimized for maximum dye adsorption. Moreover, kinetics, equilibrium modelling and thermodynamic studies were performed in order to understand the adsorption nature and mechanism of CR dye adsorption onto composites.

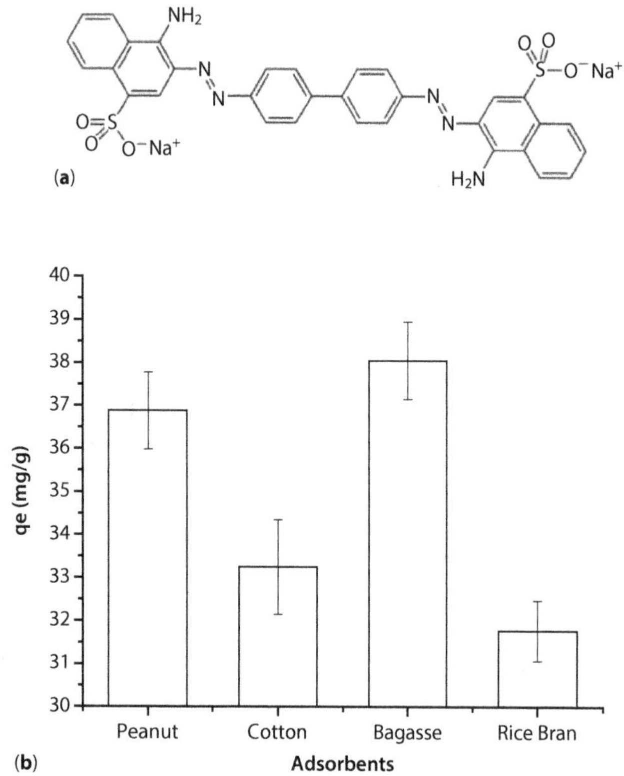

Figure 11.1 (a) Structure of Congo Red; (b) biomasses screening for Congo Red dye adsorption (sugarcane baggase, peanut hull, cotton stick, and rice bran).

11.2 Material and Methods

11.2.1 Chemical, Reagent and Instruments

The chemical and reagent, that is, HCl (37%), aniline (≥99.5%), formic acid (≥95%), pyrrole (98%), chitosan (≥85%), starch (CAS Number 9005-84-9, Molecular Weight 342.30/mole), acetic acid (≥99.99%), NaOH (50%), methanol (99.8%), ammonium persulfate (≥98%) were purchased from Sigma-Aldrich (St. Louis, Missouri, US). Commercial garde CR dye was kindly supplied by Director, Harris Dye and Chemical, Faisalabad, Pakistan. Ultra pure water with a resistivity of 18.2 MΩ cm from Milli-Q system (Millipore) was used for the preparation of solutions. The siever (OCT-DIGITAL 4527-OI), orbital shaker incubator (PA 250/25.H), analytical balance (Shimadzu, AW 220), pH meter (HI-8014 Hanna), grinder (Moulinex, France) and spectrophotometer (CE Cecil 7200, UK) were used throughout the study (otherwise stated).

11.2.2 Biomass Collection and Preparation

Sugarcanes bagasse, rice bran, peanut hull and cotton stick were collected from Shakarganj Sugar Mill Jhang, Iqbal Rice Mills, local market (Faisalabad) and former field, respectively. Collected material was washed with distilled water to remove dust and impurities and dried for a week in sunlight followed by 36 h oven drying at 60 °C. Dried biomasses were grinded in an electric ball mill, sieved and fractions of 0.25 mm were selected, screened for adsorption of CR and sugarcane bagasse was used for composites preparation.

11.2.3 Composites Preparation

The composites of bagasse were prepared with polyaniline, starch, pyrrole, chitosan aniline and chitosan pyrrole. The polyaniline/bagasse composite was prepared with slight modification following reported method [72, 108, 109]. Briefly, 5 mL aniline was dissolved in 1M HCl (50 mL), then, ammonium per disalfate was added drop wise with constant stirring. Then, 1 g of bagasse powder was mixed and stirred for 3 h and placed in refrigerator at –4 °C for 24 h, filtered and washed with water, dil HCl and methanol. Finally, prepared mass was dried at 50 °C for 48 h in an oven. For starch composite with bagasse, 6% starch solution was dissolved in 50 mL water, stirred for 15 min and 250 mL solution of ammonium per disalfate in HCl was added drop wise while keeping mixture in ice bath and 1 g of bagasse powder was added. Stirred the mixture for 30 min and placed at room temperature until the color turns dark green. Mixture was filtered and washed with water and methanol and dried at 50 °C for 48 h in an oven. For polypyrrole composite preparation, a slight modification was done in reported method [110], a 0.5 M $FeCl_3$ solution was added slowly (4 h) at room temperature in 10 mL pyrrole solution (0.2 M). Then, 1 g of bagasse powder was added and stirred the mixture for 3 h. Mixture was filtered and washed with water and methanol and dried at 50 °C for 48 h in an oven. For chitosan pyrrole composite with bagasse preparation, chitosan (1.0 g) was dissolved in 20 mL dilute acetic acid solution of 2.5% (v/v). Then, 10 mL $CdCl_2$ with concentration of 5 g/L was added into the chitosan solution and stirred for 60 min [111] and then, 10 mL pyrrole solution (0.2 M) was added. Finally, 2.0 g bagasse powder was added and stirred for 1 h and placed overnight at room temperature. The obtained mixture was filtered and washed with water and methanol and dried at 50 °C for 48 h in an oven. For chitosan aniline composite with bagasse, chitosan (1.0 g) was dissolved in 20 mL dilute acetic acid solution of 2.5% (v/v). Then, 10 mL $CdCl_2$ with concentration of 5 g/L was added into the chitosan solution and stirred for 60 min and then 5 mL aniline was dissolved in 1M HCl (50 mL) and added in chitosan solution, mixed thoroughly and then, 2.0 g bagasse powder was added and stirred for 1 h and placed overnight at room temperature. The obtained mixture was filtered and washed with water and methanol and dried at 50 °C for 48 h in an oven.

11.2.4 Dye Solution Preparation

The dye stock solution of was prepared by adding 1 g dye in 1000 mL ultrapure water. The working concentrations (10–200 mg/L) were prepared by dilution of stock solution. The solution was scanned from 190–900 nm for λ_{max} measurement (505 nm) of CR dye.

11.2.5 Adsorption Experiments

The adsorption experiments were conducted in 250 mL in Erlenmeyer flasks in temperature controlled rotary shaker at 120 rpm. Different process variables, that is, pH (4–12), composites dose (0.05–0.30 g), contact time (5–120 min), dye initial concentration (5–200 mg/L) and temperature (30–60 °C) were investigated. The pH of the medium was adjusted using 0.1N solutions of NaOH/HCl. To study individual effect parameters, one variable was varied, while others keep fixed. The CR dye residual concentration was determined by measuring absorbance at 505 nm. The linearity of dye concentration was checked by measuring the absorbance at λ_{max} and the amount of dye adsorbed onto unit mass of adsorbent was calculated by concentration difference method (Eqn. 11.1). [112].

$$qe = \frac{(C0 - C)x\left(\frac{v}{1000}\right)}{m} \tag{11.1}$$

where, q_e is adsorption capacity (mg/g), C_0 is the initial concentration of dye (mg/L), C is the concentration of dye at time "t" (mg/L), V is the volume of solution (mL) and m is the adsorbent dose (g).

11.3 Results and Discussion

11.3.1 Screening of Adsorbents

Different biomasses were screened for CR adsorption, that is, rice bran, peanut hull, cotton stick and sugarcane baggase. Screening of biomass was conducted using 0.1 g adsorbents, temperature 30 °C, dye initial concentration 50 mg/L, contact time 1 h and shaking speed 120 rpm and responses, thus obtained are shown in Figure 11.1(b). Among four types of adsorbents, sugarcane bagasse showed better efficiency of the adsorption of CR dye. Peanut hull, cotton stick, sugarcane bagasse and rice bran showed the adsorption capacities of 36.88, 33.25, 38.05, and 31.77 (mg/g), respectively. Therefore, composites of polyaniline, starch, pyrrole, chitosan aniline and chitosan pyrrole were prepared with bagasse and subjected to detail study for the adsorption of CV dye. Previous studies revealed that composites performed better for the adsorption of pollutant from wastewater versus individual agricultural-based adsorbents since composites provide reinforcement and compatible matrix and resultantly, offer excellent adsorption properties [113].

11.3.2 Effect of pH

The CV dye adsorption capacities of native biomass and composites of polyaniline, starch, polypyrrole, chitosan aniline and chitosan pyrrole was studied in the pH range of 4–12 (Figure 11.2(a)). The CV dye adsorption behavior was found variable. However, at any pH value, the composites showed better adsorption capacities. Ploypyrrole, polyaniline and starch composites showed higher CV adsorption at pH 4, whereas native bagasse and chitosan aniline furnished highest CV adsorption at pH 5 and chitisan pyrrole reponse was good at pH 6. At all other pH values, the CV adsorption capacities for all adsorbents were low, which revealed that pH range of 4–6 is best for the adsorption of CV onto composites under investigation, that is, the adsorption capacities were recorded to be 24.98, 23.85, 23.0, 18.36, 17.18 and 16.33 (mg/g) at pH 4, whereas these values were 21.35, 20.92, 18.13, 19.24, 23.01 and 17.93 (mg/g) at pH 6 and decreased to 13.2, 13.7, 11.79, 10.79, 10.8, and 9.99 (mg/g) at pH 12 for polypyrrole, starch, polyaniline, chitosan aniline, chitosan pyrrole composites and native bagasse biomass, respectively. Since pH affect the surface binding-sites and chemistry of aqueous solution and initial pH of the solution can affect the degree of ionization of the dye and charge on adsorbent surface, which are responsible for variable CV adsorption on adsorbent as a function of pH. The adsorbents used for CV adsorption are composed biopolymers, which have different functional groups and are responsible for dye binding on adsorbent surface. Thus, the CR adsorption onto composites may be explained on the basis of electrostatic interaction and chemical reaction between the adsorbent and dye ions. At lower pH, strong electrostatic attraction may exist between the positively charged surface of adsorbent

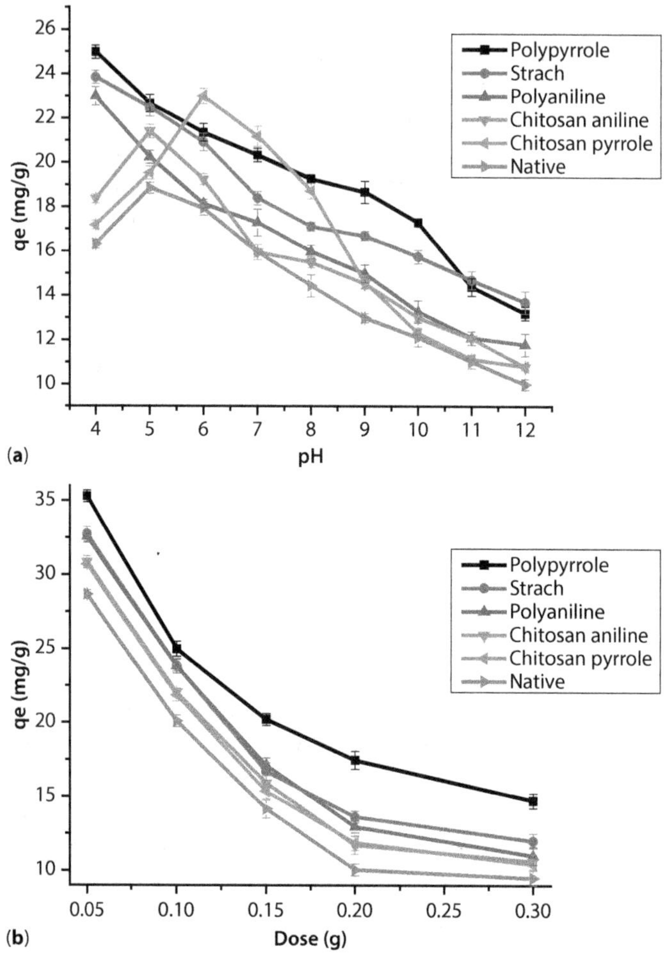

Figure 11.2 CR dye adsorption on composites (bagasse with polyaniline, starch, pyrrole, chitoson/aniline and chitoson/pyrrole) and baggase native rice biomass; A: Effect of pH; B: Effect of adsorbent dose.

and the negatively charged sulfonic acid group of CR dye. At higher pH, the number of positively charged sites decreases and negatively charged sites increases. A negatively charged surface does not favor the adsorption of dye anions due to electrostatic repulsion. At the same time, the presence of excess hydroxyl ions at higher pH conditions might compete with the dye anions for the adsorption sites. At higher pH hydroxyl ions may compete with the dye anions and hence a slow reduction in dye uptake was observed. However, a small amount of dye adsorption under alkaline condition is correlated to chemisorption on dye ions onto adsorbent [49, 78]. Secondly, the functional groups are involved in the dye adsorption process and pH affects the functional groups and resultantly, the forces between adsorbate and adsorbent may change [114, 115]. Similar trend for the adsorption of dyes on composites have been reported previously, that is, adsorption of cationic dye on magnetic κ-carrageenan/PVA nano-composite hydrogels [116], adsorption of acid violet 19 dye onto poly-aniline-Fe_2O_3 magnetic nano-composite [117], removal of reactive dye using sewage sludge derived activated carbon [118], crystal violet and Light yellow 7GL onto glutamic acid modified chitosan magnetic composite [38], methyl orange onto epichlorohydrin magnetic alginate beads [119], brilliant blue FCF onto organosilane functionalized Fe_3O_4 composite [120] and cationic dyes onto humic acid-immobilized amine modified polyacrylamide/bentonite composite [121]. So for, present investigation revealed that bagasse composites with polypyrrole, polyaniline, chitosan aniline and chitosan pyrrole and starch higher potential for CR adsorption under acidic pH range.

11.3.3 Effect of Composites Dose

The effect of composites doses were checked in the range of 0.05–0.3 g/100 mL of dye solution. Overall, trend of all types of adsorbent was similar for CV adsorption and 0.05 mg composite doses furnished higher CV dye adsorption. Composites showed significantly higher CV adsorption versus native biomass. The composites adsorption capacities were found in following order; polypyrrole > starch > polyaniline > chitosan aniline > chitosan pyrrole > native biomass (Figure 11.2(b)). At 0.05 g doses, the adsorption capacities were 35.31, 32.74, 32.56, 30.86, 30.69 and 28.66 (mg/g) and these values reduced to 14.71, 12, 10.99, 10.6, 10.41, and 9.49 (mg/g) at 0.30 g adsorbent dose for polypyrrole, starch, polyaniline, chitosan aniline, chitosan pyrrole and native biomass, respectively. It is reported that at at higher adsorbent dose, the surface area may decreased due to aggregation of adsorbent particles and resultantly, the availability of active sites become limited and dye binding efficiency is decreased [34, 35, 115, 122]. The increased adsorbent dose at constant dye concentration and volume lead to saturation of available active sites because most of the binding sites are blocked due to aggregates formation and low surface area [123, 124]. For effective adsorption, equilibrium is established between adsorbate and adsorbent. Other than nature of functional group on different adsorbents, the competition of the dye ions for limited available binding sites, electrostatic interactions, overlapping or aggregation of adsorption sites, low surface area, interference between binding sites and lower ions contact at higher adsorbent densities might be the reasons for low dye adsorption at higher adsorbent dose [34, 35, 114, 122, 125].

11.3.4 Effect of Contact Time

The effect of contact time was studied in the range of 5–120 min for all types of adsorbent and responses, thus obtained are shown in Figure 11.3(a). It was observed the CV dye adsorption onto native and composites was very fast initially, slowed down after 30 min and equilibrium shortly after 50–60 min. At equilibrium, the adsorption capacities were found out to be 35.55, 32.55, 32.7, 30.67, 30.83 and 28.96 (mg/g) for polypyrrole, starch, polyaniline, chitosan aniline, chitosan pyrrole and native biomass, respectively. Previously, it was reported that the native agricultural waste biomasses attained equilibrium after longer contact time [12, 14, 112, 114, 115, 126, 127] and those of composites attained the equilibrium in shorter duration [35, 128] and similar trend observed in present investigation and equilibrium was reached in shorter duration with very fast adsorption initially followed by the little slower adsorption till equilibrium reached within 50–60 min for different composites. Further increase in contact time did not increase the adsorption and the lower adsorption rate at the latter stage may be due to the difficulty faced by dye ions to occupy the remaining vacant active sites and intraparticle diffusion process [35, 128]. This behavior of adsorption as function of contact time may be explained as the sites were freely available to bind ions and in second phase, already exhausted sites may repel the coming ions as well as the concentration gradient between adsorbate and adsorbent also decreased with time [129, 130]. Similar finding have been document previously that with the passage of time the adsorption process decreased [35, 128].

11.3.5 Effect of Initial Concentration

The CV dye concentration in the range of 25 to 200 mg/L was studied for the evaluation of dye initial concentration effect on adsorption using native biomass, polyaniline, starch, chitosan pyrrole, polypyrrole and chitosan aniline composites and results thus, obtained are shown in Figure 11.3(b). By increasing the dye initial concentration, the adsorption capacities increased and reached maximum for the initial concentration of 175 mg/L and beyond this concentration the effect was insignificant. The adsorption capacities of 20.31, 19.77, 19.14, 16.02, 19.92, and 17.19 (mg/g) were observed for 25 mg/L initial dye concentration, whereas these values

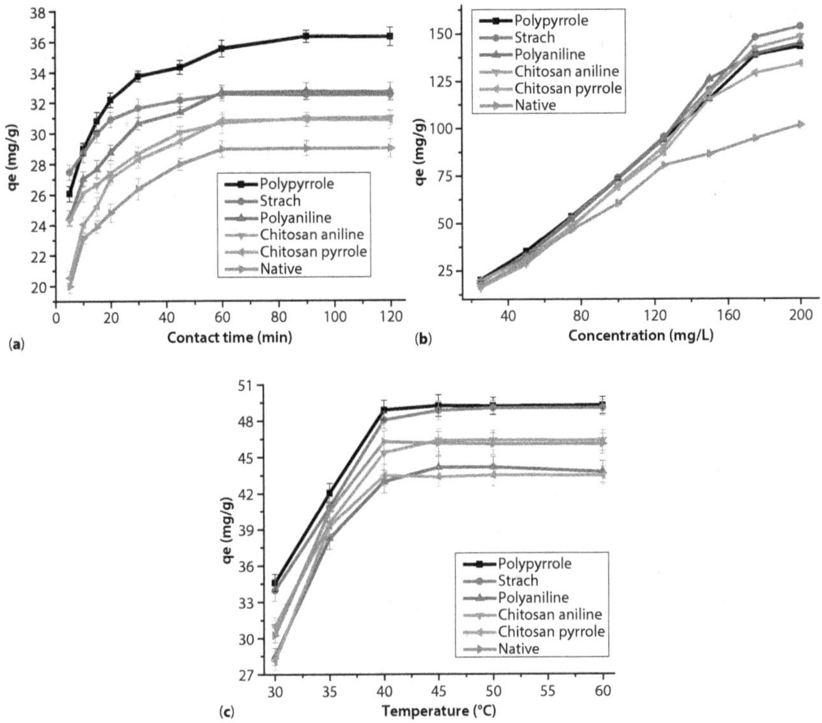

Figure 11.3 CR dye adsorption on composites (bagasse with polyaniline, starch, pyrrole, chitoson/aniline and chitoson/pyrrole) and baggase native rice biomass; A: Effect of contact time; B: Effect of CR initial concentration; and C: Effect of temperature.

reached to 138.54, 147.85, 139.11, 142.05, 129.02 and 94.4 (mg/g) for 175 mg/L dye initial concentration in case of for polypyrrole, starch, polyaniline, chitosan aniline, chitosan pyrrole and native biomass, respectively and beyond this concentration there was no considerable change in adsorption for all types of adsorbents. Previously, similar adsorption trends using different composites have been reported by researchers [37–46, 60, 65, 66]. The enhanced dye adsorption at higher concentration was due to the driving forces to overcome mass transfer resistance of dye ions between the liquid and solid phases [35, 122, 128]. At extremely higher initial concentration, the available binding sites saturated and the adsorption depends on the initial concentration [112] and the probability of interaction of ions with binding sites becomes limited, which restricted the adsorption process. Therefore, for effective adsorption, the ions in solution that interact with the binding sites are important that depends upon initial concentration. Therefore, the adsorption sites were unsaturated initially and at higher concentration the competition between ions and available binding sites increased and hence, the complexation of ions was difficult, which reduced the adsorption of dye slightly [35, 122, 128]. Therefore, the concentration of both adsorbate and adsorbent are important, which must be in equilibrium for effective adsorption [115]. In present investigation, it was observed that composites showed significantly higher dye adsorption versus native biomass and adsorption trend was in following order; starch > chitosan aniline > polyaniline = polypyrrole > chitosan pyrrole > native biomass.

11.3.6 Effect of Temperature

The effect of temperature was studied in the range of 30–60 °C on CV adsorption on to native biomass, polyaniline, starch, chitosan pyrrole, polypyrrole and chitosan aniline composites and results, thus recorded are shown in Figure 11.3(c). Removal of CV showed temperature dependent behavior initially and at higher temperature the dye adsorption was insignificant. The CV adsorption capacities were recorded to be 34.62, 33.99, 28.49, 31.02, 28.14 and 30.28 (mg/g) at 30 °C, whereas these values increased to 49.22, 48.83, 44.14, 46.37, 43.36 and 46.13 (mg/g) for 45 °C for for polypyrrole, starch, polyaniline, chitosan aniline, chitosan pyrrole and native biomass, respectively and at 60 °C, the adsorption capacities were 49.22, 49.02, 43.75, 46.33, 43.44 and 46.06 (mg/g) for polypyrrole, starch, polyaniline, chitosan aniline, chitosan pyrrole and native biomass, respectively, which revealed that the adsorption of dye remained constant and adsorbent showed stability at higher temperature. Adsorption data trend indicates an endothermic process up to certain temperature and then the adsorption was independent at higher temperature, which revealed stability of composites. The favorable effect of moderate temperature on dye adsorption might be due to re-orientation of cell wall components of adsorbent, the ionization of chemical moieties on the cell wall, increased rate of adsorbate diffusion across the external boundary layer and inside the pores of the adsorbate particles because liquid viscosity decreases as temperature increased [122, 131]. The stability of dye adsorption at extreme higher temperatures, that is, 60 °C might be attributed to stable nature of composite, which hold the adsorbed dye ions strongly, however, the studies undertaken using native biomasses showed that at higher temperature, the adsorption was decreased and author's correlated it with damaging of active sites, the weakening of adsorptive forces between active binding sites of the adsorbent and the adsorbate species [34, 35, 114, 132]. It is also reported that earlier adsorbed ions on a surface tend to desorbs from the surface at elevated temperatures [133]. Other researchers also observed a similar temperature effect on adsorption [132, 134, 135] and observed that adsorption increased up to certain temperature and then, decreased. Composite of based on different moieties also showed similar trend for dyes adsorption as a function of temperature, that is, polypyrrole composite [73], polyaniline nano composite [74] and chitosan-cotton composite [136].

11.3.7 Kinetic Study

Bagasse native biomass, polyaniline, starch, chitosan pyrrole, polypyrrole and chitosan aniline composites data was subjected to pseudo-first-order and pseudo-second-order kinetic models. These models are useful for the determination of rate of adsorption for the removal of pollutant form aqueous solution to the adsorbent and deduce the relation between amounts of dye adsorbed as a function of time. Pseudo-first-order kinetic model differential equation is shown in Eqn. 11.2 and linear form is given in Eqn. 11.3.

$$d_{qt} / d_t = k_1 (q_e - q_t) \tag{11.2}$$

$$\log(q_e - q_t) = log q_e - \frac{k_1}{2.303t} \tag{11.3}$$

where, q_e is the adsorption efficiency (at equilibrium), q_t is the adsorption at time t, and k_1 is the rate constant of pseudo first order.

Pseudo-second-order kinetic model differential form is shown in Eqn. 11.4, whereas linear form is presented in Eqn. 11.5.

$$d_{qt} / d_t = k_2 (q_e - q_t)^2 \tag{11.4}$$

$$t / q_t = 1 / k_2 q_e^2 + 1 / q_e \tag{11.5}$$

where, q_e is the adsorption efficiency (at equilibrium), q_t is the adsorption efficiency at time t and k_2 is the rate constant of pseudo-second-order kinetic model.

The data obtained in case of kinetic models for native and composite adsorbents is plotted and plots are shown in Figure 11.4(a–f) and Figure 11.5(a–f), respectively, for pseudo-first-order and pseudo-second-order kinetics models. The q_e estimated from experimental data did not correlate with the pseudo-first-order kinetic model for CV adsorption on

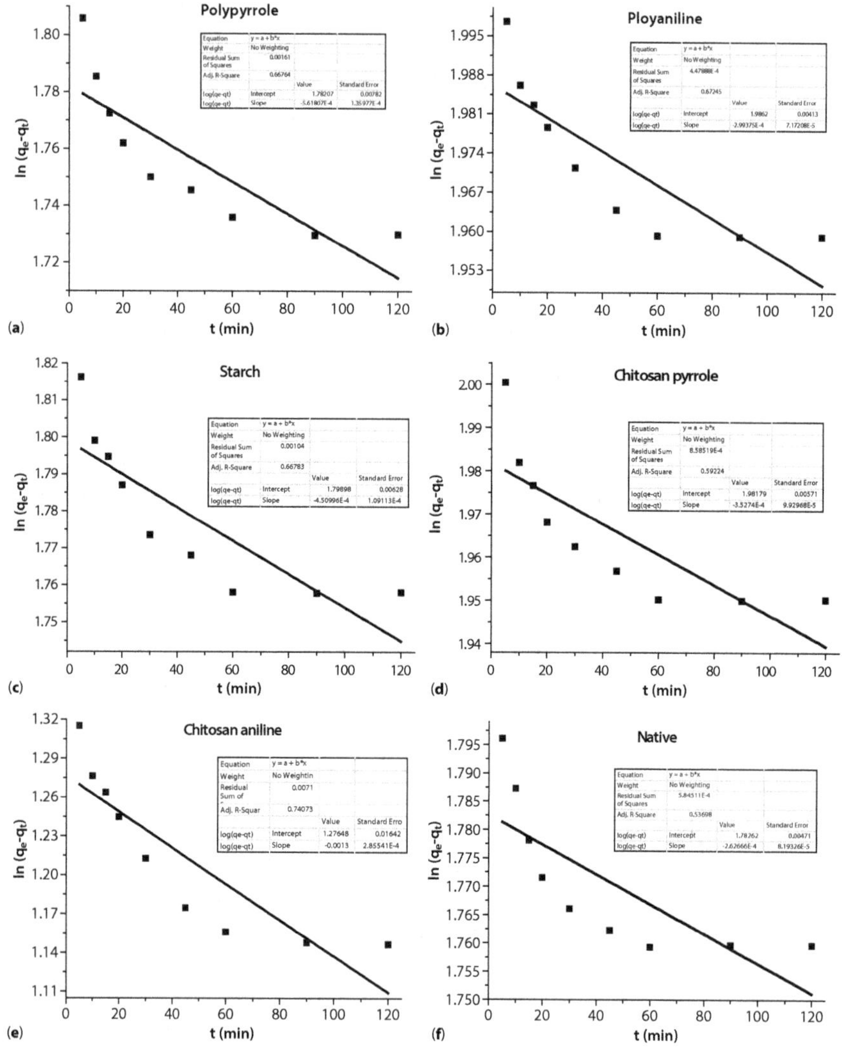

Figure 11.4 (A–F) Pseudo-first-order plots (log) for CR dye adsorption on composites (baggase with polyaniline, starch, pyrrole, chitoson/aniline and chitoson/pyrrole) and native bagasse biomass.

to native and composite adsorbents and the R^2 values (0.66, 0.67, 0.66, 0.59, 0.74 and 0.53 for polypyrrole, polyaniline, starch, chitosan pyrrole, polyaniline, chitosan aniline, and native) were also not acceptable, leading to the assumption that the CR data did not fit well with the pseudo-first order kinetics model. In case of pseudo second order kinetic model, the R^2 values (0.99, 0.99, 0.99, 0.99, 0.99 and 0.99 for native, starch, chitosan aniline,

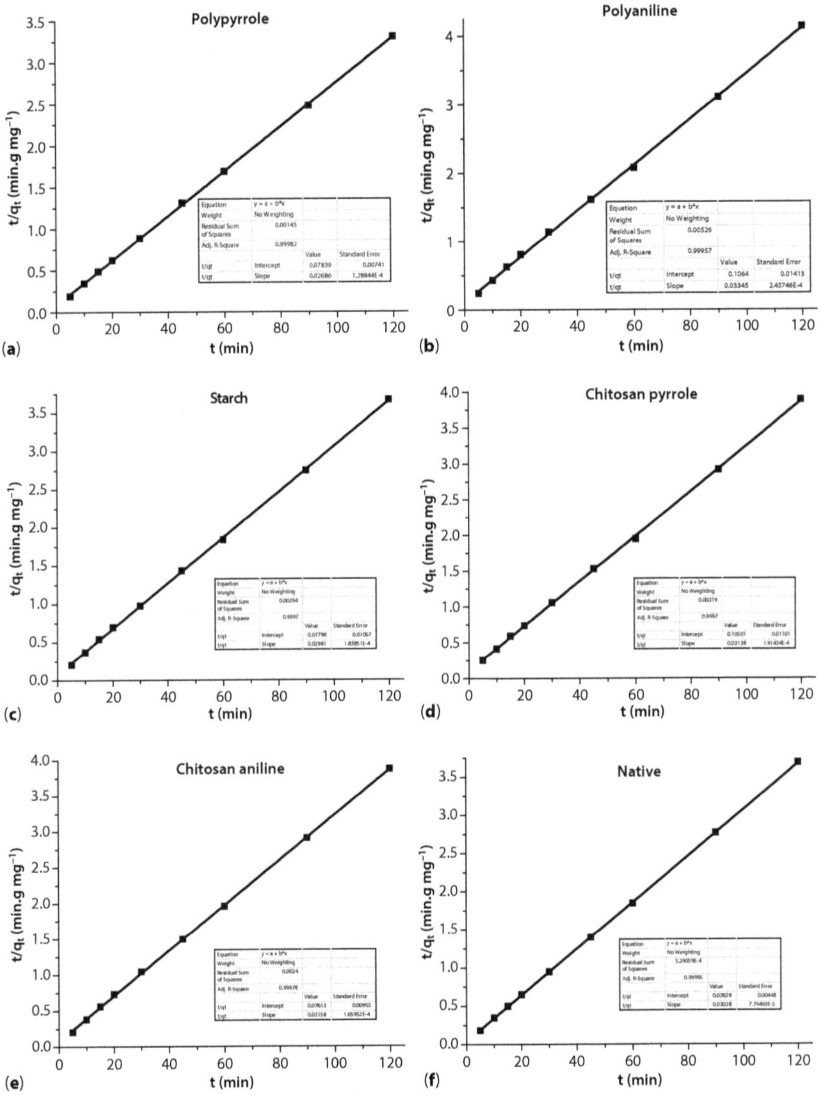

Figure 11.5 (A–F) Pseudo-second-order plots for CR dye adsorption on composites (baggase with polyaniline, starch, pyrrole, chitoson/aniline and chitoson/pyrrole) and native bagasse biomass.

polyanilne, polypyrrole and chitosan pyrrole, respectively) were also found to be enough high and the q_e values determined experimentally, coincided well with the modelled values, which revealed that CR dye adsorption on to native and polypyrrole, chitosan pyrrole starch, polyaniline and chitosan aniline composites followed pseudo-second-order kinetic model. Similar kinetics trends have also been reported previously for different dyes adsorption onto composites such as chitosan supported CNT-MNPs [43], magnetic chitosan/poly (vinyl alcohol) hydrogel beads [37], chitosan/polyurethane [47], polyurethane/chitosan composite foams [47], glutamic acid modified chitosan magnetic composite [38], chitosan/oil palm [52], ethylenediamine-modified magnetic chitosan [60], chitosan/montmorillonite [49], chitosan/activated clay [51], chitosan/bentonite [50], ferrofluid modified peanut husks [39], and magnetic N-lauryl chitosan nanocomposite [66].

11.3.8 Intraparticle Diffusion Model

During adsorption process different steps are involved, that is, movement of dye molecules toward the biosorbent surface, the adsorbed species diffuses on surface followed by interior pores diffusion (intraparticle diffusion) and then, interacts with binding sites. The possibility of intrapaticle diffusion was examined using the intraparticle diffusion model, which assumes that during the adsorption process, the adsorbed amount is proportional to the square root of the contact time. The relation for interaparticle diffusion model is shown in Eqn. 11.6 [137].

$$q_t = K_d t^{1/2} \tag{11.6}$$

where, K_d is the intraparticle diffusion constants, which gives an idea about the thickness of the boundary layer and plots between q_t and $t^{0.5}$ for native biomass, polyaniline, starch, chitosan pyrrole, polypyrrole and chitosan aniline composites are shown in Figure 11.6(a–f), respectively, for the contact time changing from 20 to 120 min. For all types of adsorbents, linear curves are obtained, which revealed direct relationship of qe with contact time, which revealed that that the CR dye uptake varies in proportional to $t^{0.5}$. Good linearization of dye adsorption data observed for adsorbents under investigation revealed the intraparticle diffusion is the rate-limiting step. The R^2 values for native biomass, chitosan aniline, chitosan pyrrole, starch, polyaniline, polypyrrole were recorded to 0.72, 0.86, 0.76, 0.82, 0.83, and 0.81, respectively.

11.3.9 Isotherm Modelling

Isotherm models are helpful in understanding the adsorption mechanism, which assume the adsorption of adsorbate as a function of equilibrium

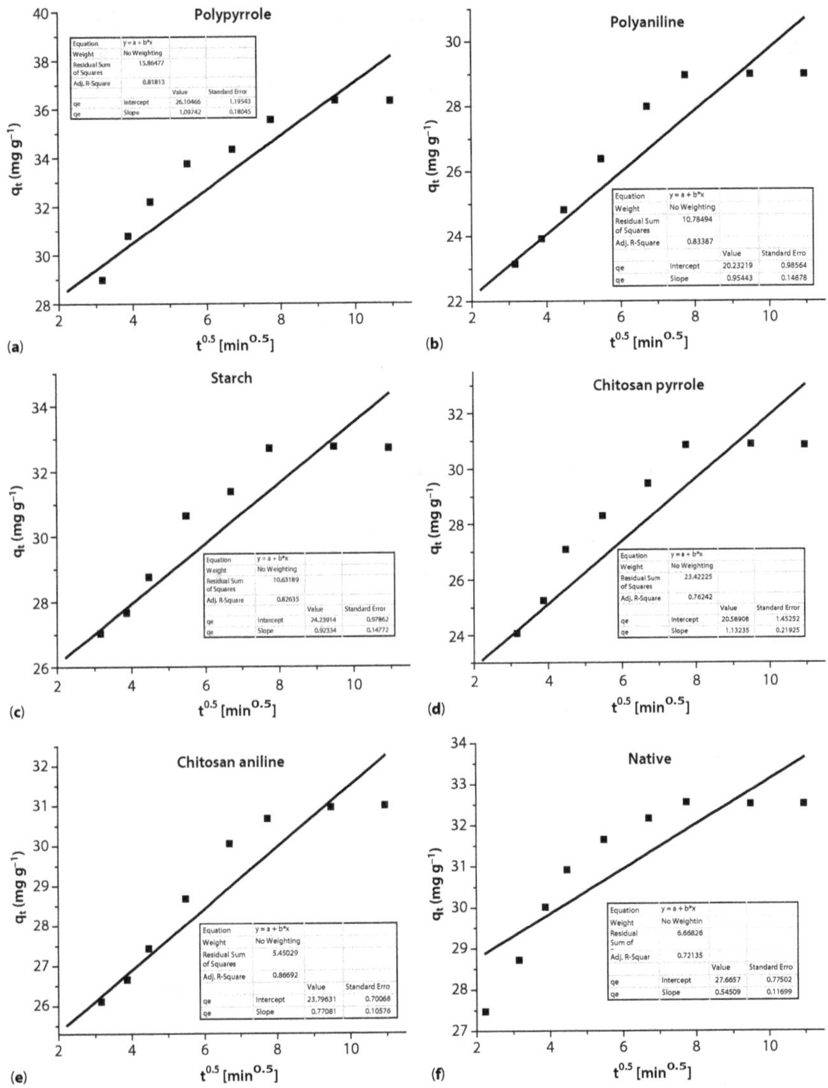

Figure 11.6 (A–F) Intraparticle diffusion model plots for CR adsorption on composites (bagasse with polyaniline, starch, pyrrole, chitoson/aniline and chitoson/pyrrole) and native bagasse biomass.

concentration. Langmuir adsorption model assumes that adsorption occurs at specific homogeneous adsorption sites within the adsorbent and intermolecular forces decrease rapidly with the distance from the adsorption surface [112, 114, 115]. The model further based on the assumption that all the adsorption sites are energetically identical and adsorption occurs on a structurally similar binding site. The Langmuir adsorption isotherm is represented in Eqn. 11.7 and linear form of this isotherm is expressed in Eqn. 11.8.

$$q_e = \frac{q_{max} \times b \times C_e}{1 + b \times C_e} \tag{11.7}$$

$$\frac{C_e}{q_e} = \frac{C_e}{q_{max}} + \frac{1}{q_{max} \times b} \tag{11.8}$$

where, $q_{\max}$ (mg/g) is maximum adsorption capacity and b (L/mg) is Langmuir constants related to the heat of adsorption.

Freundlich isotherm model can be used to describe the sorption on heterogeneous surfaces as well as a multilayer sorption. It assumes that the uptake of adsorbate ion occurs on a heterogeneous adsorbent surface. The Freundlich isotherm model is empirical in nature which assumes that the stronger binding sites are occupied first and that the binding strength decreases with increasing degree of site occupation. The non liner and linear forms of Freundlich isotherm model are shown in Eqns. 11.9 and 11.10, respectively.

$$q_e = K_f C_e^{1/n} \tag{11.9}$$

$$\log q_e = logK_F + \frac{1}{n} \log C_e \tag{11.10}$$

where, K_f and n are the Freundlich model constants designated as adsorption capacity and adsorption intensity, respectively, which can be obtained from slope and intercept of the plot of logq_e versus logC_e. C_e is the residual concentration of solute in solution (mg/L), q_e is the amount of adsorbate adsorbed by a unit mass of adsorbent at equilibrium (mg/g). The value of n in the range of 1–10 is another benchmark to assess the adsorbent–adsorbate interaction.

The Freundlich and Langmuir isotherm models plots are shown in Figures 11.7(a–f) and Figure 11.8(a–f) for native biomass, polyaniline, starch, pyrrole, chitosan aniline and chitosan pyrrole composites. The R^2 and equilibrium experimental data showed that the R^2 value for the Freundlich model were enough high and acceptable to explain the CR dye equilibrium adsorption data, whereas the R^2 values in case of Langmuir

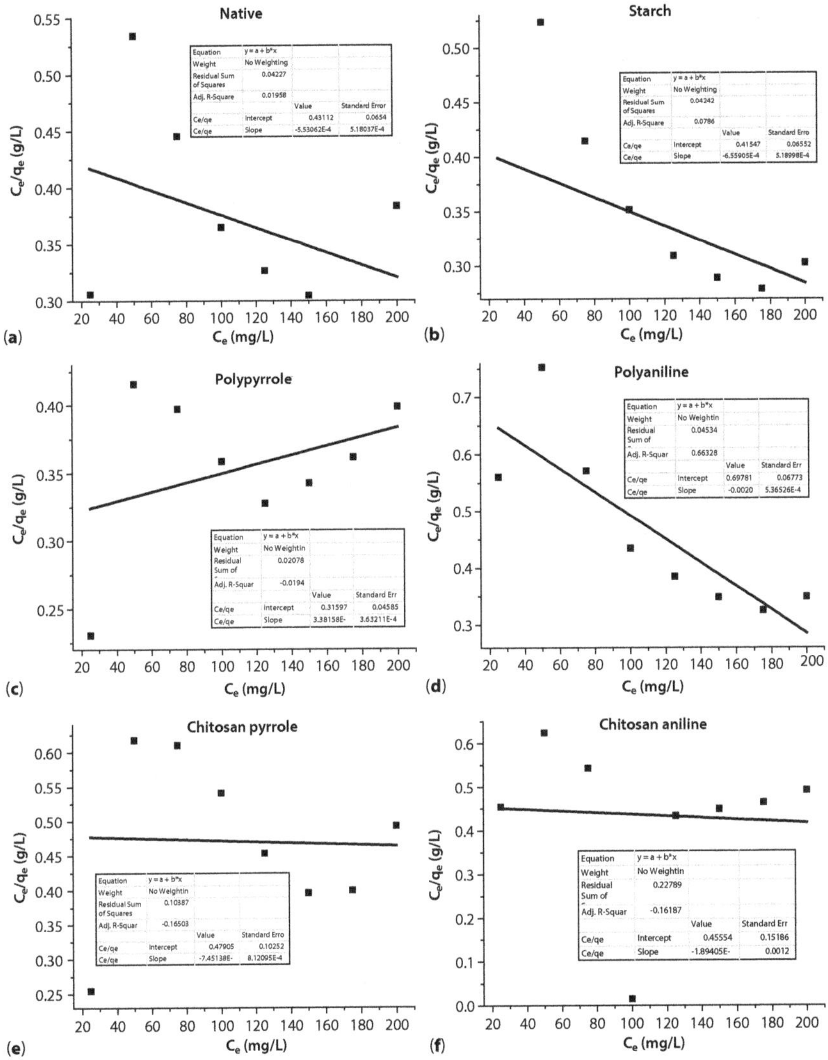

Figure 11.7 (A–F) Langmuir isotherm model curves (Ce/qe (g/L) versus Ce (mg/L)) CR adsorption on composites (bagasse with polyaniline, starch, pyrrole, chitoson/aniline and chitoson/pyrrole) and native bagasse biomass.

isotherm model were not appreciable high since this isotherm is unable to explain the CR dye adsorption onto native biomass, polyaniline, starch, pyrrole, chitosan aniline and chitosan pyrrole composites. The Freundlich isotherm model described the adsorption on heterogeneous surface and is not restricted to the formation of monolayer. The values of R^2 in case of

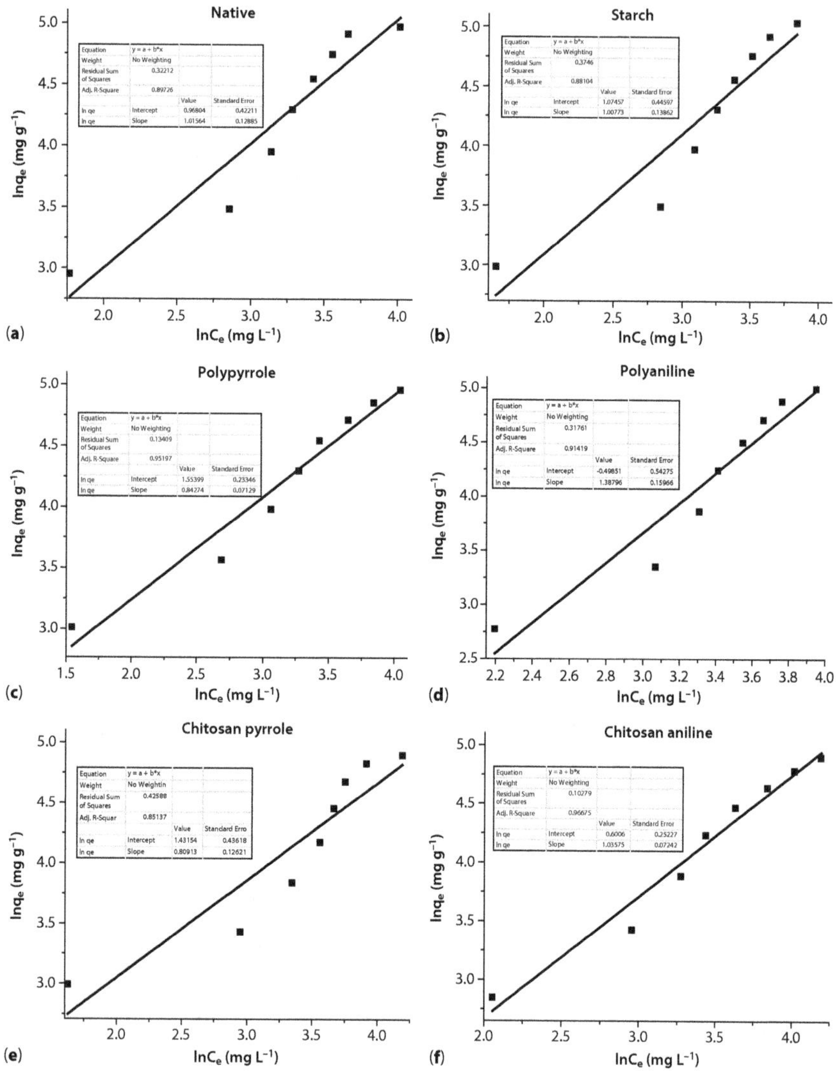

Figure 11.8 (A–F) Freundlich isotherm model curves CR adsorption on composites (bagasse with polyaniline, starch, pyrrole, chitoson/aniline and chitoson/pyrrole) and native bagasse biomass.

Freundlich isotherm model were greater than those obtained for Langmuir model and also q_e values were close to experimentally determined values, which indicate that the Freundlich isotherm model explained well CR adsorption on to native and composites adsorbents.

11.3.10 Thermodynamic Study

The thermodynamic parameters, that is, Gibbs free energy (ΔG^0), enthalpy (ΔH^0) and entropy (ΔS^0), were calculated using relations shown in Eqns. 11.11–11.13 [138] in order to understand the nature of the CR dye adsorption on to native and composite adsorbents.

$$K_c = \frac{q_e}{C_e} \tag{11.11}$$

$$\Delta G^0 = -RTlnK_c \tag{11.12}$$

$$lnK_c = \frac{\Delta S^0}{R} - \frac{\Delta H^0}{RT} \tag{11.13}$$

where, K_c, q_e and C_e are representing the equilibrium constant, adsorption at equilibrium (mg/L) and dye concentration at equilibrium (mg/L). Arrhenius plot was used to calculate the ΔH^0 and ΔS^0 values [139]. The thermodynamic results calculated for CR dye adsorption for native biomass, polyaniline, starch, polypyrrole, chitosan aniline and chitosan pyrrole composites are shown in Table 11.4 and Arrhenius plot are shown in Figure 11.9. The native $\Delta G°$ values recorded for native biomass, polyaniline, polypyrole, starch and chitosan pyrole composites, respectively indicate that the CR adsorption process was spontaneous in nature. $\Delta H°$ values for native biomass, polyaniline, starch, polypyrrole, chitosan aniline and chitosan pyrrole composites were positive, which revealed the endothermic nature of CR dye adsorption. This was also in line with effect of temperature study that CR dye adsorption onto different composites and native biomass increased with temperature and then, became stable at higher temperature. The positive values $\Delta S°$ for native biomass and composites suggests that there was an increase in randomness at the solid/solution interface and similar trend has been reported for dye adsorption [140]. These finding implies that the CR dye adsorption process was energetically stable, spontaneous and CR adsorbed onto composites by chemical interaction [141] and composites are efficient absorbents and could possibly be used for dyes adsorption from wastewater [111].

To date, chitosan along with other composites have been prepared and employed for the adsorption of pollutants. Different kinds of substances have been used to prepare composites, that is, montmorillonite, polyurethane, activated clay, bentonite [50], poly vinyl alcohol, poly vinyl chloride [142], oil palm ash [52] and perlite [143] and agricultural waste biomasses

Table 11.4 Thermodynamics parameters of congo red dye adsorption on native and composites adsorbents.

Adsorbents	Temperature K	ΔG° kJ/mol	ΔH° kJ/mol	ΔS° J/mol.K
Native	303	−26584.69		
	308	−27062.76		
	313	−27540.82	2386.11	95.6132
	318	−28018.89		
	323	−28496.95		
	333	−29453.09		
Starch	303	−42663.13		
	308	−43435.11		
	313	−44207.08	4118.55	154.395
	318	−44979.06		
	323	−45751.03		
	333	−49420.05		
Polyaniline	303	−21634.06		
	308	−22023.95		
	313	−22413.84	1993.49	77.9787
	318	−22803.74		
	323	−23193.63		
	333	−23973.42		
Polypyrrole	303	−45229.81		
	308	−46047.16		
	313	−46864.51	4301.597	163.47
	318	−47681.86		
	323	−48499.21		
	333	−50133.91		

(*Continued*)

Table 11.4 Cont.

Adsorbents	**Temperature K**	$\Delta G°$ **kJ/mol**	$\Delta H°$ **kJ/mol**	$\Delta S°$ **J/mol.K**
Chitosan pyrole	303	−20450.75		
	308	−20818.41		
	313	−21186.08	1829.66	73.5327
	318	−21553.74		
	323	−21921.40		
	333	−22656.73		
Chitosan aniline	303	−27245.52		
	308	−27736.68		
	313	−28227.85	2518.99	98.2327
	318	−28719.01		
	323	−29210.17		
	333	−30192.50		

[34]. The composite adsorption efficiency depends upon various factors. For example pH, as it can either form gel or dissolve depending on the pH values [144]. So far, by preparing composites with suitable agent can enhance the adsorption capacities of composite. It is reported that chitosan/bentonite composites had a much lower surface area as compared to bentonite [145], more protons and protonation was observed when chitosan combined with oil palm ash, which can increase the electrostatic attractions between the negatively charged dyes and the positively charged active binding sites on the adsorbent surface [52]. Chitosan/kaolin/-Fe_2O_3 composites revealed pores and pleats on the surface which provided active sites for dye entrapment [142]. Present investigation also revealed that composites showed good efficiencies for the adsorption of MG dye and the adsorption capacities were in line with reported composites, that is, 470.1 mg/g Congo Red—magnetic chitosan/poly (vinyl alcohol) hydrogel beads [37], 180, 375.4 and 217.3 (mg/g) Methylene Blue, Crystal violet and Light yellow 7GL—glutamic acid modified chitosan magnetic composite [38], 29.83 mg/g Methyl—magnetic γ-Fe_2O_3 crosslinked chitosan composite [40], 73.4, 1.24, 69.2, 9.43, 68.1, 43 and 59 (mg/g) Acridine Orange, Amido

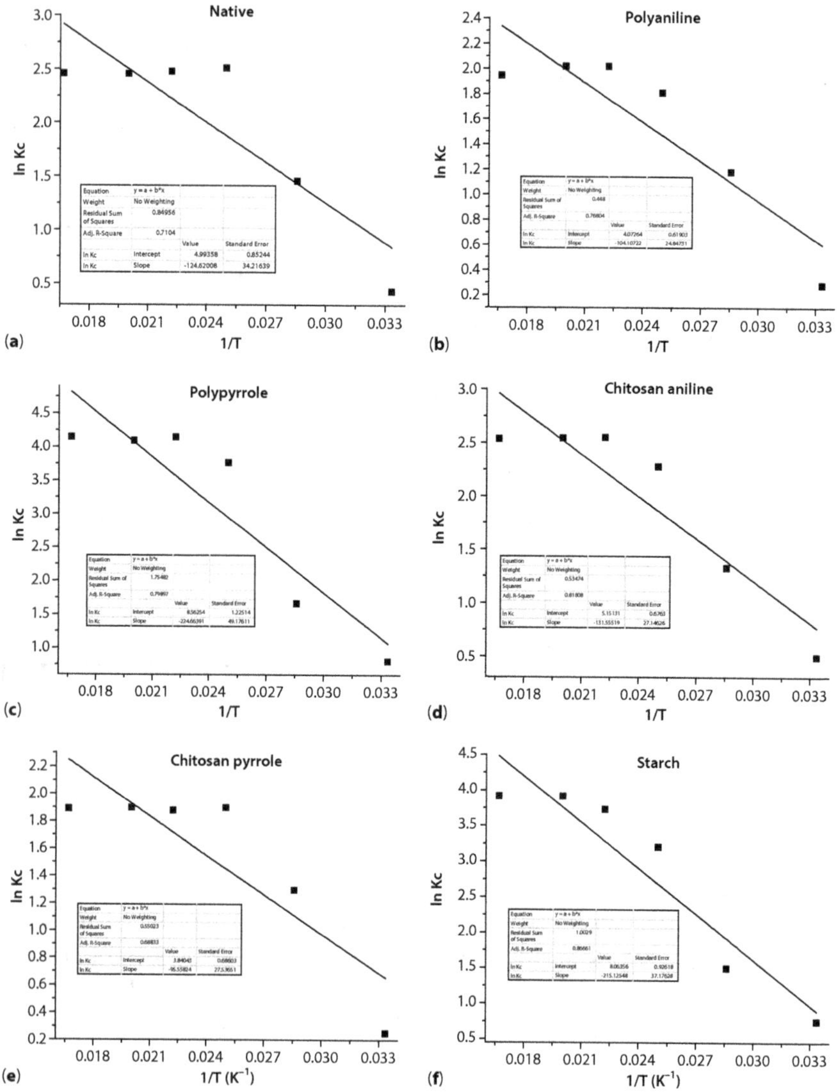

Figure 11.9 ln Kc versus 1/T (Arrhenius plot) for thermodynamic study.

Black 10B, Bismark Brown, Congo Red, Crystal Red, Malachite Green, Safranin-O—magnetically modified spent coffee grounds [42], 95.3, 86.1, 80.9 and 71.4 (mg/g) Bismarck Brown, Safranin-O, Crystal Violet, Acridine Orange—ferrofluid modified peanut husk [39], 1017 and 1214 (mg/g) Acid orange 10 and Acid orange 7—ethylenediamine-modified magnetic chitosan nanoparticles [60], 809.9 mg/g Acid red 18—chitosan suppoted

CNT-MNPs [43], 267 mg/g Remazol red 198—magnetic N-lauryl chitosan nanocomposite [66], 262 mg/g Methylene Blue—ionic liquid-coated Fe_3O_4@chitosan@graphene oxide [48], 748.50 mg/g Acid Red 337—amphoteric chitosan/gelatin composite microspheres [48], 330 mg/g methylene blue—chitosan/activated clay [51], 435 mg/g Malachite Green—chitosan/bentonite [50], 909.1 mg/g Reactive blue 19—chitosan/oil palm [52], 243.9 mg/g Acid Orange10—polypyrrole composite [73] and 143.59 mg/g Reactive Dye—polyaniline nano composite [74]. The higher adsorption capacities of composites have been correlated with modification and stability of active sites on the surfaces since combination of two types of material provide reinforcement and stability in case of good compatibility [113] and resultantly, the physico-chemical characteristics of composites are enhanced that offer excellent adsorption capacities versus native/un-modified biomasses/adsorbents [112, 146–152]. Hence, composites (polyaniline, starch, polypyrrole composites with RB biomass) showed excellent affinity for MG dye adsorption and could be used for the adsorption of dyes from textile wastewater.

11.4 Conclusion

The composites of sugarcane bagasse with polyaniline, starch, pyrrole, chitosan aniline and chitosan pyrrole were prepared and employed successfully for the adsorption of Crystal Violet dye and maximum dye adsorption was achieved at 180 mg/L CR initial concentration, 40 °C temperature, 60 min contact time, 0.05 g adsorbent doses and 2–6 pH's (variable for different composites). Pseudo-second-order kinetic model, Freundlich isotherm and intraparticle diffusion model fitted well to the CR dye adsorption data. CR dye adsorption process was endothermic, spontaneous and energetically stable. In view of enhanced dye adsorption, the composites could possibly be used for dyes removal from textile wastewater.

References

1. Bhatt, R., Sreedhar, B., Padmaja, P., Adsorption of chromium from aqueous solutions using crosslinked chitosan–diethylenetriaminepentaacetic acid. *Int. J. Biol. Macromol.*,74, 458–466, 2015.
2. Cataldo, S., Gianguzza, A., Milea, D., Muratore, N., Pettignano A., Pb(II) adsorption by a novel activated carbon alginate composite material. A kinetic and equilibrium study. *Int. J. Biol. Macromol.* , 92, 769–778, 2016.

3. Zhang, L., Luo, H., Liu, P., Fang, W., Geng, J., A novel modified graphene oxide/chitosan composite used as an adsorbent for Cr(VI) in aqueous solutions. *Int. J. Biol. Macromol.*, 87, 586–596, 2016.
4. Petcu, A.R., Lazar, C.A., Rogozea, E.A., Olteanu, N.L., Meghea, A., Mihaly, M., Nonionic microemulsion systems applied for removal of ionic dyes mixtures from textile industry wastewaters. *Sep. Purif. Technol.*, 158, 155–159, 2016.
5. Ghaly, A., Ananthashankar, R., Alhattab, M., Ramakrishnan, V., Production, characterization and treatment of textile effluents: a critical review. *J. Chem. Eng. Process Technol.*, 2014, 2014.
6. Chandran, D., A review of the textile industries waste water treatment methodologies. *Int. J. Sci. Eng. Res.*,, 7, 392–403, 2016.
7. Klemola, K., Pearson, J., Lindstrom-Seppä, P., Evaluating the toxicity of reactive dyes and dyed fabrics with the HacaT cytotoxicity test. *Autex Res. J.*, 7, 217–223, 2007.
8. Hayat, H., Mahmood, Q., Pervez, A., Bhatti, Z.A., Baig, S.A., Comparative decolorization of dyes in textile wastewater using biological and chemical treatment. *Sep. Purif. Technol.*, 154, 149–153, 2015.
9. Iqbal, M., Vicia faba bioassay for environmental toxicity monitoring: a review. *Chemosphere*, 144, 785–802, 2016.
10. Iqbal, M., Bhatti, I.A., Gamma radiation/H_2O_2 treatment of a nonylphenol ethoxylates: degradation, cytotoxicity, and mutagenicity evaluation. *J. Hazard. Mat.*, 299, 351–360, 2015.
11. Babarinde, A., Ogundipe, K., Sangosanya, K.T., Akintola, B.D., Elizabeth Hassan, A-.O., Comparative study on the biosorption of Pb(II), Cd(II) and Zn(II) using Lemon grass (*Cymbopogon citratus*): kinetics, isotherms and thermodynamics. *Chem. Int.*, 2, 89–102, 2016.
12. Babarinde, A., Onyiaocha, G.O., Equilibrium sorption of divalent metal ions onto groundnut (Arachis hypogaea) shell: kinetics, isotherm and thermodynamics. *Chem. Int.*, 2, 37–46, 2016.
13. Benabdallah, N.K., Harrache, D., Mir, A., de la Guardia, M., Benhachem, F.-Z., Bioaccumulation of trace metals by red alga Corallina elongata in the coast of Beni Saf, west coast, Algeria. *Chem. Int.*, 3, 220–231, 2017.
14. Iqbal, M., Khera, R.A., Adsorption of copper and lead in single and binary metal system onto Fumaria indica biomass. *Chem. Int.*, 1, 157b–163b, 2015.
15. Jafarinejad, S., Control and treatment of sulfur compounds specially sulfur oxides (SOx) emissions from the petroleum industry: a review. *Chem. Int.*, 2, 242–53, 2016.
16. Jafarinejad, S., Recent developments in the application of sequencing batch reactor (SBR) technology for the petroleum industry wastewater treatment. *Chem. Int.*, 3(3), 2017
17. Jamal M.A., Muneer M., Iqbal, M., Photo-degradation of monoazo dye blue 13 using advanced oxidation process. *Chem. Int.*, 1, 12–6, 2015.
18. Legrouri, K., Khouya, E., Hannache, H., El Hartti, M., Ezzine, M., Naslain, R., Activated carbon from molasses efficiency for Cr(VI), Pb(II) and Cu(II) adsorption: A mechanistic study. . *Chem. Int.*, 3, 301–10, 2017.

19. Majolagbe, A.O., Adeyi, A.A., Osibanjo, O., Vulnerability assessment of groundwater pollution in the vicinity of an active dumpsite (Olusosun), Lagos, Nigeria. *Chem. Int.*, 2, 232–41, 2016.
20. Majolagbe, A.O., Adeyi, A.A., Osibanjo, O., Adams, A.O., Ojuri, O.O., Pollution vulnerability and health risk assessment of groundwater around an engineering Landfill in Lagos, Nigeria. *Chem. Int.*, 3, 58–68, 2017.
21. Ngobiri, N., Okorosaye-Orubite, K., Adsorption and corrosion inhibition characteristics of two medicinal molecules. *Chem. Int.*, 3, 185–194, 2017.
22. Ogundipe, K.D., Babarinde, A., Comparative study on batch equilibrium biosorption of Cd(II), Pb(II) and Zn(II) using plantain (Musa paradisiaca) flower: kinetics, isotherm, and thermodynamics. *Chem. Int.*, 3, 135–149, 2017.
23. Pervaiz, M., Butt, K.M., Raza, M.A., Rasheed, A., Ahmad, S., Adnan, A., *et al.*, Extraction and applications of aluminum hydroxide from bauxite for commercial consumption. *Chem. Int.*, 1, 99–102, 2015.
24. Peter, U.C., Chinedu, U., Model prediction for constant area, variable pressure drop in orifice plate characteristics in flow system. *Chem. Int.* 2, 80–88, 2016.
25. Qureshi, K., Ahmad, M., Bhatti, I., Iqbal, M., Khan, A., Cytotoxicity reduction of wastewater treated by advanced oxidation process. *Chem. Int.*, 1, 53–59, 2015.
26. Sayed, M., Efficient removal of phenol from aqueous solution by the pulsed high-voltage discharge process in the presence of H2O2. *Chem. Int.*, 1, 81–86, 2015.
27. Ukpaka, C., Development of model for bioremediation of crude oil using moringa extract. *Chem. Int.*, 2, 19–28, 2016.
28. Ukpaka, C., Predictive model on the effect of restrictor on transfer function parameters on pneumatic control system. *Chem. Int.*, 2, 128–135, 2016.
29. Ukpaka, C., Empirical model approach for the evaluation of pH and conductivity on pollutant diffusion in soil environment. *Chem. Int.*, 2, 267–278, 2016.
30. Ukpaka, C., BTX degradation: the concept of microbial integration. *Chem. Int.* 3, 8–18, 2016.
31. Ukpaka, C., Izonowei, T., Model prediction on the reliability of fixed bed reactor for ammonia production. *Chem. Int.*, 3, 46–57, 2017.
32. Ukpaka, C., Ukpaka, C., Characteristics of groundwater in Port-Harcourt local Government area. *Chem. Int.*, 2, 136–144, 2016.
33. Ukpaka, C.P., Igwe, F.U., Modeling of the velocity profile of a bioreactor: the concept of biochemical process. *Chem. Int.*, 3, 258–267, 2017.
34. Mushtaq, M., Bhatti, H.N., Iqbal, M., Noreen, S., Eriobotrya japonica seed biocomposite efficiency for copper adsorption: isotherms, kinetics, thermodynamic and desorption studies. *J. Environ. Manag.*, 176, 21–33, 2016.
35. Rashid, A., Bhatti, H.N., Iqbal, M., Noreen, S., Fungal biomass composite with bentonite efficiency for nickel and zinc adsorption: a mechanistic study. *Ecol. Eng.*, 91, 459–471, 2016.

36. Sargın, İ., Arslan, G., Kaya, M., Efficiency of chitosan–algal biomass composite microbeads at heavy metal removal. *React. Funct. Polym.*, 98, 38–47, 2016.
37. Zhu, H.-Y., Fu, Y.-Q., Jiang, R., Yao, J., Xiao, L., Zeng, G-.M., Novel magnetic chitosan/poly (vinyl alcohol) hydrogel beads: Preparation, characterization and application for adsorption of dye from aqueous solution. *Biores. Technol.*, 105, 24–30, 2012.
38. Yan, H., Li, H., Yang, H., Li, A., Cheng, R., Removal of various cationic dyes from aqueous solutions using a kind of fully biodegradable magnetic composite microsphere. *Chem. Eng. J.*, 223, 402–411, 2013.
39. Safarik, I., Safarikova, M., Magnetic fluid modified peanut husks as an adsorbent for organic dyes removal. *Physics Procedia*, 9, 274–8, 2010.
40. Zhu, H.-Y., Jiang, R., Xiao, L., Li, W., A novel magnetically separable γ-Fe2O 3/crosslinked chitosan adsorbent: preparation, characterization and adsorption application for removal of hazardous azo dye. *J. Hazard. Mater.*, 179, 251–257, 2010.
41. Chang, P.R., Zheng, P., Liu, B., Anderson, D.P., Yu, J., Ma, X., Characterization of magnetic soluble starch-functionalized carbon nanotubes and its application for the adsorption of the dyes. *J. Hazard. Mater.*, 186, 2144–2150, 2011.
42. Safarik, I., Horska, K., Svobodova, B., Safarikova, M., Magnetically modified spent coffee grounds for dyes removal. *Europ. Food Res. Technol.*, 234, 345–350, 2012.
43. Wang, S., Zhai, Y.-Y., Gao, Q., Luo, W-.J., Xia, H., Zhou, C.-G., Highly efficient removal of acid red 18 from aqueous solution by magnetically retrievable chitosan/carbon nanotube: batch study, isotherms, kinetics, and thermodynamics. *J. Chem. Eng. Data*, 59, 39–51, 2013.
44. Ge, F., Ye, H., Li, M-.M., Zhao, B-.X., Efficient removal of cationic dyes from aqueous solution by polymer-modified magnetic nanoparticles. *Chem. Eng. J.*, 198, 11–17, 2012.
45. Mahmoodi, N.M., Magnetic ferrite nanoparticle–alginate composite: Synthesis, characterization and binary system dye removal. *J. Taiwan Instit. Chem.*, 44, 322–330, 2013.
46. Li, D-.P., Zhang, Y-.R., Zhao, X-.X., Zhao, B-.X., Magnetic nanoparticles coated by aminoguanidine for selective adsorption of acid dyes from aqueous solution. *Chem. Eng. J.*, 232, 425–433, 2013.
47. Lee, H.C., Jeong, Y.G., Min, B.G., Lyoo, W.S., Lee, S.C., Preparation and acid dye adsorption behavior of polyurethane/chitosan composite foams. *Fibers Polym.*, 10, 636–642, 2009.
48. Li, L., Duan, H., Wang, X., Luo, C., Fabrication of novel magnetic nanocomposite with a number of adsorption sites for the removal of dye. *Int. J. Biol. Macromol.*. 78, 17–22, 2015.
49. Wang, L., Wang, A., Adsorption characteristics of Congo Red onto the chitosan/montmorillonite nanocomposite. *J. Hazard. Mater.*, 147, 979–985, 2007.
50. Ngah, W.S.W., Ariff, N.F.M., Hanafiah, M.A.K.M., Preparation, characterization, and environmental application of crosslinked chitosan-coated bentonite

for tartrazine adsorption from aqueous solutions. *Water, Air, Soil Pollut.*, 206, 225–236, 2010.

51. Chang, M.-Y., Juang, R-.S., Adsorption of tannic acid, humic acid, and dyes from water using the composite of chitosan and activated clay. *J. Colloid Interf. Sci.*, 278, 18–25, 2004.
52. Hasan, M., Ahmad, A., Hameed, B., Adsorption of reactive dye onto cross-linked chitosan/oil palm ash composite beads. *Chem. Eng. J.*, 136, 164–172, 2008.
53. Kumar, R., Ansari, M.O., Barakat, M., Adsorption of brilliant green by surfactant doped polyaniline/MWCNTs composite: Evaluation of the kinetic, thermodynamic, and isotherm. *Ind. Eng. Chem. Res.*, 53, 7167–7175, 2014.
54. Du, H., Chen, W., Cai, P., Rong, X., Feng, X., Huang, Q., Competitive adsorption of Pb and Cd on bacteria–montmorillonite composite. *Environ. Pollut.*, 218, 168–175, 2016.
55. Zarrini, K., Rahimi, A.A., Alihosseini, F., Fashandi, H., Highly efficient dye adsorbent based on polyaniline-coated nylon-6 nanofibers. *J. Cleaner Prod.*, 142, 3645–3654, 2017.
56. Hou, H., Zhou, R., Wu, P., Wu, L., Removal of Congo red dye from aqueous solution with hydroxyapatite/chitosan composite. *Chem. Eng. J.*, 211–212, 336–342, 2012.
57. Yang, Y., Song, S., Zhao, Z., Graphene oxide (GO)/polyacrylamide (PAM) composite hydrogels as efficient cationic dye adsorbents. *Colloids Surf. A: Physicochem. Eng. Aspects*, 513, 315–324, 2017.
58. Yao, Y., Miao, S., Liu, S., Ma, L.P., Sun, H., Wang, S., Synthesis, characterization, and adsorption properties of magnetic Fe3O4@graphene nanocomposite. *Chem. Eng. J.*, 184, 326–332, 2012.
59. Liu, J., Yu, H., Liang, Q., Liu, Y., Shen, J., Bai, Q., Preparation of polyhedral oligomeric silsesquioxane based cross-linked inorganic-organic nanohybrid as adsorbent for selective removal of acidic dyes from aqueous solution. *J. Colloid Interf. Sci.*, 497, 402–412, 2017.
60. Zhou, L., Jin, J., Liu, Z., Liang, X., Shang, C., Adsorption of acid dyes from aqueous solutions by the ethylenediamine-modified magnetic chitosan nanoparticles. *J. Hazard. Mater.*, 185, 1045–1052, 2011.
61. Du, Z., Zhang, Y., Li, Z., Chen, H., Wang, Y., Wang, G., *et al.*, Facile one-pot fabrication of nano-Fe 3 O 4/carboxyl-functionalized baker's yeast composites and their application in methylene blue dye adsorption. *Appl. Surf. Sci.*, 392, 312–320, 2017.
62. Sivashankar, R., Sathya, A., Sivasubramanian, V., Synthesis of magnetic biocomposite for efficient adsorption of azo dye from aqueous solution. *Ecotoxicol. Environ. Safety*, 121, 149–153, 2015.
63. Tzvetkov, G., Kaneva, N., Spassov, T., Room-temperature fabrication of core-shell nano-ZnO/pollen grain biocomposite for adsorptive removal of organic dye from water. *Appl. Surf. Sci.*, 400, 481–491, 2017.

64. Shoukat, S., Bhatti, H.N., Iqbal, M., Noreen, S., Mango stone biocomposite preparation and application for crystal violet adsorption: a mechanistic study. *Microporous Mesoporous Mat.*, 239, 180–189, 2017.
65. Fungaro, D., Yamaura, M., Carvalho, T., Adsorption of anionic dyes from aqueous solution on O zeolite from fly ash-iron oxide magnetic Π nanocomposite. *J. Atom. Mol. Sci.*, 2, 305–316, 2011.
66. Debrassi, A., Baccarin, T., Demarchi, C.A., Nedelko, N., Ślawska-Waniewska, A., Dłużewski, P., *et al.*, Adsorption of Remazol Red 198 onto magnetic N-lauryl chitosan particles: equilibrium, kinetics, reuse and factorial design. *Environ. Sci. Pollut. Res.*, 19, 1594–1604, 2012.
67. Arshadi, M., Faraji, A., Amiri, M., Mehravar, M., Gil, A., Removal of methyl orange on modified ostrich bone waste–a novel organic–inorganic biocomposite. *J. Colloid Interf. Sci.*, 446, 11–23, 2015.
68. Manohar, R., Shrivastava, V., Adsorption removal of Carcinogenic acid violet19 dye from aqueous solution by polyaniline-Fe2O3 magnetic nanocomposite. *J. Mat. Environ. Sci.*, 6, 11–21, 2015.
69. Zhu, H., Zhang, M., Liu, Y., Zhang, L., Han, R., Study of congo red adsorption onto chitosan coated magnetic iron oxide in batch mode. *Desalination Water Treat.*, 37, 46–54, 2012.
70. Sui, K., Li, Y., Liu, R., Zhang, Y., Zhao, X., Liang, H., *et al.*, Biocomposite fiber of calcium alginate/multi-walled carbon nanotubes with enhanced adsorption properties for ionic dyes. *Carbohydr. Polym.*, 90, 399–406, 2012.
71. Elkady, M., Ibrahim, A.M., El-Latif, M.A., Assessment of the adsorption kinetics, equilibrium and thermodynamic for the potential removal of reactive red dye using eggshell biocomposite beads. *Desalination*. 278, 412–423, 2011.
72. Tahir, N., Bhatti, H.N., Iqbal, M., Noreen, S., Biopolymers composites with peanut hull waste biomass and application for Crystal Violet adsorption. *Int. J. Biol. Macromol.*, 94, 210–220, 2016.
73. Palanisamy, P., Agalya, A., Sivakumar, P., Polypyrrole composite-A potential material for the removal of acid dyes. *Asian J. Chem.*, 25, 5891, 2013.
74. Baseri, J.R., Palanisamy, P., Sivakumar, P., Polyaniline nano composite for the adsorption of reactive dye from aqueous solutions: equilibrium and kinetic studies. *Asian J. Chem.*, 25, 4145, 2013.
75. Jothirani, R., Kumar, P.S., Saravanan, A., Narayan, A.S., Dutta, A., Ultrasonic modified corn pith for the sequestration of dye from aqueous solution. *J. Ind. Eng. Chem.*, 39, 162–175, 2016.
76. Fosso-Kankeu, E., Mittal, H., Mishra, S.B., Mishra, A.K., Gum ghatti and acrylic acid based biodegradable hydrogels for the effective adsorption of cationic dyes. *J. Ind. Eng. Chem.*, 22, 171–178, 2015.
77. Debnath, S., Ballav, N., Maity, A., Pillay, K., Development of a polyaniline-lignocellulose composite for optimal adsorption of Congo red. *Int. J. Biol. Macromol.*. 75, 199–209, 2015.
78. Wang, J., Zhou, Q., Song, D., Qi, B., Zhang, Y., Shao, Y., *et al.*, Chitosan–silica composite aerogels: preparation, characterization and Congo red adsorption. *J. Sol-Gel Sci. Technol.*, 76, 501–509, 2015.

79. Feng, T., Xu, L., Adsorption of Congo Red from aqueous solution onto chitosan/rectorite composite. *Adv. Mater. Res.: Trans. Tech. Publ.*, 438–441, 2013.
80. Shweta, K., Jha, H., Rice husk extracted lignin–TEOS biocomposites: effects of acetylation and silane surface treatments for application in nickel removal. *Biotechnol. Reports,* 7, 95–106, 2015.
81. Aftab, K., Akhtar, K., Jabbar, A., Batch and column study for Pb-II remediation from industrial effluents using glutaraldehyde–alginate–fungi biocomposites. *Ecol. Eng.*, 73, 319–325, 2014.
82. Khan, T.A., Nazir, M., Ali, I., Kumar, A., Removal of Chromium (VI) from aqueous solution using guar gum–nano zinc oxide biocomposite adsorbent. *Arabian J. Chem.*, 10, S2388–S2398, 2017.
83. Akram, M., Bhatti, H.N., Iqbal, M., Noreen, S., Sadaf, S., Biocomposite efficiency for Cr(VI) adsorption: Kinetic, equilibrium and thermodynamics studies. *J. Environ. Chem. Eng.*, 5, 400–411, 2017.
84. Gopalakannan, V., Viswanathan, N., One pot synthesis of metal ion anchored alginate–gelatin binary biocomposite for efficient Cr (VI) removal. *Int. J. Biol. Macromol.*. 83, 450–459, 2016.
85. Aytas, S., Turkozu, D.A., Gok, C., Biosorption of uranium (VI) by bi-functionalized low cost biocomposite adsorbent. *Desalination*. 280, 354–362, 2011.
86. Kousalya, G., Gandhi, M.R., Sundaram, C.S., Meenakshi, S., Synthesis of nano-hydroxyapatite chitin/chitosan hybrid biocomposites for the removal of Fe (III). *Carbohydr. Polym.*, 82, 594–599, 2010.
87. Gopalakannan, V., Periyasamy, S., Viswanathan, N., One pot eco-friendly synthesis of highly dispersed alumina supported alginate biocomposite for efficient chromium (VI) removal. *J. Water Process Eng.*, 10, 113–119, 2016.
88. Pereira, F., Sousa, K., Cavalcanti, G., Fonseca, M., de Souza, A.G., Alves, A., Chitosan-montmorillonite biocomposite as an adsorbent for copper (II) cations from aqueous solutions. *Int. J. Biol. Macromol.*. 61, 471–478, 2013.
89. Kahraman, H.T., Development of an adsorbent via chitosan nano-organoclay assembly to remove hexavalent chromium from wastewater. *Int. J. Biol. Macromol.*. 94, 202–209, 2017.
90. Ding, C., Cheng, W., Sun, Y., Wang, X., Novel fungus-Fe 3 O 4 bio-nanocomposites as high performance adsorbents for the removal of radionuclides. *J. Hazard. Mater.*, 295, 127–137, 2015.
91. Singha, A., Guleria, A., Chemical modification of cellulosic biopolymer and its use in removal of heavy metal ions from wastewater. *Int. J. Biol. Macromol.*. 67, 409–417, 2014.
92. Googerdchian, F., Moheb, A., Emadi, R., Lead sorption properties of nano-hydroxyapatite–alginate composite adsorbents. *Chem. Eng. J.*, 200, 471–479, 2012.
93. Gandhi, M.R., Viswanathan, N., Meenakshi, S., Preparation and application of alumina/chitosan biocomposite. *Int. J. Biol. Macromol.*. 47, 146–154, 2010.
94. Varghese, L.R., Das, N., Removal of Hg (II) ions from aqueous environment using glutaraldehyde crosslinked nanobiocomposite hydrogel modified by

TETA and β-cyclodextrin: optimization, equilibrium, kinetic and ex situ studies. *Ecol. Eng.*, 85, 201–211, 2015.
95. Das, D., Vimala, R., Das, N., Removal of Ag (I) and Zn (II) ions from single and binary solution using sulfonated form of gum arabic-powdered mushroom composite hollow semispheres: equilibrium, kinetic, thermodynamic and ex-situ studies. *Ecol. Eng.*, 75, 116–122, 2015.
96. Saber-Samandari, S., Saber-Samandari, S., Gazi, M., Cellulose-graft-polyacrylamide/hydroxyapatite composite hydrogel with possible application in removal of Cu (II) ions. React. Funct. Polym., 73, 1523–1530, 2013.
97. Choo, C.K., Kong, X.Y., Goh, T.L., Ngoh, G.C., Horri, B.A., Salamatinia, B., Chitosan/halloysite beads fabricated by ultrasonic-assisted extrusion-dripping and a case study application for copper ion removal. *Carbohydr. Polym.*, 138, 16–26, 2016.
98. Kumar, I.A., Viswanathan, N., Development of multivalent metal ions imprinted chitosan biocomposites for phosphate sorption. *Int. J. Biol. Macromol.* In press, 2017, https://doi.org/10.1016/j.ijbiomac.2017.02.100.
99. Golie, W.M., Upadhyayula, S., An investigation on biosorption of nitrate from water by chitosan based organic-inorganic hybrid biocomposites. *Int. J. Biol. Macromol.*. 97, 489–502, 2017.
100. Nazari, M., Halladj, R., Adsorptive removal of fluoride ions from aqueous solution by using sonochemically synthesized nanomagnesia/alumina adsorbents: An experimental and modeling study. *J. Taiwan Instit. Chem.*, 45, 2518–2525, 2014.
101. Pandi, K., Viswanathan, N., Synthesis of alginate bioencapsulated nano-hydroxyapatite composite for selective fluoride sorption. *Carbohydr. Polym.*, 112, 662–667, 2014.
102. Soni, U., Bajpai, J., Singh, S.K., Bajpai, A., Evaluation of chitosan-carbon based biocomposite for efficient removal of phenols from aqueous solutions. *J. Water Process Eng.*, 16, 56–63, 2017.
103. Prabhu, S.M., Meenakshi, S., Synthesis of metal ion loaded silica gel/chitosan biocomposite and its fluoride uptake studies from water. *J. Water Process Eng.*, 3, 144–150, 2014.
104. Davila-Rodriguez, J.L., Escobar-Barrios, V.A., Rangel-Mendez, J.R., Removal of fluoride from drinking water by a chitin-based biocomposite in fixed-bed columns. *J. Fluorine Chem.*, 140, 99–103, 2012.
105. Prabhu, S.M., Elanchezhiyan, S.S., Lee, G., Meenakshi, S., Defluoridation of water by Tea-bag model using La 3+ modified synthetic resin@ chitosan biocomposite. *Int. J. Biol. Macromol.*. 91, 1002–1009, 2016.
106. Vecino, X., Devesa-Rey, R., Moldes, A., Cruz, J., Formulation of an alginate-vineyard pruning waste composite as a new eco-friendly adsorbent to remove micronutrients from agroindustrial effluents. *Chemosphere*, 111, 24–31, 2014.
107. Wu, S., Zhao, X., Li, Y., Zhao, C., Du, Q., Sun, J., *et al.*, Adsorption of ciprofloxacin onto biocomposite fibers of graphene oxide/calcium alginate. *Chem. Eng. J.*, 230, 389–395, 2013.

108. Kanwal, F., Rehman, R., Mahmud, T., Anwar, J., Ilyas, R., Isothermal and thermodynamical modeling of chromium (III) adsorption by composites of polyaniline with rice husk and saw dust. *J. Chilean Chem. Soc.*, 57, 1058–63, 2012.
109. Bhatti, H.N., Jabeen, A., Iqbal, M., Noreen, S., Naseem, Z., Adsorptive behavior of rice bran-based composites for malachite green dye: Isotherm, kinetic and thermodynamic studies. *J. Mol. Liq.*, 237, 322–333, 2017.
110. Shofiyani, A., Narsito, N., Santosa, S.J., Noegrohati, S., Zahara, T.A., Sayekti, E., Cadmium adsorption on chitosan/chlorella biomass sorbent prepared by ionic-imprinting technique. *Indonesian J. Chem.*, 15, 163–171, 2015.
111. Palanisamy, P., Agalya, A., Sivakumar, P., Polymer composite—a potential biomaterial for the removal of reactive dye. *J. Chem.*, 9, 1823–1834, 2012.
112. Manzoor, Q., Nadeem, R., Iqbal, M., Saeed, R., Ansari, T.M., Organic acids pretreatment effect on *Rosa bourbonia* phyto-biomass for removal of Pb (II) and Cu (II) from aqueous media. *Biores. Technol.*, 132, 446–452, 2013.
113. Rangabhashiyam, S., Selvaraju, N., Evaluation of the biosorption potential of a novel Caryota urens inflorescence waste biomass for the removal of hexavalent chromium from aqueous solutions. *J. Taiwan Instit. Chem.*, 47, 59–70, 2015.
114. Nadeem, R., Manzoor, Q., Iqbal, M., Nisar, J., Biosorption of Pb (II) onto immobilized and native Mangifera indica waste biomass. *J. Ind. Eng. Chem.*, 35, 184–195, 2016.
115. Ullah, I., Nadeem, R., Iqbal, M., Manzoor, Q., Biosorption of chromium onto native and immobilized sugarcane bagasse waste biomass. *Ecol. Eng.*, 60, 99–107, 2013.
116. Mahdavinia, G.R., Massoudi, A., Baghban, A., Shokri, E., Study of adsorption of cationic dye on magnetic kappa-carrageenan/PVA nanocomposite hydrogels. *J. Environ. Chem. Eng.*, 2, 1578–1587, 2014.
117. Shrivastava, M.R.P.V., Adsorption removal of carcinogenic acid violet19 dye from aqueous solution by polyaniline-Fe2O3 magnetic nano-composite. *J. Mater. Environ. Sci.*, 6, 11–21, 2015.
118. Reddy, S.S., Kotaiah, B., Reddy, N.S.P., Velu, M., The removal of composite reactive dye from dyeing unit effluent using sewage sludge derived activated carbon. *Turkish J. Eng. Environ. Sci.*, 30, 367–373, 2007.
119. Rocher, V., Bee, A., Siaugue, J-.M., Cabuil, V., Dye removal from aqueous solution by magnetic alginate beads crosslinked with epichlorohydrin. *J. Hazard. Mater.*, 178, 434–439, 2010.
120. Wu, Z., Wu, J., Xiang, H., Chun, M-.S., Lee, K., Organosilane-functionalized Fe3O4 composite particles as effective magnetic assisted adsorbents. *Colloids Surf. A: Physicochem. Eng. Aspects*, 279, 167–174, 2006.
121. Anirudhan, T., Suchithra, P., Adsorption characteristics of humic acid-immobilized amine modified polyacrylamide/bentonite composite for cationic dyesin aqueous solutions. *J. Environ. Sci.*, 21, 884–891, 2009.
122. Tahir, M.A., Bhatti, H.N., Iqbal, M., Solar Red and Brittle Blue direct dyes adsorption onto Eucalyptus angophoroides bark: Equilibrium, kinetics and thermodynamic studies. *J. Environ. Chem. Eng.*, 4, 2431–2439, 2016.

123. Ofomaja, A.E., Ho, Y-.S., Equilibrium sorption of anionic dye from aqueous solution by palm kernel fibre as sorbent. *Dyes Pigm.*, 74, 60–66, 2007.
124. Shukla, A., Zhang, Y-.H., Dubey, P., Margrave, J., Shukla, S.S., The role of sawdust in the removal of unwanted materials from water. *J. Hazard. Mater.*, 95, 137–152, 2002.
125. Iqbal, J., Cecil, F., Ahmad, K., Iqbal, M., Mushtaq, M., Naeem, M., *et al.*, Kinetic study of Cr (III) and Cr (VI) biosorption using Rosa damascena phytomass: a rose waste biomass. *Asian J. Chem.*, 25, 2099–2103, 2013.
126. Aksu, Z., Isoglu, I.A., Use of agricultural waste sugar beet pulp for the removal of Gemazol turquoise blue-G reactive dye from aqueous solution. *J. Hazard. Mater.*, 137, 418–430, 2006.
127. Adesola, B., Ogundipe, K., Sangosanya, K.T., Akintola, B.D., Oluwa, A., Hassan, E., Comparative study on the biosorption of Pb(II), Cd(II) and Zn(II) using Lemon grass (*Cymbopogon citratus*): Kinetics, isotherms and thermodynamics. . *Chem. Int.*, 2, 89–102, 2016.
128. Mushtaq, M., Bhatti, H.N., Iqbal, M., Noreen, S., Eriobotrya japonica seed biocomposite efficiency for copper adsorption: Isotherms, kinetics, thermodynamic and desorption studies. *J. Environ. Manag.*, 176, 21–33, 2016.
129. Gupta, A., Balomajumder, C., Simultaneous adsorption of Cr(VI) and phenol onto tea waste biomass from binary mixture: Multicomponent adsorption, thermodynamic and kinetic study. *J. Environ. Chem. Eng.*, 3, 785–796, 2015.
130. Pandey, S., Mishra, S.B., Organic–inorganic hybrid of chitosan/organoclay bionanocomposites for hexavalent chromium uptake. *J. Colloid Interf. Sci.*, 361, 509–520, 2011.
131. Al-Qodah, Z., Biosorption of heavy metal ions from aqueous solutions by activated sludge. *Desalination*, 196, 164–176, 2006.
132. Rajalakshmi, R., Subhashini, S., Lalitha, P., Usefulness of activated carbon prepared from industrial wastes in the removal of nickel from aqueous solution. *J. Chem.*, 6, 361–370, 2009.
133. Zubair, A., Bhatti, H.N., Hanif, M.A., Shafqat, F., Kinetic and equilibrium modeling for Cr (III) and Cr (VI) removal from aqueous solutions by Citrus reticulata waste biomass. *Water, Air, Soil Pollut.*, 191, 305–318, 2008.
134. Aksu, Z., Açıkel, Ü., Kabasakal, E., Tezer, S., Equilibrium modelling of individual and simultaneous biosorption of chromium (VI) and nickel (II) onto dried activated sludge. *Water Res.*, 36, 3063–3073, 2002.
135. Alpat, S., Alpat, S.K., Çadirci, B.H., Özbayrak, Ö., Yasa, İ., Effects of biosorption parameter: kinetics, isotherm and thermodynamics for Ni (II) biosorption from aqueous solution by Circinella sp. *Elect. J. Biotechnol*, 13, 4–5, 2010.
136. Jabli, M., Baouab, M.H.V., Roudesli, M.S., Bartegi, A., Adsorption of acid dyes from aqueous solution on a chitosan-cotton composite material prepared by a new pad-dry process. *J. Eng. Fiber Fabr.*, 6, 1–12, 2011.
137. Weber, W.J., Morris, J.C., Kinetics of adsorption on carbon from solution. *J. Sanit. Eng. Div.*, 89, 31–60, 1963.

138. Tan, I.A.W., Ahmad, A.L., Hameed, B.H., Adsorption of basic dye on high-surface-area activated carbon prepared from coconut husk: Equilibrium, kinetic and thermodynamic studies. *J. Hazard. Mater.*, 154, 337–346, 2008.
139. Gholipour, M., Hashemipour, H., Mollashahi, M., Hexavalent chromium removal from aqueous solution via adsorption on granular activated carbon: adsorption, desorption, modeling and simulation studies. *J. Eng. Appl. Sci.*, 6, 10–18, 2011.
140. Yadav, S., Srivastava, V., Banerjee, S., Weng, C-.H., Sharma, Y.C., Adsorption characteristics of modified sand for the removal of hexavalent chromium ions from aqueous solutions: Kinetic, thermodynamic and equilibrium studies. *Catena.*, 100, 120–127, 2013.
141. Wu, Y., Luo, H., Wang, H., Wang, C., Zhang, J., Zhang, Z., Adsorption of hexavalent chromium from aqueous solutions by graphene modified with cetyltrimethylammonium bromide. *J. Colloid Interf. Sci.*, 394, 183–191, 2013.
142. Zhu, H-.Y., Jiang, R., Xiao, L., Adsorption of an anionic azo dye by chitosan/kaolin/γ-Fe 2 O 3 composites. *Appl. Clay Sci.*, 48, 522–526, 2010.
143. Kalyani, S., Priya, J.A., Rao, P.S., Krishnaiah, A., Removal of copper and nickel from aqueous solutions using chitosan coated on perlite as biosorbent. *Sep. Sci. Technol.*, 40, 1483–1495, 2005.
144. Chiou, M-.S., Ho, P-.Y., Li, H-.Y., Adsorption of anionic dyes in acid solutions using chemically cross-linked chitosan beads. *Dyes Pigm.*, 60, 69–84, 2004.
145. Zhang, A-.C., Sun, L-.S., Xiang, J., Song, H., Peng, F., Su, S., *et al.*, Removal of elemental mercury from coal combustion flue gas by bentonite-chitosan and their modifier. *J. Fuel Chem. Technol.*, 37, 489–495, 2009.
146. Demirbas, A., Heavy metal adsorption onto agro-based waste materials: A review. *J. Hazard. Mater.*, 157, 220–229, 2008.
147. Veli, S., Alyüz, B., Adsorption of copper and zinc from aqueous solutions by using natural clay. *J. Hazard. Mater.*, 149, 226–233, 2007.
148. Weng, C.-H., Tsai, C-.Z., Chu, S-.H., Sharma, Y.C., Adsorption characteristics of copper(II) onto spent activated clay. *Sep. Purif. Technol.*. 54, 187–197, 2007.
149. Aljlil, S.A., Alsewailem, F.D., Adsorption of Cu & Ni on Bentonite Clay from Waste Water. *Athens J. Nat. Formal Sci.*, 1, 21–30, 2014.
150. Eloussaief, M., Jarraya, I., Benzina, M., Adsorption of copper ions on two clays from Tunisia: pH and temperature effects. *Appl. Clay Sci.*, 46, 409–413, 2009.
151. Musso, T., Parolo, M., Pettinari, G., Francisca, F.. Cu (II) and Zn (II) adsorption capacity of three different clay liner materials. *J. Environ. Manag.*, 146, 50–58, 2014.
152. Zacaroni, L.M., Magriotis, Z.M., Cardoso, M.D.G., Santiago, W.D., Mendonça J.G., Vieira S.S., *et al.* Natural clay and commercial activated charcoal: Properties and application for the removal of copper from cachaça. *Food Control.*, 47, 536–544, 2015.

12

A Review on the Removal of Nitrate from Water by Adsorption on Organic–Inorganic Hybrid Biocomposites

Wondalem Misganaw Golie, Kaisar Ahmad and Sreedevi Upadhyayula*

Department of Chemical Engineering, Indian Institute of Technology Delhi, New Delhi, India

Abstract

Increasing nitrate concentration in water resources due to increased anthropological activities is a potential risk to public health worldwide. Among the unit operations involving water treatment, adsorption process is an economical and environmentally friendly treatment technique with ease of operation, simplicity of design, allows nitrate recovery, and feasibility to implement in field conditions and in point of use treatment systems. Various conventional and nonconventional adsorbents have been used for the removal of nitrate from water. In this chapter, a list of various adsorbents from literature is surveyed and their adsorption capacities under various operational conditions are compared. This survey showed that chitosan-based adsorbents, especially organic–inorganic hybrid biocomposites, have good potential for the nitrate removal from water.

Keywords: Biosorption, biocomposite, nitrate removal, fixed-bed

12.1 Introduction

A wide variety of chemicals generated from different sources and discharged into the environment in potentially harmful concentrations and have been considered as a growing water pollution problem [1, 2]. Nitrate

**Corresponding author*: sreedevi@chemical.iitd.ac.in

Shahid-ul-Islam (ed.) Advanced Materials for Wastewater Treatment, (433–478)

is possibly one of the most widespread chemical contaminants in water resources imposing a serious threat to drinking water supplies and promoting eutrophication and has become a prime concern on a global scale [3–5]. Although nitrate is a common nitrogenous compound, anthropologic activities contributed immeasurably to the imbalances in the nitrogen cycle and resulted in alarmingly increased levels of nitrate in drinking water sources and other water bodies [5, 6]. Agriculture is a major contributor of nitrate to surface waters and groundwater pollution due to the excessive and long-term use of chemical fertilizers and animal manures on cropland [7, 8]. More than 80 pounds of nitrogen per acre per year may leach into groundwater beneath irrigated lands, usually as nitrate [7]. Other sources which have led to the higher nitrate contamination of water sources include chemical industries, waste from explosive industries, agricultural and urban runoff, intensive animal husbandry, dairy operations, discharge from wastewater and sewage, septic tank systems, leachate from landfills, and automobiles emissions [6].

12.1.1 Risks Associated to High Level of Nitrate in Water

Almost 90% of the rural population uses groundwater for domestic purposes and people in rural areas are at the highest risk from excess fertilizer use [9]. The epidemiological studies show that drinking water is the main source of nitrate and intake of water with high nitrate concentrations has considerable health effects [3]. The toxicity of nitrate to humans is mainly attributable to its in vivo reduction to nitrites. Ingested nitrate is reduced to nitrite, which binds with hemoglobin to form methemoglobin by oxidation of ferrous iron (Fe^{2+}) in hemoglobin to ferric form (Fe^{3+}). This prevents or reduces the ability of blood to transport oxygen and leads methemoglobinaemia, the so-called "blue-baby syndrome" which is dangerous especially in infants causes cyanosis, asphyxia and consequently to death, at higher concentrations [3–14]. This is the primary adverse health effect associated with human exposure to high level of nitrate in water. Moreover, the reduction of nitrate to nitrite in the alimentary canal and nitrite readily reacts with amide and amino compounds to generate N-nitroso compounds (NOC), most of which are potentially carcinogenic for humans [10–17]. Other diseases associated with intake of high concentration of nitrate in water are hypertension, increased infant mortality, central nervous system birth defects, diabetes, weakness, mental depression, headache, diarrhea, spontaneous abortions, and changes to the immune system [3, 10]. High nitrate concentrations in water bodies contribute to eutrophication and this can lead to the development of anoxic conditions by depleting dissolved oxygen, increase in plant and animal biomass, and affects the biodiversity. It has a great environmental, societal, and economic impact on the commercial and recreational fisheries [19–21].

The increased demand for water purity in both industrial and domestic use leads to a strict regulation of the water pollutants. The World Health Organization (WHO) and US Environmental Protection Agency (USEPA) set the Maximum Contaminant Level (MCL) of nitrate ions in drinking water to be 50 mg/L and 45 mg/L as NO_3^-, respectively [10, 16, 22] and many countries have formulated their own MCLs according to their domestic conditions.

12.1.2 Technologies for the Removal of Nitrate from Water

The removal of nitrate from water is necessary to meet regulated concentrations in order to protect human health and the environment and has been a subject of extensive academic and industrial research. Nitrate cannot be easily removed from water by coagulation and filtration due to its high stability and solubility, as well as its low potential for co-precipitation. Numerous technologies are currently available to remove nitrate from water. Ion exchange, reverse osmosis, and electrodialysis are the most widely used methods, however; nitrate is not converted into harmless compounds but only concentrated and produce a brine waste that still requires treatment [23–27]. Moreover; reverse osmosis and electrodialysis are an energy intensive treatment and these methods have high capital operation costs. Biological denitrification is commonly used for the treatment of municipal and industrial wastewater [22–30]. It can treat large volumes of high concentration nitrate and it can reduce nitrate to nontoxic forms. However, this process has several disadvantages since it is difficult to handle and leads to the production of undesirable by-products [22]. Biological denitrification relies on bacteria. Therefore, Substrate addition is necessary and downstream treatment is usually required to remove contamination. Moreover, the startup time takes weeks so it is not suitable for intermittent operations [23–30]. Chemical denitrification [31–36] is another method to remove nitrate from water. However, this method has several limitations, such as generation of byproducts, such as ammonia and sludge which need post treatment, high operational cost, pH constraints, lack of full scale systems, and materials are not reusable [32–33]. Electrochemical reduction [37–44], and catalytic reduction [45–62] are the recent evolving methods in which nitrate is converted into harmless end products (nitrogen gas) without generating secondary wastes. However, both reduction methods are not without limitations. In electrochemical nitrate reduction, electrode fouling, operational difficulty, and extensive energy requirement are the main challenges. In catalytic reduction, generation of ammonia and use of precious metals as active metal for catalysis and hydrogen gas as reducing agent increases the capital and operating cost of the reduction process.

Adsorption technology, the focus of this thesis, is an economical and environmentally friendly treatment technique with ease of operation, simplicity of design, allows nitrate recovery, and feasibility to implement in field conditions and in point of use treatment systems. Furthermore, this process can remove multiple pollutants from water and wastewater [63–67]. Table 12.1 gives the summary on the advantages and disadvantages of currently available nitrate removal technologies.

Table 12.1 Nitrate removal.

Technology	Advantage	Disadvantage
Ion exchange	Stable, fast removal, easily automated, cost-effective, low-maintenance, nondestructive	Requires treatment or disposal of brine waste, suitable only for small – and medium-scale operations, may produce toxic by-products, not good for waters with high total dissolved solids (i.e., groundwater only)
Reverse osmosis	Good for waters with high total dissolved solids content, no by-products, nondestructive	Expensive – energy intensive and large capital costs, inefficient nitrate removal, requires treatment or disposal of concentrate
Electrodialysis	Good for waters with high total dissolved solids content, no by-products, nondestructive	Expensive – energy intensive and large capital costs, inefficient nitrate removal, requires treatment or disposal of concentrate
Chemical	High removal, zero-valent iron for in-situ treatment, efficient, destructive method	pH sensitive, no reuse of material, High operational cost, production of by-products.

(Continued)

Table 12.1 Cont.

Technology	Advantage	Disadvantage
Electrochemical	Selective toward nontoxic by-product (i.e., N_2), destructive method	Electrode fouling, operational difficulty, energy intensive
Catalytic	Selective toward nontoxic by-product (i.e., N_2), efficient, destructive method	Catalyst fouling, expensive metals, reductant such as H_2
Biological	Produces a nontoxic by-product, suitable for large applications, can handle high concentrations, destructive method	Electron donor and carbon source required, requires downstream treatment (e.g., ozonation), high-maintenance, long start-up times, pH and temperature sensitive, requires biomass waste disposal.
Adsorption	Economical, ease of operation, simplicity of design	Requires saturated/spent adsorbent disposal. Removal efficiency varies with different adsorbents.

12.2 Adsorbents for the Removal of Nitrate from Water

Adsorption technology has been found as an economical successful method for the removal of various types of inorganic anions [23, 24, 63, 64, 68–70]. Adsorbents used for water treatment are either of natural origin or synthetic. Various types of carbonaceous adsorbents have investigated for the removal of nitrate from water [71–78]. However, activated carbon is quite expensive and has low affinity toward inorganic pollutants. Moreover, processing of activated carbon is quite expensive and has negative environmental impact. On the other hand, various kinds of materials, such as activated waste sepiolite [64], silica materials [78], surfactant modified bentonite [79], layered double hydroxides (LDHs) [80–85], zeolites [86–91], modified sugar cane bagasse [92], and agricultural wastes [93–97] have been used for the removal nitrate from water. However, modification and production process of such materials consumes great quantity of energy and chemicals which intern involving environmental costs.

Recently, great attention has been driven toward the natural materials and their derivative products, such as cellulose, lignin, starch, alginate, gelatin, chitin, and chitosan for the removal of aquatic pollutants [98–102]. Among these biomaterials, chitosan has received wide attention in the environmental engineering field as effective biosorbents for the removal of various aquatic pollutants. Chitosan is a natural amino polysaccharide produced from N-deacetylation of chitin, a major component present in crustacean shells from crabs, shrimps, and insects, and is the second most abundant natural amino polysaccharide after cellulose [98–107]. The unique features of chitosan include, being readily available, low cost, non-toxic, biodegradability, biocompatibility, and reactive as well as adsorptive properties, resulting from the presence of high contents amino and hydroxyl functional groups in the polymer chain [98–108].

However, chitosan in its original form is not suitable for practical use due to its sensitivity to wide range of pH and relatively low sorption performance. Moreover, chitosan has poor morphological properties and low mechanical stability, which cause a significant pressure drop which would affect filtration, possible obstruction, and is difficult to regenerate in the packed column operation, and finally, these properties restrict its field applications [109]. Chitosan needs certain modifications to overcome the abovementioned limitations to increase the stability of the material in acid media (good mechanical properties, etc.), and to enhance the sorption performance, including sorption capacity, selectivity, good diffusion and hydrodynamic behavior [23, 90, 110, 111]. The easy separation from real effluent after treatment and the cost effectiveness of the sorbents preparation are also important. Thus, chitosan modification by introducing the desired properties in physical, chemical, and mechanic properties is key issues in order to enhance its contaminant sorption capacity, to improve its affinity for the specific ions, to change the selectivity series for ion sorption, and to alter the optimum pH range. Modification of chitosan molecule by grafting, cross-linking, or functionalization for forming composites leads to the formation of chitosan derivatives with superior properties, [24, 108–114].

Chatterjee and Woo [115] conducted batch adsorption experiments for the removal of nitrate using chitosan hydrobeads. The same authors [116] reported the adsorption of nitrate onto cross-linked and conditioned chitosan beads. A. Sowmya and Meenakshi [117] studied of the removal of nitrate ion from water using quaternized chitosan beads (QCB) which is produced through the reaction of cross-linked chitosan beads and glycydyl trimethyl ammonium chloride (GTMAC). The same authors [118] investigated the application of protonated cross-linked chitosan beads (PCB) and carboxylated cross-linked chitosan beads (CCB) for the removal of nitrate from water. Jiang et al. [119] synthesized Fe_3O_4/ZrO_2/chitosan composite and studied its ability to adsorb nitrate. Hu et al. [120] synthesized granular chitosan-Fe^{3+} complex through precipitation method and studied its performance on nitrate adsorption from water. On the other hand, Rajeswari et al. [121] investigated the removal of nitrate ions using polyethylene glycol/chitosan and polyvinyl alcohol/chitosan composites in a series of batch experiments. Moreover, C. Feng et al. [122] prepared granular chitosan-Fe(III)-Al(III) complex and tested its application for removal of nitrate water. Recently, Vahdatpoora et al. [123] prepared a series of chitosan/Zeolite Y/Nano Zirconium oxide (CTS/ZY/Nano ZrO_2) nanocomposites and demonstrated its application for the removal of remove nitrate from water.

Higher nitrate adsorption capacity, high nitrate selectivity, economic feasibility, availability, reusability, and good chemical stability and mechanical integrity are desirable characteristics of adsorbent to be effective in water treatment system for practical applications. The recent report [124] on the investigation of biosorption of nitrate from water using chitosan-based organic–inorganic biocomposites indicated that inorganics materials, such as bentonite, titanium oxide, and alumina are an attractive immobilization inorganic materials for chitosan due to their environmental benignity, low cost as well as chemical and mechanical stability. In this report, chitosan-based organic–inorganic hybrid biocomposites, such as chitosan/bentonite, chitosan/ titanium oxide, and chitosan/ alumina (ChBT, ChTi, and ChAl respectively) were prepared and characterized. Stability of ChBT, ChTi, and ChAl increased with increase in cross-linker and inorganic dosage. Batch adsorption studies were conducted and the operating parameters were optimized and the amounts of nitrate adsorbed on ChBT, ChTi, and ChAl were 35.68 and 43.62, and 45.38 mg/g, respectively. Adsorption capacities increased with the increase in temperature from 283 K to 313 K and decreased above 313 K. Also, the amounts of nitrate adsorbed on the three adsorbents were found to decrease at higher cross-linker dosage. Among the three adsorbents, ChAl has shown highest stability and performance in all operating conditions. A typical preparation method of metal-organic composites for the case of ChAl is shown in Figure 12.1.

Equilibrium studies were carried out in the temperature range of 283 to 313 K and found that the Freundlich isotherm fitted well to the experimental data with the highest correlation coefficient (R^2) and least chi-square (χ_2) values for all the adsorbents used. Results from thermodynamic study revealed that the magnitude of $\Delta G°$ highly dependent of the type of

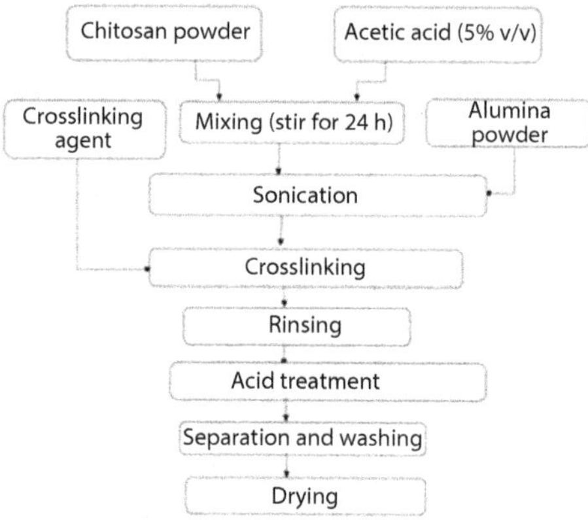

Figure 12.1 Preparation of ChAl composite.

isotherm constants from which equilibrium constant derived. The values of thermodynamic parameters show that adsorption process was spontaneous and endothermic for all adsorbents used with increased randomness at the solid liquid interface. Mechanism of nitrate adsorption process was physisorption. The kinetics of adsorption process was better described by a pseudo-second-order equation. A continuous nitrate adsorption study on ChBT, ChTi, and ChAl biocomposites was conducted in a fixed-bed column at temperature of 298 ± 2 K and at the solution pH of 6.8 ± 0.2. Nitrate removal efficiency increased with increase in bed height and decreased with increase in influent nitrate concentration and flow rate. The breakthrough time increased with increase in bed height. Thomas, Yoon-Nelson, Clarks, and BDST models described the adsorption process well. The best performance for nitrate removal from water was obtained using ChAl biocomposite.

Table 12.2 shows summary some of the results, as well as operation conditions used, reported in the literature for nitrate removal from water by adsorption process. Some values are not indicated in the table because that information was not available in the original reference. Literatures indicated that chitosan is a promising adsorbent material for the removal of nitrate from water. However, chitosan his not without limitations and needs modification to improve its chemical and mechanical stability, as well as to enhance its adsorption capacity and reusability. Cost and environmental considerations are critical issues and one should not ignore these during modification of chitosan. Therefore, there is a need to explore an alternative low-cost, environmental friendly and effective modification method.

Table 12.2 Summary of adsorbents for nitrate removal.

Adsorbent	Temp (°C)	pH	Time (min)	Adsorption capacity (mg/g)	Adsorption isotherm	Thermodynamics nature	Kinetics model	Ref.
Sepiolite by activated HCl	30	6.7	5	38.16	Freundlich	–	pseudo-second-order	[64]
$ZnCl_2$ Activated carbon	35	6.58	60	41.2	Langmuir	Endothermic and spontaneous	pseudo-second-order	[71]
Commercial activated carbon	10	–	10 min	1.22	Langmuir	–	–	[72]
Wheat straw charcoal	15	–	10	1.10	–	–	–	[72]
Mustard straw charcoal	15	–	10	1.30	–	–	–	[72]
Activated carbon cloth (ACC)	25	7	60	125.85	Langmuir	–	–	[73]
$ZnCl_2$ treated coconut granular activated carbon	25	3–10	–	10.2	Freundlich	Endothermic and spontaneous	pseudo-second-order	[75]
Functionalized CNT	25	7	1440	142.86	Langmuir and Freundlich	–	pseudo-second-order	[76]
Bamboo powder charcoal	10		120	1.25	Langmuir	–	–	[77]
Bamboo charcoal	30	2–4	–	6.45	Langmuir	–	–	[77]
Chemically modified biochar		4.64	60	28.21	Langmuir	Endothermic and spontaneous	Pseudo-second-order	[78]

(*Continued*)

Table 12.2 Cont.

Adsorbent	Temp (°C)	pH	Time (min)	Adsorption capacity (mg/g)	Adsorption isotherm	Thermodynamics nature	Kinetics model	Ref.
Ammonium-functionalized mesostructured silica	25	<8	60	46.5	Freundlich	Exothermic and spontaneous	–	[78]
HDTMA modified QLD-bentonite	30	5.4	1020	14.76	–	–	–	[79]
Zn-Al Chloride LDHs	25	6	40	41.77	–	Endothermic and spontaneous	–	[81]
Mg-Al-Cl hydrotalcite	25 ± 2	6	40	41.78	–	Endothermic and spontaneous	–	[82]
Ca-Al-Cl hydrotalcite	–	6	40	37.65	–	Endothermic and spontaneous	Pseudo-first-order	[83]
Layered double hydroxides	21	~8.5	240	20–35	Langmuir	–	–	[84]
Calcined hydrotalcite-type compounds	25	–	1440	61.7–147.0	Langmuir	–	–	[85]
Functionalized zeolite	20	–	120	11.4	Freundlich	–	pseudo-second order	[86]
Surfactant-modified zeolite	25	5–6	1440	6.07	Langmuir	–	–	[89]
Chitosan coated zeolite	4–20	7	4320	37.2–45.88	–	–	–	[90]

Cetylpyridinium bromide (CPB) modified zeolite	25	–	90	9.36	Langmuir, Freundlich and D-R	Exothermic and spontaneous	pseudo-second order	[91]
Modified sugar cane bagasse	25	–	30	18,502	Freundlich	Endothermic and non-spontaneous	pseudo-second order	[92]
Amine-grafted corn cob	–	6.5	–	49.9	Freundlich	–	Elovich model	[93]
Amine-grafted coconut copra	–	6.5	–	59.0	Freundlich	–	Elovich model	[93]
Sugarcane bagasse	30	–	2880	87.42	Langmuir	–	–	[94]
Rice hull	30		1440	74.4	Langmuir	–	–	[95]
Pine bark anion exchanger	30		1440	65.72	Langmuir	–	–	[95]
Raw wheat residue	25		20	1.24	Freundlich	–	pseudo-second-order	[96]
Modified wheat residue	25	–	20	128.96	Freundlich	–	pseudo-second-order	[96]
Zr(IV)-loaded sugar beet pulp	30	6.0	–	63	–	–	–	[97]
Zr(IV)-loaded sugar beet pulp	30	6.0	–	63	–	–	–	[97]
Chitosan hydrobeads	30	5	1440	92.1	Langmuir	Exothermic and spontaneous	pseudo-second order	[115]
Conditioned cross-linked chitosan beads	30	5	1440	104.0	Langmuir	–	–	[116]
Chitosan beads	30	5	1440	90.7	Langmuir	–		[116]

(Continued)

Table 12.2 Cont.

Adsorbent	Temp (°C)	pH	Time (min)	Adsorption capacity (mg/g)	Adsorption isotherm	Thermodynamics nature	Kinetics model	Ref.
Quaternary amine modified chitosan beads	30	6–9	30	24	Freundlich	Exothermic and spontaneous	pseudo-second order	[117]
Functionalized chitosan beads	30	5–7	120	113.1	Freundlich	Exothermic and spontaneous	pseudo-second-order	[118]
Fe_3O_4/ZrO_2/chitosan	25	3	1440	89.3	Langmuir	–	pseudo-first-order	[119]
Granular chitosan-Fe^{3+} complex	50	5.0	120	8.35	Langmuir-Freundlich	Exothermic and spontaneous	pseudo-second-order	[120]
PEG/chitosan	30	3	40	50.68	Freundlich	Endothermic and spontaneous	pseudo-second-order	[121]
PVA/chitosan	30	3	40	35.03	Freundlich	Endothermic and spontaneous	pseudo-second-order	[121]
Granular chitosan-Fe(III)-Al(III) complex	15	3	–	58.45	Freundlich	–	pseudo-first-order	[122]
Chitosan /Zeolite Y/Nano ZrO2 nanocomposite	35	3	60	23.58	Freundlich	–	pseudo-second-order	[123]
Chitosan/titanium oxide	30	6	180	70.94	Freundlich	Endothermic and spontaneous	pseudo-second-order	[124]
Chitosan/alumina	30	6	180	82.04	Freundlich	Endothermic and spontaneous	pseudo-second-order	[124]

12.3 Models for Adsorption Process

Adsorption is a surface phenomenon and widely used fundamental physicochemical process in practice to remove substances from fluid phases (gases or liquids). When the fluid and solid surface comes into contact, the solutes accumulated on the surface of the solid [65, 125–127]. The solute that accumulated on the surface of the solid is called adsorbate, whereas, the solid on which the adsorption occurs is known as adsorbent [65, 68, 125–127]. Interaction between the solute and solid surface can be by physical attraction (physisorption) or chemical reaction (chemisorption). Physisorption is reversible and the force of attraction existing between adsorbate and adsorbent are weak Van der Waal forces and it has low enthalpy of adsorption i.e. ΔH adsorption is less than 40 kJ/mol [125–127]. Physisorption takes place with formation of multilayer of adsorbate on adsorbent and it is not site-specific. Conversely, the force of attraction in chemisorption is an ionic or covalent bond formation between the adsorbate molecules and the atoms of the functional groups of the adsorbent. The chemisorption is irreversible, site specific with formation of monolayer of adsorbate on adsorbent, and has high enthalpy of adsorption. Typically, adsorption equilibrium is not established instantaneously. The mass transfer from the solution to the adsorption sites within the adsorbent particles is constrained by mass transfer resistances that determine the time required to reach the state of equilibrium [65, 125–127].

In water treatment, adsorption has been proved as an efficient removal process for a multiplicity of solutes [65]. Adsorption process can be conducted in batch and continuously flow fixed-bed dynamic system. Batch adsorption studies provide useful information on the application and investigation of characteristics of a particular adsorbent for the removal of specific contaminant for preliminary screening of the systems before running more costly tests. In practice, application of adsorption system is usually performed in a continuously flow fixed-bed dynamic system and in order to treat large volume treated of water using a defined quantity of adsorbent [23, 24, 125–129].

12.3.1 Batch Adsorption Models

Evaluation of equilibrium adsorption isotherms, kinetics, and thermodynamics parameters has a vital role in systematically understanding the adsorption process of various pollutants in water solutions [65, 126, 130].

12.3.1.1 Adsorption Isotherms and Models

Adsorption isotherm is of fundamental importance in the adsorption process to describe the association between the equilibrium concentration of adsorbate in the solution and solid phase concentrations at constant temperature [65, 68, 110]. The knowledge of adsorption equilibrium data provides the basis for assessing the adsorption processes and helps to understand the nature of interaction between adsorbate and adsorbent, to investigate the adsorption properties, to understand the adsorption mechanism, to evaluate the potential applicability of biosorbents, and to predict maximum adsorption capacity [65, 68, 110, 131, 132]. The adsorption isotherm indicates how the adsorbent distribute between the liquid phase and solid phase when the adsorption process reaches an equilibrium state. These are information is of importance in the design of design and optimization of adsorption system used for removal of pollutants from aqueous solution [133, 134]. There are a number of batch adsorption equilibrium isotherm models which have been used to explain the adsorption of solute on to the adsorbent. Some of them are Langmuir, Freundlich, the Sips, the Redkich-Peterson, Temkin, Redlich-Peterson and Dubbinin-Radushkevitch isotherm models [110, 130, 135–137].

12.3.1.2 Langmuir Isotherm

Langmuir (1918) originally introduced a model for the homogeneous and monomolecular adsorption of gases on the energetically homogeneous surfaces [137, 138]. The basic assumptions of theoretical Langmuir sorption isotherm model underlying that the adsorbed species are attached to the surface at definite localized sites and all the adsorption sites are energetically identical. Moreover, the model assumes that there is no interaction between neighboring adsorbed adsorbate molecules. This model is now widely used for the characterization of adsorption of ions from solution. The Langmuir isotherm is expressed as follows [138]:

$$q_e = \frac{q_m K_L C_e}{1 + K_L C_e} \tag{12.1}$$

The linear form of Langmuir equation is given as:

$$\frac{C_e}{q_e} = \frac{1}{K_L q_m} + \frac{C_e}{q_m} \tag{12.2}$$

where, q_e is adsorption capacity at equilibrium (mg/g), C_e is concentration of nitrate in the solution at equilibrium (mg/L), q_m is the maximum adsorption capacity (mg/g), and K_L is the Langmuir constant related to the energy of adsorption represents the degree of adsorption affinity the adsorbate (L/mg). The parameters K_L and q_m can obtained from the slope and intercept of the plot of C_e/q_e against C_e. To predict the favorability of adsorption, the Langmuir isotherm is expressed in terms of a dimensionless constant separation factor or equilibrium parameter, R_L [139].

$$R_L = \frac{1}{1+K_L C_O} \tag{12.3}$$

where, C_o = the highest initial adsorbate concentration, b = the Langmuir's adsorption constant (L/mg). The value of R_L indicates the type of the isotherm to be either unfavorable ($R_L > 1$), linear ($R_L = 1$), favorable ($0 < R_L < 1$) or irreversible ($R_L = 0$) [134, 139].

12.3.1.3 Freundlich Isotherm

Freundlich adsorption isotherm model [140] is an empirical relation between liquid phase and solid phase solute concentrations, and this is applicable for multilayer adsorption with non-uniform distribution of adsorption heat and affinities over the heterogeneous surface. Unlike the Langmuir isotherm, the Freundlich isotherm describes the mathematical relationship for non-ideal and reversible adsorption process. The Freundlich empirical equation is expressed as [140]:

$$q_e = K_f C_e^{1/n} \tag{12.4}$$

A linear form of this expression is:

$$lnq_e = lnK_f + \frac{1}{n} lnC_e \tag{12.5}$$

where, C_e is concentration of solute in the liquid phase at equilibrium (mg/L), q_e is amount of solute adsorbed on the surface of adsorbent (mg/g). K_f (mg/g) (L/mg)$^{1/n}$ and n (dimensionless) are the Freundlich constants calculated from the intercept and slope of linear plots of Eqn. (12.5). The value of K_f constant is an indication of theoretical adsorption capacity. The value of n is an indicates the favorability and intensity adsorption process and surface heterogeneity of the system [134]. If $1 < n < 10$, this indicates favorable adsorption, and $n = 1$ characterizes a linear adsorption phenomenon [134, 140–142].

12.3.1.4 Temkin Isotherm

The Temkin isotherm assumes that the fall in the heat of adsorption is linear rather than logarithmic as stated by Freundlich. The heat of sorption of all the molecules in the layer would decrease linearly with coverage due to adsorbate/adsorbent interactions [137]. The Temkin isotherm has been used in the following form:

$$q_e = \frac{RT}{b}\ln(AC_e) \tag{12.6}$$

A linear form of the Temkin isotherm can be expressed as:

$$q_e = \frac{RT}{b}\ln(A) + \frac{RT}{b}\ln(C_e) \tag{12.7}$$

$$\frac{RT}{b} = B$$

This model can be linearized as follows:

$$q_e = B\ln(A) + B\ln(C_e) \tag{12.8}$$

where A and B are the Temkin isotherm constant (L/g) and heat of sorption (J/mol) respectively. R is the gas constant (J/mol.k), b is the Temkin isotherm constant linked to the energy parameter, B. The constants B and A can calculated from the slope and intercept linear plot of q_e against $\ln(C_e)$ respectively [137].

12.3.1.5 Dubinin–Radushkevich (D–R) Isotherm

The Dubinin–Radushkevish (DR) isotherm assumes that sorption has a multilayer character with heterogeneity of surface energies. The adsorption data was applied to the D–R model to distinguish between physical and chemical adsorption. The Dubinin–Radushkevich model can be expressed as follows [135, 137, 143]:

$$q_e = q_m \exp\left[-\beta_{\mathrm{DR}}\left(RT\ln\left(1+\frac{1}{C_e}\right)\right)^2\right] \tag{12.9}$$

Eqn. (12.9) is expressed in short form as follows:

$$q_e = q_m \exp((-\beta_{DR}\varepsilon^2)) \tag{12.10}$$

Linear form of D–R isotherm equation is given as:

$$\ln q_e = \ln q_m - \beta_{DR}\varepsilon^2 \tag{12.11}$$

where q_m is the adsorption capacity (mg/g) and β_{DR} is the activity coefficient related t1o adsorption energy (mol^2/kJ^2) and ε is Dubinin–Radushkevich isotherm constant (Polanyi potential). The values of β_{DR} and q_o were computed from the slope and intercept of the plot $\ln(q_e)$ against ε^2, respectively. Polanyi potential, ε can be calculated by the equation:

$$\varepsilon = RT\ln\left(1+\frac{1}{C_e}\right) \tag{12.12}$$

where T is the temperature (K) and R is the gas constant (kJ/ mol.K). The mean adsorption energy (E) is defined when one mole of ion is transferred from infinity in solution to the surface of the adsorbent. It is calculated from the β_{DR} value using the following equation:

$$E = \frac{1}{\sqrt{2\beta_{DR}}} \tag{12.13}$$

If the value of E lies between 8 and 16 kJ/mol the sorption process is a chemisorption one, while values of below 8 kJ/mol indicate a physical adsorption process [143–145].

12.3.1.6 Sips Isotherm

The Sips isotherm is derived from the limiting behavior of the Langmuir and Freundlich isotherms [146, 147]. At low adsorbate concentrations, this model reduces to the Freundlich isotherm, while at high concentrations it determines a monolayer adsorption capacity characteristic like the Langmuir isotherm. The Sips model equation is expressed as [146–148]:

$$q_e = \frac{q_m K_S C_e^{\frac{1}{n_s}}}{1 + K_S C_e^{\frac{1}{n}}} \tag{12.14}$$

$$\frac{1}{q_e} = \frac{1}{q_m K_S}\left(\frac{1}{C_e}\right)^{1/n_s} + \frac{1}{q_m} \tag{12.15}$$

where K_S (L/mg) and q_m (mg/g) are the Sips equilibrium constant and maximum adsorption capacity values and n_s is the dimensionless heterogeneity factor. If the value of n_s is between 0 and 1, the system is described as heterogeneous. If $n_s = 1$, the Sips equation reduces to the Langmuir equation and it indicates a homogeneous adsorption process [146–148].

12.3.1.7 Redlich–Peterson Isotherm

The Redlich–Peterson model is a three-parameter isotherm, which syndicates elements of the Langmuir and Freundlich isotherms in a single equation and compensates the imprecisions [146–148]. The adsorption mechanism is unique and does not follow ideal monolayer adsorption characteristics. Redlich-Peterson model is given as:

$$q_e = \frac{K_{RP} C_e}{1 + \alpha_{RP} C_e^{\beta_{RP}}} \tag{12.16}$$

where K_{RP} is the Redlich–Peterson adsorption capacity constant (L/g), a_{RP} is the R-P isotherm constant and β_{RP} is the exponent between 0 and 1. If $\beta_{RP} = 1$, the R-P equation reduces to the Langmuir isotherm equation and if $\beta_{RP} = 0$, it is closer to the Freundlich equation [146–148].

12.3.1.8 Thermodynamic Parameters

Thermodynamic considerations are important in order to predict the feasibility of the process and to understand the nature of the adsorption [149]. The free energy change (ΔG^0) is a critical factor for determining the degree of spontaneity of the adsorption process; a higher negative value of ΔG^0 indicates a more energetically favorable adsorption. In the adsorption study, it is essential to establish the adsorption mechanisms (i.e., either chemical or physical). Thermodynamic parameters of adsorption, including the Gibbs energy change (ΔG^0), the enthalpy change (ΔH^0), and the entropy change (ΔS^0) provide information concerning the type and mechanism of the adsorption process [65, 131]. The adsorption enthalpy in the case of physisorption is mostly lower than 40 kJ/mol. Chemisorption is based on chemical reactions between the adsorbate and the surface sites, and the interaction energies are therefore in the order of magnitude of reaction enthalpies (>40 kJ/mol). It has to be noted that the differentiation between physisorption and chemisorption is widely arbitrary and the boundaries are fluid [65]. It is noticeable that evaluation of $\Delta G°$, $\Delta H°$, and $\Delta S°$ is directly affected by the thermodynamic equilibrium constant (K_C; dimensionless). In adsorption thermodynamics, the equilibrium constant K_C can be derived from the constants of various adsorption isotherms (i.e., Langmuir and Freundlich) or the partition coefficient. The different estimation approaches might lead to great variations in the thermodynamic parameters; therefore, the most appropriate approaches should be recommended.

Changing in temperature will change the equilibrium capacity of the adsorbent for a particular adsorbate. At equilibrium conditions, free energy of adsorption process is defined as follows:

$$\Delta G^o = -RTlnK_C \tag{12.17}$$

where, $\Delta G°$ is the standard free energy change (kJ/mol), R is the ideal gas constant (8.314 J/mol.K), and T is temperature (K) and K_C is equilibrium constant.

The Gibbs Helmholz equation is expressed as:

$$\left(\frac{\partial\left(\frac{\Delta G^0}{T}\right)}{\partial T}\right)_P = \frac{-\Delta H^0}{T^2} \tag{12.18}$$

Substituting Eqn. (12.17) into Eqn. (12.18), the van't Hoff equation is obtained as follows [149]:

$$\left(\frac{\partial lnK_C}{\partial T}\right)_P = \frac{\Delta H^0}{T^2} \tag{12.19}$$

where $\Delta H°$ is the change in the standard enthalpy. Integrating Eqn. (12.19) for T at a constant pressure, the following equation is attained:

$$-RTlnK_C = \Delta H^0 - TRY \tag{12.20}$$

where Y is the integration constant. The relationship between the standard Gibbs free surface energy change, standard enthalpy change and standard entropy change ($\Delta S°$) is as follows:

$$\Delta G^0 = \Delta H^0 - T\Delta S^0 \tag{12.21}$$

From Eqns. (12.17), (12.19), and (12.21) the following equation is attained:

$$lnK_C = \frac{\Delta S^o}{R} - \frac{\Delta H^o}{RT} \tag{12.22}$$

where $\Delta H°$ is the standard enthalpy change (kJ/mol) and $\Delta S°$ is the standard entropy change (kJ/mol.K). The enthalpy change ($\Delta H°$) and the entropy change ($\Delta H°$) can be determined from the slope and intercept of a plot of ln K_C against $1/T$ of Eqn. (12.22).

12.3.1.9 Adsorption Kinetics

Adsorption process is a time dependent process and adsorption equilibrium is not established instantaneously [65]. This phenomenon can be described by adsorption kinetics [134]. In adsorption process, there are a number of steps of mass transfer and adsorption rate is typically determined by slow mass transfer processes from the liquid to the solid phase [65, 150–152]. Understanding the kinetics of the sorption process the different rate equation is used to fit the experimental data for any kind of sorption. Pseudo-first-order, pseudo-second-order, and intraparticle diffusion models are most commonly used models in the kinetic study of adsorption process [150–153].

Legergren proposed the simplest empirical kinetic model at the end of 19th century [152, 153]. The model is associated with the kinetics of one-site adsorption governed by the rate of the surface reaction. The equation is given as:

$$\frac{dq_t}{dt} = k_1(q_e - q_t) \tag{12.23}$$

where, q_t and q_e are amount the solid phase concentrations (mg/g) at time and at equilibrium, respectively and k_1 is equilibrium rate constant of pseudo-first-order adsorption (1/min).

After integration with the initial condition $q_t = 0$ and $t = 0$; we obtain the linearized form of this equation:

$$ln(q_e - q_t) = lnq_e - k_1t \tag{12.24}$$

The value of the sorption rate (k_1) can be calculated from linear regression analysis or from the slope of the linear plot of experimental data in $\ln(q_e - q_t)$ vs. t.

The pseudo-second-order equation is the most widely used rate equation for the sorption of a solute from a liquid phase. The experimental adsorption data can also be tested for pseudo-second-order kinetics equation. The commonly represented pseudo-second-order adsorption equation [152–154] is:

$$\frac{dq_t}{dt} = k_2(q_e - q_t)^2 \tag{12.25}$$

Integrating the Eqn. (12.25) for the boundary conditions of $t = 0$ to $t = t$ and $q_t = 0$ to $q_t = q_t$ gives:

$$\frac{1}{q_e - q_t} = \frac{1}{q_e} + k_2 t \tag{12.26}$$

Eqn. (12.26) can be rearranged to obtain Eqn. (12.27), the expression used to calculate the amount of adsorbate on the surface of the adsorbents at any time, *t*.

$$q_t = \frac{q_{ek_2}^2 t}{1 + q_e t} \tag{12.27}$$

Eqn. (12.27) is linearized form to obtain the following expression:

$$\frac{t}{q_t} = \frac{1}{k_2 q_e^2} + \frac{t}{q_e} \tag{12.28}$$

When t approaches to 0, Eqn. (12.25) can be simplified to Eqn. (12.29) which is the expression to calculate the initial adsorption rate, *h*, (mg/g min).

$$h = k_2 q_e^2 \tag{12.29}$$

Substituting Eqn. (12.29) in to Eqn. (12.28), the linearized pseudo-second-order kinetic equation can be expressed as follows:

$$\frac{t}{q_t} = \frac{1}{h} + \frac{t}{q_e} \tag{12.30}$$

where q_t is the amount of adsorbate on the surface of the adsorbents (mg/g) at any time, t; k_2 is the pseudo-second-order rate constant (g/mg min); and q_e is the amount of anions adsorbed at equilibrium (mg/g) and h is the initial sorption rate. The value of q_e (1/ slope), k (slope^2/intercept), and h (1/intercept) of the pseudo-second-order equation can be found out experimentally by plotting t/q_t against t.

In adsorption process, there are multiple steps involving transport of the solute molecules from the aqueous phase to the surface of the solid particulate, followed by diffusion of the solute molecules into the pore interiors [150, 152]. The most commonly applied pore-diffusion model is the intra-particle diffusion model [150]:

$$q_t = k_d\left(t^{\frac{1}{2}}\right) + C \tag{12.31}$$

where, k_d is intraparticle diffusion rate constant (mg/g min$^{1/2}$). The k_d is the slope of straight line portions of the plot of q_t vs. $t^{1/2}$. If the plot yields straight line passing through the origin and the slope gives the rate constant k_d, intra-particle diffusion is the rate controlling step of adsorption process. If there is a deviation of the straight line from the origin indicates that this process is not the only controlling step [99, 150].

12.4 Column Study

Although batch adsorption studies provide useful information on the application and investigation of characteristics of a particular adsorbent for the removal of specific waste constituents, experimental are only limited to investigate the characteristics of the adsorbent and preliminary screening of the systems before running more costly tests. Results obtained from equilibrium data cannot easily applied directly to continuous process and may not give accurate scale up information in the column operation system [23, 24, 128, 129]. Therefore, continuous column experiments are also necessary to obtain practical operational information with respect to the application of a particular adsorbent for the removal of contaminants from water [155]. In practice, application of adsorption system is usually carried out in the fixed bed adsorption mode. In this process, adsorbent is continuously in contact with fresh solution, which results in maximum loading of the sorbent at constant solute concentration and therefore efficient utilization of adsorbent. Continuous process is simple to operate and can be controlled easily, can be easily scaled up from a laboratory-scale procedure, and it is economical process [156, 157]. In addition, large volume of water can be continuously treated using a defined quantity of adsorbent. Moreover; continuous adsorption study can be used to identify the feasibility of the adsorbent under continuous loading [128, 158]. Further, in order to obtain basic engineering data for the design of the application process, it is essential to study the continuous flow system [159].

Adsorption within the frequently used fixed-bed adsorbers is not only a time dependent but also a spatial-dependent process. The dependence on time (t) and space (z) is referred to as adsorption dynamics or column dynamics [65].

$$q = F(t,z) \tag{12.32}$$

$$C = F(t,z) \tag{12.33}$$

The adsorption equilibrium is the basis of all adsorption models. To predict adsorption dynamics, information about adsorption equilibrium as well as about adsorption kinetics is necessary.

12.4.1 Breakthrough Curve Analysis

The performance of fixed-bed adsorption column is described by the concept of the breakthrough curve [156]. A breakthrough curve is obtained by plotting normalized concentration, defined as the ratio of effluent adsorbate concentration (C_t) to the influent adsorbate concentration (C_0) as a function of time of treatment or volume of effluent for a given bed volume [18, 156]. The breakthrough time and the shape of the breakthrough curve are important characteristics for determining the operation and the dynamic response of an adsorption column. These determine the operation life span of the bed and regeneration time and thus directly affect the feasibility and economics of the adsorption process, [129, 160, 161]. The profile of breakthrough curves and parameters are dependent on column operating conditions such as influent concentration, flow rate, and depth of bed. Experimental determination the effects of these parameters on breakthrough curves is very important to evaluate the performance of the column and to and to develop scale up procedures [162]. Parameters and profile of breakthrough curve determine the operation life span of the bed and regeneration time. A plot breakthrough curve is expressed in terms of normalized concentration, defined as the ratio of effluent adsorbate concentration (C_t) to the influent adsorbate concentration (C_0) as a function of time or volume of effluent for a given bed volume [18, 156]. Time equivalent to total or stoichiometric capacity and time equivalent to usable capacity (breakthrough time) are given in equations (1) and (2) respectively [163]

$$t_t = \int_{t=0}^{t=\text{total}} \left(1 - \frac{C_t}{C_0}\right) dt \tag{12.34}$$

$$t_b = \int_{t=0}^{t_b} \left(1 - \frac{C_b}{C_0}\right) dt \tag{12.35}$$

where t_t is exhaustion time, t_b is breakthrough time at which $C_t = C_b$ (for the present system, $C_b = 50$ mg/L). The total adsorbed nitrate quantity (q_{total}) in the column for a given influent concentration and flow rate can be calculated using Eqn. (12.36) [156].

$$q_{\text{total}} = QC_0 \int_{t=0}^{t=\text{total}} \left(1 - \frac{C_t}{C_0}\right) dt \tag{12.36}$$

where Q is the volumetric flow rate (L/h), C_0 is the influent nitrate concentration (mg/L) and t_{total} is the total flow time (min), respectively.

The equilibrium nitrate adsorption capacity of the column, q_e (mg/g), can be calculated using Eqn. (12.37)

$$q_e = \frac{q_{\text{total}}}{M} \tag{12.37}$$

where M is the amount of the adsorbent packed in the column (g). The total amount of nitrate entering the column, (W_{total}) in grams can be found out with the help of Eqn. (12.38)

$$W_{\text{total}} = C_0 Q t_e \tag{12.38}$$

where t_e = exhaustion time (h). The total percentage removal of nitrate (R) can also be found from the ratio of total adsorbed quantity (q_{total}) to the total amount sent to column (W_{total}) as [163]:

$$R(\%) = \frac{q_{\text{total}}}{W_{\text{total}}} \tag{12.39}$$

The total effluent volume (V_e) and breakthrough volume (V_b) can be estimated as follows:

$$V_e = Qt_e \tag{12.40}$$

$$V_b = Qt_b \tag{12.41}$$

where, V_e is total effluent volume up to exhaustion time (L), total effluent volume up to breakthrough time (L), Q is volumetric flow rate (L/h) t_e is exhaustion time (h), and t_b is breakthrough time (h).

The mass transfer zone (MTZ) is the length of the adsorption zone in the column, shows the efficiency in the use of adsorbents in the column can be calculated as follows [163, 164]:

$$MTZ = H\left(\frac{t_e - t_b}{t_e}\right) \tag{12.42}$$

where, H is total depth of the adsorbent in the column (cm), t_e is exhaustion time (h), and t_b is breakthrough time (h).

The time of contact between the liquid phase and the solid adsorbent is called Empty Bed Contact Time (EBCT) which measures the critical depth and the contact time for an adsorbent. It can be calculated using Eqn. (12.43).

$$EBCT = \left(\frac{V}{Q}\right) \tag{12.43}$$

where V is volume of fixed bed (L) and Q flow rate (L/h).

12.4.2 Models of Column Studies

Successful design of a continuous fixed bed column adsorption process requires prediction of breakthrough curve parameters for the effluent under given specific operating conditions. Several such models have been used in the literature to characterize the performance of fixed bed for the removal contaminants from water and wastewater are presented here [128, 156, 162, 165].

12.4.2.1 Adams-Bohart Model

Adams–Bohart model [166] based on the surface reaction theory assumes that equilibrium is not instantaneous and explains that the kinetics is controlled by external mass transfer in the initial part of adsorption. This model is also based on the postulate that the adsorption rate is proportional to the residual capacity of adsorbent and the solute concentration. This model established the fundamental equations describing the relationship between C_t/C_0 and t in a continuous system. The Adam's–Bohart model is used for the description of the initial part of the breakthrough curve. The mass transfer rates obey the following equations:

$$\frac{\partial q}{\partial t} = -k_{AB} q C_b \tag{12.44}$$

$$\frac{\partial C_b}{\partial Z} = -\frac{k_{AB}}{U_O} q C_b \tag{12.45}$$

Some assumptions are made for the solution of these differential equation systems: When the differential equation systems solved, the following equation is obtained with parameters k_{AB} and N_0 assuming that the concentration field is considered to be low and when $t \to \infty$, $q \to N_0$,

$$\ln\left(\frac{C_t}{C_0}\right) = k_{AB} C_0 t - \frac{k_{AB} N_0 H}{U_0} \tag{12.46}$$

where, C_0 and C_t are the influent and effluent adsorbate concentration (mg/L), N_0 is the saturation concentration (mg/L), k_{AB} is the kinetic constant (L/mg .h), U_0 is the linear velocity calculated by dividing the flow rate by the column section area (cm/h), H is the bed depth of column (cm), and N_0 is the saturation concentration (mg/L). The values of N_0 and k_{AB} can be obtained from the intercept and slope the linear plot of $\ln(C_t/C_0)$ against time (t).

12.4.2.2 Thomas Model

Description of the dynamic behavior of adsorption of adsorbate onto the adsorbent is important for successful design of a continuous fixed-bed adsorption column process. Thomas model is one of the most commonly and a widely used kinetic model in the to predict the performance of column operations [167]. This model can be applied to evaluate the adsorption rate constant and maximum uptake of the adsorbate. This model can be applied to evaluate the adsorption rate constant and maximum uptake of the adsorbate [167]. The model assumes that the adsorption is described by a pseudo second-order reaction rate principle which reduces a Langmuir isotherm at equilibrium. The model has the following form [165–168]:

$$\frac{C_t}{C_0} = \frac{1}{1 + exp\left[k_{\text{Th}}\left(\frac{q_0 M - C_0 V}{Q}\right)\right]} \tag{12.47}$$

The above equation can be linearized to give,

$$\ln\left(\frac{C_0}{C_t} - 1\right) = \frac{k_{\text{Th}} C_0 M}{Q} - k_{\text{Th}} C_0 \frac{V}{Q} \tag{12.48}$$

$$\ln\left(\frac{C_0}{C_t} - 1\right) = \frac{k_{\text{Th}} C_0 M}{Q} - k_{\text{Th}} C_0 t \tag{12.49}$$

where, k_{Th} is the Thomas rate constant (L/h.mg); q_0 is the equilibrium adsorbate uptake per g of the adsorbent (mg/g); C_0 is the influent adsorbate concentration (mg/L); C_t is the effluent concentration at time t (mg/L); M is the mass of adsorbent (g), V is the throughput volume (L), and Q is the flow rate (L/h). The value of C_t/C_0 is the ratio of outlet and influent adsorbate concentrations. The values of k_{Th} and q_o were determined from the intercept and slope of the linear plot of $\ln[(C_o/C_t)-1]$ against time (t).

12.4.2.3 Yoon and Nelson Model

The Yoon-Nelson model was derived based on the assumption that the rate of decrease in the probability of adsorption of adsorbate molecule is proportional to the probability of the adsorbate adsorption and the adsorbate breakthrough on the adsorbent [169]. Yoon-Nelson kinetic model is one of the simplest model used to predict the exhaustion time and the behavior of adsorption process for a given adsorbate concentration. The Yoon and Nelson equation applied to a single component system is expressed as [156, 169]:

$$\frac{C_t}{C_e} = \frac{1}{1 + \exp[k_{\text{YN}}(\tau - t)]} \tag{12.50}$$

The linearized form of the Yoon and Nelson model is as follows:

$$\ln\left(\frac{C_t}{C_0 - C_t}\right) = k_{\mathrm{YN}}t - \tau k_{\mathrm{YN}} \tag{12.51}$$

where k_{YN} is the rate velocity constant (L/h), τ is the time in required for 50% adsorbate breakthrough (h). The values of τ and k_{YN} can be determined from the intercept and slope of the plot linear plot of ln $[C_t/(C_0 - C_t)]$ against sampling time (t), respectively.

The derivation of Eqn. (12.40) was based on the definition that 50% breakthrough occurs at $t = \tau$. Therefore, if the breakthrough curve is symmetrical then the adsorption bed should be completely saturated at $t = 2\tau$. Accordingly, the following equation can be obtained for a given bed [156, 170, 171]

$$q_0 = \frac{1}{2}\frac{C_0 Q(2\tau)}{M} = \frac{C_0 Q\tau}{M} \tag{12.52}$$

This equation also permits one to determine the adsorption capacity of the column (q_0) as a function of inlet adsorbate concentration (C_0), flow rate (Q), adsorbent quantity in the column (M) and 50% breakthrough time (τ) by using the Yoon–Nelson model [156, 170, 171].

12.4.2.4 Clark Model

Clark [172] defined a new model by combining the concept of Freundlich equation and the mass transfer. The model is used for the simulation of breakthrough curves and the equation is as follows [165, 170–173]:

$$U_0 \frac{dC_b}{dZ} = k_a (C_b - C_e) \tag{12.53}$$

where k_a is the mass transfer coefficient (min^{-1}). Clark resolved this system and obtained the following solution:

$$\left[\frac{C_0^{n-1}}{1 + \left[\left(\frac{C_0^{n-1}}{C_b^{n-1}} - 1\right)e^{rt_b}\right]e^{-rt}}\right]^{\frac{1}{n-1}} = C_t \tag{12.54}$$

$$\frac{C_t}{C_0} = \left(\frac{1}{1 + Ae^{-rt}} \right)^{\frac{1}{n-1}} \quad (12.55)$$

$$A = \left(\frac{C_0^{n-1}}{C_b^{n-1}} - 1 \right) e^{rt_b} \quad (12.56)$$

$$r = (n-1)\frac{k_T}{U_0} v \quad (12.57)$$

The migration velocity of the concentration front in the bed can be determined from the relationship below,

$$v = \frac{U_0 C_0}{N_0 + C_0} \quad (12.58)$$

where k_T is the mass transfer coefficient in (h^{-1}), v is the migration velocity of the concentration front in the bed (cm/h), U_0 is the gas flow rate (cm/h) and N_0 is the adsorptive capacity (mg adsorbate /L adsorbent).

The linearized form of Clark equation is:

$$\ln\left[\left(\frac{C_0}{C} \right)^{n-1} - 1 \right] = -rt + lnA \quad (12.59)$$

where C_0 is the influent adsorbate concentration (mg/L); C_t is the effluent concentration at time t (mg/L), C_b is the outlet concentration at breakthrough, t_b is the breakthrough time (h), A and r are Clark parameters. n is Freundlich parameter which can be evaluated from batch equilibrium experiments. The values of the Clark parameters, A and r are determined from the slope and the intercept of the linear plots of $\ln((C_0/C_t)^{n-1} - 1)$ versus t of the experimental data., enabling the prediction of the breakthrough curve according to the relationship between C/C_0 and t in Eqn. (12.55) [165, 172].

12.4.2.5 The Wolborska Model

The Wolborska model is also used for the description of adsorption dynamics using mass transfer where diffusion mechanisms are involved in the range of the low concentration breakthrough curve [173, 174]. The mass transfer in the fixed bed sorption is described by the following equations:

$$\frac{\partial C}{\partial t} = U_0\left(\frac{\partial C}{\partial z}\right) + \left(\frac{\partial q}{\partial t}\right) = D\left(\frac{\partial^2 q}{\partial^2 z}\right) \tag{12.60}$$

$$\frac{\partial q}{\partial t} = -v\left(\frac{\partial q}{\partial z}\right) = \beta_a(C_b - C_s) \tag{12.61}$$

where C is the adsorbate concentration in the liquid phase (g/L). t is time (h), U_0 is the linear flow rate (cm/h), q is the adsorbate concentration in the solid phase (mg/g), C_s is the adsorbate concentration at the solid/liquid interface (mg/L), D is the axial diffusion coefficient (cm^2/h), v is the migration rate (cm/h), and βa is the kinetic coefficient of the external mass transfer (h^{-1}).

Assuming that $C_S << C_b$, $v << U_0$ and axial diffusion is negligible, Wolborska solved the above differential equations and found the following solution in the original variables,

$$\ln\left(\frac{C_t}{C_0}\right) = \frac{\beta_a C_0}{N_0}t - \frac{\beta_a}{U_0}H \tag{12.62}$$

For column dynamics with axial dispersion [173, 174]:

$$\beta_a = \frac{U_0^2}{2D}\left(\sqrt{1+\frac{4\beta_0 D}{U_0^2}} - 1\right) \tag{12.63}$$

where, β_0 is the external mass transfer coefficient with a negligible axial dispersion coefficient D. Wolbraska observed that in short beds or at high flow rates of solution through the bed, the axial diffusion is negligible and $\beta a = \beta_0$. The expression of the Wolborska solution is equivalent to the Adams–Bohart relation if the coefficient k_{AB} is equal to β_a/N_0. By plotting $\ln(C_t/C_0)$ versus t of the experimental data, the resulting curve would be linear should the Wolborska model solution fit the data.

12.4.2.6 Bed Depth Service Time (BDST) Model

The BDST model gives relationship between bed height (H) and service time (t) in terms of process concentrations and other adsorption parameters.

The BDST model constants can be for scaling up of the process for other flow rates and concentrations without additional tests. It is used to evaluate the column performance for any bed length using the data of known bed depth. The modified version of the equation used in this evaluation is given as [175]:

$$t = \left(\frac{N_o H}{C_o u}\right) - \left(\frac{1}{K_\alpha C_o}\right) \ln\left(\frac{C_o}{C_t} - 1\right) \tag{12.64}$$

where, C_0 and C_t (mg/L) are the influent and effluent nitrate concentration, K_a is the kinetic constant BDST model (L/mg.h), u is the linear velocity (cm/h) calculated by dividing the flow rate by the column cross-section area, H is the bed depth of column (cm), and N_0 is the saturation concentration (mg/L). The values of N_0 and K_a can be calculated from the intercept and slope the linear plot of time (t) against bed depth of column (H).

12.5 Conclusion

Excess concentration of nitrate water can lead to plausible health problems to the public and affects biodiversity due to eutrophication. The currently available nitrate removal techniques are found to be costly, inefficient, and generate secondary pollution by generating byproducts. Adsorption technique is a cost-effective and feasible method for the removal of nitrate from water. Higher nitrate adsorption capacity, high nitrate selectivity, economic feasibility, availability, reusability, and good chemical stability and mechanical integrity are desirable characteristics of adsorbent to be effective in water treatment system for practical applications. Several adsorbent materials, such as carbonaceous, oxides and minerals, modified agricultural and industrial wastes have been widely used for the removal of nitrate from water. However, modification and production process of such materials consumes great quantity of energy and chemicals which intern involving environmental costs. Therefore, there is a need for low cost, eco-friendly, and sustainable materials for the removal of nitrate from water. Chitosan has received wide attention in the environmental engineering field as an effective adsorbent for the removal of several water contaminants. However, chitosan in its original form is not suitable for practical use due to its poor chemical and mechanical stability. Moreover, chitosan has very low density and needs a large volume of fix-bed for continuous adsorption system. These properties restrict its field application.

Therefore, chitosan needs certain modifications to improve its chemical and mechanical stability, as well as to enhance its adsorption capacity and reusability. Cost and environmental considerations are critical issues and one should not ignore these during modification of chitosan. Therefore, there is a need to explore an alternative low-cost, environment friendly and effective modification method. Inorganics materials, such as bentonite, titanium oxide, and alumina are an attractive immobilization inorganic materials for chitosan due to their environmental benignity, low cost as well as chemical and mechanical stability.

The application and performance of chitosan-based organic–inorganic hybrid biocomposites were investigated. In future, these studies can be extended to real water and wastewater. Most of the studies reported in the literature on the removal of nitrate from water using various adsorbents focused on batch equilibrium studies. However, batch studies are only limited to investigate the characteristics of the adsorbent and the results obtained cannot easily applied directly to continuous process. Therefore, continuous column experiments are also necessary to obtain practical operational information with respect to the application. Moreover, most of the studies on the application of various materials investigated for the removal of nitrate from synthetic water. In future, these studies can be extended to real water and wastewater.

Nomenclatures

A	Temkin isotherm constant	(L/g)
a	Total interfacial area of the particles	(cm^2)
B	Temkin isotherm constant related to heat of sorption	(J/mol)
B_t	Biot number	–
b	Temkin isotherm constant linked to the energy parameter	–
C_a	Equilibrium adsorbate concentration on the adsorbent	(mg/L)
C_b	Breakthrough concentration	(mg/L)
C_e	Equilibrium concentration of adsorbate	(mg/L)
C_o	initial concentration of adsorbate	(mg/L)

C_s	Adsorbate concentration at the solid/liquid interface	(mg/L)
D	Axial diffusion coefficient	(cm^2/h)
D_i	Intraparticle diffusion	(cm^2/min)
d_p	Mean particle diameter	(cm)
E	sorption energy	(kJ/mol)
EBCT	Empty Bed Contact Time	(h)
H	Total depth of the bed	(cm)
k_1	Pseudo-first-order rate constant	(1/min)
k_2	Pseudo-second-order rate constant	(g/mg min)
k_{AB}	Adams-Bohart model kinetic constant	(L/mg .h)
k_d	Intraparticle diffusion rate constant	($min^{0.5}$)
K_D	Adsorption distribution coefficient	(L/g)
K_f	Freundlich constant	$(mg/g)\ (L/mg)^{1/n}$
k_f	initial external mass transfer coefficient	(cm/min)
K_L	Langmuir constant	(L/mg)
K_{RP}	The Redlich-Peterson adsorption capacity	(L/g)
K_P	Partition coefficient	–
K_S	The Sips equilibrium constant	(L/mg)
k_{Th}	Thomas rate constant	(L/h.mg)
k_{YN}	Yoon-Nelson rate velocity constant	(h^{-1})
K_α	Kinetic constant BDST model	(L/mg.h)
M	Mass of biosorbent	(g)
M_S	Biosorbent mass concentration in the solution	(g/cm^3)
MTZ	The mass transfer zone	(cm)
N_o	Adsorption capacity	(mg/L)
n	Freundlich constant related to adsorption intensity	–
n_s	The Sips heterogeneity factor	–

q_e	Equilibrium adsorption capacity	(mg/g)
$q_{e,m}$	Model equilibrium adsorption capacity	(mg/g)
q_m	Langmuir maximum adsorption capacity	(mg/g)
q_t	Adsorption capacity at any time, t	(mg/g)
q_0	Maximum equilibrium nitrate adsorption capacity from Thomas model	(mg/g)
q_{total}	Total amount of nitrate adsorbed	(mg)
R	Gas constant	(J/mol.K)
R^2	Correlation coefficient	–
R_L	Separation factor or equilibrium parameter	–
R	Total percentage removal of nitrate	(%)
t_b	Breakthrough time	(h)
t_e	Exhaustion time	(h)
T	Contact time	(min)
T	Temperature	(K)
U_0	Linear velocity	(cm/h)
V	Migration velocity of the concentration front in the bed	(cm/h)
V	Volume of the bed	(L)
V_b	Breakthrough volume	(L)
V_e	Total effluent volume	(L)
V	Volume of the solution	(L)
W_{tot}	Total amount of nitrate sent to the column	(mg)
ΔG^o	Standard free energy change	(kJ/mol)
ΔH^o	Standard enthalpy change	(kJ/mol)
ΔS^o	Standard entropy change	(kJ/mol.K)
a_{RP}	The R-P isotherm constant	–
β_a	External mass transfer coefficient	(h^{-1})
β_o	External mass transfer coefficient with a negligible axial dispersion coefficient	(h^{-1})

β_{RP}	The R-P isotherm parameter	–
ε	Polanyi sorption potential	–
χ^2	Chi-square	–
ρ	Density of water	(g/L)
ρ_{app}	Apparent density of biosorbent	(g/cm^3)
τ	Time in required for 50% adsorbate breakthrough	(h)

References

1. Velizarov, S., Crespo, J.G., Reis, M.A., Removal of inorganic anions from drinking water supplies by membrane bio/processes. *Rev. Environ. Sci. Biotechnol.*, 3(2004), 361–380, 2004.
2. Petrović, M., Gonzalez, S., Barceló, D., Analysis and removal of emerging contaminants in wastewater and drinking water. *TrAC – Trends Anal. Chem.*, 22(August 2015), 685–696, 2003.
3. Fewtrell, L., Drinking-water nitrate, methemoglobinemia, and global burden of disease: A discussion. *Environ. Health Perspect.*, 112(14), 1371–1374, 2004.
4. Spalding, R.F., Exner, M.E., Occurrence of nitrate in groundwater—a review. *J. Environ. Qual.*, 22, p. 392, 1993.
5. Blackwood, B., Halloran, P.O., Porter, S., Nitrate removal from drinking water by point of use ion exchange. *J. Res. Nurs.*, 1–10, 2010.
6. Fields, S., Global nitrogen: cycling out of control. *Environ. Health Perspect.*, 112(10), 556–563, 2004.
7. Harter, T., Agricultural impacts on groundwater nitrate. *Southwest hydrol. (Nitrates Groundw. Spec. Issue)*, 8(4), , p. 22, 2009.
8. Carpenter, S., Caraco, N.F., Correll, D.L., Howarth, R.W., Sharpley, A.N., Smith, V.H., Nonpoint Pollution of Surface Waters with Phosphorus and Nitrogen. *Issues Ecol.*, 4(3), 1–12, 1998.
9. Bouchard, D.C., Williams, M.K., Surampalli, R.Y., Nitrate contamination of groundwater: sources and potential health effects. *Am. Water Work. Assoc.* (September), p. 6, 1992.
10. WHO, *Guidelines for drinking-water quality*, Fourth., 1. 2011.
11. Samatya, S., Kabay, N., Yüksel, Ü., Arda, M., Yüksel, M., Removal of nitrate from aqueous solution by nitrate selective ion exchange resins. *React. Funct. Polym.*, 66(11), 1206–1214, 2006.
12. Majumdar, D., The blue baby syndrome. *Resonance*(October), 20–30, 2003.
13. Ward, M.H., deKok, T.M., Levallois, P., Brender, J., Gulis, G., Nolan, B.T., VanDerslice, J., Workgroup report: Drinking-water nitrate and health–recent findings and research needs. *Environ. Health Perspect.*, 113(11), 1607–1614, 2005.

14. Hunter, H., Comly, M.D., Cyanosis in infants caused by nitrates in well water. *J. Am. Med. Assoc.*, 129, 112–116, 1945.
15. Kuo, H.-W., Wu, T.-N., Yang, C.-Y., Nitrates in drinking water and risk of death from rectal cancer in Taiwan. *J. Toxicol. Environ. Health. A*, 70(20), 1717–1722, 2007.
16. Morales-suarez-varela, M.M., Llopis-gonzalez, A., Tejerizo-perez, M.L., Impact of nitrates in drinking water on canc. *Eur. J. Epidemiol.*, 11, 15–21, 1995.
17. Chang, C.C., Chen, C.C., Wu, D.C., Yang, C.Y., Nitrates in drinking water and the risk of death from rectal cancer: does hardness in drinking water matter? *J. Toxicol. Environ. Health. A*, 73(19), 1337–1347, 2010.
18. Lohumi, N., Gosain, S., Jain, A., Gupta, V.K., Verma, K.K., Determination of nitrate in environmental water samples by conversion into nitrophenols and solid phase extraction–spectrophotometry, liquid chromatography or gas chromatography–mass spectrometry. *Anal. Chim. Acta*, 505(2), 231–237, Mar. 2004.
19. Khandare, H.W., Scenario of Nitrate contamination in groundwater: its causes and prevention introduction. *Int. J. ChemTech Res.*, 5(4), 1921–1926, 2013.
20. CENR, An Assesment of Coastal Hypoxia and Eutrophication in U.S. Waters. 1–82, 2003.
21. Bohlen, P.J., Hendrix, P.F., Exotic earthworm invasions in north america: ecological and policy implications. *Bioscience*, 52(8), 669–682, 2002.
22. Sharma, S. K., Sobti, R.C., Nitrate removal from ground water: a review. *E-Journal Chem.*, 9(4), 1667–1675, 2012.
23. Bhatnagar, A., Sillanpää, M., A review of emerging adsorbents for nitrate removal from water. *Chem. Eng. J.*, 168(2), 493–504, Apr. 2011.
24. Loganathan, P., Vigneswaran, S., Kandasamy, J., Enhanced removal of nitrate from water using surface modification of adsorbents–a review.*J. Environ. Manage.*, 131, 363–74, 2013.
25. Kapoor, B.A., Viraraghavan, T., Nitrate removal from drinking water-review. *J. Environ. Eng.* (April), 371–380, 1997.
26. Centi, G., Perathoner, S., Remediation of water contamination using catalytic technologies. *Appl. Catal. B Environ.*, 41(1–2), 15–29, Mar. 2003.
27. Jensen, V.B., Darby, J.L., Drinking water treatment for nitrate with a focus on tulare lake basin and salinas valley groundwater. *Cent. Watershed Sci. Univ. California, Davis http//groundwaternitrate.ucdavis.edu*, 2012.
28. Soares, M.I.M., Biological denitrification of groundwater. *Water Air Soil Pollut.*, 123, 183–193, 2000.
29. Dahab, M.F., Lee, Y.W., Dahab, F., Woon, Y., All use subject to JSTOR terms and conditions nitrate using removal biological from supplies denitrification water. *Water Pollut. Control Fed.*, 60(9), 1670–1674, 2015.
30. Shrimali, M., Singh, K.P., New methods of nitrate removal from water. *Environ. Pollut.*, 112(3), 351–359, 2001.

31. Fu, F., Dionysiou, D.D., Liu, H., The use of zero-valent iron for groundwater remediation and wastewater treatment: A review. *J. Hazard. Mater.*, 267, 194–205, 2014.
32. Ziajahromi, S., Zand, A.D., Khanizadeh, M., Nitrate removal from water using synthesis nanoscale zero-valent iron (NZVI). 105–110, 2012.
33. Yang, G.C.C., Lee, H.L., Chemical reduction of nitrate by nanosized iron: kinetics and pathways. *Water Res.*, 39, 884–894, 2005.
34. Hsu, J.-C., Liao, C.-H., Wei, Y.-L., Nitrate removal by synthetic nanoscale zero-valent iron in aqueous recirculated reactor. *Sustain. Environ. Res.*, 21(6), 353–359, 2011.
35. Huang, Y.H., Zhang, T.C., Effects of low pH on nitrate reduction by iron powder. *Water Res.*, 38, 2631–2642, 2004.
36. Vorlop, K., Tacke, T., 1st Steps towards noble-metal catalyzed removal of nitrate and nitrite from drinking-water. *Chem. Ing.Tech.*, 61(10), 836–837, 1989.
37. Li, M., Feng, C., Zhang, Z., Yang, S., Sugiura, N., Treatment of nitrate contaminated water using an electrochemical method. *Bioresour. Technol.*, 101(16), 6553–7, Aug. 2010.
38. Pérez, G., Ibáñez, R., Urtiaga, A.M., Ortiz, I., Kinetic study of the simultaneous electrochemical removal of aqueous nitrogen compounds using BDD electrodes. *Chem. Eng. J.*, 197, 475–482, 2012.
39. Dortsiou, M., Katsounaros, I., Polatides, C., Kyriacou, G., Electrochemical removal of nitrate from the spent regenerant solution of the ion exchange. *Desalination*, 248(1–3), 923–930, 2009.
40. Duca, M., Koper, M.T.M., Powering denitrification: the perspectives of electrocatalytic nitrate reduction. *Energy Environ. Sci.*, 9726–9742, 2012.
41. Ghanim, A.N., Ajjam, S.K., Kinetic modelling of nitrate removal from aqueous solution during electrocoagulation. *Civ. Environ. Res.*, 3(7), 64–73, 2013.
42. Figueiredo, M.C., Compounds on platinum surfaces Electrocatalytic reduction of nitrogen containing compounds on platinum surfaces."
43. Peel, J.W., Reddy, K.J., Sullivan, B.P., Bowen, J.M., Electrocatalytic reduction of nitrate in water. 37. 2512–2519, 2003.
44. Peel, J.W., Reddy, K.J., Sullivan, B.P., Bowen, J.M., Electrocatalytic reduction of nitrate in water. *Water Res.*, 37(3), 2512–2519, 2003.
45. Vorlop, K., Tacke, T., Sellb, M., Development of catalysts for a selective nitrate and nitrite removal from drinking water. *Catal. Today*, 17, 21–30, 1993.
46. Epron, F., Gauthard, F., Pinéda, C., Barbier, J., Catalytic reduction of nitrate and nitrite on Pt–Cu/Al2O3 catalysts in aqueous solution: role of the interaction between copper and platinum in the reaction. *J. Catal.*, 198(2), 309–318, Mar. 2001.
47. Prüsse, U., Vorlop, K., Supported bimetallic palladium catalysts for water-phase nitrate reduction. *J. Mol. Catal. A Chem.*, 173, 313–328, 2001.
48. Chaplin, B.P., Reinhard, M., Schneider, W.F., Schu, C., Shapley, J.R., Strathmann, T.J., Werth, C.J., Critical review of Pd-based catalytic treatment of priority contaminants in water. *Environ. Sci. Technol.*, 2012.

49. Garron, A., Epron, F., Use of formic acid as reducing agent for application in catalytic reduction of nitrate in water. *Water Res.*, 39(13), 3073–81, Aug. 2005.
50. Chen, Y.-X., Zhang, Y., Chen, G.-H., Appropriate conditions or maximizing catalytic reduction efficiency of nitrate into nitrogen gas in groundwater. *Water Res.*, 37(10), 2489–95, May 2003.
51. Geng, B., Zhu, Y., Jin, Z., Li, T., Kang, H., Wang, S., Studies on catalytic reduction of nitrate in groundwater. *Front. Environ. Sci. Eng. China*, 1(3), 357–361, 2007.
52. Pintar, A., Catalytic processes for the purification of drinking water and industrial effluents. in *Catalysis Today*, 2003, 77(4), 451–465.
53. Rashid, S., Shen, C., Chen, X., Li, S., Chen, Y., Wen, Y., Liu, J., Enhanced catalytic ability of chitosan–Cu–Fe bimetal complex for the removal of dyes in aqueous solution. *RSC Adv.*, 5, 90731–90741, 2015.
54. Sá, J., Vinek, H., Catalytic hydrogenation of nitrates in water over a bimetallic catalyst. *Appl. Catal. B Environ.*, 57(4), 247–256, May 2005.
55. Jin, R., Gao, W., Chen, J., Zeng, H., Zhang, F., Liu, Z., Guan, N., Photocatalytic reduction of nitrate ion drinking water by using metal-loaded MgTiO3-TiO2 composite semiconductor catalyst. *J. Photochem. Photobiol. A Chem.*, 162(2–3), 585–590, 2004.
56. Dodouche, I., Barbosa, D.P., Rangel, M.D.C., Epron, F., Palladium-tin catalysts on conducting polymers for nitrate removal. *Appl. Catal. B Environ.*, 93, 50–55, 2009.
57. Pintar, A., Batista, J., Levec, J., Integrated ion exchange / catalytic process for e $ cient removal of nitrates from drinking water. 56, 1551–1559, 2001.
58. Epron, F., Gauthard, F., Barbier, J., Influence of oxidizing and reducing treatments on the metal – metal interactions and on the activity for nitrate reduction of a Pt-Cu bimetallic catalyst. *Appl. Catal. A Gen.*, 237, 253–261, 2002.
59. Vorlop, K., Kinetic investigation of the catalytic nitrate reduction: construction of the test reactor system. 199–202, 1999.
60. Gašparovičová, D., Králik, M., Hronec, M., Vallušová, Z., Vinek, H., Corain, B., Supported Pd-Cu catalysts in the water phase reduction of nitrates: functional resin versus alumina. *J. Mol. Catal. A Chem.*, 264, 93–102, 2007.
61. Gao, W., Jin, R., Chen, J., Guan, X., Zeng, H., Zhang, F., Guan, N., Titania-supported bimetallic catalysts for photocatalytic reduction of nitrate. *Catal. Today*, 90, 331–336, 2004.
62. Soares, O.S.G.P., Órfão, J.J.M., Pereira, M.F.R., Bimetallic catalysts supported on activated carbon for the nitrate reduction in water: Optimization of catalysts composition. *Appl. Catal. B Environ.*, 91(1–2), 441–448, Sep. 2009.
63. Arya, V., Philip, L., Adsorption of pharmaceuticals in water using Fe3O4 coated polymer clay composite. *Microporous Mesoporous Mater.*, 232, 273–280, 2016.
64. Oztürk, N., Bektaş, T.E., Nitrate removal from aqueous solution by adsorption onto various materials. *J. Hazard. Mater.*, 112, 155–162, 2004.

65. Worch, E., *Adsorption technology in water treatment*. Walter de Gruyter GmbH & Co. KG, Berlin/Boston, 2012.
66. Malika, C., Kenza, A., Yasmine, A.O., Abdeltif, A., Aicha, B., Removal of nitrate from drinking water by adsorption using ion exchange resin. *Desalin. Water Treat.*, 24(February), 109–116, 2010.
67. Dabrowski, A., Adsorption–from theory to practice. *Adv. Colloid Interface Sci.*, 93, 135–224, 2001.
68. Rashed, M.N., Adsorption technique for the removal of organic pollutants from water and wastewater. *Org. Pollut. - Monit. Risk Treat.*, 167–194, 2013.
69. M. E. R. 2, Mirna Habuda-Stanić, A.F. 1,*, A review on adsorption of fluoride from aqueous solution. *Materials (Basel).*, Materials(7), 317–6366, 2014.
70. Xie, Y., Li, S., Wang, F., Liu, G., Removal of perchlorate from aqueous solution using protonated cross-linked chitosan. *Chem. Eng. J.*, 156, 56–63, 2010.
71. Demiral, H., Gündüzoğlu, G., Removal of nitrate from aqueous solutions by activated carbon prepared from sugar beet bagasse. *Bioresour. Technol.*, 101, 1675–1680, 2010.
72. Mishra, P.C., Patel, R.K., Use of agricultural waste for the removal of nitrate-nitrogen from aqueous medium. *J. Environ. Manage.*, 90(1), 519–522, 2009.
73. Afkhami, A., Madrakian, T., Karimi, Z., The effect of acid treatment of carbon cloth on the adsorption of nitrite and nitrate ions. *J. Hazard. Mater.*, 144, 427–431, 2007.
74. Le Cloirec, P., Adsorption onto activated carbon fiber cloth and electrothermal desorption of volatile organic compound (VOCs): a specific review. *Chinese J. Chem. Eng.*, 20(3), 461–468, 2012.
75. Bhatnagar, A., Ji, M., Choi, Y., Jung, W., Lee, S., Kim, S., Lee, G., Suk, H., Kim, H., Min, B., Kim, S., Jeon, B., Kang, J., Removal of nitrate from water by adsorption onto zinc chloride treated activated carbon. *Sep. Sci. Technol.*, 43(4), 886–907, 2008.
76. Ahmadzadeh Tofighy, M., Mohammadi, T., Nitrate removal from water using functionalized carbon nanotube sheets. *Chem. Eng. Res. Des.*, 90(11), 1815–1822, 2012.
77. Mizuta, K., Matsumoto, T., Hatate, Y., Nishihara, K., Nakanishi, T., Removal of nitrate-nitrogen from drinking water using bamboo powder charcoal. *Bioresour. Technol.*, 95, 255–257, 2004.
78. Saad, R., Hamoudi, S., Belkacemi, K., Adsorption of phosphate and nitrate anions on ammonium-functionnalized mesoporous silicas. *J. Porous Mater.*, 15, 315–323, 2008.
79. Xi, Y., Mallavarapu, M., Naidu, R., Preparation, characterization of surfactants modified clay minerals and nitrate adsorption. *Appl. Clay Sci.*, 48(1–2), 92–96, 2010.
80. Goh, K.H., Lim, T.T., Dong, Z., Application of layered double hydroxides for removal of oxyanions: a review. *Water Res.*, 42, 1343–1368, 2008.

81. Islam, M., Patel, R., Synthesis and physicochemical characterization of Zn/Al chloride layered double hydroxide and evaluation of its nitrate removal efficiency. *Desalination*, 256(1–3), 120–128, 2010.
82. Islam, M., Patel, R., Nitrate sorption by thermally activated Mg/Al chloride hydrotalcite-like compound. *J. Hazard. Mater.*, 169, 524–531, 2009.
83. Islam, M., Patel, R., Physicochemical characterization and adsorption behavior of Ca/Al chloride hydrotalcite-like compound towards removal of nitrate. *J. Hazard. Mater.*, 190(1–3), 659–668, 2011.
84. Hosni, K., Srasra, E., Nitrate adsorption from aqueous solution by MII-Al-CO3 layered double hydroxide. *Inorg. Mater.*, 44(7), 742–749, 2008.
85. Soías-Viciana, M.M., Ureña-Amate, M.D., González-Pradas, E., García-Cortéz, M.J., López-Teruel, C., Nitrate removal by calcined Hydrotalcite-type compounds. *Clays Clay Miner.*, 56(1), 2–9, 2008.
86. Onyango, M.S., Masukume, M., Ochieng, A., Otieno, F., Functionalised natural zeolite and its potential for treating drinking water ontaining excess amount of nitrate. *Water SA*, 36(5), 655–662, 2010.
87. Wang, S., Peng, Y., Natural zeolites as effective adsorbents in water and wastewater treatment. *Chem. Eng. J.*, 156, 11–24, 2010.
88. Kanawade, S.M., Use of natural zeolite for treating drinking water containing excess amount of nitrate. *Int. J. Multidiscip. Res. Dev.*(April), 336–345, 2015.
89. Schick, J., Caullet, P., Paillaud, J.L., Patarin, J., Mangold-Callarec, C., Batch-wise nitrate removal from water on a surfactant-modified zeolite. *Microporous Mesoporous Mater.*, 132(3), 395–400, 2010.
90. Arora, M., Eddy, N.K., Mumford, K.A., Baba, Y., Perera, J.M., Stevens, G.W., Surface modification of natural zeolite by chitosan and its use for nitrate removal in cold regions. *Cold Reg. Sci. Technol.*, 62(2–3), 92–97, 2010.
91. Zhan, Y., Lin, J., Zhu, Z., Removal of nitrate from aqueous solution using cetylpyridinium bromide (CPB) modified zeolite as adsorbent. *J. Hazard. Mater.*, 186(2–3), 1972–1978, 2011.
92. Schwantes, D., Gonçalves Jr., A.C., Schons, D.C., Veiga, T.G., Diel, R.C., Schwantes, V., Nitrate adsorption using sugar cane bagasse physicochemically changed. *J. Agric. Environ. Sci.*, 4(1), 51–59, 2015.
93. Kalaruban, M., Loganathan, P., Shim, W.G., Kandasamy, J., Ngo, H.H., Vigneswaran, S., Enhanced removal of nitrate from water using amine-grafted agricultural wastes. *Sci. Total Environ.*, 565, 503–510, 2016.
94. Orlando, U.S., Baes, A.U., Nishijima, W., Okada, M., A new procedure to produce lignocellulosic anion exchangers from agricultural waste materials. *Bioresour. Technol.*, 83, 195–198, 2002.
95. Orlando, U.S., Baes, A. U., Nishijima, W., Okada, M., Preparation of agricultural residue anion exchangers and its nitrate maximum adsorption capacity. *Chemosphere*, 48, 1041–1046, 2002.
96. Wang, Y., Gao, B.Y., Yue, W.W., Yue, Q.Y., Adsorption kinetics of nitrate from aqueous solutions onto modified wheat residue. *Colloids Surfaces A Physicochem. Eng. Asp.*, 308, 1–5, 2007.

97. Guo, C., Zhou, L., Lv, J., Effects of expandable graphite and modified ammonium polyphosphate on the flame-retardant and mechanical properties of wood flour-polypropylene composites. *Polym. Polym. Compos.*, 21(7), 449–456, 2013.
98. Michalak, I., Chojnacka, K., Witek-Krowiak, A., State of the art for the biosorption process – a review. *Appl. Biochem. Biotechnol.*, 170, 1389–1416, 2013.
99. Perju, M.M., Dragan, E.S., Removal of azo dyes from aqueous solutions using chitosan based composite hydrogels. *Ion Exch. Lett.*, 3(January 2010), 7–11, 2010.
100. Gadd, G.M., Biosorption: Critical review of scientific rationale, environmental importance and significance for pollution treatment. *J. Chem. Technol. Biotechnol.*, 84(1), 13–28, 2009.
101. Fomina, M., Gadd, G.M., Biosorption: current perspectives on concept, definition and application. *Bioresour. Technol.*, 160, 3–14, 2014.
102. Kuyucak, N., Volesky, B., Biosorbents for recovery of metals from industrial solutions. *Biotechnol. Lett.*, 10(October), 137–142, 1988.
103. Rinaudo, M., Chitin and chitosan: properties and applications. *Prog. Polym. Sci.*, 31, 603–632, 2006.
104. Rocha, L.S., Almeida, A., Nunes, C., Henriques, B., Coimbra, M.A., Lopes, C.B., Silva, C.M., Duarte, A.C., Pereira, E., Simple and effective chitosan based films for the removal of Hg from waters: Equilibrium, kinetic and ionic competition. *Chem. Eng. J.*, 300, 217–229, 2016.
105. Ravi Kumar, M.N., A review of chitin and chitosan applications. *React. Funct. Polym.*, 46, 1–27, 2000.
106. Wan Ngah, W.S., Teong, L.C., Hanafiah, M.A.K.M., Adsorption of dyes and heavy metal ions by chitosan composites: A review. *Carbohydr. Polym.*, 83(4), 1446–1456, 2011.
107. Bhatnagar, A., Sillanpää, M., Applications of chitin – and chitosan-derivatives for the detoxification of water and wastewater – A short review. *Adv. Colloid Interface Sci.*, 152(1–2), 26–38, 2009.
108. Kyzas, G., Bikiaris, D., Recent modifications of chitosan for adsorption applications: a critical and systematic review. *Mar. Drugs*, 13, 312–337, 2015.
109. Rangel-mendez, J.R., Escobar-barrios, V.A., Davila-rodriguez, J.L., Potosino, I., Cientifica, D.I., Chitin based biocomposites for removal of contaminants from water : a case study of fluoride adsorption. *Biopolymers* (October), 8–180, 2010.
110. Liu, X., Zhang, L., Removal of phosphate anions using the modified chitosan beads: Adsorption kinetic, isotherm and mechanism studies. *Powder Technol.*, 277, 112–119, 2015.
111. Wang, J., Chen, C., Chitosan-based biosorbents: Modification and application for biosorption of heavy metals and radionuclides. *Bioresour. Technol.*, 160, 129–141, 2014.

112. Loganathan, P., Vigneswaran, S., Kandasamy, J., Naidu, R., Defluoridation of drinking water using adsorption processes. *J. Hazard. Mater.*, 248–249, 1–19, 2013.
113. Viswanathan, N., Meenakshi, S., Enriched fluoride sorption using alumina/chitosan composite. *J. Hazard. Mater.*, 178 (1–3), 226–232, 2010.
114. Samatya, S., Yüksel, Ü., Yüksel, M., Kabay, N., Removal of fluoride from water by metal ions (Al^{3+}, La^{3+} and ZrO^{2+}) loaded natural zeolite. *Sep. Sci. Technol.*, 42 (October 2015), 2033–2047, 2007.
115. Chatterjee, S., Woo, S.H., The removal of nitrate from aqueous solutions by chitosan hydrogel beads. *J. Hazard. Mater.*, 164, 1012–1018, 2009.
116. Chatterjee, S., Lee, D.S., Lee, M.W., Woo, S.H., Nitrate removal from aqueous solutions by cross-linked chitosan beads conditioned with sodium bisulfate. *J. Hazard. Mater.*, 166, 508–513, 2009.
117. Sowmya, A., Meenakshi, S., An efficient and regenerable quaternary amine modified chitosan beads for the removal of nitrate and phosphate anions. *J. Environ. Chem. Eng.*, 1(4), 906–915, 2013.
118. Sowmya, A., Meenakshi, S., Effective removal of nitrate and phosphate anions from aqueous solutions using functionalised chitosan beads. *Desalin. Water Treat.*, 52(13–15), 2583–2593, 2014.
119. Jiang, H., Chen, P., Luo, S., Tu, X., Cao, Q., Shu, M., Synthesis of novel nano-composite Fe3O4/ZrO2/chitosan and its application for removal of nitrate and phosphate. *Appl. Surf. Sci.*, 284, 942–949, 2013.
120. Hu, Q., Chen, N., Feng, C., Hu, W., Nitrate adsorption from aqueous solution using granular chitosan-Fe^{3+} complex. *Appl. Surf. Sci.*, 347, 1–9, 2015.
121. Rajeswari, A., Amalraj, A., Pius, A., Adsorption studies for the removal of nitrate using chitosan/PEG and chitosan/PVA polymer composites. *J. Water Process Eng.*, 9, 123–134, 2016.
122. Hu, Q., Chen, N., Feng, C., Hu, W., Zhang, J., Liu, H., Nitrate removal from aqueous solution using granular chitosan-Fe(III)–Al(III) complex: kinetic, isotherm and regeneration studies. *J. Taiwan Inst. Chem. Eng.*, 0, 1–10, 2016.
123. Teimouri, A., Nasab, S.G., Vahdatpoor, N., Habibollahi, S., Salavati, H., Chermahini, A.N., Chitosan /Zeolite Y/Nano ZrO2 nanocomposite as an adsorbent for the removal of nitrate from the aqueous solution. *Int. J. Biol. Macromol.*, 93, 254–266, 2016.
124. Golie, W.M., Upadhyayula, S., An investigation on biosorption of nitrate from water by chitosan based organic-inorganic hybrid biocomposites. *Int. J. Biol. Macromol.*, 97, 489–502, 2017.
125. Weber, W.J., Adsorption processes. *Pure Appl. Chem.*, 37, 375–392, 1974.
126. Thomas, W.J., Crittenden, B., *Adsorption technology and design*. Elsevier Science & Technology Books, 1998.
127. McKay, C.P., Elemental composition, solubility, and optical properties of Titan's organic haze. *Planet. Space Sci.*, 44(8), 741–747, 1996.

128. Golie, W.M., Upadhyayula, S., Continuous fixed-bed column study for the removal of nitrate from water using chitosan/alumina composite. *J. Water Process Eng.*, 12, 58–65, 2016.
129. Agrawal, P., Bajpai, A.K., Dynamic column adsorption studies of toxic Cr(VI) ions onto iron oxide loaded gelatin nanoparticles. *J. Dispers. Sci. Technol.*, 32(February), 1353–1362, 2011.
130. Tran, H.N., You, S.-J., Chao, H.-P., Thermodynamic parameters of cadmium adsorption onto orange peel calculated from various methods: A comparison study. *J. Environ. Chem. Eng.*, 4(3), 2671–2682, 2016.
131. Milonjić, S.K., A consideration of the correct calculation of thermodynamic parameters of adsorption. *J. Serbian Chem. Soc.*, 72(12), 1363–1367, 2007.
132. Giles, C.H., Smith, D., Huitson, A., A general treatment and classification of the solute adsorption isotherm. I. Theoretical. *J. Colloid Interface Sci.*, 47(3), 755–765, 1974.
133. Kulkarni, R.M., Srinikethan, G., V. S. K, Equilibrium and kinetic studies for the adsorption of cadmium ion on Zeolite 4A. *J Biochem Tech*, 3(2012), 158–160, 2013.
134. Ratnamala, G.M., Shetty, K.V., Srinikethan, G., Removal of Remazol Brilliant Blue dye from dye-contaminated water by adsorption using red mud : Equilibrium , kinetic , and thermodynamic studies. *Water Air Soil Pollut*, 223(2012), 6187–6199, 2012.
135. Das, G., Pradhan, N.C., Madhu, G.M., Preetham, H.S., Removal of cadmium from aqueous streams by zeolite synthesized from fly ash. *J. Mater. Environ. Sci.*, 4(3), 410–419, 2013.
136. Grassi, M., Kaykioglu, G., Belgiorno, V., Lofrano, G., Emerging compounds removal from wastewater. *Green Chem. Sustain.*, 15–38, 2012.
137. Dada, O., Olalekan, A.O., Olatunya, A.P., Dada, A.M., Langmuir , Freundlich, Temkin and Dubinin – Radushkevich isotherms studies of equilibrium sorption of Zn^{2+} unto phosphoric acid modified rice husk. *IOSR J. Appl. Chem.*, 3(1), 38–45, 2012.
138. Langmuir, I., The constitution and fundamental properties of solids and liquids. *J. Am. Chem. Soc.*, 38, 2221–95, 1916.
139. Viswanathan, N., Meenakshi, S., Synthesis of Zr(IV) entrapped chitosan polymeric matrix for selective fluoride sorption. *Colloids Surfaces B Biointerfaces*, 72(October), 88–93, 2009.
140. Freundlich, H.M.F., Adsorption in solution. *Z. Phys. Chem.*, 57, 385–470, 1906.
141. Subash, N., Krishna Prasad, R., Kinetics and mass transfer models for sorption of titanium industry effluent in activated carbon. *Desalin. Water Treat.*, 57(16), 7254–7261, 2016.
142. Mohan, S.V., Karthikeyan, J., Removal of lignin and tannin colour from aqueous solution by adsorption onto activated charcoal. *Environ. Pollut.*, 97(1), 183–187, 1997.

143. Chabani, M., Amrane, A., Bensmaili, A., Kinetics of nitrates adsorption on Amberlite IRA 400 resin. *Desalination*, 206, 560–567, 2009.
144. Dawodu, F.A., Akpomie, G.K., Ogbu, I.C., Isotherm modeling on the equilibrium sorption of cadmium (II) from solution by agbani clay. *Int. J. Multidiscip. Sci. Eng.*, 3(Ii), 9–14, 2012.
145. A. A. Z. Vassilis J. Inglezakisa, Heat of adsorption, adsorption energy and activation energy in adsorption and ion exchange systems. *Desalin. Water Treat.*, 39(November), 149–157, 2012.
146. Foo, K.Y., Hameed, B.H., Insights into the modeling of adsorption isotherm systems. *Chem. Eng. J.*, 156(1), 2–10, 2010.
147. Kumara, N.T.R.N., Hamdan, N., Petra, M.I., Tennakoon, K.U., Ekanayake, P., Kumara, N.T.R.N., Hamdan, N., Petra, M.I., Tennakoon, K.U., Ekanayake, P., Equilibrium isotherm studies of adsorption of pigments extracted from kuduk-kuduk (*Melastoma malabathricum* L.) pulp onto TiO_2 nanoparticles. *J. Chem.*, 2014, 1–6, 2014.
148. Allen, S.J., Mckay, G., Porter, J.F., Adsorption isotherm models for basic dye adsorption by peat in single and binary component systems. *J. Colloid Interface Sci.*, 280, 322–333, 2004.
149. P.S., Chowdhury, S., Insight Into Adsorption Thermodynamics. *J. Hazard. Mater.*, 162, p. 440, 2003.
150. Ho, Y.S., Ng, J.C.Y., McKay, G., Kinetics of pollutant sorption by biosorbents: review. *Sep. Purif. Rev.*, 29(2), 189–232, 2000.
151. Plazinski, W., Rudzinski, W., Plazinska, A., Theoretical models of sorption kinetics including a surface reaction mechanism: A review. *Adv. Colloid Interface Sci.*, 152(1–2), 2–13, 2009.
152. Qiu, H., Lv, L., Pan, B., Zhang, Q., Zhang, W., Zhang, Q., Critical review in adsorption kinetic models. *J. Zhejiang Univ. Sci. A*, 10(5), 716–724, 2009.
153. Ho, Y.S., McKay, G., Sorption of dye from aqueous solution by peat. *Chem. Eng. J.*, 70, 115–124, 1998.
154. Ho, Y.S., Mckay, G., A comparison of chemisorption kinetic models applied to pollutant removal on various sorbents. *Process Saf. Environ. Prot.*, 76(November), 332–340, 1998.
155. Suksabye, P., Thiravetyan, P., Nakbanpote, W., Column study of chromium(VI) adsorption from electroplating industry by coconut coir pith. *J. Hazard. Mater.*, 160(1), 56–62, 2008.
156. Z¨umriye Aksu, F.G., S¸ eyda S¸en C¸ a˘gatay, Continuous fixed bed biosorption of reactive dyes by dried Rhizopus arrhizus: Determination of column capacity. *J. Hazard. Mater.*, 143, 362–371, 2007.
157. Cruz-Olivares, J., P??rez-Alonso, C., Barrera-D??az, C., Ure??a-Nu??ez, F., Chaparro-Mercado, M.C., Bilyeu, B., Modeling of lead (II) biosorption by residue of allspice in a fixed-bed column. *Chem. Eng. J.*, 228, 21–27, 2013.
158. Dorado, A.D., Gamisans, X., Valderrama, C., Solé, M., Lao, C., Cr(III) removal from aqueous solutions: a straightforward model approaching of the adsorption in a fixed-bed column. *J. Environ. Sci. Health. A. Tox. Hazard. Subst. Environ. Eng.*, 49(March 2016), 179–86, 2014.

159. Maji, S.K., Pal, A., Pal, T., Adak, A., Modeling and fixed bed column adsorption of As(III) on laterite soil. *Sep. Purif. Technol.*, 56(3), 284–290, 2007.
160. Elangovan, R., Philip, L., Chandraraj, K., Biosorption of hexavalent and trivalent chromium by palm flower (Borassus aethiopum) . *Chem. Eng. J.*, 141(1–3), 99–111, 2008.
161. Chen, I., Kan, C., Futalan, C.M., M. J. C, Lin, S., Tsai, W.C., Wan, M., Batch and fixed bed studies : removal of copper (II) using chitosan-coated kaolinite beads from aqueous solution. *Sustain. Environ. Res.*, 25(2), 73–81, 2015.
162. Das Gupta, M., Loganathan, P., Vigneswaran, S., Adsorptive removal of nitrate and phosphate from water by a purolite ion exchange resin and hydrous ferric oxide columns in series. *Sep. Sci. Technol.*, 47(October), 1785–1792, 2012.
163. Yagub, M.T., Sen, T.K., Afroze, S., Ang, H.M., Fixed-bed dynamic column adsorption study of methylene blue (MB) onto pine cone. *Desalin. Water Treat.*, 55(September 2015), 1026–1039, 2015.
164. Adak, A., Bandyopadyay, M., Pal, A., Fixed bed column study for the removal of crystal violet (C. I. Basic Violet 3) dye from aquatic environment by surfactant-modified alumina. *Dye. Pigment.*, 69, 245–251, 2006.
165. Xu, Z., Cai, J.-G., Pan, B.-C., Mathematically modeling fixed-bed adsorption in aqueous systems. *J. Zhejiang Univ. A (Applied Phys. Eng.*, 14(3), 155–176, 2013.
166. Bohart, G.S., Adams, E.Q., Some aspects of the behavior of charcoal with respect to chlorine. *J. Franklin Inst.*, 189, p. 669, 1920.
167. Thomas, H.C., Heterogeneous ion exchange in a flowing system. *J. Am. Chem. Soc.*, 66(2), 1664–1666, 1944.
168. Mathialagan, T., Viraraghavan, T., Adsorption of cadmium from aqueous solutions by perlite. *Hazard. Mater.*, 94(August 2013), 291–303, 2002.
169. Yoon, Y.H., Nelson, J.H., Application of gas adsorption kinetics. I. A theoretical model for respirator cartridge service life. *Am. Ind. Hyg. Assoc. J.*, 45, 509–516, 1984.
170. Lin, S.H., Juang, R.S., Wang, Y.H., Adsorption of acid dye from water onto pristine and acid-activated clays in fixed beds. *J. Hazard. Mater.*, 113(1–3), 195–200, 2004.
171. Öztürk, N., Kavak, D., Adsorption of boron from aqueous solutions using fly ash: batch and column studies. *J. Hazard. Mater.*, 127(1–3), 81–88, 2005.
172. Clark, R.M., Evaluating the cost and performance of field-scale granular activated carbon systems. *Environ. Sci. Technol.*, 21(6), 573–580, 1987.
173. Aksu, Z., Gönen, F., Biosorption of phenol by immobilized activated sludge in a continuous packed bed: prediction of breakthrough curves. *Process Biochem.*, 39, 599–613, 2004.
174. Wolborska, A., Adsorption on activated carbon of p-nitrophenol from aqueous solution. *Water Res.*, 23(1), 85–91, 1989.
175. Kanadasan, G., Mashitah, M.D., Vadivelu, V.M., Fixed bed adsorption of methylene blue by using palm oil mill effluent waste activated sludge. *3rd IWA Asia Pacific Young Water Prof. Conf. 2010*, 1–8, 2010.

13

Nitrate Removal and Nitrogen Sequestration from Polluted Waters Using Zero-Valent Iron Nanoparticles Synthesized under Ultrasonic Irradiation

Mohammadreza Kamali[1], Maria Elisabete Costa[2*] and Isabel Capela[3]

[1]Department of Environment and Planning, Center for Environmental and Marine Studies, CESAM, Aveiro Institute of Materials, CICECO, University of Aveiro, Portugal
[2]Department of Materials and Ceramics Engineering, Aveiro Institute of Materials, CICECO, University of Aveiro, Aveiro, Portugal
[3]Department of Environment and Planning, Center for Environmental and Marine Studies, CESAM, University of Aveiro, Aveiro, Portugal

Abstract

The main purpose of this chapter is to address the conditions in which improvements on the surface area and surface reactivity of zero-valent iron nanosized particles (ZVI NPs) can be achieved. A synthesis method based on liquid-phase reduction (LPR) under selected conditions was adopted. Transmission electron microscopy (TEM), scanning electron microscopy (SEM), X-ray diffraction (XRD) and gas adsorption (using BET isotherm) confirmed the changes on the size of the ZVI NPs, when ultrasonic irradiation (UI) was used. The NPs prepared with and without UI were tested in a removal procedure of nitrate from contaminated waters. A L-8 Taguchi experimental design was utilized to investigate the efficiency of the synthesized NPs. Nitrate concentrations were determined by ion chromatography and considered as response data. The results indicated that the reactivity of the NPs was significantly higher for NPS prepared under UI. The investigation of the mechanisms involved in nitrate reduction demonstrated that ammonium ions were the dominant end-product. Furthermore ammonium ions could be precipitated as struvite ($NH_4MgPO_4{\cdot}6H_2O$), a valuable fertilizer, and also

**Corresponding author*: Elisabete.costa@ua.pt

Shahid-ul-Islam (ed.) Advanced Materials for Wastewater Treatment, (479–506)

a good candidate for the removal of environmental contaminants such as heavy metals.

Keywords: Zero valent iron nanoparticles, ultrasonic irradiation, nitrate decontamination, struvite

13.1 Introduction

Several pathways are responsible for nitrate contamination of water resources such as industrial effluents and agricultural diffuse sources [1]. The concentration of nitrate in polluted waters is generally high enough to constitute a major challenge for the environment (e.g., eutrophication of aquifers) and human health [2, 3] and also to produce direct toxic effects on aquatic organisms [4].

Within European Union, the recommended limit for nitrate concentration in drinking freshwaters to prevent the eutrophication is 25 mg/L [5]. So, adopting efficient and cost effective methods for the treatment of nitrate contaminated waters is of high importance.

So far, some physicochemical [6] and biological [7, 8] methods have been utilized for the removal of nitrate from the contaminated waters. Most of these methods have not been widely adopted on full-scale applications due to economic and technical difficulties [9]. Table 13.1 presents the results of some recent studies on the chemical and biological methods applied for the removal of nitrate from the polluted waters and wastewaters as well as the main mechanisms involved.

In this regard, the application of nanoscale zero-valent Iron (nZVI) particles for the treatment of various organic and inorganic pollutants has considerably increased in recent years thanks to the developments achieved in the field of the synthesis and characterization of nanosized materials [10, 11].

There are two main type of methods for the synthesis of nanomaterials. Top-down methods such as ball milling [20] can convert the bulk materials to nanosized building blocks. However, bottom-up techniques such as microemulsion [21], inert gas condensation [22], chemical vapor condensation [23], liquid flame spray [24], gas-phase [12, 13] and liquid-phase reduction [14, 15] are the dominant methods used for the synthesis of nanosized zero-valent iron particles (nZVI). This particular type of metal nanoparticle is gaining a growing interest for applications aimed to the decontamination of polluted waters as is the case of nitrate removal from contaminated effluents [29, 30]. It offers significant advantages as compared to other competing single nanoparticles (NPs) or NP-based

Table 13.1 Recent applications of different methods for the removal of nitrate from polluted waters.

Method	Nitrate initial concentration (mg/L)	Operational conditions				Removal efficiency	Mechanisms	Ref.
		Initial pH	Temperature (°C)	Time (min)	Material dosage			
Biochar[1]	50	4.64	Room	60	2 g/L	28.21 mg/g[2]	Adsorption	[12]
Al-Fe_2O[3]	25 (mg of N/L)	2	26 ± 1	60	40 g[4]	18.9%	Reduction	[13]
Al-Fe_2O[5]	25 (mg of N/L)	12	26 ± 1	60	40 g[6]	60.3%	Reduction	[13]
Bipolar Si/BDD[7]	350	5.5 ± 0.2	25 ± 3	120	*[8]	100%		[14]
Granular activated carbon[9]	200	6.5	–	2	–	10 mg/g[2]	Adsorption	[15]
Granular activated carbon						68% to 60%[10]	Adsorption	[15]
Electrocoagulation (Al electrode)	4.3	10.08	Room	30	*[11]	60.6%	Reduction	[16]
Electrocoagulation (Fe electrode)	4.3	10.08	Room	30		69.7%	Reduction	[16]
A520E ion exchange resin	20 (mg of N/L)	–	24 ± 1	120	3.0 g/L	84%	Adsorption	[17]
Acidogenic liquid[12]	25 (mg of N/L)		25	60	–	~70%	Reduction	[18]
Vegetated floating beds	150	–	23/19[13]	7 (days)	–	~100%	Reduction	[19]

[1] Developed from sugarcane bagasse.
[2] Nitrate/ adsorbent concentration.
[3] With an iron content of 20%.
[4] 0.8–2 mm particles in a column packed with Al-Fe alloys.
[5] With an iron content of 20%.
[6] 0.8–2 mm particles in a column packed with Al-Fe alloys.
[7] Bipolar Boron-Doped Diamond on silicone substrate.
[8] BDD was used as cathode and anode material (with an active surface of 70 cm^2).
[9] In a continuous fixed bed pilot reactor.
[10] From nitrate polluted groundwaters.
[11] BDD was used as cathode and anode material (with an active surface of 70 cm^2).
[12] Fabricated using simulated food waste composed of 35%, 45%, 16%, and 4% of rice, cabbage, pork, and tofu, respectively, by weight.
[13] Day and night, respectively.

composites including zero-valent copper NPs [31], zinc hydroxide NPs [32], nano-copper-loaded titania [33], zero-valent iron supported on treated activated carbon [34], titanium electrode [35], Chitosan/Zeolite Y/Nano ZrO_2 nanocomposite [36], SiO_2-FeOOH-Fe nanostructures [37], Fe^0/Pd/Cu nanocomposite [38] and alumina supported nano zero-valent zinc [39]. Low toxic effects [40], high efficiency [41] and lower cost are the attributes in favor of nZVI as decontamination agents.

However, there is a need of cost effective methods for the synthesis of such nanomaterials with enhanced properties. Sonochemistry as a simple, fast and environmentally friendly method [42] is able to promote the synthesis of nanostructures mainly because it favors the formation of a large number of nuclei in the precipitating medium. Sonochemistry is mediated through a hot spot mechanism, where the powerful ultrasound irradiation (UI) (20 KHz – 10 MHz) causes cavitation which involves the formation, growth, and instantaneous implosive collapse of bubbles in a liquid [43]. The potential of UI will be here exploited in terms of its ability to enhance the specific surface area of the synthesized nanoscale materials, aiming at enhanced surface activity. As reported in literature, very often nanosized particles have incipient crystallinity. From the point of view of pratical applications implying the use and reuse of nanoparticles in polluted solutions, robust nanoparticles with dissolution endurance are highly desired. In this regard particles crystallinity becomes an issue of interest. In order to improve the amorphous or incipient crystalline structure of NPs, high temperature annealings are often used which normally favor the growth or strong agglomeration of the NPs, hence resulting in a substantial loss of surface area. So, investigating a reliable and cost effective methodology that might ensure both characteristics to the synthesized particles, that is, high crystallinity and large specific surface area, will be of primordial importance.

Taguchi method, also called robust design method, is a statistical method developed by Genichi Taguchi who formulated a special set of design guidelines for factorial experiments with a large potential of applications [44]. This method was firstly developed in order to improve the quality of manufactured products. More recently, it has been also successfully applied in some scientific areas such as materials science and engineering [45] and biology [46, 47]. Taguchi's design of experiments allows to identify the main parameters affecting a certain production process and their interplay with a minimum number of experiments [48]. The orthogonal array (OA) is the basis for the arrangement of the experimental variables. OA allows to design the experiment with a minimum number of test runs while keeping the accuracy of the experiment [49].

In the context of a nitrate removal process, the end-products are also of high importance in terms of environmental and economic considerations. In the present study, in order to recover the nitrogen from nitrate polluted waters, nitrate removal mechanisms were studied when nanoscale zero-valent Iron prepared under ultrasonic irradiation ($nZVI^{UI}$) particles were utilized. The results suggested an eco-friendly process for the recovery of ammonium magnesium phosphate hexahydrate (AMP), struvite ($NH_4MgPO_4{\cdot}6H_2O$), as a valuable product which can be used as a fertilizer containing nitrogen (N) and phosphorous (P) [50].

13.2 Materials and Methods

13.2.1 Experimental

13.2.1.1 Reagents

Ferric chloride ($FeCl_3.6H_2O$) (99%), sodium borohydride ($NaBH_4$) (≥98.0%) were purchased from Sigma-Aldrich and used for the NZVI synthesis. Potassium nitrate (KNO_3) (99.0%) was also purchased from Riedel-de Haën and used to prepare a nitrate stock solution. Magnesium chloride hexahydrate ($MgCl_2.6H_2O$) (>99.0%) and sodium phosphate dibasic (Na_2HPO_4) (99.95%) were also purchased from Sigma-Aldrich for the synthesis of struvite. Hydrochloric acid (HCl, 37%) and sodium hydroxide (NaOH) (≥98.0%) were purchased from JMGS, Portugal and used for pH control. All the solutions were prepared with deionized water.

13.2.1.2 Synthesis Protocol

Liquid-phase reduction is one of the most popular methods for the synthesis of nanoscale zero-valent Iron particles, being highly cited in the recent decades [51–61]. The synthesis of the zero-valent iron nanomaterials ($nZVI^{LPR}$ and $nZVI^{UI}$) was carried out in a 500 mL three-open neck flask reactor. The central neck was housed with a mechanical overhead stirrer (IKA Eurostar 40), operating at 250 rpm. 0.1 M solution of sodium borohydride was prepared by dissolving an appropriate amount of $NaBH_4$ in 100 ml of deionized water. This solution was added to an ethanolic solution of $FeCl_3$, using a Watson Marlow (120U/R) peristaltic pump which ensured a constant addition rate, at ambient temperature. After delivering the whole borohydride solution, the resulting mixture was stirred in the reactor for more 10 minutes. The synthesis involving ultrasonic irradiation was carried out in a 100 mL glass beaker, sonicated under an ultrasound probe (Hielscher, UP200s W, 24 kHz). Sonication amplitude was adjusted to 40% providing a theoretical dissipation rate of 80 W. The resulting precipitated particles were washed with absolute ethanol to remove water and impurities. Magnet collection was used for separating the precipitates from the liquid solution. Then the collected particles were dried overnight at 50 °C, in air atmosphere.

For the crystallization of struvite, equimolar amounts of Na_2HPO_4 and $MgCl_2.6H_2O$ were added to a three-open neck flask reactor containing NH_4^+ (as a result of nitrate reduction mechanism). Various pHs were adopted as initial pH to investigate the effects of initial pH on struvite crystallization. The solution was kept under magnetic stirring for 1 h at 250 rpm. The resulting white powder was collected using vacuum filtration, then washed with deionized water several times and finally dried at 50 °C for 12 h.

13.2.2 Characterization

X-ray diffraction (XRD, Rigaku, Geigerflex, Japan) analysis was performed using a scan rate of 2° 2θ/min from 30° to 110° 2θ, to access the crystalline nature of the precipitated products. The crystallite size and lattice strain of the particles was determined through Scherrer equation (Eqn. (13.1)):

$$d = K\lambda / \beta\cos\theta \tag{13.1}$$

where d is the average crystallite size, K is a dimensionless shape factor, λ is the wavelength of the X-ray used, β is the full-width at half maximum intensity (FWHM) of the sample, and θ is the Bragg angle. To access particle morphology scanning electron microscopy (SEM) was carried out with Hitachi S4100 coupled to energy dispersive X-ray spectroscopy (EDS). In the case of nanosized particles, morphology analysis was carried out by transmission electron microscopy (TEM, Hiatchi H9000na, Japan) being SAED images acquired as well. Nitrogen adsorption-desorption isotherms of the synthesized particles were measured using a Micromeritics equipment, Gemini V2 (USA). The specific surface areas of the particles were calculated based on BET equation.

The stability of the nanoparticles prepared under UI against dissolution at various pHs was also evaluated. 10 mg of the nanoparticles were dispersed in 100 mL of deionized water previously purged with Argon gas for 2 hrs to minimize the dissolved oxygen (DO). The initial pH of the suspension was adjusted to different values, from 1.5 to ~7. The suspension was sonicated for 10 min under an ultrasonic cleaning bath (Branson,

1510E-MT) with a theoretical dissipation rate of 70. The suspension was then shaked for 30 min using a platform shaker (Edmund Bühler, KL2) fixed at 200 rpm. After vacuum filtration of the suspension, the collected solids particles were dried in an oven at 100 °C for 4 h. The weight of the dried particles was used to determine the amount of the dissolved materials under the various pHs. Regarding the yields of the two synthesis methods (with and without application of UI), the dried products resulting from each process were waited and the results used for estimating the respective yields.

13.2.3 Taguchi Design and Reactivity Analysis

A Taguchi L8 statistical design (Table 13.2) including 4 reaction time levels (30 min, 60 min, 90 min, and 120 min) and 2 levels for NMs ($nZVI^{LPR}$ and $nZVI^{UI}$) with no initial pH control was utilized with the aim of comparing the reactivity of the materials prepared under different conditions regarding nitrate removal, with and without assistance of ultrasonic irradiation. L and 8 are referring to Latin square and the number of experiments, respectively [48]. Signal-to-noise (S/N) ratio (larger is better) was used to optimize the specific surface area. "Signal" and "noise" represent the desirable and undesirable values for the response data, respectively [48]. Eqn. (13.2) was employed to calculate the S/N ratio where Y_i is the response data of the Taguchi statistical design, and n is the number of experiments [62, 63].

Table 13.2 L8 orthogonal array of experiments.

Run	Reaction time (min)	Sample identification	Mean nitrate removal	S/N (dB) nitrate removal	Mean nitrite production	S/N (dB) nitrite production
1	30	$nZVI^{LPR}$	87.42	38.83	0.26	−11.64
2	30	$nZVI^{UI}$	92.25	39.30	0.24	−12.42
3	60	$nZVI^{LPR}$	88.30	38.92	0.34	-9.37
4	60	$nZVI^{UI}$	93.10	39.38	0.18	−14.61
5	90	$nZVI^{LPR}$	91.73	39.25	0.36	-8.91
6	90	$nZVI^{UI}$	94.44	39.50	0.18	−14.87
7	120	$nZVI^{LPR}$	93.55	39.42	0.33	-9.48
8	120	$nZVI^{UI}$	95.31	39.58	0.17	−15.56

$$\frac{S}{N}\left[dB\right] = -10\log\frac{1}{n}\left(\sum_{i=1}^{n}(Y_i)^2\right) \tag{13.2}$$

Batch tests for nitrate reduction were conducted in a three-necked flask. The reactor was filled with 450 mL of nitrate solution (200 mg of nitrate per liter of solution). Then, 50 mL of NZVI suspension containing 0.6 g of nanoparticles was added to the reactor and continuously stirred with a mechanical stirrer (IKA Eurostar 40) at 200 rpm. Before being added to the reactor containing nitrate, the nanoparticles suspension was sonicated in an ultrasonic cleaning bath (Branson, 1510E-MT) for 10 min. Aliquots of the suspension were withdrawn periodically from the reactor, filtered and then immediately analyzed. Nitrate, nitrite and ammonium ion concentrations were measured by ion chromatography using a Dionex, ICS-3000 ion chromatography system, USA. The concentrations of nitrite and ammonium ions are evaluated because these compounds are envisaged as possible products from nitrate removal reactions, depending on the operating mechanism. Hence, in order to investigate the products from the nitrate reduction process using nano zero-valent Iron, the NO_3^- solution after being treated for 12o min with nZVI prepared under UI was sampled and analyzed by Raman spectroscopy using a T64000 Jobin Yvon SPEX spectrometer, with an Ar laser ($\lambda = 532$ nm) as excitation font. The spectrum was obtained in a back-scattering geometry, between 0 and 4000 cm^{-1}. For struvite precipitation, the remaining solution was filtered using vacuum filtration, and then added under stirring with stoichiometric amounts of Na_2HPO_4 and $MgCl_2.6H_2O$ solutions. Appropriate amounts of 1% NaOH solution were added when pH adjustments were required. The solution was stirred for 30 min and then the white precipitate readily formed was collected using vacuum filtration. The resulting dried powder was characterized as stated in 13.3.2.2.

13.3 Results and Discussion

13.3.1 Characterization

Figure 13.1 presents the XRD pattern of both type of particles synthesized in the present work, that is, particles prepared under liquid-phase reduction ($nZVI^{LPR}$) or under the combination of liquid-phase reduction and ultrasonic irradiation (LPR^{UI}). As observed peaks of different intensities are detected at various 2θ positions, which can be assigned to zero-valent

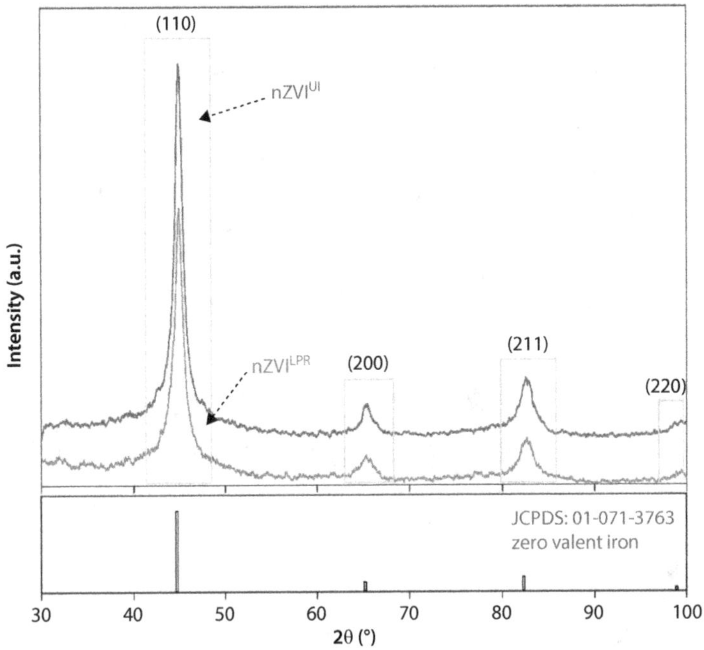

Figure 13.1 XRD patterns of the samples prepared under liquid-phase reduction (nZVI^{LPR}) and under a combination of liquid-phase reduction-ultrasonic irradiation (nZVI^{UI}).

iron according to JCPDS No. 07-71-3763 [64]. The crystallite size calculated for both type of particles were quite similar, that is, 13 and 12 nm for nZVI^{LPR} and LPR^{UI} respectively, thus indicating that the application of UI did not penalize the crystallinity of nZVI particles. This may reflect the benefits of local temperature increase of the reaction media induced by ultrasonic irradiation [11].

Figure 13.2 (a–d) presents the TEM images of the synthesized zero-valent iron particles (with and without UI). Both types of particles present a spherical shape with a nanometric size, and appear aligned as chains which are randomly arranged as aggregates (Figure 13.2(a) and (b)). Such shapes for nZVI have been also reported by previous studies [26, 64–66]. It could be also noticed that the average diameter of the particles obtained under UI is smaller than that of the particles not submitted to UI (Figure 13.2 (c) and (d)). Furthermore the SAED patterns of the two type of particles samples presented as insets in the Figure 13.2 ((c) and (d)) do not exhibit significant differences, thereby indicating that the crystallinity of precipitated particles is not affected by UI. This reasoning is in line with the crystallite size results previously commented.

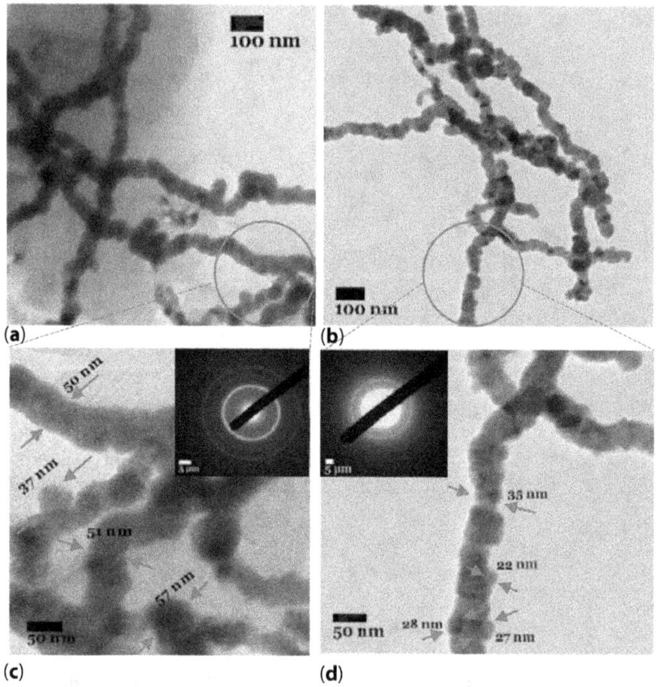

Figure 13.2 TEM images and SAED patterns of the prepared samples without (a and c) and with (b and d) ultrasonic irradiation.

The values of the specific surface area of the particles $nZVI^{LPR}$ and $nZVI^{UI}$ are 21 and 26 m^2/g, respectively. These results indicate that nZVI particles prepared under UI have a lower size than the particles prepared without UI. Also it was observed that the yield of $nZVI^{UI}$ process is higher (17%) than that of $nZVI^{LPR}$. Altogether these results suggest that UI favored the formation of a larger number of nuclei which growth could not proceed to the same extent as that reached by the less nuclei formed under normal precipitation conditions (LPR) [43, 67, 68]. The higher nucleation rate is probably induced by the high-speed jets or shock waves produced by collapsing bubbles that result from ultrasonic cavitation in the reaction media [11, 42]. Regarding the positive impact of UI on the precipitation yield, it was also reported by Barbosa *et al.* (2013) [69] when producing nano-hydroxyapatite by a wet chemical method. Considering that UI enhances the molecular movement in the liquid medium, the likelihood of species collision in the reaction medium increases as well under UI, hence favoring the mass transport necessary for particles growth and ultimately the precipitation yield achieved during the precipitation experiment, that is, 17% higher yield for the synthesis of the particles under ultrasonic irradiation in this study [67, 68].

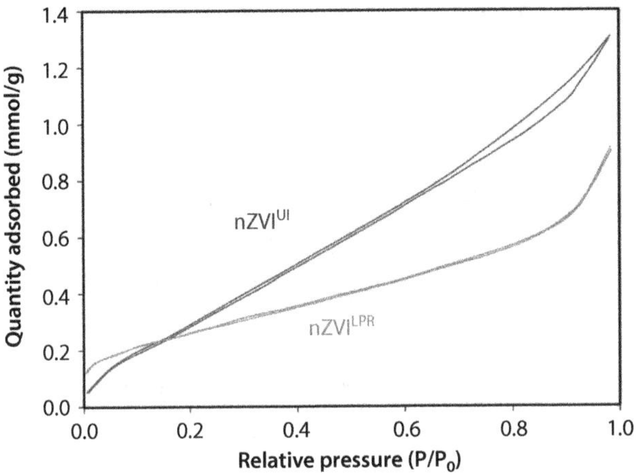

Figure 13.3 Nitrogen Adsorption-desorption isotherms of the $nZV^{I\ and}\ nZVI^{LPR}$ samples.

Figure 13.3 shows the adsorption-desorption isotherms of the prepared samples. As observed the particles precipitated under UI present a hysteretic adsorption-desorption isotherm of type IV (according to IUPAC classification) which indicates these particles as being mesoporous. The hysteresis free isotherm of the particles prepared in absence of UI belongs to the type II, which is characteristic of non-porous materials. Also the amount of adsorbed nitrogen is larger in the case of the mesoporous nanoparticles, reflecting their larger specific surface area calculated based on BET equation. The role of UI as porosity enhancer was already reported and attributed to the ability of ultrasonic waves for inhibiting particle pore blocking due to micro-jet, shock-wave and micro-streaming phenomena occurring during UI [70].

13.3.2 Reactivity of nZVI

13.3.2.1 Statistical Analysis

Tables 13.2 and 13.3 present the values of the responses, that is, of the particles reactivity, for each individual set of experiments and the overall results, respectively, expressed in terms of nitrate removal percentage, and of the calculated S/N ratios corresponding to the designed Taguchi L8 orthogonal array. The optimal level of each factor is that corresponding to the highest S/N ratio [62]. Among three existing S/N ratios (smaller is better, higher is better and nominal is better), the reactivity of the particles was analyzed in terms of larger is better S/N quality characteristic. As shown in Table 13.2, the optimal values of reactivity were exhibited by the

Table 13.3 Signal to noise ratio, factor effect and ranking for each factor.

Factor/ level	Reaction time (min)	S/N (dB) nitrate removal	S/N (dB) nitrite production	NMs	S/N (dB) nitrate removal	S/N (dB) nitrite production
	30	39.07	−12.03	$nZVI^{LPR}$	39.11	−9.85
	60	39.15	−11.99	$nZVI^{UI}$	39.44	−14.37
	90	39.38	−11.89			
	120	39.15	−12.52			
E_f		4.59	0.63		3.53	4.52
Rank		1	2		2	1

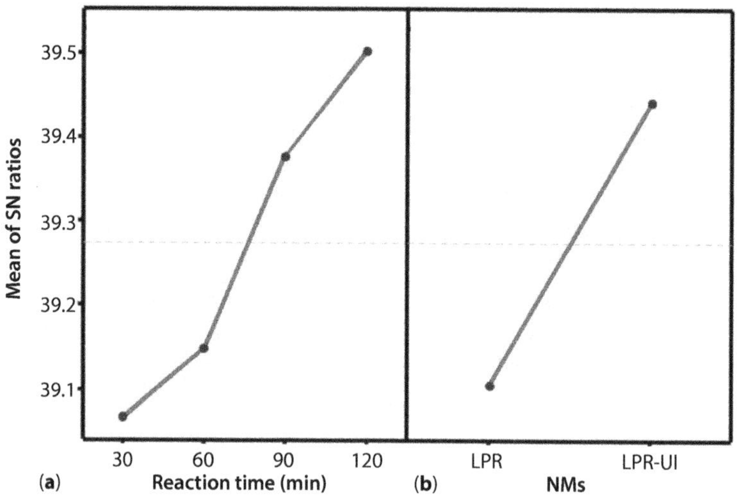

Figure 13.4 S/N ratios of the reaction time effect on the nitrate removal (a) and the S/N ratios belonging to different types of nanomaterials used including $nZVI^{LPR}$ and $nZVI^{UI}$ (b). The circles represent the optimal levels of the factors.

nanoparticles that were obtained under UI [70]. The variation of optimal values with the reaction time is depicted in Figure 13.4. In this figure, the observed nitrate removal percentages corresponding to the experimental variables including reaction time (in 4 levels) and the type of the nanomaterials (in 2 levels) and the S/N ratios belonging to different sets of the experiments are presented. Also, the concentration of the nitrite produced

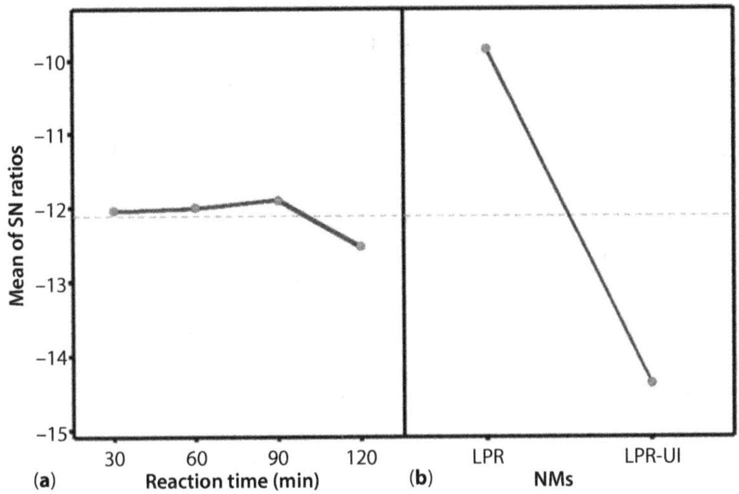

Figure 13.5 S/N ratios of effects of the reaction time (min) (a) and the types of nanomaterials (including $nZVI^{LPR}$ and $nZVI^{UI}$) (b) on the nitrite production as a possible end-product of the nitrate removal by nano zero-valent iron particles.

as a result of nitrate removal process and the effects of the studied experimental variables including reaction time and the type of the nanomaterials were investigated as presented in Table 13.2 and Figure 13.5. It should be stated that in the actual conditions of the experiments in which the pH was not controlled, the concentration of the nitrite produced in the reaction media was always less than 1% of the nitrate removed. In Table 13.3, the highest effect factor (E_f) is a parameter defined as the difference between the maximum and the minimum S/N ratio. E_f is a statistically important output that qualifies the relative importance of the factors. The values of E_f are presented in Table 13.3.

According to the results of the Taguchi design, there are statistically differences between the different levels of the reaction time and also between the types of the nanomaterials. Both reaction time and the type of nanomaterials have significant effects on the nitrate removal efficiency (E_f = 4.59 and 3.53 for the reaction time and the type of the NMs, respectively), as response data. The output of the Taguchi experimental design thus confirm that the nanomaterials prepared under ultrasonic irradiation present a better performance toward nitrate removal as compared to the nanomaterials prepared under a conventional liquid-phase reduction process.

From Figure 13.4 and Table 13.2 it is evident that the application of ultrasonic irradiation during nZVI synthesis could improve the nitrate removal from the contaminated waters. The better performance of ($nZVI^{UI}$) is attributed to their larger area for the involved reactions, whereas a smaller area will be available in the case of $nZVI^{LPR}$ particles, hence explaining the observed differences in NO_3^- removal.

Hwang et al [71] achieved a nitrate removal of ~50% using nZVI particles having a specific surface area of 45.4 m²/g after 3 hr of reaction. In the present study, using the same weight ratio of nZVI particles to nitrate as the authors did (6.7), a higher nitrate removal, that is, 92% could be already obtained just after 30 min. In addition to the differences in the synthesis conditions, which affect the crystal structure and morphology of the nanomaterials, the pre-sonication applied to the nanomaterials before the nitrate removal test may also be a factor contributing for the higher removal efficiency reached in this study. The effects of such pre-treatments have been also referred by Wang et al [64] in a process of bromate removal from the solution using nZVI particles.

13.3.2.2 Nitrate Removal Reaction: Mechanisms and Pathways

(1) Nitrate Removal Mechanisms under Uncontrolled pH

Following Yang & Lee (2005) [72], the main end-products of the reduction process of nitrate by zero-valent Iron are suggested to be nitrogen gas and ammonium ion according to the following reaction pathways:

$$NO_3^- + 4Fe^0 + 10H^+ \longrightarrow 4Fe^{2+} + NH_4^+ + 3H_2O \qquad (13.3)$$

$$2NO_3^- + 5Fe^0 + 12H^+ \longrightarrow 5Fe^{2+} + N_{2(g)} + 6H_2O \qquad (13.4)$$

$$10Fe^0 + 6NO_3^- + 3H_2O \longrightarrow 5Fe_2O_3 + 6OH^- + 3N_{2(g)} \qquad (13.5)$$

In addition to the above mentioned pathways, the eqns. (13.6) and (13.7) may account for the production of nitrite in the reaction solution [73, 74].

$$Fe^0 + NO_3^- + 2H^+ \longrightarrow Fe^{2+} + NO_2^- + H_2O \qquad (13.6)$$

$$Fe^0 + NO_3^- + 2H_3O^+ \longrightarrow Fe^{2+} + NO_2^- + 3H_2O \qquad (13.7)$$

For accessing the effective mechanisms involved in the removal of nitrate using $nZVI^{UI}$, Raman spectroscopy of the nitrate solution (in the actual condition of uncontrolled pH) after 2 hrs of reaction was performed at room temperature, being the results shown in Figure 13.6.

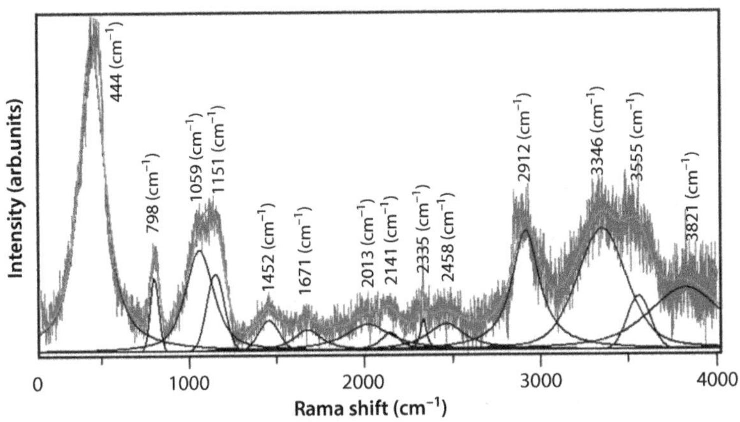

Figure 13.6 Raman Spectrum of the nitrate solution after 120 min of treatment with nZVI ^{UI}.

The sharp peak around 400–450 cm^{-1} can be attributed to hematite (α-Fe_2O_3) [72, 75]. Other iron compounds may also form in the reaction media. According to Cowan (1997) [76] the vibration at 798 cm^{-1} is attributed to Fe-$^{18}O_2$. Also, a peak appearing around 1160 cm^{-1} is assigned to δ-FeOOH [77]. It may be thus concluded that reduction processes of the metallic iron to different iron oxides is taking place.

As the amount of nitrite formed in the reaction medium was very low (corresponding to ~1% of nitrate removed as measured by ion chromatography), it can be suggested that the conversion of nitrate to nitrite is not the dominant mechanism accounting for the removal of nitrate from solution, although a chemical reduction mechanism is confirmed. Hence, it can be anticipated that other mechanisms which lead to the release of N_2 or to ammonia as chemical reduction products are also taking place [72]. This can be confirmed by the Raman spectrum (Figure 13.6) where the peaks at 3332 and 1452 cm^{-1} are assigned to NH_4^+ stretch and NH_4^+ deformation, respectively [78]. A medium intensity band can also observed around 1450 cm^{-1} [79, 80]. In addition, peaks associated to the presence of nitrite in the solution are also identified in the Raman spectrum. In this regard, the peak at 1056 cm^{-1} corresponds to NO_2^- [81]. The formation of N_2 as an end-product of the nitrate reduction by nano zero-valent Iron is also confirmed by the presence of the relevant Raman shift around 2325 [82] and 2331 cm^{-1} [83, 84]. There are also Raman bands around 3555, 3385 and 3821 cm^{-1} which can be assigned to water stretching vibrations [85, 86]. In addition, a peak at 2912.5 cm^{-1} is attributed to the HOH stretching vibration of water hydrogen bonded to the anions in the solution [87, 88]. According to these results reported in this section, it can be thus concluded that the various nitrate removal mechanisms underlying the eqns. (13.3)–(13.7) are at play. However nitrite is not a dominant product; hence eqns. (13.6) and (13.7) are contributing less for the overall nitrate conversion into new products.

(2) Nitrate Removal Mechanisms under Initial Acidic pH

Recent studies have reported that when the pH decreases from 7 to 3, the removal of nitrate and generation of NH_4^+ in the reaction medium increases [89]. Batch experiments under the same conditions used for the reactivity tests (3.2) were performed except the initial pH which was adjusted to an acidic pH value (pH = 4) at which the iron nanoparticles were experimentally confirmed to not dissolve. The concentrations of nitrite, nitrate and ammonium ions were measured periodically, being the results presented in Figure 13.7 ((a) and (b)).

As observed in Figure 13.7(a), the removal of nitrate, increases rapidly with time, especially during the first 5 minutes of the reaction when the pH (4) increases to 6.8. In fact, ~54% of nitrate is removed during this period. After that nitrate removal still keeps increasing but at a much lower rate, reaching a plateau of ~95.5% after 60 minutes. The pH variation is now relatively weak, changing from 6.8 to the final value of 7.2 that is reached after 30 min of reaction. In Figure 13.7(b) the curves describing the concentration of nitrite and ammonium ions produced during the removal process of nitrate are depicted. As observed the concentration of NO_2 – is slightly higher (it corresponds to ~4% of nitrate conversion) than that measured under uncontrolled pH condition (<1% of nitrate conversion). It can be anticipated according to Eqns. 13.6 and 13.7 that under sufficient amount of H^+ ions the formation of nitrite can take place. However, nitrite can also be converted to NH_4^+ according to the following equation [90]:

$$NO_2^- + 3Fe^0 + 8H^+ \longrightarrow 3Fe^{2+} + NH_4^+ + 2H_2O \qquad (13.8)$$

This may be the main cause of the slight reduction of NO_2^- concentration between 10 min to 20 min of the reaction. However, nitrogen mass balance shows that the production of ammonium ion is the predominant mechanism of the nitrate removal by the synthesized nano zero-valent iron under initial acidic pHs. As shown in the Figure 13.7(b), the concentration of ammonium ions has a high increase rate during the initial period, following a variation trend similar to that exhibited by the nitrate removal and accounting for about 87% of the reaction products while nitrite is ~4% of the nitrate removal products. This also suggests that N_2 could also account for the balance.

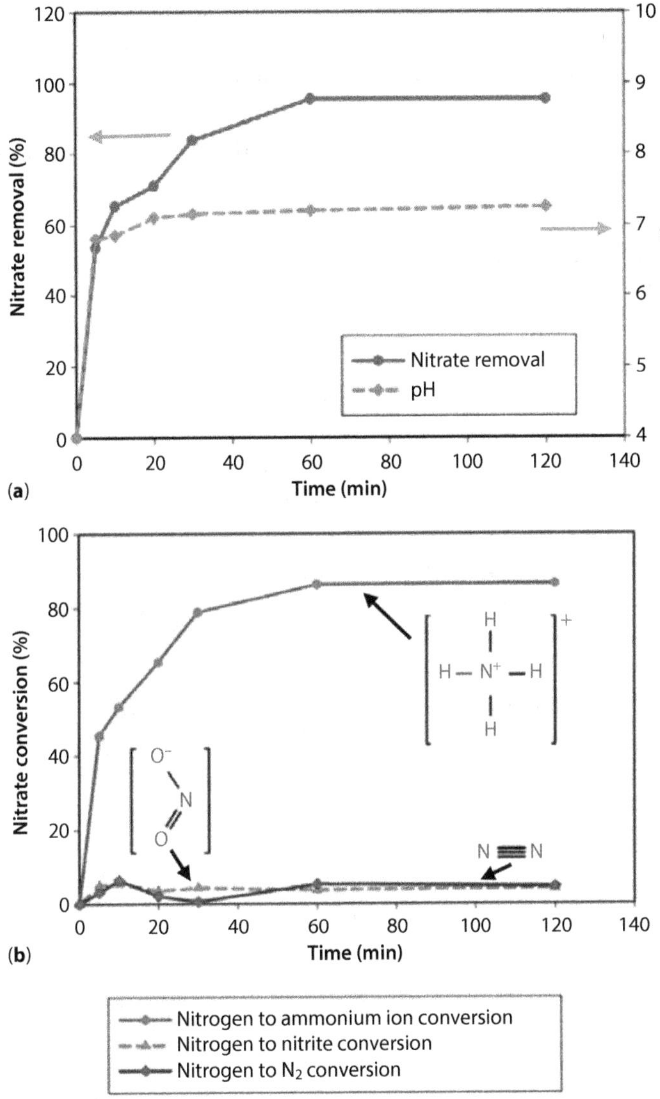

Figure 13.7 Nitrate removal by the sample prepared under ultrasonic irradiation (nZVI^{UI}) (a). (b) indicates the percentage of the nitrogen converted to the final products. The lines and represent the conversion (%) of the initial nitrate to ammonium ion and to nitrite, respectively, measured in the reaction solution. The line represents the conversion (%) of the initial nitrate to nitrogen which was calculated based on the shares of other end-products of the reaction.

It is known that ammonium ions can induce some toxic effects on the environment leaving organisms [91–93]. Hence, it is worthy to explore environmentally friendly methods for immobilizing this compound.

Struvite ($MgNH_4PO_4 \cdot 6H_2O$) has been widely studied in the literature due to its importance for agriculture applications and natural processes, medicine, and environment protection [94]. Through the precipitation of struvite, essential elements such as nitrogen and phosphorous can be recovered from the treatment medium for further applications. The formation of struvite can take place in the presence of ammonium ion (NH_4^+) through the following equation [95]:

$$Mg^{2+} + NH_4^+ + HPO_4^{2+} \longrightarrow MgNH_4PO_4 + H^+ \tag{13.9}$$

According to Eqn. 13.9, the crystallization of struvite results in the release of H^+ to the medium. So, the low pHs inhibit the formation of struvite [95]. J. Wang, Song, Yuan, Peng, & Fan (2006) [96] indicated that equimolar ratio of Mg/N/P (1:1:1) is considered an optimum value for the crystallization of struvite. Also, it has been mentioned in the literature that pH values between 8.5 to 9.25 are optimum for the crystallization of struvite [96, 97].

So, the initial pH for the crystallization of struvite in the present process of nitrate removal was adjusted to 8.5. Equimolar amounts of Na_2HPO_4 and $MgCl_2.6H_2O$ were added to the solution resulting from nitrate reduction. The pH of the solution decreased slightly, to 8.2 probably as a result of the release of H^+ ions in the medium. However, when the initial pH was set to 5 (acidic environment), or to 12 (basic environment) no struvite precipitation took place. The inhibition of the struvite crystallization at pH 12 may be related to the fact that, at this pH the ammonium ions are present as NH_3 [98].

The XRD pattern of the obtained white precipitate during struvite synthesis process is shown in Figure 13.8(a). The XRD peaks and their intensities matched the reference values of pure struvite (JCPDS: 04-010-2894) as also stated by the recent studies [99]. The EDX pattern of the prepared materials is also presented in Figure 13.8(b) confirming the presence of N, Mg, O, and P elements in the composition of the material.

Figure 13.9 shows the SEM pictures of the obtained struvite crystals. The SEM images clearly show that the obtained crystals are predominantly elongated rectangular bars of different lengths (typically 10 to 20 microns) which is a morphology also observed by Zhang, Ding, & Ren (2009) [100].

The present results thus confirm that struvite may be readily obtained by a simple precipitation process at room temperature.

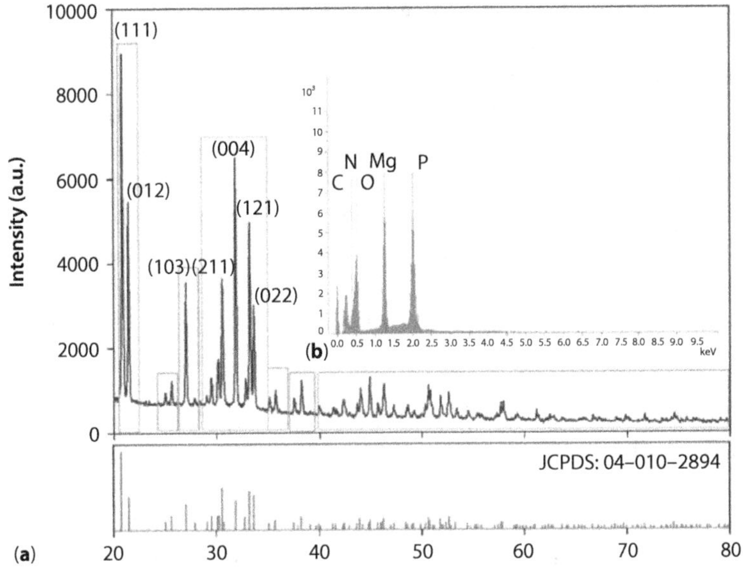

Figure 13.8 XRD (a) and EDX (b) patterns of the synthesized struvite.

Figure 13.9 SEM images of the struvite prepared in this study.

13.4 Conclusion

This study was performed in order to study the effects of the ultrasonic irradiation on the properties and reactivity of the synthesized ZVI nanomaterials. The combination of sonochemistry and liquid-phase reduction under certain synthesis conditions resulted in promoting lower sized NMs which higher surface area under UI, while maintaining particle crystallinity. The experimental data based Taguchi approach for nitrate reduction reflect the effects of the enhanced properties on the reactivity of the nanomaterials obtained under ultrasonic irradiation which account for ~96% of nitrate removal from solution after 30 min of reaction time. Chemical reduction mechanisms leading to the release of ammonium ions as reduction reaction products was confirmed as the dominant mechanism accounting for nitrate removal. The precipitation of struvite was proposed in this study to recover the nitrogen from nitrate polluted waters. The higher synthesis yield of the sample prepared under ultrasonic irradiation compared with the materials synthesized from non-irradiated solutions can also satisfy economic considerations.

Acknowledgments

Thanks are due, for the financial support to CESAM (UID/AMB/50017), to CICECO-Aveiro Institute of Materials, POCI-01-0145-FEDER-007679 (FCT Ref. UID/CTM/50011/2013), to FCT/MEC through national funds, and the co-funding by the FEDER, within the PT2020 Partnership Agreement and Compete 2020. Thanks are also due to FCT for the doctoral scholarship Nº SFRH/BD/103695/2014 for the first author.

References

1. Arauzo, M., Martínez-Bastida, J.J., Environmental factors affecting diffuse nitrate pollution in the major aquifers of central Spain: groundwater vulnerability vs. groundwater pollution. *Environ. Earth Sci.*, 73(12), 8271–8286, 2015.
2. Sutton, M., Howard, C., Erisman, J., Bleeker, B.G.A., Grennfelt, P., van Grisven, H., B. Grizzetti, *European nitrogen assessment (ENA)*. Cambridge university press, 2011.
3. Camargo, J.A., Alonso, A., Salamanca, A., Nitrate toxicity to aquatic animals: a review with new data for freshwater invertebrates. *Chemosphere*, 58(9), 1255–1267, 2005.
4. Baker, J.A., Gilron, G., Chalmers, B.A., Elphick, J.R., Evaluation of the effect of water type on the toxicity of nitrate to aquatic organisms. *Chemosphere*, 168, 435–440, 2017.
5. European Commission, Nitrates Directive (91/676/EEC) Status and trends of aquatic environment and agricultural practice, Development guide for Member States' reports. Directorate General for Environment, Brussels, 2000.

6. Majlesi, M., Mohseny, S.M., Sardar, M., Golmohammadi, S., Sheikhmohammadi, A., Improvement of aqueous nitrate removal by using continuous electrocoagulation/electroflotation unit with vertical monopolar electrodes. *Sustain. Environ. Res.*, 2–5, 2016.
7. Rajeswari, M., Seenivasagan, R., Prabhakaran, P., Rajakumar, S., Ayyasamy, P.M., Lab scale process on the removal of nitrate in ground water enriched with denitrifying bacterium and starch as a sole carbon source. *J. Water Process Eng.*, 13, 189–195, 2016.
8. Wang, B., Liu, S., Li, F., Fan, Z., Removal of nitrate from constructed wetland in winter in high-latitude areas with modified hydrophyte biochars. *Korean J. Chem. Eng.*, 34(3), 717–722, 2017.
9. Kamali, M., Gameiro, T., Costa, M.E., Capela, I., Anaerobic digestion of pulp and paper mill wastes – an overview of the developments and improvement opportunities. *Chem. Eng. J.*, 298, 162–182, 2016.
10. Kamali, M., Gomes, A.P.D., Khodaparast, Z., Seifi, T., Review on recent advances in environmental remediation and related toxicity of engineered nanoparticles. *Environ. Eng. Manag. J.*, 15, 923–934, 2016.
11. Xu, H., Zeiger, B.W., Suslick, K.S., Sonochemical synthesis of nanomaterials. *Chem. Soc. Rev.*, 42, 2555–2567, 2013.
12. Divband Hafshejani, L., Hooshmand, A., Naseri, A.A., Mohammadi, A.S., Abbasi, F., Bhatnagar, A., Removal of nitrate from aqueous solution by modified sugarcane bagasse biochar. *Ecol. Eng.*, 95, 101–111, 2016.
13. Xu, J., Pu, Y., Qi, W.K., Yang, X.J., Tang, Y., Wan, P., Fisher, A., Chemical removal of nitrate from water by aluminum-iron alloys. *Chemosphere*, 166, 197–202, 2017.
14. Ghazouani, M., Akrout, H., Jomaa, S., Jellali, S., Bousselmi, L., Enhancing removal of nitrates from highly concentrated synthetic wastewaters using bipolar Si/BDD cell : optimization and mechanism study. *JEAC*, 783, 28–40, 2016.
15. Abolghasem Alighardashi, S.S., Removal of nitrate from synthetic aqueous solution and groundwater in a continuous pilot system using chemical activated carbon derived from walnut shell. *Water Pract. Technol.*, 11, 784–795, 2016.
16. Jo, E.Y., Park, S.M., Yeo, I.S., Cha, J.D., Lee, J.Y., Kim, Y.H., Lee, T.K., Park, C.G., A study on the removal of sulfate and nitrate from the wet scrubber wastewater using electrocoagulation. 57(17), 7833–7840, 2016.
17. Nur, T., Shim, W.G., Loganathan, P., Vigneswaran, S., Kandasamy, J., Nitrate removal using Purolite A520E ion exchange resin : batch and fixed-bed column adsorption modelling. *Int. J. Environ. Sci. Technol.*, 12, 1311–1320, 2015.
18. Zhang, H., Jiang, J., Li, M., Yan, F., Gong, C., Wang, Q., Biological nitrate removal using a food waste-derived carbon source in synthetic wastewater and real sewage. *J. Environ. Manag.*, 166(3), 407–413, 2016.
19. Bartucca, M.L., Mimmo, T., Cesco, S., Del Buono, D., Nitrate removal from polluted water by using a vegetated floating system. *Sci. Total Environ.*, 542, 803–808, 2016.

20. Li, S., Yan, W., Zhang, W., Solvent-free production of nanoscale zero-valent iron (nZVI) with precision milling. *Green Chem.*, 11, 1618–1626, 2009.
21. Zolfaghari, Z., Tavasoli, A., Tabyar, S., Pour, A.N., Enhancement of bimetallic Fe-Mn/CNTs nano catalyst activity and product selectivity using microemulsion technique. *J. Energy Chem.*, 23, 57–65, 2014.
22. Xiang, X., Xia, F., Zhan, L., Xie, B., Preparation of zinc nano structured particles from spent zinc manganese batteries by vacuum separation and inert gas condensation. *Sep. Purif. Technol.*, 142, 227–233, 2015.
23. Kim, J.-C., Lee, J.-W., Park, B.-Y., Choi, C.-J., Characteristics of Fe/SiO2 nanocomposite powders by the chemical vapour condensation process. *J. Alloys Compd.*, 449(1–2), 258–260, 2008.
24. Makel, J.M., Keskinen, H., Forsblom, T., Keskinen, J., Generation of metal and metal oxide nanoparticles by liquid flame spray process. *J. Mater. Sci.*, 39, 2783–2788, 2004.
25. Chen, S., Zhang, Y., Han, W., Wellburn, D., Liang, J., Liu, C., Synthesis and magnetic properties of Fe_2O_3-TiO_2 nano-composite particles using pulsed laser gas phase evaporation-liquid phase collecting method. *Appl. Surf. Sci.*, 283, 422–429, 2013.
26. Nurmi, J.T., Tratnyek, P.G., Sarathy, V., Baer, D.R., Amonette, J.E., Pecher, K., Wang, C., Linehan, J.C., Matson, D.W., Penn, R.L., Driessen, M.D., Characterization and properties of metallic iron nanoparticles: spectroscopy, electrochemistry, and kinetics. *Environ. Sci. Technol.*, 39, 1221–1230, 2005.
27. Efecan, N., Shahwan, T., Eroğlu, A.E., Lieberwirth, I., Characterization of the uptake of aqueous Ni2+ ions on nanoparticles of zero-valent iron (nZVI). *Desalination*, 249, 1048–1054, Dec. 2009.
28. Hwang, Y., Lee, Y.-C., Mines, P.D., Huh, Y.S., Andersen, H.R., Nanoscale zero-valent iron (nZVI) synthesis in a Mg-aminoclay solution exhibits increased stability and reactivity for reductive decontamination. *Appl. Catal. B Environ.*, 147, 748–755, 2014.
29. Peng, L., Liu, Y., Gao, S., Chen, X., Xin, P., Dai, X., Evaluation on the nanoscale zero valent iron based microbial denitrification for nitrate removal from groundwater. *Nat. Publ. Gr.*, 1–11, 2015.
30. Liu, Y., Li, S., Chen, Z., Megharaj, M., Naidu, R., Influence of zero-valent iron nanoparticles on nitrate removal by Paracoccus sp., *Chemosphere*, 108, 426–432, 2014.
31. Lucchetti, R., Onotri, L., Clarizia, L., Di Natale, F., Di Somma, I., Andreozzi, R., Marotta, R., Removal of nitrate and simultaneous hydrogen generation through photocatalytic reforming of glycerol over "in situ" prepared zero-valent nano copper/P25. *Appl. Catal. B Environ.*, 202, 539–549, 2017.
32. R. Kamaraj, A. Pandiarajan, S. Jayakiruba, M. Naushad, and S. Vasudevan, Kinetics, thermodynamics and isotherm modeling for removal of nitrate from liquids by facile one-pot electrosynthesized nano zinc hydroxide, *J. Mol. Liq.*, 215, 204–211, 2016.

33. R. Lucchetti, A. Siciliano, L. Clarizia, D. Russo, I. Di Somma, F. Di Natale, M. Guida, R. Andreozzi, and R. Marotta, Sacrificial photocatalysis: removal of nitrate and hydrogen production by nano-copper-loaded P25 titania. A kinetic and ecotoxicological assessment, *Environ. Sci. Pollut. Res.*, 24(6), 5898–5907, 2017.
34. A. M. E. Khalil, O. Eljamal, T. W. M. Amen, Y. Sugihara, and N. Matsunaga, Optimized nano-scale zero-valent iron supported on treated activated carbon for enhanced nitrate and phosphate removal from water, *Chem. Eng. J.*, 309, 349–365, 2017.
35. X. Ma, M. Li, C. Feng, W. Hu, L. Wang, and X. Liu, Development and reaction mechanism of efficient nano titanium electrode: Reconstructed nanostructure and enhanced nitrate removal efficiency, *J. Electroanal. Chem.*, 782, 270–277, 2016.
36. A. Teimouri, S. G. Nasab, N. Vahdatpoor, S. Habibollahi, H. Salavati, and A. N. Chermahini, Chitosan/Zeolite Y/Nano ZrO_2 nanocomposite as an adsorbent for the removal of nitrate from the aqueous solution, *Int. J. Biol. Macromol.*, 93, 254–266, 2016.
37. B. Ensie and S. Samad, Removal of nitrate from drinking water using nano SiO2-FeOOH-Fe core-shell, *Desalination*, 347, 1–9, 2014.
38. H. Liu, M. Guo, and Y. Zhang, Nitrate removal by Fe0/Pd/Cu nano-composite in groundwater, *Environ. Technol.*, 35(7), 917–924, 2014.
39. H. B. Ahmad, Y. Abbas, M. Hussain, N. Akhtar, T. M. Ansari, M. Zuber, K. M. Zia, and S. A. Arain, Synthesis and application of alumina supported nano zero valent zinc as adsorbent for the removal of arsenic and nitrate, *Korean J. Chem. Eng.*, 31(2), 284–288, 2014.
40. M. Bhuvaneshwari, D. Kumar, R. Roy, S. Chakraborty, A. Parashar, A. Mukherjee, N. Chandrasekaran, and A. Mukherjee, Toxicity, accumulation, and trophic transfer of chemically and biologically synthesized nano zero valent iron in a two species freshwater food chain, *Aquat. Toxicol.*, 183, 63–75, 2017.
41. F. Fu, D. D. Dionysiou and H. Liu, The use of zero-valent iron for groundwater remediation and wasterwater treatment: a review, *J. Hazard. Mater.*, 267, 194–205, 2014.
42. J. H. Bang and K. S. Suslick, Applications of Ultrasound to the Synthesis of Nanostructured Materials, *Adv. Mater.*, 22, 1039–1059, 2010.
43. R. Feng, Y. Zhao, C. Zhu, and T.J. Mason, Enhancement of ultrasonic cavitation yield by multi-frequency sonication, *Ultrason. Sonochem.*, 9, 231–236, Oct. 2002.
44. R. S. Rao, C. G. Kumar, R. S. Prakasham, and P. J. Hobbs, The Taguchi methodology as a statistical tool for biotechnological applications: A critical appraisal, *Biotechnol. J.*, 3(4), 510–523, 2008.
45. S. Naghibi, M. A. Faghihi Sani, and H. R. Madaah Hosseini, Application of the statistical Taguchi method to optimize TiO2 nanoparticles synthesis by the hydrothermal assisted sol-gel technique, *Ceram. Int.*, 40(3), 4193–4201, 2014.

46. U. Özdemir, B. Özbay, I. Özbay, and S. Veli, Application of Taguchi L32 orthogonal array design to optimize copper biosorption by using Spaghnum moss, *Ecotoxicol. Environ. Saf.*, 107, 229–235, 2014.
47. G. Al, U. Özdemir, and Ö. Aksoy, Cytotoxic effects of Reactive Blue 33 on Allium cepa determined using Taguchi's L8 orthogonal array, *Ecotoxicol. Environ. Saf.*, 98, 36–40, 2013.
48. N. Daneshvar, A. R. Khataee, M. H. Rasoulifard, and M. Pourhassan, Biodegradation of dye solution containing Malachite Green: Optimization of effective parameters using Taguchi method, *J. Hazard. Mater.*, 143(1–2), 214–219, 2007.
49. Y. J. Kim, J. Heo, K. S. Park, and S. Kim, Proposition of novel classification approach and features for improved real-time arrhythmia monitoring, *Comput. Biol. Med.*, 75, 190–202, 2016.
50. S. A. Lobanov and V. Z. Poilov, Treatment of wastewater to remove ammonium ions by precipitation, *Russ. J. Appl. Chem.*, 79(9), 1473–1477, 2006.
51. Z. Yang, H. Xu, C. Shan, Z. Jiang, and B. Pan, Effects of brining on the corrosion of ZVI and its subsequent As(III/V) and Se(IV/VI) removal from water, *Chemosphere*, 170, 251–259, 2017.
52. X. Guo, Z. Yang, H. Dong, X. Guan, Q. Ren, X. Lv, and X. Jin, Simple combination of oxidants with zero-valent-iron (ZVI) achieved very rapid and highly efficient removal of heavy metals from water, *Water Res.*, 88, 671–680, 2016.
53. B. Casentini, F. T. Falcione, S. Amalfitano, S. Fazi, and S. Rossetti, Arsenic removal by discontinuous ZVI two steps system for drinking water production at household scale, *Water Res.*, 106, 135–145, 2016.
54. P. Devi and A. K. Saroha, Simultaneous adsorption and dechlorination of pentachlorophenol from effluent by Ni-ZVI magnetic biochar composites synthesized from paper mill sludge, *Chem. Eng. J.*, 271, 195–203, 2015.
55. Z. Zhang, J. Liu, X. Cao, X. Luo, R. Hua, Y. Liu, X. Yu, L. He, and Y. Liu, Comparison of U(VI) adsorption onto nanoscale zero-valent iron and red soil in the presence of U(VI)-CO_3/Ca-U(VI)-CO_3 complexes, *J. Hazard. Mater.*, 300(6), 633–642, 2015.
56. L. Liang, X. Guan, Y. Huang, J. Ma, X. Sun, J. Qiao, and G. Zhou, Efficient selenate removal by zero-valent iron in the presence of weak magnetic field, *Sep. Purif. Technol.*, 156, 1064–1072, 2015.
57. Y. Lubphoo, J. M. Chyan, N. Grisdanurak, and C. H. Liao, Nitrogen gas selectivity enhancement on nitrate denitrification using nanoscale zero-valent iron supported palladium/copper catalysts, *J. Taiwan Inst. Chem. Eng.*, 57, 143–153, 2014.
58. Y. Georgiou, K. Dimos, K. Beltsios, M. A. Karakassides, and Y. Deligiannakis, Hybrid polysulfone-Zero Valent Iron. membranes: Synthesis, characterization and application for AsIII remediation, *Chem. Eng. J.*, 281, 651–660, 2015.
59. M. A. Salam, O. Fageeh, S. A. Al-Thabaiti, and A. Y. Obaid, Removal of nitrate ions from aqueous solution using zero-valent iron nanoparticles supported on high surface area nanographenes, *J. Mol. Liq.*, 212, 708–715, 2015.

60. X. Zhou, B. Lv, Z. Zhou, W. Li, and G. Jing, Evaluation of highly active nanoscale zero-valent iron coupled with ultrasound for chromium(VI) removal, *Chem. Eng. J.*, 281, 155–163, 2015.
61. R. Fu, Y. Yang, Z. Xu, X. Zhang, X. Guo, and D. Bi, The removal of chromium (VI) and lead (II) from groundwater using sepiolite-supported nanoscale zero-valent iron (S-NZVI), *Chemosphere*, 138, 726–734, 2015.
62. R. Norouzbeigi and S. Majdabadi Farahani, Modified combustion synthesis of Nano-NiFe2O4: Optimization using Taguchi experimental design, *J. Magn. Magn. Mater.*, 384, 289–295, 2015.
63. K. Do Kim, D. N. Han, and H. T. Kim, Optimization of experimental conditions based on the Taguchi robust design for the formation of nano-sized silver particles by chemical reduction method, *Chem. Eng. J.*, 104(1–3), 55–61, 2004.
64. Q. Wang, S. Snyder, J. Kim, and H. Choi, Aqueous ethanol modified nanoscale zerovalent iron in bromate reduction: synthesis, characterization, and reactivity, *Environ. Sci. Technol.*, 43(9), 3292–3299, 2009.
65. L. Li, M. Fan, R. C. Brown, J. (Hans) Van Leeuwen, J. Wang, W. Wang, Y. Song, and P. Zhang, Synthesis, Properties, and Environmental Applications of Nanoscale Iron-Based Materials: A Review, *Crit. Rev. Environ. Sci. Technol.*, 36, 405–431, 2006.
66. H. Woo, J. Park, S. Lee, and S. Lee, Effects of washing solution and drying condition on reactivity of nano-scale zero valent irons (nZVIs) synthesized by borohydride reduction, *Chemosphere*, 97, 146–152, Feb. 2014.
67. E. B. Flint and K. S. Suslick, The temperature of cavitation, *Science*, 253, 1397–1399, 1991.
68. A. Shui, W. Zhu, L. Xu, D. Qin, and Y. Wang, Green sonochemical synthesis of cupric and cuprous oxides nanoparticles and their optical properties, *Ceram. Int.*, 39, 8715–8722, 2013.
69. M. C. Barbosa, N. R. Messmer, T. R. Brazil, F. R. Marciano, and A. O. Lobo, The effect of ultrasonic irradiation on the crystallinity of nano-hydroxyapatite produced via the wet chemical method, *Mater. Sci. Eng. C. Mater. Biol. Appl.*, 33, 2620–2625, 2013.
70. H. Li, M. Zhu, Y. Pang, H. Du, and T. Liu, Influences of ultrasonic irradiation on the morphology and structure of nanoporous Co nanoparticles during chemical dealloying, *Prog. Nat. Sci. Mater. Int.*, 26(6), 562–566, 2016.
71. Y.-H. Hwang, D.-G. Kim, and H.-S. Shin, Effects of synthesis conditions on the characteristics and reactivity of nano scale zero valent iron, *Appl. Catal. B Environ.*, 105(1–2), 144–150, Jun. 2011.
72. G. C. C. Yang and H. L. Lee, Chemical reduction of nitrate by nanosized iron: Kinetics and pathways, *Water Res.*, 39(5), 884–894, 2005.
73. C. P. Huang, H. W. Wang, and P. C. Chiu, Nitrate reduction by metallic iron, *Water Res.*, 32, 2257–2264, 1998.
74. S. Choe, Y. Chang, K. Hwang, and J. Khim, Kinetics of reductive denitrification by nanoscale zero-valent iron, *Chemosphere*, 41, 1307–1311, 2000.

75. D. L. a. de Faria, S. V. Silva, and M. T. de Oliveira, Raman microspectroscopy of some iron oxides and oxyhydroxides, *J. Raman Spectrosc.*, 28(February), 873–878, 1997.
76. P. J. A. Cowan, *Inorganic Biochemistry: An Introduction*, 2nd ed. New York: Wiley-VCH, 1997.
77. J. T. Keiser, C. W. Brown, and R. H. Heidersbach, Characterization of the passive film formed on weathering steels, *Corros. Sci.*, 23(3), 251–259, 1983.
78. S. E. W. N.B. Colthup, L.H. Daly, *Introduction to Infrared and Raman Spectroscopy*, Second. New York: Academic press, 1975.
79. A. Ianoul, T. Coleman, and S. A. Asher, UV resonance Raman spectroscopic detection of nitrate and nitrite in wastewater treatment processes, *Anal. Chem.*, 74(6), 1458–1461, 2002.
80. H. A. Szymanski, *Raman Spectroscopy: Theory and Practice*. Springer Science & Business Media, 2012.
81. S. Higuchi, W. Treatment, and E. Division, Determination of nitrite ion in waste and treated waters by resonance Raman spectrometry, *Water Res.*, 14, 747–752, 1980.
82. C. Klinke, R. Kurt, J.-M. Bonard, and K. Kern, Raman spectroscopy and field emission measurements on catalytically grown carbon nanotubes, p. 6, 2005.
83. N. Huang, M. Short, J. Zhao, H. Wang, H. Lui, M. Korbelik, and H. Zeng, Full range characterization of the Raman spectra of organs in a murine model, *Opt. Express*, 19(23), p. 22892, 2011.
84. R. Ganoe and R. J. DeYoung, Remote Sensing of Dissolved Oxygen and Nitrogen in Water Using Raman Spectroscopy, *NASA Rep.*, 2013.
85. R. L. Frost, A. López, R. Scholz, and Y. Xi, Vibrational spectroscopy of the borate mineral priceite : implications for the molecular structure, *Spectrosc. Lett.*, 48, 101–106, 2015.
86. B. A. Kolesov, Raman spectra of single H2O molecules isolated in cavities of crystals, *J. Struct. Chem.*, 47(1), 21–34, 2006.
87. R. L. Frost, W. N. Martens, and K. L. Erickson, Thermal decomposition of the hydrotalcite thermogravimetric analysis and hot stage raman spectroscopic study, 82, 603–608, 2005.
88. R. Li, Z. Jiang, Y. Guan, H. Yang, and B. Liu, Effects of metal ion on the water structure studied by the Raman O-H stretching spectrum, *J. Raman Spectrosc.*, 40(9), 1200–1204, 2009.
89. D. Kim, Y. Hwang, H. Shin, and S. Ko, Kinetics of Nitrate Adsorption and Reduction by Nano-scale Zero Valent Iron (NZVI): Effect of Ionic Strength and Initial pH, 20, 175–187, 2016.
90. Y. H. Huang and T. C. Zhang, Kinetics of Nitrate Reduction by Iron at Near Neutral pH, *J. Environ. Eng.*, 128(7), 604–611, 2002.
91. R. Esteban, I. Ariz, C. Cruz, and J. F. Moran, Review: Mechanisms of ammonium toxicity and the quest for tolerance, *Plant Sci.*, 248, 92–101, 2016.

92. Moreno-Marín, F., Vergara, J. J., Pérez-Llorens, J. L., Pedersen, M. F., and Brun, F. G , Interaction between ammonium toxicity and green tide development over seagrass meadows: A laboratory study, *PLoS One*, 11(4), 2016.
93. W. L. Pan, I. J. Madsen, R. P. Bolton, L. Graves, and T. Sistrunk, Ammonia/Ammonium Toxicity Root Symptoms Induced by Inorganic and Organic Fertilizers and Placement, *Agron. J.*, 108(6), p. 2485, 2016.
94. V. V. Vol'khin, D. A. Kazakov, G. V. Leont'eva, Y. V. Andreeva, E. A. Nosenko, and M. Y. Siluyanova, Synthesis of struvite (MgNH4PO4·6H2O) and its use for sorption of nickel ions, *Russ. J. Appl. Chem.*, 88(12), 1986–1996, 2015.
95. S. H. Lee, R. Kumar, and B. H. Jeon, Struvite precipitation under changing ionic conditions in synthetic wastewater: Experiment and modeling, *J. Colloid Interface Sci.*, 474, 93–102, 2016.
96. J. Wang, Y. Song, P. Yuan, J. Peng, and M. Fan, Modeling the crystallization of magnesium ammonium phosphate for phosphorus recovery, *Chemosphere*, 65(7), 1182–1187, 2006.
97. N. O. Nelson, R. L. Mikkelsen, and D. L. Hesterberg, Struvite precipitation in anaerobic swine lagoon liquid: Effect of pH and Mg:P ratio and determination of rate constant, *Bioresour. Technol.*, 89(3), 229–236, 2003.
98. S. Körner, S. K. Das, S. Veenstra, and J. E. Vermaat, The effect of pH variation at the ammonium/ammonia equilibrium in wastewater and its toxicity to Lemna gibba, *Aquat. Bot.*, 71(1), 71–78, 2001.
99. I. Merino-Jimenez, V. Celorrio, D. J. Fermin, J. Greenman, and I. Ieropoulos, Enhanced MFC power production and struvite recovery by the addition of sea salts to urine, *Water Res.*, 109, 46–53, 2017.
100. T. Zhang, L. Ding, and H. Ren, Pretreatment of ammonium removal from landfill leachate by chemical precipitation, *J. Hazard. Mater.*, 166(2–3), 911–915, 2009.

Index

Acetone, 181
Acrylamide, 205
Acrylonitrile, 205
Actinomycete, 183, 193
Activated carbon, 175
Activated sludge, 171, 178, 179, 183, 185–189, 192, 193, 197, 210, 211, 214, 215, 219
Additives, 35
Adsorbents, 35
Adsorption, 59, 265
  Freundlich, 361–362
    Langmuir, 361–362
Adsorption experiments, 402
Adsorption kinetics, 71
Aerobic degradation, 171, 198
Algal biosrbents, 103–106
Allochtonous plants,
  adventitious, 228
  cultivated, 228
  established plants, 228
  invasive plants, 228
  naturalized plants, 228
  non-cultivated, 228
  sub-spontaneous plants, 228
Ammonium ions, 479, 486, 494, 496, 498
Amoebae, 178, 218
Anaerobic degradation, 171, 189
Aquatic weeds, 173
Aqueous media, 31
Archaebacterial, 175
Arsenic Concentration in Water, 2–3
Arsenic Toxicity and Mechanism, 6–10
  alternation in the gene expression, 9
  arsenic impairs pathways of glucose catabolism,
  binding to sulfhydryl group, 7–8
  oxidative stress, 6–7
  replacement of phosphate group, 8–9
Ash content, 317
Aspergillus, 108, 110–111, 119
ATP, 190, 191
Auxiliaries, 29

Bacillus, 111, 113–114
Bacteria, 111
Bacteria biosorbent, 111–114
BET (Brunauer, Emmett, and Teller) method, 320
Bioaugmentation, 207, 109, 211
Biochar,
  for dye removal, 321–322
  for fluoride removal, 322–323
  for heavy metal removal, 320–321
  for other pollutant removal, 323–324
  for persistent organic pollutants (POPs) removal, 323
  for soil amendment, 324
  overview, 313–314
  plant-based, 316–320
  preparation methods, 315–316
  surface area, 320
Biocomposites, 439, 440, 464
Biodegradation, 40, 42
Biodegradable, 207
Biodiversity, 223, 227, 229, 238, 250, 251

Biological oxidation, 175, 185, 213, 196
Biological treatment, 177, 179–181, 196, 205
Biomass collection and preparation, 401
Biomasses, 146
Bioremediation, 40, 43
Biosorbants, 438, 439, 446
Biosorption mechanism, 114–120
Blue-baby syndrome, 434
BNR, 196
BOD, 175, 179, 181, 183, 187–189, 190, 193, 195, 200, 210, 213–215
Bulk density, 317
Bunch dyeing, 31

C. vulgaris, 104
Calamine soil, 225, 226, 249
Catalytic reduction, 435, 436
Cavitation, 482, 488,
Cell permeability, 40
Chemical constitution, 35
Chemical oxidants,
  chlorine, 332–333, 336
  ferrate(VI), 331–332, 334–337, 339–340, 347, 357–359, 376
  ozone, 332–333, 336, 338
Chemical oxidation technology, 331–332, 378
Chemical, reagent and instruments, 400
Chemoautotroph, 200
Chitin, 181
Chitosan, 433, 438–440, 442–444, 463
Chitosan composites, 105, 108–109, 111–112, 114, 118, 121
Chlorine, 172, 175
Cholera, 172, 174
Ciliate, 178, 180, 212, 214, 218
Clay, 102, 104, 108, 114
Clay minerals, 139
Coagulants, 35
Coagulation, 204–206
Coliforms, 181, 182, 218
Colorants, 35
Colour, 36
Composites preparation, 401
Conditions for lipid degradation, 276
Constant dyeing, 31
Contamination, 35, 223–227, 230–233, 236, 249
Continuous dyeing, 31
Coupled semiconductor, 334, 371–372
Covalent binding, 269
CPC, 36
Cr(VI) adsorption, 104, 108, 111, 113, 114
CTAB, 36
Cyanobacterial, 181, 218

Deamination, 184, 187
Decaying speed, 43
Degradation and decolorization of industrial dyes, 279
Degradation of aromatic compounds, 296
Degradation of lipids, 291
Degradation of miscellaneous compounds, 292
Denitrifying, 187–189, 218
Detoxification, 39
Detoxification of arsenic, 10–12
  adsorption and recent advancement in adsorption technology, 15–16
  antioxidants agents, 10, 11
  arsenic remediation technologies, 12–15
  chelating agents, 11, 12
Detritus, 203
Disinfection agent, 39
Dissolved salts, 37
Doping,
  anion doping, 369, 371
  cation doping, 369–371
Dudnin-Radushkevich isotherm model, 135

Dye decolorization with free ligninolytic enzymes, 280
Dye degrading microorganisms, 43
Dye removal, 58
Dye removal by immobilized ligninolytic enzymes, 286
Dye removal, biochar for, 321–322
Dye solution preparation, 402
Dye wastewaters, 42
Dyes, 35, 37
Dye-sensitization, 334, 370, 373–374

Ecology, 223, 224, 238, 249–251
EDTA, 173
EDX (Energy Dispersive X-Ray), 320
Effect of composites dose, 405
Effect of contact time, 406
Effect of initial Concentration, 406
Effect of pH, 403
Effect of temperature, 408
Effective degradation, 37
Elemental analysis, 320
Emerging pollutants, 332–333, 340, 360, 362, 376, 378
Encapsulation, 268
Endangered, 173
Endemic, 223–227, 238–251
Endoenzymes, 179, 180, 189, 217
Entrapment, 267
Environmental, 223, 226, 227, 230, 234–238, 249–251
Environmental aplications of ligninolytic enzymes,
Environmentally friendly, 43
Enzyme immobilization strategies, 265
Enzymes sources, 276
Epidemiological, 434
Escherichia coli, 112
ETC, 173
Ethanol, 178, 188, 199
Eukaryotes, 175
Eutrophication, 434, 463, 173, 174, 191, 194

Exhaust dyeing, 31
Exoenzymes, 180, 217
Exotic, 224, 228–230, 238, 251
Exponential growth, 175
Exposure of arsenic in human body, 3–4

Fabrics printing, 30
Factors affecting adsorption, 62
Facultative Anerobic Bacteria, 189
Families,
  amaranthaceae, 239
  apiaceae, 230, 239, 242, 249
  asteraceae, 230, 236, 239, 249,
  boraginaceae, 236, 240,
  brassicaceae, 236, 240, 249
  calceolariaceae, 247, 249
  equisetaceae, 242, 249
  euphorbiaceae, 242
  hypericaceae, 243, 249
  juncaceae, 243, 244, 249
  lamiaceae, 230, 236, 244, 249
  marsilaceae, 244, 249
  plantaginaceae, 245, 249
  plumbaginaceae, 245, 249
  poaceae, 230, 236, 246, 249
  polygonaceae, 246, 249
  potamogetonaceae, 247, 249
  ranunculaceae, 247, 249
  scrophulariaceae, 236, 247, 249
  solanaceae, 247, 249
Fast pyrolysis, 315
Fermantative microbial community, 182
Ferrate(VI),
  characterization, 338
  kapp, 340–342
  kinetics, 331, 334, 339–341
  redox potential, 335
  stoichiometry, 331, 334, 341, 348–349, 378
Fertilizer industry, 174
Filtration, 203, 213
Flagellates, 178, 218

Flocculation, 35, 204–206, 221
Fluoride removal, biochar for, 322–323
Freundilich isotherm model, 134
FTIR (Fourier Transform Infrared Spectroscopy), 320
Fulvic acid, 173
Ffunctional groups, 103, 104, 107
Fungi, 106,
Fungi biosorpbents, 106–111

Gasification, 315
Gastroenteritis, 173
Geochemical background, 225
Global distribution, 235, 238–249
Glucose, 180, 188, 202, 210
Growth media, 40

Hair complexion, 37
Heavy metal, 225
Heavy metal removal, biochar for, 320–321
Hepatitis, 172, 174
High pressure, 41
Hydraulic retention time, 192
Hydrolytic microbial community, 179
Hydrothermal carbonization (HTC), 315–316
Hyperaccumulator, 226, 234–237, 239, 246,
Hypochlorite, 172

Immobilization, 264
Immobilized Lac, 274
Immobilized LiP, 271
Immobilized MnP, 273
Industrial wastes, 159
Insecticides, 172
Invasive, 228–230, 238, 249–251
Ion exchangers, 35
Isotherm modelling, 412
Isotherm models, 133

Kinetic study, 409
Kinetics of adsorption, 136

Langmuir isotherm model, 133
LAS, 36
Liquid-phase reduction, 479, 483, 486–487, 491, 497

MCRT, 179, 186, 207
Metabolism and excretion of arsenious compounds, 4–6
Metalliferous soils, 225
Metal-organic, 439
Methaemoglobin, 174
Methanogens, 184
Microbiological disintegration, 39
Microbiological processes, 37
Mineralization, 40
MLVSS, 207
Modified clay minerals, 140
Moisture content, 317

Nanoparticles, 479–480, 482, 484, 486, 489–490, 494, 37
Native, 223, 227–230, 236, 238–251
Natural clay minerals, 139
Natural clay minerals along with reducing agents, 140
Nematode, 177, 178, 180, 186, 187
Nitrate, 433–441, 445, 447, 456, 461, 463–464, 479–481, 483–486, 489–498
Nitrates, 173, 177, 178, 187–190
Nitrification, 183, 185–188, 195, 220
Nitrobacter, 183, 185
Nitrogen sequestration, 479
Nitrogenous aromatics, 39
Nitrous oxide, 182, 189, 196
Nocardioforms, 183, 194
Nondegenerateable pollutants, 180
Nonionic surfactants, 36
Nutrients, 180, 190, 192–194, 218

Organic–inorganic, 433, 439, 464
Orthogonal array, 482, 485, 489
Osmotic stress, 37

OTAB, 36
Oxidizing agents, 37
Ozonation, 39

PAO, 195
Partially permeable membrane, 37
Pathogen, 172, 174, 218
Penicillium chrysogenum, 107, 119, 120
Peroxide, 172
Persistent organic pollutants (POPs) removal, biochar for, 323
Pesticides, 172
pH, 317
PHB, 195, 196
Phenolics, 172
Photo electrochemical cells, 30
Photocatalysis,
  mechanism, 334, 361–363
  photocatalyst, 331, 334, 360, 367–368, 371, 374–376, 378
  reactor, 334, 363–367
  semiconductor, 360–361, 368–374
Photocatalytic degradation, 37
physical intetraction forecs, 103
Physicochemical treatments, 40
Phytoremediation, 231–233, 235–238, 249–251
Phytotechnologies,
  hydraulic control, 233
  phytodegradation, 230, 232, 233
  phytoextraction, 233–235
  phytostabilization, 226, 227, 232
  phytovolatilization, 232, 233
  rhizofiltration, 230, 233
Plant-based biochar,
  physico-chemical characterization, 316–320
Plyvinyl alcohol, 172
Polio, 172, 174
Pollutant removal, 323–324
Polluted waters, 479–480, 483, 498
Polyaniline as adsorbent, 74
Polyphosphate, 191–193, 196
Potential technologies, 29
Precipitation, 175, 194–196
Prokaryotes, 175
Properties of immobilized enzymes,
Proteases, 180
Protozoa, 180, 186, 187, 215, 218
Pseudo-first-order kinetics, 136
Pseudomonas, 111–114, 118
Pseudo-second-order kinetics, 137
Pyrolysis, 315

Radiation sources, 37
Radio-active pollutants, 172
Raman spectroscopy, 486, 492
RAS, 196, 197
Retention Time, 179, 192, 195, 214
Reverse osmosis, 36
Rotifers, 178, 180, 186, 218

Saccharomyces, 178
Saccharomyces cerevisiae, 107, 109–110
Saprophytic microorganisms, 208
Sargassum sp., 105, 106
Screening of adsorbents, 402
SDS, 36
Sedimentation, 174, 204, 206, 213, 216
Sediments, 172, 174, 218
Self-immobilization, 270
SEM (Scanning Electron Microscopy), 320
Serpentine soil, 225, 226, 235–237, 249
Sewage, 172–173, 204, 218, 220
Slow pyrolysis, 315
Soda ash, 172
Sodium alginate, 104, 106, 113
Sodium hydroxide, 172
Sodium silicates, 172
Soil erosion, 174
Soil treatment, biochar for, 324
Solubilizing agents, 31–32
Solution pH, 137
Sonochemistry, 482, 497

SRB, 198–200
Starch, 172, 180, 193, 201, 210
Stoichiometric formula, 104
Stritification, 172
Struvite, 479–480, 483, 484, 486, 496–498
Substantial barriers, 33
Suspended solids, 172, 174, 193, 203, 204
Synthesis of ferrate(VI),
- dry thermal, 334, 336, 338
- electrochemical, 334, 336–337, 339
- wet chemical, 334, 336, 338

Synthetic dyes, 31

Taguchi, 479, 482, 485, 489, 491, 498,
Teichoic acid, 190, 191
Teichuronic acid, 191
Textile effluent treatment, 29
Thermal pollutants, 172, 218
Thermodynamics of adsorption, 135
Tolerance, 236, 243, 244, 248, 250
Treatment tool, 43
Tricholoma lobayense, 104
Turbidity, 35

Ultrafiltration, 36
Ultrasonic irradiation, 479–480, 483, 485–488, 491, 495, 497–498
Uranium adsorption, 104–105, 108–111, 120
Urea, 184, 185, 213

Winkler test, 201

Xenobiotics and industrial effluents, 295

Zero-valent iron, 479–480, 482–483, 486–487, 491–494

Also of Interest

Check out these published volumes in the Advanced Materials Series

Advanced Magnetic and Optical Materials
Edited by Ashutosh Tiwari, Parameswar K. Iyer, Vijay Kumar and Hendrik Swart
Published 2016. ISBN 978-1-119-24191-1

Advanced Surfaces for Stem Cell Research
Edited by Ashutosh Tiwari, Bora Garipcan and Lokman Uzun
Published 2016. ISBN 978-1-119-24250-5

Advanced Electrode Materials
Edited by Ashutosh Tiwari, Filiz Kuralay and Lokman Uzun
Published 2016. ISBN 978-1-119-24252-9

Advanced Molecularly Imprinting Materials
Edited by Ashutosh Tiwari and Lokman Uzun
Published 2016. ISBN 978-1-119-33629-7

Intelligent Nanomaterials (2nd edition)
Edited by Tiwari, Yogendra Kumar Mishra, Hisatoshi Kobayashi and Anthony P. F. Turner
Published 2016. ISBN 978-1-119-24253-6

Advanced Composite Materials
Edited by Ashutosh Tiwari, Mohammad Rabia Alenezi and Seong Chan Jun
Published 2016. ISBN 978-1-119-24253-6

Advanced Surface Engineering Materials
Edited by Ashutosh Tiwari, Rui Wang, and Bingqing Wei
Published 2016. ISBN 978-1-119-24244-4

Advanced Ceramic Materials
Edited by Ashutosh Tiwari, Rosario A. Gerhardt and Magdalena Szutkowska
Published 2016. ISBN 978-1-119-24244-4

Advanced Engineering Materials and Modeling
Edited by Ashutosh Tiwari, N. Arul Murugan and Rajeev Ahuja
Published 2016. ISBN 978-1-119-24246-8

Advanced 2D Materials
Ashutosh Tiwari and Mikael Syväjärvi
Published 2016. ISBN 978-1-119-24249-9

Advanced Materials Interfaces
Edited by Ashutosh Tiwari, Hirak K. Patra and Xumei Wang
Published 2016. ISBN 978-1-119-24245-1

Advanced Bioelectronics Materials
Edited by Ashutosh Tiwari, Hirak K. Patra and Anthony P.F. Turner
Published 2015. ISBN 978-1-118-99830-4

Graphene
An Introduction to the Fundamentals and Industrial Applications
By Madhuri Sharon and Maheswar Sharon
Published 2015. ISBN 978-1-118-84256-0

Advanced Theranostic Materials
Edited by Ashutosh Tiwari, Hirak K. Patra and Jeong-Woo Choi
Published 2015. ISBN: 978-1-118-99829-8

Advanced Functional Materials
Edited by Ashutosh Tiwari and Lokman Uzun
Published 2015. ISBN 978-1-118-99827-4

Advanced Catalytic Materials
Edited by Ashutosh Tiwari and Salam Titinchi
Published 2015. ISBN 978-1-118-99828-1

Graphene Materials
Fundamentals and Emerging Applications
Edited by Ashutosh Tiwari and Mikael Syväjärvi
Published 2015. ISBN 978-1-118-99837-3

DNA Engineered Noble Metal Nanoparticles
Fundamentals and State-of-the-art-of Nanobiotechnology
By Ignác Capek
Published 2015. ISBN 978-1-118-07214-1

Advanced Electrical and Electronics Materials
Process and Applications
By K.M. Gupta and Nishu Gupta
Published 2015. ISBN: 978-1-118-99835-9

Advanced Materials for Agriculture, Food and Environmental Safety
Edited by Ashutosh Tiwari and Mikael Syväjärvi
Published 2014. ISBN: 978-1-118-77343-7

Advanced Biomaterials and Biodevices
Edited by Ashutosh Tiwari and Anis N. Nordin
Published 2014. ISBN 978-1-118-77363-5

Biosensors Nanotechnology
Edited by Ashutosh Tiwari and Anthony P. F. Turner
Published 2014. ISBN 978-1-118-77351-2

Advanced Sensor and Detection Materials
Edited by Ashutosh Tiwari and Mustafa M. Demir
Published 2014. ISBN 978-1-118-77348-2

Advanced Healthcare Materials
Edited by Ashutosh Tiwari
Published 2014. ISBN 978-1-118-77359-8

Advanced Energy Materials
Edited by Ashutosh Tiwari and Sergiy Valyukh
Published 2014. ISBN 978-1-118-68629-4

Advanced Carbon Materials and Technology
Edited by Ashutosh Tiwari and S.K. Shukla
Published 2014. ISBN 978-1-118-68623-2

Responsive Materials and Methods
State-of-the-Art Stimuli-Responsive Materials and Their Applications
Edited by Ashutosh Tiwari and Hisatoshi Kobayashi
Published 2013. ISBN 978-1-118-68622-5

Other Scrivener books edited by Ashutosh Tiwari
Nanomaterials in Drug Delivery, Imaging, and Tissue Engineering
Edited by Ashutosh Tiwari and Atul Tiwari
Published 2013. ISBN 978-1-118-29032-3

Biomedical Materials and Diagnostic Devices Devices
Edited by Ashutosh Tiwari, Murugan Ramalingam, Hisatoshi Kobayashi and Anthony P.F. Turner
Published 2012. ISBN 978-1-118-03014-1

Intelligent Nanomaterials (first edition)
Processes, Properties, and Applications
Edited by Ashutosh Tiwari Ajay K. Mishra, Hisatoshi Kobayashi and Anthony P.F. Turner
Published 2012. ISBN 978-0-470-93879-9

Integrated Biomaterials for Biomedical Technology
Edited by Murugan Ramalingam, Ashutosh Tiwari, Seeram Ramakrishna and Hisatoshi Kobayashi
Published 2012. ISBN 978-1-118-42385-1

www.ingramcontent.com/pod-product-compliance
Lightning Source LLC
LaVergne TN
LVHW011532200926
848484LV00001B/248